Chasing Down Religion

In the Sights of History and the Cognitive Sciences

Edited by
Panayotis Pachis and Donald Wiebe

SHEFFIELD UK   BRISTOL CT

Published by Equinox Publishing Ltd.

UK: Office 415, The Workstation, 15 Paternoster Row, Sheffield, South Yorkshire S1 2BX

USA: ISD, 70 Enterprise Drive, Bristol, CT 06010

www.equinoxpub.com

First published as *Chasing Down Religion: In the Sights of History and the Cognitive Sciences, Essays in Honor of Luther H. Martin* by Editions Barbounakis, Thessaloniki, Greece, in 2010.

© Panayotis Pachis, Donald Wiebe and contributors 2010

All rights reserved. No part of this publication may be reproduced or transmitted in any form or by any means, electronic or mechanical, including photocopying, recording or any information storage or retrieval system, without prior permission in writing from the publishers.

British Library Cataloguing-in-Publication Data

A catalogue record for this book is available from the British Library.

ISBN-13: 978-1-78179-207-0 (paperback)
Library of Congress Cataloging-in-Publication Data
Chasing down religion : in the sights of history and the cognitive sciences / edited by Panayotis Pachis and Donald Wiebe.
     pages cm
 Includes bibliographical references and index.
 ISBN 978-1-78179-207-0 (paperback)
 1. Religion. 2. Religion--History. 3. Religions. 4. Psychology, Religious. 5. Cognitive science. I. Pachis, Panayotis. II. Wiebe, Donald, 1943-
 BL48.C487 2015
 200.9--dc23
                    2014041557

Printed and bound in the UK by Lightning Source UK Ltd., Milton Keynes and Lightning Source Inc., La Vergne, TN

IV

Ἀ[πό]λλων εἶπεν ὅ μοι Λύκιος... πρός δε σε] καὶ τόδ' ἄνωγα,
τὰ μὴ πατέουσιν ἅμαξαι τὰ στείβειν, ἑτέρων δ' ἴχνια μὴ καθ'
ὁμὰ δίφρον ἐλᾶ]ν μηδ' οἷμον ἀνὰ πλατύν, ἀλλὰ κελεύθους
ἀτρίπτου]ς, εἰ καὶ στεινοτάτην ἐλάσεις.

Lycian Apollo said to me: "...I bid you: tread a path which
carriages do not trample; do not drive your chariot upon
the common tracks of others, nor along a wide road,
but on unworn paths, though cour course be more narrow".

(Callimachus, *Aetia fragm.* 1,22-28
[circa 250 BCE]
[text, translation
and notes by C.A. Trypanis, LCT]).

CONTENTS

CONTRIBUTORS X
ABBREVIATIONS XVIII
LUTHER H. MARTIN PUBLICATIONS XX

PREFACE XXXVI

Ancient and modern approaches to the representation of supernatural beings: Dio Chrisostom (Oration 12) and Dan Sperber (explaining culture) compared
Beck Roger 01

Gnosis in European religious History
Berner Ulrich 09

The first shall be last: The Gospel of Mark after the first century
Braun Will 31

Comparative religion scholars in debate: Theology vs History in letters addressed to Ugo Bianchi
Casadio Giovanni 49

Why did Greeks and Romans pray aloud? Antropomorphism, 'Dumb Gods' and human cognition
Chalupa Aleš 69

Reflections on the origins of religious thought and behavior
Geertz W. Armin 81

Why is it better to be a plant than an animal? Cognitive poetics and ascetic ideals in the book of Thomas the Contender (Nhc II,7)
Gilhus Sælid Ingvild 99

Miracles, Memory and Meaning: A cognitive
approach to Roman myths
Griffith B. Alison 117

'Whatever story sings, the arena displays for you'.
Performance, narrative and myth in Graeco-Roman
discourse
Heever van den Gerhard 133

Religion and violence: A psychoanalytic inquiry
Hewitt Aileen Marsha 151

Disciplinary Clans
Hrotic M. Steven 175

The social capital of religious communities in
the age of globalization
Kippenberg G. Hans 191

Towards a cognitive Historiography –
Frequently posed objections
Lisdorf Anders 209

How science and Religion are more like theology and
commonsense explanations than they are like each
other: a cognitive account
McCauley N. Robert 217

The Transmission of Historical cognition
McCorkle W. William Jr. 237

Religion before 'Religion'?
McCutcheon T. Russell 251

The discourse of a myth: Diodorus Siculus and the
Egyptian *theologoumena* during the Hellenistic age
Pachis Panayotis 270

The History of Religions and evolutionary models:
Some reflections on framing a mediating vocabulary
Paden E. William 293

Religion and Modern culture
Petrou S. Ioannis 307

Do Relics do? Making a place for 'presence' in the
comparative study of Relics
Robinson Douglas 325

Mithras in the magical papyri. Religio-Historical
reflections on various magical texts
Sanzi Ennio 339

The use of Egyptian tradition in Alexandria
of the Hellenistic and Roman periods: self-display
and identity of rulers, ideology and further political
propaganda
Savvopoulos Kyriakos 351

Buddhist Hymns and Medieval plainsong:
Some reflections on the links between neuroscience,
music and religion
Trainor Kevin and Clark Anne 373

Citations of Biblical texts in Greek, Jewish and
Christian inscriptions of late antiquity: a case
of religious demarcation
Tsalampouni Ecaterini 397

Memorable religions: transmission, codification,
and change in divergent Melanesian contexts
Whitehouse Harvey 413

Recovering 'religious experience' in the
explanation of religion
Wiebe Donald 441

Can the study of religion be scientific?
Xygalatas Dimitris 459

Bibliography	473
Index of Authors	539
Index of Writers	543
General Index	551

CONTRIBUTORS

BECK ROGER was Lecturer at the University of Manitoba between 1963 and 1964. In 1964 he started his career at the University of Toronto, Erindale College and Department of Classics. He is Professor Emeritus since 1998. He was Secretary of the Classical Association of Canada between 1977 and 1979. His research interests are Mithraism and religion in the Roman Empire; ancient astrology and astronomy; Petronius and the ancient novel.

BERNER ULRICH is Professor of Religious Studies at the University of Bayreuth, Germany, faculty of cultural sciences. His main interests are European History of Religions, African Christianity, and Method and Theory in the Study of Religion. He is a member of the Advisory Board for SAPERE (Scripta Antiquitatis Posterioris ad Ethicam Religionemque pertinentia).

BRAUN WILLI is Professor at the Department of History and Classics/Program of Religious Studies and Director of the Program of Religious Studies Faculty of Arts at the University of Alberta Edmonton, Canada. He is the author of *Feasting and Social Rhetoric in Luke 14 (2005[1995])*. He is the editor of *Rhetoric and Reality in Early Christianities (2005)*. He is co-editor with R. McCutcheon of *Guide to the Study of Religion* (2000) and also *Introducing Religion: Essays in Honor of Jonathan Z. Smith* (2008). He is the author of numerous articles on the above topics.

CASADIO GIOVANNI is Professor Ordinarius of History of Religions at the University of Salerno. Past member of the executive committee of the European Association for the Study of religions (EASR) as Publications officer. Member of the editorial board of the second edition of Eliade's *Encyclopedia of Religion*. Member of the advisory board of several international series and journals (including *Archaeus* and *Vergilius*). 109 publications on ancient Mediterranean religions and religious historiography. Books: *Storia del culto di Dioniso in Argolide* (1994), *Vie gnostiche all' immortalità* (1997), *Il vino dell' anima* (1999), *Ugo Bianchi: Una vita per la storia delle religioni* (ed. 2002).

CHALUPA ALEŠ is Assistant Professor at the Department for the Study of Religions at Masaryk University, Brno, Czech Republic. His field of research is ancient religions, especially Roman religion, Graeco-Roman magic and Mithraism. In recent years, he has also focused on the cognitive science of religion and its possible contribution to the methodology of the study of ancient religious traditions.

CLARK ANNE works on the Christian tradition in the Middle Ages. Her Ph.D. is from Columbia University. Her research focuses on styles of piety, questions of gender, the role of the body, and women's religious life. She has published *Elisabeth of Schönau: The Complete Works* (Paulist Press, 2000) and *Elisabeth of Schönau: A Twelfth-Century Visionary* (University of Pennsylvania Press, 1992). She has also published articles on Elisabeth of Schönau, Hildegard of Bingen, Gertrude of Helfta, women's monastic communities, and cognitive theory in the study of religion. Her current research is on the cult of the Virgin Mary in the later Middle Ages.

GEERTZ W. ARMIN Dr. Phil., is Professor in the History of Religions at the Department of the Study of Religion, Chair of the Religion, Cognition and Culture Research Unit (RCC) and Section Leader in the MIND*Lab* coalition, Aarhus University, Denmark. His interests include Indigenous religions, Native American religions, method and theory in the study of religion, the cognitive science of religion, contemporary spirituality. His publications range from the cognitive science of religion to method and theory in the study of religions and the religions of indigenous peoples. He has published articles and chapters on evolutionary theory, atheism, the neurobiology of prayer and introductions to the cognitive science of religion.

GILHUS INGVILD SÆLID is Professor in the history of religions at the University of Bergen (1988-). Research interests: late antiquity and Gnosticism, new religious movements, and theories and methods in the history of religions. Vice Dean (1988) and Dean of the Faculty of Arts, University of Bergen (1989-90) and member of the board of the Norwegian Research Council for Social Sciences and Humanities (1993-96). Member of the board of the University of Bergen (1999-), Deputy Chairman of the board of the Norwegian Research Council (2003-), Chairman of the board of Chr. Michelsen Institute. Development Studies and Human Rights (2001-), member of the board of the Bergen Festival (1999-) and member of the Executive Commitee IAHR (1999-2004). Deputy General Secretary of the Executive Committee IAHR (2005-2010). Member of the the Royal Norwegian Society of Sciences and Letters in Trondheim and of the Norwegian Academy of Science and Letters in Oslo.

GRIFFITH B. ALISON is a senior Lecturer at the University of Canterbury in Christchurch, New Zealand and the chief editor of the *Electronic Journal of Mithraic Studies*. She is the author of numerous articles on the worship of Mithras in Rome and also on Roman religion and its alternatives.

HEWITT AILEEN MARSHA is Professor of Ethics and Contemporary Theology in the Faculty of Divinity, and cross-appointed to the Centre for the Study of Religion at the University of Toronto. She also teaches in the Arts Faculty, where she is the co-ordinator of Ethics, Society and the Law, a Trinity College undergraduate Arts programme. Professor Hewitt is the author of several articles and books in the areas of ethics, critical social theory, feminism and liberation theology. She is also a practising psychoanalyst.

HROTIC M. STEVEN. His primary interests concern universal patterns of human behavior as rooted in our evolutionarily constrained cognition. This has led him to study primate behavior, religion, and implicit cultural schema. Following a year of fieldwork, his current research investigates how academic thought and practice, idealized as perfectly objective and impersonal, may nevertheless reflect biases based in our biological heritage.

KIPPENBERG G. HANS was chair for theory and history of religions at the University of Bremen from 1989 to 2004. Since 1998 he has been a fellow of the Max-Weber Kolleg, Erfurt. He served as editor of *NUMEN: International Review for the History of Religions* from 1989 to 2000. He published *Discovering Religious History in the Modern Age* (2002) and was in charge of a critical edition of Max Weber's section on «Religion» in *Wirtschaft und Gesellschaft*, published uncer the authentic title *Religiöse Gemeinschaften* in 2001. Together with Martin Riesebrodt, he edited the volume *Webers «Religious systemetik»* (2001). In 2006, together with Tilman Seidensticker, he published *The 9/11 Handbook: Annotated Translated and*

Interpretation of the Attackers' Spiritual Manual. His book *Gewalt als Gottesdienst: Religionskriege der Globalisiering* appeared in 2008.

LISDORF ANDERS received his PhD "The Dissemination of Divination in Roman Republican Times – A Cognitive Approach" from the University of Copenhagen in 2007. His research interests are how the cognitive science of religion can help understand the large scale patterns of history.

McCAULEY N. ROBERT, William Rand Kenan Jr. University Professor, Director, Center for Mind, Brain, and Culture, Emory University. His research focuses on philosophy of science (especially philosophy of psychology), cognitive science of religion, naturalized epistemology. *Selected Publications: Rethinking Religion: Connecting Cognition and Culture* (Cambridge 1990) and *Bringing Ritual to Mind: Psychological Foundations of Cultural Forms* (Cambridge 2002), both with E. Thomas Lawson. Editor of *The Churchlands and Their Critics* (Blackwell 1996). Research articles in *Philosophy of Science, Synthese, Philosophical Psychology, Consciousness and Cognition, Theory and Psychology, Method and Theory in the Study of Religion, Journal of the American Academy of Religion,* and *History of Religions.*

Fellowships and Honors: grants or fellowships from the American Council of Learned Societies, British Academy, National Endowment for the Humanities, Lilly Endowment, American Academy of Religion, Templeton Foundation, and Council for Philosophical Studies; Emory Williams Distinguished Teaching Award, 1996; Massée-Martin/NEH Distinguished Teaching Professor (1994-1998, inaugural appointment); President, Society for Philosophy and Psychology (1997-1998).

McCORKLE Lee is an evolutionary anthropologist interested in ritual, language and communication. He earned his PhD in 2007 from the Institute of Cognition and Culture at the Queen's University, Belfast. He is an Assistant Professor of Psychology and Social Sciences at Tiffin University and the author of *Ritualizing the Disposal of the Deceased: From Corpse to Concept (2010).*

McCUTCHEON T. RUSSELL is Professor at the University of Alabama. His areas of interest include the history of scholarship on myths and rituals, the history of the publicly-funded academic study of religion as practiced in the U.S., secularism, as well as the relations between the classification "religion" itself and the rise of the nation-state. In the fall of 2007 he stepped down after having served three years as the Executive Secretary, Treasurer, and webmaster for the *North American Association for the Study of Religion (NAASR)*. In November of 2005 he was elected as the President of the Council of Societies for the Study of Religion (CSSR). The Council ceased existence at the end of 2009. He has also published widely on teaching and edited classroom and references resources. He is the founder and series editor for Equinox Publishing's anthology series, *"Critical Categories in the Study of Religion"*. His most recent editing project is a new monograph series, also with Equinox Press, UK, entitled *"Religion in Culture: Studies in Social Contest and Construction"*.

PACHIS PANAYOTIS is Associate Professor of religious Studies at the Aristotle University of Thessaloniki, Greece. His research focuses on the religions of the Graeco-Roman era and the method and theory in the study of religion. He has published numerous articles and books related to the above topics. He is co-editor of *Theoretical Frameworks for the Study of Graeco-Roman Religions* (2002) and *Hellenisation, empire and globalization: Lessons from Antiquity* (2004). He recently published: *Imagistic Traditions in the Graeco-Roman World: A Cognitive Modeling of History of Religious Research*, Vanias Editions, Thessaloniki 2009, *Η λατρεία της Ίσιδας και του Σάραπι. Από την τοπική στην οικουμενική κοινωνία*, Barbounakis Editions, Thessaloniki 2010, *Θρησκεία και οικονομία στην αρχαία ελληνική κοινωνία*, Barbounakis Editions, Thessaloniki 2010, *Οι «ανατολικές λατρείες» της ελληνορωμαϊκής εποχής. Συμβολή στην ιστορία και τη μεθοδολογία της έρευνας*, Barbounakis Editions, Thessaloniki 2010, *Religion and politics in the Graeco-Roman world. Redescribing the Isis-Sarapis cult*, Barbounakis Editions, Thessaloniki 2010.

PADEN E. WILLIAM has been a member of the Department of Religion, University of Vermont from 1965 until his retirement in the Spring of 2009. He served as Chair from 1972-78 and 1990-2005. He has been a visiting scholar at Wolfson College, Oxford (1999, 1992), and spent time as a research fellow and lecturer in Japan (1999, 1992). Publications include the books, *Religious Worlds: The Comparative Study of Religion* (Beacon Press,

1988; ²1994), and *Interpreting the Sacred: Ways of Viewing Religion* (Beacon Press, 1992; ²2003), both translated in numerous foreign language editions.

PETROU IOANNIS is Professor of Sociology and Ethics in the School of Theology of the Aristotle University of Thessaloniki. He has participated in two European Programmes: "European Identity, Welfare State, Religion", as scientific supervisor of the Greek Research Group, and "Material/Module for the Training of Specialized Executives in the Approach of Problems of Intercultural Communication in the Field of Local Community (Neighborhood/ Municipality/Prefecture)", as coordinator. He was a member of the European Commission Think Tank in the area of European Cultural Dimension. He is a member of the Environmental Committee of the Aristotle University as well as of the interdisciplinary network "Medical Practice and Law" and of the Bioethics Association. His recent books are: *Multiculturality and religious freedom* (2005), *Social Theory and Modern Culture* (2005), *Sociology* (2007) (in Greek).

ROBINSON DOUGLAS is a former student and teaching assistant of Luther Martin at the University of Vermont where he graduaded Phi Beta Kappa with honors in Religion in 2009. Douglas's research on method and theory and relics has resulted in numerous undergraduate conference presentations and grant funding. His honors thesis was entitled *Strategies of Ritual Presences: Relics in Cross-Cultural Perspective*.

SANZI ENNIO (Roma 1965), under the guidance of Ugo Bianchi, studied History of Religions at University of Rome «La Sapienza», graduating in 1991 with the thesis *Giove di Doliche. Dimensione sociale di organizzazione cultuale*. He recieved his PhD degree in 1997 with the thesis, *Soteriologia, escatologia e cosmologia nel culto di Mithra, di Iside e Osiride, e di Iuppiter Dolichenus. Osservazioni storico-comparative* under the guidance of Ugo Bianchi and Giulia Sfameni Gasparro. He has also published Misteri, *Soteriologia, Dualismo. Ricerche storico-religiose* (Roma 1995), *I culti orientali nell' Impero romano. Un' antologia di fonti* (Cosenza, 2003), *Cultos orientais e Magia no mundo elenístico-romano. Modelos e perspectivas metodológicas* (Fortaleza, 2006), with Carla Sfameni, *Magia e Culti orientali. Per la storia religiosa della Tarda Antichità* (Cosenza 2009), and several articles on the religious phenomena of the «Second Hellenism» under the title *Oriental Cults and Magic*. He has taught at the University of Messina and is working on Coptic sources for the fields of research mentioned above.

SAVVOPOULOS KYRIAKOS is Lecturer of archaeology and history at the Alexandria Center for Hellenistic Studies and research fellow at the Alexandria and the Mediterranean Research Center of the Bibliotheca Alexandrina. He received his PhD from the Faculty of Archaeology, Leiden University (2009). His thesis concerns the contribution of the Egyptian tradition in culture, identity and public life of Alexandria in the Hellenistic and Roman periods, as an attempt to give a better understanding in Greco-Egyptian interaction.

TRAINOR KEVIN specializes in South Asian religious traditions. His area of research includes Theravada Buddhist traditions, especially Buddhism in Sri Lanka. Other research interests include gender analysis and ritual studies. He completed an M. Div. degree in Church History at Union Theological Seminary in New York City and a Ph.D. in Religion at Columbia University; his B.A. in Religion is from Colgate University. His publications include: *Embodying the Dharma: Buddhist Relic Veneration in Asia*, edited with David Germano (SUNY Press, 2004), *Buddhism: The Illustrated Guide* (Oxford University Press, 2004), *Relics, Ritual, and Representation in Buddhism: Rematerializing the Sri Lankan Theravada Tradition* (Cambridge University Press, 1997), "Constructing a Buddhist Ritual Site: Stupa and Monastery Architecture", in *Unseen Presence: The Buddha and Sanchi*, edited by Vidya Dehejia (Marg Publications, 1996), "In the Eye of the Beholder: Nonattachment and the Body in Subha's Verse (Therigatha 71)", *Journal of the American Academy of Religion* 61 (1993): 57-79, and "When Is a Theft Not a Theft? Relic Theft and the Cult of the Buddha's Relics in Sri Lanka", *Numen* 39, (1992): 1-26.

TSALAMPOUNI ECATERINI is a Lecturer in the New Testament, School of Pastoral and Social Theology, Faculty of Theology, Aristotle University of Thessaloniki. She obtained her PhD in 1999 at the Faculty of Theology of the Aristotle University of Thessaloniki. Title: "*Macedonia in the New Testament era*". She teaches courses on NT exegesis and theology and the Greco-Roman world and the NT. Research interests: Greco-Roman world and the New Testament, Greco-Roman epigraphy and early Christianity. Member of EABS and SBL.

WHITEHOUSE HARVEY is Head of the School of Anthropology and Museum Ethnography, Director of the Centre for Anthropology and Mind, and a Fellow of Magdalen College. After carrying out two years of field research on a 'cargo cult' in New Britain, Papua New Guinea in the late eighties, he developed a theory of 'modes of religiosity' that has been the subject of extensive critical evaluation and testing by anthropologists, historians, archaeologists, and cognitive scientsts. In recent years, he has focused his energies on the development of collaborative programs of research on cognition and culture.

WIEBE DONALD is Professor at Trinity College, University of Toronto. His primary areas of research interest are philosophy of the social sciences, epistemology, philosophy of religion, the history of the academic and scientific study of religion, and method and theory in the study of religion. He is the author of *Religion and Truth: Towards and Alternative Paradigm for the Study of Religion* (1981), *The Irony of Theology and the Nature of Religious Thought* (1991), *Beyond Legitimation: Essays on the Problem of Religious Knowledge* (1994), and *The Politics of Religious Studies: The Continuing Conflict with Theology in the Academy* (1999). He has edited several books and sets of congress proceedings, and edits the series TORONTO STUDIES IN RELIGION for Peter Lang Press. In 1985 Dr. Wiebe, with Luther H. Martin and E. Thomas Lawson, founded the North American Association for the Study of Religion which became affiliated to the IAHR in 1990; he twice served as President of that Association (1986-87, 1991-92).

XYGALATAS DIMITRIS is a social anthropologist, specializing in ritual and cognition. He received his PhD. from the Institute of Cognition and Culture, Queen's University, Belfast. He has conducted extensive fieldwork in Greece, Bulgaria, Spain and Mauritius, and held post-doctoral positions at the Department for the Study of Religion at Aarhus University and the Program in Hellenic Studies at Princeton University. He is currently a research fellow at MindLab, Aarhus University Hospital.

ABBREVIATIONS

AJP	American Journal of Philology
ASS.	Archivio Storico per la Sicilia Oruentale
BSRA	Bulletin de la Société Royale d' Alexandrie. Alexandrie
BO	Beiträge zur Orientslistik
CASA	Cronache di Archeologia e di Storia di Arte
CIMRM	Vermaseren, M.J. (ed.), 1956-1960. Corpus Inscriptionum et Monumentorum Religionis Mithriacae, vols. I-II., The Hague.
CP	Classical Philology
CQ	Classical Quarterly
EPRO	M. J. Vermaseren (ed.), 1961—1990. Études Préliminaires aux Religions Orientales dans l' Empire Romain. Leiden: Brill.
GRBS	Greek, Roman and Byzantine Studies
HThR	Harvard Theological Review
HSCPh	Harvard Studies in Classical Philology
JDAI	Jahrbuch des Deutchen Archäologischen Instituts
JRS	Journal of Roman Studies
JSQ	Jewish Studies Quarterly
NScS	Notizie degli Acavi di Antichità. Roma Accad. Nazionale di Lincei
PGM:	K. Preisendanz (ed.),1972-1974. Papyri Magicae Graecae^3, Leipzig.
RAC	Reallexikon für Antike und Christentum. Stuttgart, 1950 -
RE	A. Pauly –G. Wissowa. Realencyclopädei der klassischen Altertumswissenschaft

RICIS	L. Bricault (ed.), 2005. *Recueil des inscriptions concernant les cultes isiaques* (Mémoires de l'Académie des Inscriptions et Belles-Lettres 31), vols. I-III, Paris.
ZPE	Zeitschrift für Papyrologie und Epigraphik.

LUTHER H. MARTIN
PUBLICATIONS

BOOKS:

1. *Essays on Jung and the Study of Religion,* ed. with James Goss, Lanham, MD: University Press of America, 1985.
2. *Hellenistic Religions: An Introduction.* New York: Oxford University Press, 1987.
   Brno: Masaryk University Press, Czech translation with revisions and a new Introduction, 1997.
   Thessaloniki: Vanias Press, Greek translation with revisions and a new Introduction, 2004.
3. *Technologies of the Self: A Seminar with Michel Foucault,* ed. with Huck Gutman and Patrick H. Hutton,
   Amherst: The University of Massachusetts Press, 1988.
   London: Tavistock Press, 1988.
   Tokyo: Iwanami Shoten (trans. Hajime Tamura), 1990.
   Torino: Boringhieri Editore (trans. Saverio Marchignoli), 1992.
   Frankfurt: S. Fischer Verlag (trans. Michael Bischoff), 1993.
   Seoul: Dongmoonsun, 1998.
4. *Religious Transformations and Socio-Political Change,* ed., Berlin: Mouton de Gruyter, 1993.
5. *Rationality and the Study of Religion,* ed. with Jeppe Sinding Jensen. Aarhus: Aarhus University Press, 1997; paperback edition, London: Routledge, 2003.
6. *The Study of Religion during the Cold War, East and West,* ed. with I. Dolezalová and D. Papoušek. New York: Peter Lang Press, 2001.
7. *Theoretical Frameworks for the Study of Graeco-Roman Religions,* ed. with Panayotis Pachis. Thessaloniki: University Studio Press, 2003

8. *Theorizing Religions Past: Archaeology, History, and Cognition,* ed. with Harvey Whitehouse, Walnut Creek, CA: AltaMira Press, 2004.
9. *Hellenisation, Empire and Globalisation: Lessons from Antiquity,* ed. with Panayotis Pachis. Thessaloniki: Vanias Press, 2004.
10. *Imagistic Traditions in the Graeco-Roman World,* ed. with Panayotis Pachis. Thessaloniki: Vanias Press, 2009.
11. *Past Minds: Studies in Cognitive Historiography,* ed. with Jesper Sørensen, London: Equinox, 2010.

SPECIAL ISSUES OF JOURNALS EDITED:

1. *History, Historiography, and the History of Religions, Historical Reflections/Réflexions Historiques* 20.3 (1994).
2. *Comparativism Old and New, Method & Theory in the Study of Religion* 8.1 (1996).
3. *The Definition of Religion in Social-scientific/Historical Research, Historical Reflections/Réflexions Historiques,* 25.3 (1999).
4. *Cognitive Science and the Study of Religion, Method & Theory in the Study of Religion* 16.3 (2004).
5. *Memory, Cognition and Historiography, Historical Reflections/Réflexions Historiques,* 31.2 (2005).

CHAPTERS IN BOOKS:

1. "Mythology and Cosmology in the Hellenistic World: A Methodological Inquiry", in: *Truth and Grace: Essays in Honor of Charles E. Crain,* Ira G. Zepp, Jr., ed. (Westminster, MD: Western Maryland College, 1978), Ch. 2 (n.p.)
2. "Jung as Gnostic", in: *Essays on Jung and the Study of Religion,* L Martin and J. Goss, eds. (University Press of America, 1985), 70-79.
3. "C. G. Jung and the Historian of Religion: The Case of Hellenistic Religion", in: *Essays on Jung and the Study of Religion,* L. Martin and J. Goss, eds. (University Press of America, 1985), 132-142.

4. "Identity and Self-knowledge in the Syrian Thomas Tradition", in: *Identity Issues and World Religions, Selected Proceedings of the Fifteenth Congress of the International Association for History of Religions*, V. Hayes, ed. (AASR: Australia 1986), 34-44; rpt under the title, "Technologies of the Self and Self-knowledge in the Syrian Thomas Tradition", in: *Technologies of the Self: A Seminar with Michel Foucault*, L. Martin, H. Gutman, P. Hutton, eds. (Amherst: The University of Massachusetts Press, 1988), 50-63.

5. "Fundamental Problems in the World-wide Pursuit of the Study of Religion", in: *Marburg Revisited: Institutions and Strategies in the Study of Religion* (Marburg: diagonal Verlag, 1989), 27-30.

6. "The History of Religions: A Field for Historical or Social Scientific Inquiry?", in: *Studies on Religions in the Context of Social Sciences: Methodological and Theoretical Relations*, W. Tyloch, ed. (Warsaw: Polish Society for the Science of Religions, 1990), 110-116.

7. "Genealogy and Sociology in the Apocalypse of Adam", in: *Gnosticism and the Early Christian World*, J. Goehring, C. W. Hedrick, J. T. Sanders, H. D. Betz, eds. (Sonoma, CA: Polebridge Press, 1990), 25-36.

8. "Pagan Religious Background", in: *Early Christianity: Origins and Evolution to AD 600*, Ian Hazlitt, ed. (London: SPCK/ Nashville: Abingdon, 1991), 52-64.

9. "Fate and Futurity in the Greco-Roman World", in: *Proceedings of the Boston Area Colloquium in Ancient Philosophy*, Vol. 5, J. J. Cleary and D. C. Shartin, eds. (Latham, MD: University Press of America, 1991), 291-311.

10. "Introduction", *Religious Transformations and Socio- Political Change: Eastern Europe and Latin America*, L. Martin, ed. (Berlin: Mouton de Gruyter, 1993), 1-5.

11. "The New Testament Writings as Hellenistic Religious Texts", in: *Bible in Cultural Contact*, D. Papoušek and H. Pavlincová, eds. (Brno: Czech Society for the Study of Religion, 1994), 209-213.

12. "Religion and Dream Theory in Late Antiquity", in: *The Notion of "Religion" in Comparative Research*, ed. Ugo Bianchi, (Rome: "L'erma" di Bretschneider, 1994), 369-374.

13. "Reflections on the Mithraic Tauroctony as Cult Scene", in: *Studies in Mithraism*, ed. J. R. Hinnells, Storia delle Religioni 9, (Rome: "L'erma" di Bretschneider, 1994), 217-224.

14. "The Manichaean Mission in China: Systemic or Syncretistic?" [in Chinese], in: *Shijie Zongjiao Ziliao* 93 (1993), 23-26, and [in English] in: *Religion and Modernization in China*, Dai Kangsheng, Zhang Xinying, Michael Pye, eds. (Cambridge: Roots and Branches, 1995), 187-196.

15. "Secrecy in Hellenistic Religious Communities", in: *Secrecy and Concealment in Late Antique and Islamic History of Religions*, H. Kippenberg and G. Stroumsa, eds., (Leiden: E. J. Brill, 1995), 101-121.

16. "Historicism, Syncretism, Comparativism", *Religions in Contact*, D. Papoušek and I. Doleželová, eds. (Brno: Czech Society for the Study of Religion, 1996), 31-37.

17. "Self and Power in the Thought of Plotinus", in: *Człowiek i Wartości*, A. Komendera, ed. (Cracow: Wydawnictwo Naukowe WSP, 1997), 91-99.

18. "Akin to the Gods or Simply One to Another: Comparison with respect to Religions in Antiquity", in: *Vergleichen und Verstehen in der Religionswissenschaft*, H.-J. Klimkeit, ed. (Wiesbaden: Harrassowitz, 1997, 147-159.

19. "Prayer in Greco-Roman Religions" (with L. Alderink), in: *Prayer from Alexander to Constantine: A Critical Anthology*, M. Kiley, ed. (London: Routledge, 1997), 123-127.

20. "Rationality and Relativism in History of Religions Research", in: *Rationality and the Study of Religion*, Jeppe S. Jensen and L. Martin, eds. (Aarhus: Aarhus University Press, 1997); paperback ed.(London: Routledge, 2003), 145-156.

21. "Secular Theory and the Academic Study of Religion", in: *Secular Theories on Religion. A Selection of Recent Academic Perspectives*, T. Jensen and M. Rothstein, eds. (Copenhagen: The Museum Tusculanum Press, 2000), 137-148; Greek translation: *Journal of the Theological Faculty of the Aristotle University of Thessaloniki*, 1999 (2001), 163-175.

22. "Of Religious Syncretism, Comparative Religion and Spiritual Quests", in: Special issue of *Method & Theory in the Study of Religion* 12 (2000)/*Method and Theory in the IAHR: Collected Essays from the XVIIth Congress of the International Association for the History of Religions, Mexico City 1995*, A. Geertz and R. McCutcheon, eds. (Leiden: E. J. Brill, 2000), 277-286.
23. "Greek and Roman Philosophy and Religion", in: *The Early Christian World*, 2 vols, Philip F. Esler, ed. (London: Routledge, 2000), 1: 53-79.
24. "La helenización del pensamiento judeochristiano", in: *La genealogía del christianismo¿ origen de Occidente?*, Herbert Frey, ed. (Mexico City: Conaculta, 2000), 92-110.
25. "Comparison", *Guide to the Study of Religion*, W. Braun and R. McCutcheon, eds. (London: Cassell Academic, 2000), 45-56; rpt. in Greek, *Egheiridio Threskeiologias*, D. Xygalatas, trans. Thessaloniki: Vanias, 2004, pp. 93-111.
26. "The Academic Study of Religion during the Cold War: The Western Perspective", in: *The Study of Religion during the Cold War, East and West*, ed. with I. Dolezalová, L. H. Martin and D. Papoušek, eds. (New York: Peter Lang Press, 2001), 209-223; rpt. in: *What is Religious Studies? A Reader in Disciplinary Formation*, S. J. Sutcliffe, ed. (London: Equinox, *in press*).
27. "Ritual Competence and Mithraic Ritual", in: *Religion as a Human Capacity: A Festschrift in Honor of E. Thomas Lawson*, T. Light and B. Wilson, eds. (Leiden: E. J. Brill, 2004), 295-263.
28. "History, Historiography, and Christian Origins: The Case of Jerusalem", *Redescribing Christian Origins*, R. Cameron and M. Miller, eds. (Leiden: E. J. Brill, 2004), 263-273.
29. "Redescribing Christian Origins: Historiography or Exegesis?", *Redescribing Christian Origins*, R. Cameron and M. Miller, eds. (Leiden: E. J. Brill, 2004), 475-481.
30. "New approaches to the Study of Syncretism" (with Anita Leopold), *New Approaches to the Study of Religion*, 2 vols., P. Antes, A. Geertz and R. Warne, eds. (Berlin: Walter de Gruyter, 2004), 2: 93-107.

31. "Petitionary Prayer: Cognitive Considerations", in: *Religion im kulturellen Diskurs. Festschrift für Hans G. Kippenberg zu seinem 65. Geburtstag/ Religion in Cultural Discourse. Essays in Honor of Hans B. Kippenberg on Occasion of His 65th Birthday*, B. Luchesi and K. von Stuckrad, eds. (Berlin: Walter de Gruyter, 2004), 115-126.

32. "Towards a Cognitive History of Religions", *Unterwegs. Neue Pfade in der Religionswissenschaft. Festschrift für Michael Pye zum 65. Geburtstage/ On the Road. New Paths in the Study of Religions. Festschrift in Honour of Michael Pye on His 65th Birthday*, C. Kleine, M. Schrimpf, K. Triplett, eds. (München: Biblion, 2004), 75-82; rpt. with corrections and minor revisions in: *REVER* [*Revista de Estudos de Religico*: Pontifical Catholic University, Sãn Paulo] 5 (2005), www.pucsp/rever/rv4_2005/.

33. "Towards a Scientific History of Religions", in: *Theorizing the Past: Historical Evidence for the Modes Theory*, H. Whitehouse and L. Martin, eds. (Walnut Creek, CA: AltaMira Press, 2004), 7-14.

34. "Forward", Roger Beck, *Beck on Mithraism: Collected Works with New Essays.* (London: Ashgate, 2004), xiv-xv.

35. "The Very Idea of Globalization: The Case of Hellenistic Empire", in: *Hellenisation, Empire and Globalization: Lessons from Antiquity*, L. Martin and P. Pachis, eds. (Thessaloniki: Vanias Press, 2004), 123-139.

36. "Religion and Cognition", in: *The Routledge Companion to the Study of Religion*, J. Hinnells, ed. (London: Routledge, 2005), 473-488; revised for 2nd edition, 2010, 526-542.

37. "Performativity, Discourse and Cognition: 'Demythologizing' the Roman Cult of Mithras", in: *Rhetoric and Reality in Early Christianity,* W. Braun, ed. (Waterloo: Wilfrid Laurier University Press, 2005), 187-217.

38. "The Promise of Cognitive Science for the Historical Study of Religions, with Reference to the Study of Early Christianity", in: *Explaining Early Judaism and Christianity: Contributions from Cognitive and Social Sciences*, P. Luomanen, I. Pyysiäinen, and R. Uro, eds. (Leiden: E. J. Brill, 2007), 37-56.

39. "What Do Religious Rituals Do? (And How Do They Do It?): Cognition and the Study of Religion", in: *Introducing Religion: Essays in Honor of Jonathan Z. Smith*, R. McCutcheon and W. Braun, eds. (London: Equinox, 2008), 325-339.
40. "The Academic Study of Religion: A Theological or Theoretical Undertaking?", in: *Cultures in Contact: Essays in Honor of Professor Gregorios D. Ziakas*, P. Pachis, P. Vasiliadis, and D. Kaimakis, eds. (Thessaloniki: Vanias Press, 2008), 333-345.
41. "Can Religion Really Evolve? (And What Is It Anyway?)", in: *The Evolution of Religions: Studies, Theories, and Critiques*, J. Bulbulia, R. Sosis, E. Harris, R. Genet, C. Genet and K. Wyman, eds. (Santa Margarita, CA: The Collins Foundation Press, 2008), 349-355.
42. "The Amor and Psyche Relief in the Mithraeum of Capua Vetere: An Exceptional Case of Graeco-Roman Syncretism or an Ordinary Instance of Human Cognition?", in: *The Mystic Cults of Magna Grecia*,. P. A. Johnston and G. Casadio, eds. (Austin: University of Texas Press, 2009), 277-289.
43. "Why Christianity was Accepted by Romans But Not by Rome", in: *Religionskritik in der Antike*, U. Berner and I. Tanaseanu, eds. (Münster: LIT-Verlag, 2009), 93-107.
44. "Imagistic Traditions in the Graeco-Roman World", in: *Imagistic Traditions in the Graeco-Roman World: A Cognitive Modeling of History of Religious Research*, L. Martin and P. Pachis, eds. (Thessaloniki: Vanias, 2009), 237-247.
45. "Origins of Religion, Cognition and Culture: The Bowerbird Syndrome", in: *The Origins of Religion, Cognition and Culture*, A. Geertz, ed. (London: Equinox, *in press for 2010*).
46. "Evolution, Cognition, and History", in: *Past Minds: Studies in Cognitive Historiography*, L. Martin and J. Sørensen, eds. (London: Equinox, *in press for 2010*).
47. "The Deep History of Religious Ritual", in: *Ritual, Cognition and Culture*, A. Geertz, ed. (London: Equinox, *forthcoming*).

48. **"Past Minds: Evolution, Cognition and Biblical Studies"**, in: *Cognitive Science Approaches in Biblical Studies,* R. Uro and I. Czachesz, eds., *forthcoming.*

ARTICLES:

1. "The Nekyia of Belle de Jour", *Agora* 1 (1970), 43-49.
2. "'The Treatise on the Resurrection' (CG I,3) and Diatribe Style", *Vigiliae Christianae* 27 (1973), 277-280.
3. "Note on 'The Treatise on the Resurrection' (CG I,3) 48.36", *Vigiliae Christianae* 27, (1973): 281.
4. "The Anti-philosophical Polemic and Gnostic Soteriology in 'The Treatise on the Resurrection' (CG I, 3)", *Numen* 20 (1973), 20-37.
5. "History of the Psychological Interpretation of Alchemy", *Ambix* 22 (1975), 10-20.
6. "Josephus' Use of Heimarmene in the Jewish Antiquities XII, 171-3", *Numen* 22 (1981), 127-137.
7. "Why Cecropian Minerva? Hellenistic Religious Syncretism as System", *Numen* 30 (1983): 131-145.
8. "Hawthorne's The Scarlet Letter: A is for Alchemy?" *American Transcendental Quarterly* 58 (1985), 31-42.
9. "Those Elusive Eleusinian Mystery Shows", *Helios* 13 (1986), 17-31.
10. "Aelius Aristides and the Technology of Oracular Dreams", *Historical Reflections/Réflexions Historique* 14 (1987), 65-72.
11. "Roman Mithraism and Christianity", *Numen* 36 (1989), 2-15.
12. "The Study of Religion in its Social - Scientific Context: A Perspective on the 1989 Warsaw Conference on Methodology" (with E. Thomas Lawson and Donald Wiebe), *Method & Theory in the Study of Religion* 2 (1990), 98-101.
13. "The *Encyclopedia Hellenistica* and Christian Origins", *Biblical Theology Bulletin,* 20 (1990), 123-127.
14. "Greek Goddesses and Grain: The Sicilian Connection", *Helios* 17 (1990), 251-261.
15. "Report on the XVIth Congress of the International Association for the History of Religion" (with Rosalind I. J. Hackett and

Donald Wiebe), *Method & Theory in the Study of Religion*, 2 (1990), 250-252.

16. "Fate, Futurity, and Historical Consciousness in Western Antiquity", *Historical Reflections /Réflexions Historique* 17 (1991), 151-169.

17. "Artemidorus: Dream Theory in Late Antiquity", *The Second Century* 8 (1991), 97-98.

18. "Recent Historiography and the History of Religions", *Method & Theory in the Study of Religion*, 3 (1991), 115-120.

19. "On Declaring WAR: A Critical Comment" (with Donald Wiebe), *Method & Theory in the Study of Religion* 5 (1993), 47-52.

20. "The Academic Study of Religion in the United States. Historical and Theoretical Reflections", *Religio. Revue pro religionistiku* 1 (1993), 73-80.

21. "Religia I wlâdza polityczna. Hipotezy krytyki wybranych wzorów zaleźnoź" [Reflections on Religion, Power and their Relationship]. *Przegląd Religioznawczy* 2/168 (1993), 67-74.

22. "The Anti-Individualistic Ideology of Hellenistic Culture", *Numen* 41 (1994), 117-140.

23. "Gods or Ambassadors of God: Barnabas and Paul in Lystra", *New Testament Studies* 41 (1995), 152-156.

24. "The Discourse of (Michel Foucault's) Life: A Review Essay", *Method & Theory in the Study of Religion* 7 (1995), 57-69.

25. "The Post-Eliadean Study of Religion and the New Comparativism", Introduction to a Symposium on "The New Comparativism in the Study of Religion", L. Martin, ed. *Method & Theory in the Study of Religion* 8 (1996), 1-3.

26. "Historicism, Syncretism, and the Cognitive Alternative: A Response to Michael Pye", *Method & Theory in the Study of Religion* 8 (1996), 215-224; rpt. in A. Leopold and J. S. Jensen, eds. *Syncretism in the Study of Religion*. (London: Routledge, 2004), 286-294.

27. "Cognition, Capitalism, and Causality": Report on NAASR/ SSSR Panels, Nashville, November 1996, with D. Krymkowski, *Method & Theory in the Study of Religion* 9 (1997), 163-167.

28. "Biology, Sociology and the Study of Religion", *Religio. Revue pro religionistiku* 5 (1997), 21-35.
29. "Religion as an Independent Variable: An Exploration of Theoretical Issues" (with Daniel Krymkowski), *Method & Theory in the Study of Religion* 10, 1998, 187-198.; rpt. in Polish translation as "Religia jako Zmienna Niezalena: Rewizja Hipotezy Webera", *Ask: społeczenstwo, badania, metody* 7 (1998), 7-16.
30. "Introduction: The Definition of Religion in the Context of Social-Scientific Study", *Historical Reflections/Réflexions Historique* 25 (1999), 387-389.
31. "A Scientific Explanation for Religion", *05401* (2000), 16-18.
32. "Kingship and the Consolidation of Religio-Political Power during the Hellenistic Period", *Religio. Revue pro religionistiku* 8 (2000), 151-160; rev. version, in: *Theoretical Frameworks for the Study of Graeco-Roman Religions*, L. Martin and P. Pachis, eds. Thessaloniki: University Studio Press (2003), 89-96.
33. "History, Historiography and Christian Origins", *Studies in Religion/ Sciences religieuse* 29 (2000), 69-90.
34. Review Symposium on Russell T. McCutcheon's *Manufacturing Religion: The Discourse on sui Generis Religion and the Politics of Nostalgia*, L. Martin, ed. and "Introduction", *Culture and Religion* 1 (2000), 95-97.
35. "Comparativism and Sociobiological Theory", *Numen* 48 (2001), 290-308.
36. "To use 'Syncretism', or not to use 'Syncretism': That is the Question", *Historical Reflections/Réflexions Historique* 27 (2001), 389-400.
37. "Marcel Gauchet's *The Disenchantment of the World*", "Review Symposium on Marcel Gauchet, *The Disenchantment of the World: A Political History of Religion*, L. Martin, ed., *Method & Theory in the Study of Religion* 14 (2002), 114-120.
38. "Rituals, Modes, Memory and Historiography: The Cognitive Promise of Harvey Whitehouse", *Journal of Ritual Studies* 16 (2002), 30-33.
39. "Cognition, Society and Religion: New Approaches in the Study of Culture" [in Chinese], *Zhongguo Xueshu* [*China Scholarship*] 3

(2002), 77-104; English, rev. *Culture and Religion* 4 (2003), 207-231.

40. "'Disenchanting' the Comparative Study of Religion", *Method & Theory in the Study of Religion* 16 (2004), 36-44.

41. "History, Cognitive Science and the Problematic Study of Folk Religions: The Case of the Eleusinian Mysteries of Demeter", *Temenos: Nordic Journal of Comparative Religion* 39-40 (2003/2004), 81-99.

42. "Establishing a Beachhead: NAASR, Twenty Years Later", with Donald Wiebe, 2004, published electronically: www.as.ua.edu/naasr/ Establishing abeachhead.pdf

43. "Towards a New Scientific Study of Religion: Commentary on Scott Atran and Ara Norenzaya 'Religion's Evolutionary Landscape'", *Behavioral and Brain Sciences* 27 (2004), 744-745.

44. "The Hellenisation of Judaeo-Christian Faith or the Christianisation of Hellenic Thought?", *Religion & Theology* 12 (2005), 1-19.

45. "Aspects of 'Religious Experience' among the Graeco-Roman Mystery Religions", *Religion & Theology* 12 (2005), 349-369.

46. "Cognitive Science, Ritual, and the Hellenistic Mystery Religions", *Religion & Theology* 13 (2006), 383-395.

47. "The Roman Cult of Mithras: A Cognitive Perspective", *Religion. Revue pro religionistiku* 14 (2006), 131-146.

48. "Rozhovors s [Interview with] profesorem Lutherem H. Martinem", Aleš Chalupa, Kateřina Řepová, Radek Kundt, *Sacra aneb Rukovet religionisty* (Brno) 4 (2006), 69-78.

49. Obituary for Gary Lease (with Donald Wiebe), *Numen* 55 (2008), 337-339; rev. (with Russell McCutcheon) for *Method & Theory in the Study of Religion* 21 (2009), 109-112.

50. "Daniel Dennett's *Breaking the Spell*: A Response to Its Critics", *Method & Theory in the Study of Religion* 20.1 (2008), 61-66.

51. "The Uses (and Abuse) of the Cognitive Sciences for the Study of Religion", *CSSR Bulletin* 37 (2008), 95-98.

52. "Globalization, Syncretism, and Religion in Western Antiquity: Some Neurocognitive Considerations", *Zeitschrift für Missionswissenschaft und Religionswissenschaft* 94.1-2 (2010), 5-17.

RESEARCH EXPERIMENTS CITED

"Pilot Study: Repetition and Memory in the Transmission of Religious Doctrine", designed in collaboration with Justin Barrett and Harvey Whitehouse, carried out at The University of Vermont with the assistance of Shane Barney, 2001. Cited by Whitehouse, "Modes of Religiosity: Towards a Cognitive Explanation of the Sociopolitical Dynamics of Religion", in *MTSR* 14-3/4 (2002): 293-315, 297, n. 4.; and Whitehouse, *Modes of Religiosity: A Cognitive Theory of Religious Transmission*, Walnut Creek, CA: AltaMira Press, 2004: 97-98 and n. 1.

PUBLISHED PHOTOGRAPHS:

Fresco from the Marino Mithraeum:
Cover Plate, *The Origin of the Mithraic Mysteries*, David Ulansey, New York: Oxford University Press, 1989; *Scientific American*, December 1989, 135. *Biblical Archaeology Review* 20.5 (Sept/Oct) 1994, 44-45.

Tauroctony from the Mithraeum of the Circus Maximus, 249
Mithraeum of the "Seven Spheres", Ostia, 252 in: L. Martin, "Ritual Competence and Mithraic Ritual", in: *Religion as a Human Capacity: A Festschrift in Honor of E. Thomas Lawson*, T. Light and B. Wilson, eds. (Leiden: E. J. Brill, 2004).

REVIEWS:

1. I. M. Lewis, *Ecstatic Religion: An Anthropological Study of Spirit Possession and Shamanism*, in: *Journal of the American Academy of Religion* 40 (1972), 528-529.
2. Wayne A, Meeks, *The Writings of St. Paul*, in: *Journal of the American Academy of Religion* 40 (1972), 542-543.
3. Elaine Pagels, *The Johannine Gospel in Gnostic Exegesis*, in: *Journal of the American Academy of Religion* 43 (1975), 325-326.

4. James Hillman, *Re-visioning Psychology*, in: *The Review of Books and Religion* 4.8 (1975), 10.
5. Hwa-Sun Lie, *Der Begriff Skandalon im Neuen Testament und der Wiederkehrgedanke bei Laotse*, in *Journal of Biblical Literature* 94 (1975), 633-634.
6. Victor Turner, *Revelation and Divination in Nbembu Ritual*, in: *The Review of Books and Religion* 5 (1976), 15.
7. Bennetta Jules-Roserne, *African Apostles: Ritual and Conversion in the Church of John Marakne*, in: *The Review of Books and Religion* 5 (1976), 15.
8. C. J. Bleeker, *The Rainbow: A Collection of Studies in the Science of Religion*, in: *Journal of the American Academy of Religion* 44 (1976), 588-589.
9. Viktor E. Frankl, *The Unconscious God: Psychotherapy and Theology*, in: *Parabola* 1 (1976), 112.
10. Daniel Noel, *Seeing Castaneda - Reactions to the 'Don Juan' Writings of Carlos Castaneda*, in: *The Review of Books and Religion* 5.6 (1976), 8.
11. M.-L. von Franz, *C. G. Jung: His Myth in Our Time*, in: *Ambix* 24 (1977), 65-66.
12. Walter H. Capps (ed.), *Seeing With a Native Eye. Essays in Native American Religion*, in: *The New Review of Books and Religion* 1.8 (1977), 11-12.
13. Wendy O'Flaherty, *Critical Study of Texts*, in: *The New Review of Books and Religion*, 4.9 (1980), 14.
14. Holmes Welch and Anna Seidel (eds.), *Facts of Taoism*, in: *The New Review of Books and Religion* 10.5 (1982), 11.
15. Wayne Meeks, *The First Urban Christians*, in: *Journal of the American Academy of Religion* 51 (1983), 686.
16. Howard C. Kee, *Miracle in the Early Christian World: A Study in Sociohistorical Method*, in: *Journal of the American Academy of Religion* 53 (1985), 138.
17. Robert L. Wilken, *The Christians as the Romans Saw Them*, in: *Journal of the American Academy of Religion* 53 (1985), 319-320.

18. Terrence W. Tilley, *Story Theology*, in: *Founders Hall. St. Michaels College Alumni Publication*, October (1985), n.p.
19. S. R. F. Price, *Rituals and Power: The Roman Imperial Cult in Asia Minor*, in: *Journal of the American Academy of Religion* 54 (1986), 189-190.
20. Walter Burkert, *Greek Religion*, in: *Journal of the American Academy of Religion* 54 (1986), 766-767.
21. H. Schutz, *The Romans in Central Europe*, in: *Journal of the American Academy of Religion* 54 (1986), 799-800.
22. Reinhold Merkelbach, *Mithras*, in: *Journal of Biblical Literature* 106 (1987), 155-157.
23. Marvin S. Meyer, *The Ancient Mysteries: A Sourcebook*, in: *Critical Review of Books in Religion* (1988), 307-308.
24. Abraham J. Malherbe, *Moral Exhortation, A Greco-Roman Sourcebook*, in: *Critical Review of Books in Religion* (1988), 304-307.
25. Robert A. Segal, *The Poimandres as Myth: Scholarly Theory and Gnostic Meaning*, in: *Journal of the American Academy of Religion* 56 (1988), 816-818.
26. Walter Burkert, *Ancient Mystery Cults*, in: *Critical Review of Books in Religion* (1989), 288-290.
27. J. A. Trumbower, *Born from Above: The Anthropology of The Gospel of John*, in: *Founders Hall. St. Michaels College Alumni Publication* 14 (1992), n.p.
28. Brian E. Daley, *The Hope of the Early Church: A Handbook of Patristic Eschatology*, in: *Numen* 39 (1992), 265-267.
29. Jonathan Z. Smith, *Drudgery Divine: On the Comparison of Early Christianities and the Religions of Late Antiquity*, in: *Critical Review of Books in Religion* (1992), 275-278.
30. H. S. Versnel, *Ter Unus: Isis, Dionysus, Hermes: Three Studies in Henotheism*, in: *Critical Review of Books in Religion* (1992), 307-309.
31. E. Koskenniemi, *Der philostrateische Apollonios*, in: *Critical Review of Books in Religion* (1992), 298-300.

32. Robert Garland, *Introducing New Gods: The Politics of Athenian Religion*, in: *Method & Theory in the Study of Religion* 6 (1994), 377-382.
33. William M. Brashear, *A Mithraic Catechism from Egypt (P. Berol. 21196)*, Tyche Supplement band, in: *Numen* 41 (1994), 325-326.
34. Antoine Faivre, *The Eternal Hermes: From Greek God to Alchemical Magus*, in: *Numen* 43 (1996), 317-318.
35. Review essay: Program for the Analysis of Religion among Latinos Studies Series: A. M. Stevens-Arroyo & A. M Díaz-Stevens (eds), *An Enduring Flame: Studies on Latino Popular Religiosity*; A. M. Stevens-Arroyo & G. R. Cadena (eds), *Old Masks, New Faces: Religion and Latino Identities*; A. M. Stevens-Arroyo & A. I. Pérez y Mena (eds), *Enigmatic Powers: Syncretism with African and Indigenous Peoples' Religions Among Latinos*; A. M. Stevens-Arroyo & S. Pantoja (eds), *Discovering Latino Religion: A Comprehensive Social Science Bibliography*, in: *Numen* 45 (1998), 104-107.
36. Guy Stroumsa, *Hidden Wisdom: Esoteric Traditions and the Roots of Christian Mysticism*, in: *Numen* 45 (1998), 222-223.
37. J. Kloppenborg and S. Wilson, *Voluntary Associations in the Graeco-Roman World*, in: *Method & Theory in the Study of Religion* 11 (1999), 150-155.
38. Keith Hopkins, *A World Full of Gods: Pagans, Jews and Christians in the Roman Empire*, in: *Numen* 48 (2001), 122-124.
39. J. A. Trumbower, *Rescue for the Dead*, in: *Saint Michael's College Magazine* 2.3 (2002), 13.
40. Antonia Tripolitis, *Religions of the Hellenistic-Roman Age*, in: *The Journal of Religion* 83 (2003), 329-330.
41. M. D. Faber, *The Magic of Prayer: An Introduction to the Psychology of Faith*, in: *Studies in Religion/Sciences Religieuses* 32 (2003), 205-206.
42. Hans Kippenberg, *Discovering Religious History in the Modern Age*, in *History of Religions*, in: *History of Religions* 43.3 (2004), 251-254.

43. René Gothóni, ed. *How to do Comparative Religion? Three Ways, Many Goals.* Berlin: Walter de Gruyter, in: *Numen* 53 (2005), 227-229.
44. Cynthia White, *The Emergence of Christianity*, in: *Journal of Church and State* 50 (2008), 588-589.
45. Jerome H. Neyrey, *Give God the Glory: Ancient Prayer and Worship in Cultural Perspective*, in *Biblical Theology Bulletin* 39 (2009), 50-51.

PREFACE

We are delighted, finally, to be able to present this volume of essays to Luther Martin in recognition of his contributions to the academic study of religion not only "as a catalytic presence at the University of Vermont and its Department of Religion" – as his colleague Bill Paden put it in a citation to mark Martin's recent retirement from the University – but also as "a mover and shaker on the larger world stage" of Religious Studies. Martin's entire forty-three year teaching career, from instructor to Professor Emeritus, has been spent in the Department of Religion at the University of Vermont. The local impact of his teaching there is clearly evident in the large number of students attracted to his courses and those who have taken on graduate work in the field or have taken up a career in religious studies or related fields. His influence, however, has also been global by virtue of his capacity to draw scholars from around the world to spend sabbatical research leaves at UVM and by his active writing and publication program that drew invitations to conferences, congresses, and visiting professorships at such institutions as Queen's University, Belfast, Masaryk University in Brno, and Aristotle University in Thessaloniki.

Martin's enthusiastic engagement in what might well be called "the politics of religious studies" – his co-founding of the North American Association for the Study of Religion (and its prestigious journal, *Method and Theory in the Study of Religion*) and, later in his career, his involvement in the founding of the *International Association for the Cognitive Science of Religion* – mark major developments in the study of religion in the context of the modern research university. He has also worked diligently to connect these ventures with the work of the International Association for the History of Religions, the oldest organization of scholars and scientists engaged in the objective, non-partisan study of religion. His overall contribution to the field of Religious Studies, therefore, is remarkable not only because of the professionalism of his scholarship on Hellenistic religions, but also because of the innovative application of the cognitive (and neuro-) sciences to those traditions and his active participation in the methodological debates to encourage the objective study of religions in colleagues and universities around the world. He has, moreover, by dint of his gregarious personality and his "love for life" (food, wine, spirits, and spirited conversation) created a community of scholarship with students, colleagues, and friends interested in understanding and explaining religion.

This overriding interest in explaining religion explains the title of this volume of essays. Whether as a historian whose work, according to Martin, can be compared with the investigations of the detective who patiently and insightfully finds solutions to unresolved problems, or as a scientists whose main objective is not simple description but finding the causes for events and actions, Martin has been primarily concerned to get to the roots of the religious impulse in human existence. Like a hunter, he has been "chasing down religion" for years, holding it in the sights (crosshairs) of history and the cognitive sciences. And, perhaps like Dawkins and Dennett (and most of the contributors to this volume) he would also very much like to see the demise of uncritical studies of religion because of what he considers their detrimental effects on modern society. Unlike the former, however, he is more measured and reasonable in his public statements about religion.

The essays that comprise this volume constitute a commemoration and celebration of a long and productive career. They are not restricted to scholars of his own generation; they include contributions from his brightest undergraduate students and those of his students who have proceeded to graduate work in the field. Luther's diverse interests, and the overlapping fields and disciplines in which he has worked and taught, makes it virtually impossible to place these essays into separate categories. The essays, therefore, follow in alphabetical order, but this is not to say that there is no structure or shape to the volume for the contributors deal with the three major foci of his work – the Greco-

Roman world, cognitive science approaches to explaining religious phenomena, and methodological issues in the academic study of religions. This volume, therefore, is more than mere commemoration and celebration;

sound scholarship and critical engagement of issues in one or other of these areas of thought and research characterizes the essays published here, and they will, we believe, contribute to an ongoing discussion and debate that will ultimately lead to an explanation of religion.

We hope Luther will accept these essays as a token of the respect in which he is held by his colleagues in the religious studies community represented in our regional, national, and international societies and associations for this field, and we hope the essays herein will encourage and stimulate others in our field to continue with the project of the scientific study of religion.

Panayotis Pachis
Donald Wiebe

ANCIENT AND MODERN APPROACHES TO THE REPRESENTATION OF SUPERNATURAL BEINGS: DIO CHRYSOSTOM (ORATION 12) AND DAN SPERBER (EXPLAINING CULTURE) COMPARED[1]

ROGER BECK

In Oration 12, composed for and delivered at the Olympic Games of 97, 101 or 105 CE, Dio Chrysostom treats religion in terms of the formation and transmission of representations, both private or mental and public. The public representation on which Dio focuses is of course the great statue of Zeus which Phidias had created half a millennium before. Dio poses the question: precisely what is it that Phidias represented in that statue, and what are the mental representations that we should make or do in fact make when we view it?

Dio's approach is in ways analogous to modern cognitivist approaches, especially to Dan Sperber's theory of religions as 'epidemics' of mental and public representations. Dio also anticipates (as do other ancient sources) the modern cognitivist premise of the naturalness of religion: that is, the assumption that the representation of supernatural beings is a normal function of the human mind/brain to be studied and explained as such.

---

1. Versions of this contribution were presented at the NAASR annual meeting (Philadelphia, November 2005) and at the APA annual meeting (Chicago, January 2008). Certain background explanatory passages I have reproduced or paraphrased from my book, The Religion of the Mithras Cult in the Roman Empire (Beck 2006: esp. p. 89). There seemed little point in going out of my way to give essentially the same explanation in entirely new language.

***

First let me sketch out Sperber's paradigm of a religion, as he presents it in his *Explaining Culture: A naturalistic approach* (1996). A religion, according to Sperber, is the sum of the mental representations of its adherents, its material representations (texts, artifacts, meeting places, etc.), and its enacted representations (rituals, words spoken and received, etc.). A living religion, one might say more abstractly, is the interplay between the mental representations of its adherents and the public representations which give expression to, and in turn condition, the mental representations.

The cognitive approach thus locates a religion primarily in the representation-forming minds of those who adhere to it, whether actively or passively. As humans we all form representations of beings which do not exist in the natural world, at least in a normal, empirically testable way. Such representations are for the most part evanescent. Some representations, however, because they are conformable to the representations of others in the same sociocultural group, are preserved, fostered, and modified by the interaction of mind with mind; also of mind with the projections of mind in the actual world: spoken and written words, artistic representations, mimetic ritual, and so on.

Every thought, every mental representation is an event in the actual world: it either is, or correlates with, a brain event, a firing of neurons which is in principle observable and describable in entirely natural terms. That the mental representations of dead Zeus-worshippers are irrecoverable is beside the point; what matters is that as natural events they are empirically reconstructable, at least in principle. We give no hostages to a metaphysical realm of 'the sacred' or whatever.

Sperber treats cultures, and a fortiori religions, as epidemics of mental representations. He intends 'epidemics' literally, not metaphorically. But the human mind/brain is not a passive transmitter: output never precisely matches input. To quote: 'The most obvious lesson of recent cognitive work is that recall is not storage in reverse, and comprehension is not expression in reverse. Memory and communication transform information' (Sperber 1996: 31).

And now to the gods. Not all representations of supernatural beings are 'religious', in the sense of belonging to that domain of life which we label 'religion'. Nor do all religious representations necessarily involve supernatural beings. Nevertheless, there is a high degree of coincidence: more often than not, the mental representation of supernatural beings is a vital part of practicing a religion. Certainly, that was so for the religions current in classical antiquity. Paganism was literally unthinkable without the mental representation of the Olympian and countless other gods.

While there is little religious thought that does not in a primary or secondary way involve supernatural beings, the human mind does construct and entertain representations of innumerable supernatural and paranatural beings entirely outside the religious domain. Folk tale and fantasy literature abound with inventions of this sort, whose connection with 'religion' is tenuous or nonexistent. And that is precisely the point. Except in the degree of ontological commitment, there is no essential difference between an ancient Athenian's representation of Pallas Athena or an ancient Roman's of Jupiter Optimus Maximus and a twenty-first century person's representations of wizards, elves, orcs, and dementors à la J. R. R. Tolkien or J. K. Rowling. I do not mean to trivialize the question of ontological commitment [Sometimes called 'the Mickey Mouse problem' (Atran 2002: x)]. Certainly, the ancient Athenian and the ancient Roman committed to their patron deities in a way that we do not commit to wizards, elves, orcs, and dementors - or for that matter to Pallas Athena and Jupiter Optimus Maximus. But that problem is not of much relevance here. Our point is simply the uniform way in which representations of supernatural beings are formed in the human mind/brain.

The first benefit of the cognitive approach is thus to strip away the special status of 'religious' representations of supernatural beings, and consequently to de-mystify and de-problematize them. The ability to form mental representations of supernatural and paranatural beings is simply part of the evolved mental endowment of the species Homo sapiens[2].

As a further consequence, the cognitive approach radically redefines the 'why' questions: why religion, why the gods? Granted our natural propensity to entertain representations of the non-natural, it is not the presence of the gods in our minds that requires explanation so much as their expulsion in relatively recent times; not 'why religion?' but rather 'why religion no longer?' As the cognitivists emphasize, religion is 'natural', science is not (McCauley 2000).

---

2. On evolution and selection in the cognitive study of religion (CSR) see Atran 2002, Boyer 2001, Mithen 1996, Tooby and Cosmides 1992. On the application of CSR methods to the historical study of religions, see Whitehouse and Martin 2004

***

Now to Dio Chrysostom, Oration 12, and the subject of the speech, Phidias' great - in size, acknowledged artistry, and numinous presence - statue of Olympic Zeus. As I said in my introduction, Dio's speech poses the question: just what is it that Phidias represented in that statue, and what are the mental representations that we should make or do in fact make when we view it?

For Dio, unlike Sperber, there were objective realities to which both the statue and our mental representations of the god relate. Dio's question, then, is both normative and practical: are our representations of Olympic Zeus appropriate? Can we improve them, both privately in our minds and publicly in, for example, images and formal orations?

We of course in the secular academy must discount that would-be reality of a Zeus-out-there. We must treat the representations, public and mental, as the sole reality. When we do so, we shall find Dio an excellent and up-to-date guide to the cult of Olympic Zeus interpreted as a particular stream of representations negotiated and transmitted by its participants. As participants we must of course include the athletes and spectators of the Olympic games, although Dio in his speech is not greatly concerned with that aspect of the cult.

The climax of Dio's Olympicus is an imaginary speech by the sculptor Phidias in which he defends himself against the charge of sculpting so definitive a representation of Zeus that subsequent representations of the god, at least in the medium of the visual arts, are inconceivable. Phidias doesn't so much rebut the charge, which is of course a not-so-veiled compliment, as explain what it is that he has tried to convey in his statue, in other words the mental representations which he has tried to instantiate in the material representation. He claims to have substituted for Homer's god of storm and battle an altogether milder, more humane and universally benevolent representation of the deity.

Phidias explicates his Zeus by reference to the god's cult-titles (75-6, trans. J. W. Cohoon, Loeb edn):

> And consider whether you will not find that the statue is in keeping with all the titles by which Zeus is known. For he alone of the gods is entitled 'Father and King', 'Protector of Cities', 'God of Friendship,' and 'God of Comradeship,' and also 'Protector of Suppliants', and 'God of Hospitality', 'Giver of Increase,' and has countless other titles, all indicative of goodness.

He is addressed as 'King' because of his dominion and power; as 'Father', I think, on account of his solicitude for us and his kindness; as 'Protector of Cities' in that he upholds the law and the common weal; as 'Guardian of the Race' on account of the tie of kinship which unites gods and men; as 'God of Friendship' and 'God of Comradeship' because he brings all men together and wills that they be friends of one and other and never enemy or foe; as 'Protector of Suppliants' since he inclines his ear and is gracious to men when they pray; as 'God of Refuge' because he gives refuge from evils; as 'God of Hospitality' because we should not be unmindful even of strangers, nor regard any human being as an alien; as 'Giver of Wealth and Increase' since he is the cause of all crops and is the giver of wealth and power'.

Phidias then goes on to specify in detail how he has realized artistically each of these qualities of Zeus in his masterpiece.

For the cognitivist cult-titles are excellent data, for they have one foot in each camp. They are both mental and public representations. Each encapsulates a single specific thought about a god, a representation conjured up in and by the minds of his worshippers; but at the same time the title is in the public domain as an agreed official fact. You can, for example, interpret the significance of Zeus Xenios, 'Zeus, God of Strangers,' but you can't question the legitimacy of the title itself or the attribution of that particular function to that particular god. As a statement about cult, the proposition 'Zeus is not the God of Strangers' is false; as a statement about Zeus Xenios it's a contradiction in terms.

Dio's Phidias does not merely list Zeus' cult-titles, he interprets them. His interpretations are actually re-representations. And they are mental representations before they enter the public domain as spoken and written words. The passage illustrates well what I mean by the negotiation of representations and what Sperber means when he talks of information transformed by memory and communication. This give-and-take of representations is the stuff of religions. As an example, consider what 'Phidias' has to say about Zeus Homognios, rather sinisterly translated as Zeus the 'Guardian of the Race'. The original intent of the adjective was indeed exclusive: 'of the same family or race' - my family, my race, not yours. Dio's Phidias reinterprets it inclusively to cover not just all humanity but gods and humans together as common kin. In so doing he re-represents Zeus Homognios in a fresh universalistic way.

How do those who are 'into' religion (or who 'do religion', to use another colloquialism) go about constructing representations? Like any Greek intellectual worth his salt, Dio can't resist the question of origins and causes, and he raises it long before addressing the particular representations of Olympic Zeus. According to Dio there are two 'sources' of religion, the second of which comprises the makers of public representations: the poets, the lawgivers (who establish the institutions of religion), the visual artists, and the philosophers (39-48). Since Dio is about to speak on Phidias' statue, he focuses on the creators in the third category, the visual artists, whom he also calls the 'craftsmen' (dêmiourgoi), no doubt deliberately invoking connotations of Plato's cosmic 'demiurge' in the Timaeus. In their public representations, then, religions are the product, the inventions of the human mind, and as such susceptible to reasoned natural analysis.

Dio's first 'source' of religion is human cognition (27-37). Cognition, Dio holds, is innate and autonomous in the sense that we do not need to be taught how to use it in order to form mental representations and so to apprehend our environment correctly. Dio does not of course use the language of cognition and mental representation. In the discourse of his times he speaks of doxa and epinoia, 'opinion' and 'thought' (27), and of 'surmise' (33, hyponoêsai as a verb). But the intent is the same, and Dio's main point is unaffected: that our capacity for opinions, thoughts, and surmises is innate, not culturally acquired. On this Dio is emphatic and insistent (27):

> ...opinion and thought common to the entire human race, both Greeks and barbarians, essential and innate (anankaia kai emphytos), naturally occurring in every rational creature, without mortal teacher or mystagogue, never deceptive...

In two regards Dio would part company from a present-day cognitivist. First, in his view, our innate capacity for making representations is naturally attuned to an external reality which is not just our physical environment complete with animals, minerals, vegetables, and our conspecifics. Rather it is our environment as kosmos, an orderly whole which we naturally represent to ourselves, whether in whole or in part, as ordered by Providence - or, as one might say today, intelligent design. That at least is how normal people, Greek or barbarian, represent the world to themselves. It takes a particularly perverse or stupid person, such as an Epicurean, to distort his naturally formed representations so as to exclude Providence. Secondly, for Dio natural representations are veridical; they do not deceive. Strictly from a cognitive perspective, the making of representations should not be encumbered with considerations of objective truth and reality.

From the point of view of the Cognitive Study of Religion (CSR), Dio is most valuable and most interesting as an archaic proponent of what Tooby and Cosmides (1992) have termed the 'Integrated Causal Model' (ICM) of the acquisition, location, and transmission of cultures. The ICM maintains that the human mind/brain is endowed with systems - software running on 'wetware', as the saying goes - selected in and by the evolutionary process, which form representations modified, not created, by interaction with conspecifics in a particular society and culture. The ICM was put forward in reaction to the 'Standard Social Scientific Model' (SSSM) which maintains that cultural systems, of which religions constitute a particular form, are downloaded in a process of teaching and learning into an originally content-free human mind/brain, the metaphorical 'blank slate' on which cultures inscribe themselves.

In his insistence both on the priority of mental over public representation and on the innate ability and propensity of the human mind to form content-specific representations, I have no doubt that Dio would declare himself an ICM person.

8

GNOSIS IN EUROPEAN RELIGIOUS HISTORY

ULRICH BERNER

1. Introduction: The Gnosis-Concept as Problematic. A Look at the Research-History

Gnosis as it appears in the European history of religions is more difficult to delineate than, for example, Neoplatonism or pantheism. For the concept of Gnosis has evoked completely different meanings over the last decades (King 2005, 5-19; Pearson 2007, 8-12). On the one hand, "Gnosis" can signify a particular historical phenomenon – a current (*Strömung*) within Late Antiquity's history of religions, and comparable to Neoplatonism (Haardt 1967, 9); on the other hand, "Gnosis" can designate a general religious phenomenon – a type of religiosity comparable to pantheism (Böhlig 1989, 81). This fluctuation in use can be traced back to the influence of Hans Jonas in particular, whose 1934 analysis of late antique Gnosis employed categories of modern existential philosophy (Jonas 1934, 14-91). His description of the Gnostic attitude toward existence (*Daseinshaltung*) made it look plausible to apply the Gnosis-concept to other, even modern, phenomena[1]. Jonas himself had already made a comparison between late antique Gnosis and modern nihilism (Jonas 1963, 320-40)[2].

The Messina Gnosis-colloquium of 1996 had thought to bring about clarity in the use of the concept. The suggestion concerning terminology would distinguish "Gnosis" from "Gnosticism" – "Gnosis" would be broadly

---

1. See Jonas (1934, 140-251). On the effects of his work, see, for example, Taubes (1954) and Weiss (1955, 83-85). The concept of *Daseinshaltung* was adopted by Schenke (1965, 116-17) and Klimkeit (1986, 23), among others.
2. See Hanratty (1980, 120): "Nihilism was essentially a Gnostic response of dread and despair to the experience of existence in a godless and therefore meaningless or indifferent universe". See also Couliano (1992, 249-66).

used, and signify a general religious phenomenon whose content would be "knowledge of the divine mysteries reserved for an elite." (Bianchi 1966, XXVI). By contrast, the term "Gnosticism" would be more restricted, and with concrete content: "beginning methodologically with a certain group of systems of the Second Century A.D. which everyone agrees are to be designated with this term" (ibid.), and the particular content of Gnosticism was identified as "that which involves the idea of the divine consubstantiality of the spark that is in need of being awakened and reintegrated" (ibid, XXVII). It was assumed, in Messina, that it was possible to construct a world-history for both Gnosis and Gnosticism (for the latter, Bogomils and Cathars were mentioned, religious movements from the European Middle Ages).

The suggested terminology of Messina did not take hold, however (Tröger 2001, 16; Markschies 2001, 22-24; Marjanen 2005, 47). We see, for example, Kurt Rudolph later employing "Gnosis" to designate "a historical category to signify a particular late antique worldview" (Rudolph 1980, 65)[3]. Whether this historical phenomenon is called "Gnosis" or "Gnosticism", what we at least have is a chance to delineate Gnosis within the European history of religions; that is, as a current that runs from Late Antiquity to the European Middle Ages. In this context, Kurt Rudolph pointed to the Bogomils and the Cathars as religious movements "that were either direct continuations of late antique gnosis or else brought new value to their traditions and ideas" (Rudolph 1980, 401; Haardt 1967, 10). Thus it does not seem impossible – as a first option – henceforth to employ the concept of Gnosis as a historical category and to follow the Gnostic current through the European history of religions at least as far as the Middle Ages[4].

And yet the metaphor of the "current" does not satisfy completely; for if the history of Gnosis in Europe comes to an end in the Middle Ages, then the current runs dry. Thus in fact another definition should be posited that would permit Gnosis to follow through, within the European history of religions, right up to the present day. In this respect, the approach of Gilles Quispel is of special interest.

---

3. Elsewhere Rudolph (1992, 41) advocated that the concept "Gnosticism" be abandoned entirely. Koslowski (2006, 225) subsequently took up again the difference between Gnosis and Gnosticism.
4. Tröger (2001, 1961) follows the effect of late antique Gnosis into Islam, and from there to the present day. In this context, one may cite the Mandaens as well. See, for example, Pearson (2007, 315-16).

Quispel considered Gnosis from a phenomenological point of view, as "a basic structure of religious apperception ... [;] one possibility of human religiosity among others, that comes up again and again over time ..." (Quispel 1995, 39). It is a view that identifies Gnosis as a current throughout the entire Western history of religions, from Late Antiquity to the present. Its late antique source is found alongside Christianity and Neoplatonism (ibid., 48), from which point "some kind or other of Gnosis has accompanied the Church up to today" (ibid., 75; Quispel 1981, 432). Quispel cites Rosicrucianism and Freemasonry as examples of modern "Gnosticizing movements and sects" (that is, since Jacob Böhme) (Quispel 1995, 75). The "most important Gnosis of our century", according to Quispel, was the psychology of C.G. Jung (ibid., 76; Quispel 1980, 21-22; Smith 1988, 538-40). Adopting some Jungian categories, Quispel described Gnosis as "discovery of the self" (Quispel 1995, 11). This binding of phenomenological and psychological perspectives results in a view of Gnosis that is of particular interest for the conception of a European history of religions. According to Quispel, Gnosis has "a compensatory function in relation to Christianity" – it is, "so to speak, the shadow that won't go away and that indicates sensitive gaps in the proclamation of the Church or in Christian life" (ibid., 79)[5].

The second option, then, is a phenomenological definition that permits Gnosis to be seen as a current continuing up to the present day. This theory, however, contains the danger of a step-by-step broadening of the concept. That is, the Gnostic current could be so strong that it carries far too many phenomena along with it. Quispel himself had briefly remarked that Gnosis "lives again or lives on in present-day occult movements (theosophy, anthroposophy, the hippie movement, etc.)" (Quispel 1973, 318; Geisen 1992; Sauer 1992). In political science, the Gnosis-concept has also been used to describe phenomena as different as, for example, positivism, psychoanalysis and National Socialism[6]. It has even been extended to modern natural science (Ruyer 1978).

---

5. The metaphor of the shadow was used by Rudolph (1980, 395) and Stroumsa (1985, 487), with no mention of Quispel. A more recent parallel is Assmann's talk of Egyptian cosmotheism as a shadow of Western Christianity (Assmann 2003, 64).
6. See Voegelin (1960, 1): "Among Gnostic movements we must count such examples as progressivism, positivism, Marxism, psychoanalysis, communism, fascism, and National Socialism". See also Faber (1984), Strohm (2005, 35), and Wegener (2007).

Quispel had applied his Gnosis-concept not only to documents pertaining to the history of religions in the narrow sense, but also to the history of literature and art. He devoted special attention to the work of Hermann Hesse (Quispel 1975; Quispel 1978; Galbreath 1981, 27-29). Alongside Hesse, other novelists have been brought into connection with Gnosticism: Albert Camus, Jean-Paul Sartre and Franz Kafka (Donovan 1990), Thomas Mann (Borchers 1980), and Gottfried Benn (Weber 1983)[7]. Quispel had briefly mentioned "the highly influential gnosis of English poet and painter William Blake" (Quispel 1981, 423; 1980, 23). Next to Blake, other painters have been cited: Paul Gauguin, Piet Mondrian, Wassily Kandinsky (Pauen 1994, 118-27), and also Max Beckmann[8]. Finally, Gnostic elements have been identified in music and sport (Fontaine 1992, XV-XVII; Goodger 1982, 333). Given this background, then, the caution regarding a spread of the Gnosis-concept holds a certain plausibility (Kretschmar 1953, 429; Couliano 1984, 290-91; Filoramo 1991, XVIII; Hoheisel 1993, 71-74). In theological controversy, the Gnosis-dispute still has meaning, as we see for example in the title *Against the Protestant Gnostics* (Lee 1987). In popular scientific representations, the dispute takes on polemical traits as well, where talk is heard in German contexts of Gnosis as the "highpoint of all religions of superior knowledge (*Spitzenklasse aller Besserwisserreligionen*)" (Hartmann 1982, 22-24). In view of such evaluations, it does not seem implausible to suggest dropping the Gnosis-concept as a scientific category to describe religious movements, and replacing it with something else (Clasquin 1992, 53)[9]. This should indeed be considered, for the Gnosis-concept is also used as a self-designation in current (new) religious movements and institutions (Pearson 2007, 339-41). The overlapping of object- and metalanguage concepts may just add to the confusion.

One must ask, then, whether the Gnosis-concept is actually appropriate for designating a current in the European history of religions. An immediate

---

7. On the problem of how to configure the concept in this connection, see Galbreath (1981, 20-23).
8. According to Fischer (1972, 19), "Beckmann's worldview [can] be designated overall as Gnostic".
9. Clasquin's suggestion is to piggyback on the Hinduistic distinction between Bhakti-, Karma-, and Jnana yoga – to replace "Gnostic" by "Jnanic", while at the same time introducing the concepts of "Bhaktic" and "Dharmic". Williams (1996, 265) has suggested the expression "biblical demiurgical" as a substitute; see also Williams (2005, 56).

alternative might be the term "Dualism", used by some historians. Steve Runciman identified in Dualism the "characteristic teaching" of late antique Gnosis: "The characteristic doctrine of this new religion was Dualism. It taught that not God but Satan, the Demiurge, made the world and its wicked matter" (1960, 172). Late antique Gnosticism he saw as "a restatement of the Dualist position" (ibid.). The history of the Dualist tradition would extend beyond the Paulicians and Bogomils to the Cathars of the European Middle Ages (Obolensky 1948, 286-89); Loos 1974, 21-31; Lambert 1991, 22-23). Similar to Runciman, Arno Borst described the Cathars' teaching as Dualist rather than as part of Gnosis-history, and he related the Gnosis-concept to Late Antiquity (Borst 1992, 56-60). "Dualism" is thus specified as "the belief – most strongly cultivated in Buddhism, and most weakly in Islam – that the world and (the) Good are completely irreconcilable, for the earthly world is not only corrupted by human guilt, but was created evil right from the beginning, by the Devil" (Borst 1990, 199). According to Borst, the Dualist heresy nests "fully and ever within Christianity", so that whether or not a historical connection can be made between religious movements of Late Antiquity and the Middle Ages is no longer the deciding question (Borst 1990, 201-02)[10].

Another option is the concept of "Mysticism", seen by some researchers as practically synonymous with "Gnosis". Carl Keller, for example, views Gnosis as "largely identical with Mysticism", so that Mystics "are in a certain sense Gnostics" (Keller 1987, 74-75)[11]. Peter Koslowski, while not identifying Mysticism with (philosophical) Gnosis, yet brings the two close together when he describes both as "attempts to advance to the border of the knowable, by means of reason and feeling" (Koslowski 1988a, 11)[12]. The "close connection" between Gnosis and mysticism is stressed as well by Karl Wilhelm Tröger (Tröger 2001, 217).

This third option involves dispensing with the Gnosis-concept altogether and choosing an alternative – "Dualism", or perhaps "Mysticism". Where this

---

10. Lambert (1991, 28) sees as characteristic of the Dualist heresy that "it needs but a slim basis for re-emergence; [...] it is a form of belief that can reappear through self-ignition alone [...]". The metaphor "self-ignition" ("Selbstentzündung") is also used by P. Sloterdijk (1991, 24).
11. See Keller (1985, 72): "Gnosis is the *Ur*-form of real Christian mysticism".
12. Koslowski (1988b, 373) delineates Gnosis more closely as a "theory of total reality developing in and through time". See also Koslowski (1992, 94).

might prove unsatisfactory, however, is in each case's concentrating on a single, self-contained, essential aspect of the historical phenomenon of late antique Gnosis – either for its Dualist content or for its special type of (religious) knowledge.

In what follows, I will be considering a definition of the Gnosis-concept that in fact binds the above two – Gnosis as Dualism and Gnosis as special type of knowledge (*Erkenntnis*)[13]. I understand "Dualism" as a particular type of belief in God, in which the (good) God is not held responsible for all of Creation, but rather is partly exonerated, through the assumption of a secondary power, god, or principle, between God and humanity. And by "Gnostic knowledge" I am denoting a special type of knowledge (*Erkenntnis*) of God that blurs the difference between belief and knowledge (*Wissen*), and makes use of the narrative form of mythos[14].

This is a metalanguage definition that makes no claim to apply to all late antique texts designated "Gnostic" (either by antique authors or modern researchers). The definition allows for demarcation of late antique Gnosis from both Neoplatonism and Christianity. Thus, for example, neither Plotinus nor Origen would pass for Gnostics, since in their systems there is no Dualism in the above-stated sense[15]. Nor would Clement of Alexandria be classified as Gnostic, for in his use of the Aristotelian theory of knowledge he makes a sharp distinction between believing and knowing[16].

This concept of Gnosis can be seen as one possible definition, among many, of the essence of a historical phenomenon – that is, of late antique Gnosticism. At the same time, it can serve as a systematic category with reference to the basic problems of belief in God – the theodicy-problem (which is solved by Dualism), and the problem of the relation of belief to knowledge (which is solved by the concept of Gnosis as a special type of knowledge). The Gnosis-concept is thus phenomenological, a description of a kind of religiosity appearing in different religious traditions – for example as a

---

13. The Gnosis-models of Tröger (2001, 331) and Markschies (2001, 25-26) are both more complicated and narrower in content.
14. On this, see Koslowski's defining Gnosis as "narrative metaphysics made dynamic" (1988b, 373). See also Martin (1987, 12), who stresses the difference between Gnosis and episteme.
15. Further to this, see Armstrong (1978, 109-123), Alt (1990, 64-65), Früchtel (1994, 184-85), and Stammkötter (1992).
16. See Berner (1981, 236-37).

Christian or a Jewish Gnosis – or based on several traditions at once, with the result that tracing its origin is not always possible (or necessary)[17].

When Gnosis develops (into) institutional structures, as for example in the case of Manichaeism, then one may speak of a Gnostic religion, as distinct from Gnostic religiosity[18]. If only one of the two defining aspects is present – dualism or the special type of knowledge – then one ought not to speak of Gnosis, but rather of Gnostic elements. Again, Gnostic elements have to be distinguished from Gnostic motifs, as, for instance, individual persons' or gods' names that appear in Gnostic documents of Late Antiquity. The metaphor of the Gnostic current ought to designate the Gnostic tradition - or reception-history - that is, the effects of late antique Gnosis in the strict sense, but further, all religious phenomena that rest on a conscious attachment to late antique Gnosis. Since the metaphor of the current evokes the assumption of continuity – "underground" or not – it is inappropriate to include within the Gnostic current phenomena of Gnostic religiosity that do not belong to the Gnostic tradition- or reception-history. The phenomena, however, that do not belong within the Gnostic current may still be considered as belonging to Gnosis-history[19]. Regarding the Western history of Gnosis, it would be better to apply the metaphor of the shadow, with reference to the theory of Gilles Quispel. Also appropriate would be the metaphorical talk about the return of the repressed – with reference to the theory of Jan Assmann[20]; this would provide a description of the relationship between Gnostic religiosity and the dominant Christian tradition in the European religious history.

In my next observations on Gnosis in the European religious history, I focus on the Middle Ages – on the dispute and conflict between Roman Catholic Christianity and Catharism. The question of the historical connection between late antique Gnosis and the Gnostic movements of the Middle Ages –

---

17. On the concept of syncretism in Gnosis-research, see King (2005, 222-24). In this context also belongs the talk of the "syncretistic" and/or "parasitic" character of Gnosis; see Böhlig (1989, 9-10), Bianchi (1964, 38-39), Martin (1987, 137); and Pearson (1995, 87). Williams (1996, 93, and 1995, 57) criticizes the parasite-metaphor.
18. Martin (1987, 143) speaks with reference to the Manichaeism of a Gnostic Church. For a critique of talk of a Gnostic religion, see Keller (1985, 72) and Williams (2005, 77).
19. Tröger (2001, 206) likewise distinguishes between the "history of the effects of Gnosis in the strict sense" and "Gnosis-history". He takes the Gnosis-concept even further, however, as he includes phenomena that are "Gnostic in the broader sense", that is, that may include only one of several distinguishing Gnostic traits.
20. See Assmann (2003, 87 and 105); also Berner (2006).

that is, whether a Gnostic current carried on into the Middle Ages – will only briefly be touched upon. Instead, the central question here is whether Catharism – as per Quispel and in contradiction to Borst – can be seen as Medieval Gnosis[21]. With reference to Quispel's theory, I will be asking whether as early as the Middle Ages there was a readiness, first to perceive the "shadow" of Christianity as such and second, to attempt its integration[22].

As a working example, the above question will be put to the work of Alanus ab Insulis, theologian and poet of the twelfth century. In his theological work, Alanus attempted to refute the teaching of the Cathars; yet in his poetical work he developed ideas that in many ways reflect Gnosis. It is especially in the artistic working-out of Gnostic elements or motifs that the complexity of the European history of religions can be seen (Berner 2007). And this applies most especially in the modern case of English poet and painter William Blake, which will be presented briefly in this paper's final section.

2. Gnosis in the Religious History of Medieval Europe

2.1 From Late Antiquity to Middle Ages: A Gnostic Current?

The historical connection between late antique Gnosis and the Gnostic movements of the Middle Ages has not yet been convincingly demonstrated. The greatest difficulty lies in establishing a bridge between the Gnostics of Late Antiquity and the seventh- and eighth-century Paulicians of Asia Minor. Research has proposed different routes: the Paulician movement has been connected to Manichaeism, to Marcionism, and even to early Syriac Christianity (Obolensky 1948, 43-44; Loos 1974, 34-35; Lambert 1991, 26). It remains uncertain, however, whether Dualism as described in the ninth century by Petrus Sikeliotes, was originally a part of the Paulician worldview[23]. In addition, the warlike activities of the Paulicians – which

---

21. Borst (1992, 901) does not use the Gnosis-concept with direct reference to the Cathars, but rather applies it to the Amalricians who, in contrast to the Cathars, exhibit no traits of Dualism.
22. See Quispel (1995, 83).
23. See Petri Siculi *Sermo primus*, especially columns 1306-7. See also Rottenwöhrer (1986, 302-3). Fontaine (2005, 139-40) stresses the Dualistic character of the Paulician teaching, as does Culianu (1992, 209).

played a large part in the conflicts between Byzantines and Arabs – would not fit well with the peaceful essence of the Gnostic movements[24]. Thus it is not entirely certain whether a Gnostic current existed, in the above-defined sense, between Late Antiquity and the Middle Ages.

And yet what does appear certain is the supposition of continuity between the Paulician and Bogomil movements of the tenth century in the Balkan region. Like the Paulicians, the Bogomils rejected many elements of ecclesial Christianity, for example the revering of images and of the cross. Part of their world-view, however, displays Dualism right from the beginning, if in varying forms (Rottenwöhrer 1986, 324-31). In this respect, the report of Euthymios Zigabenos of the end of the eleventh century is of particular interest.

Satanael, the fallen son of God, created the material world and humanity, but cannot bring human beings to life. He must ask (the good) God to breathe a soul into the creature. Satanael had a hand in creation several times; thus we see Cain originating from Satanael's association with Eve. The good God took divinity and creativity away from Satanael, but left him regent of the world. Later the good God took hold of the world's progress, through his creation of the Word – identified as the Archangel Michael and at the same time as Jesus Christ – and reduced Satanael to Satan[25].

This version of the Bogomil system allows the Gnostic character of the movement to be seen clearly – Dualism as the exoneration of God, and the mythical presentation as "narrated metaphysics".

Both indications are seen as well in the Bogomil text *Interrogatio Iohannis*. The Dualism here is a bit modified, and the metaphysical narrative is more closely orientated toward the Genesis-text: the fallen Satan cannot begin his creation-work until God has granted him both permission and the opportunity[26]. Satan's creation-function is partly just a shaping capacity - that is, not really a creative capacity. Yet since it is Satan who makes humans and effects their deviation (in Paradise, which he had made for this purpose), it seems valid to speak here of Dualism: evil is imputed neither to God nor to

---

24. Beck (1993, 74) is of the opinion that the Paulicians were "certainly not a militant Sect".
25. See Euthymii Zigabeni *Panoplia dogmatica*, columns 1289-1306.
26. "Pater misertus est eius et dedit ei requiem, facere quod vult usque ad diem septimum" (text of the Vienna manuscript according to the edition of R. Reitzenstein (1929, 300).

humanity, but rather to Satan as a secondary power situated between God and human beings[27].

What appears equally certain is the supposition of continuity between the Bogomils and the Cathar movements of twelfth-century southern France (Loos 1974, 133; Beck 1993, 81-82). From the point of view of history of religions, however, it would certainly be more important to consider each of these movements in its historical context – the Bogomils as up against Eastern Christianity, and the Cathars in their opposition to Western Christianity. Regarding Dualism as a basic form of human religiosity would of course also allow for the possibility of "self-incitement" (*Selbstentzündung*) whereupon the influence of the Bogomils would assume second place.

2.2. Presentation and Critique of Catharism in the Theological Work of Alanus ab Insulis

It must first be asked whether Alanus's *Contra hæreticos* can be considered at all as a source for Cathar teachings. The credibility of such a view can be confirmed through comparison with the thirteenth-century Cathar *Liber de duobus principiis*. It is possible that Alanus systematized Cathar teaching too strongly. And yet in the *Liber de duobus principiis* one can see systemization in its attempt to ground the teachings not merely exegetically but also philosophically. In any case, it should not be presumed that Alanus gleaned his representation of Catharism from the anti-Manichaeistic writings of Augustine alone, with no knowledge of contemporary sources (Häring 1977, 330-31).

Alanus's argument against the Cathars – designated throughout as *hæretici* – begins with a presentation of Dualism, which he clearly sees as the foundation of the Cathar system[28]. The two principles are further specified as "God" and "Lucifer", and each is assigned a domain of creation[29]. Alanus's report is confirmed by a tract from the *Liber de duobus principiis*, at the

---

27. On this, see Couliano (1992, 209), who prefers to designate the Bogomils as "pseudodualists". On the Dualism-concept, see Berner (1997, 282-83).
28. "Aiunt hæretici temporis nostri quod duo sunt principia rerum, et principium lucis, et principium tenebrarum" (I, 2; PL volume 210, column 308).
29. "Principium lucis dicunt esse Deum, a quo sunt spiritualia, videlicet animæ et angeli; principium tenebrarum, Luciferum, a quo sunt temporalia" (ibid.).

beginning of which Dualism is raised, as characteristic of the whole movement, in contradiction to Monotheism[30].

In the same way, Bonacursus, a Catholic convert and ex-Cathar, begins his report on the heresy of Catharism with the presentation of Dualism. In contrast to Alanus, however, Bonacursus speaks of different versions of Cathar Dualism[31]. What is decisive, however, is Bonacursus's affirmation that what the different versions have in common is a second(ary) principle, the Devil, who participated in Creation[32]. This fundamental commonality makes it meaningful to talk of "the" Dualism of the Cathars despite individual differences.

Alanus deals systematically with the Cathar arguments for their Dualistic foundation. On the one hand, the Cathars cite Scriptures, for example the New-Testament saying that a good tree brings forth no bad fruit;[33] and on the other hand, they try to argue philosophically, for example with the axiom that the character of immutability must carry over from origin to effect[34].

After presenting the Cathars' Dualism, Alanus tries to refute their arguments on both exegetical and philosophical planes. Regarding the exegetical basis, Alanus refers to the concept of "intention", and connects the New-Testament saying to human intention, which makes a second(ary) principle between God and humanity unnecessary[35]. The philosophical argument Alanus attacks by differentiating between *causa efficiens* and *causa formalis*: God is only the *causa efficiens* of Creation; therefore the expectation

---

30. "De duobus autem principiis ad honorem patris sanctissimi volui inchoare, sententiam unius principii reprobando, quamvis hoc sit fere contra omnes religiosos" (*Liber de duobus principiis* I, 1). On this, see also Stroumsa (1992).
31. "Nam quidam illorum dicunt Deum creasse omnia elementa, alii dicunt illa elementa diabolum creasse; ..." (*Vita hæreticorum*, in PL volume 204, column 775).
32. "[...]; sententia tamen omnium est, illa elementa diabolum divisisse" (ibid.). Couliano (1992: 214-38) gives a differentiated overview of the various types of Cathar Dualism. He also posits different sources – Bogomilism and Origenism. Couliano did not take into consideration the presentation of Alanus.
33. See Alanus ab Insulis, *De fide Catholica contra hæreticos*, I, 2 (PL volume 210, column 308).
34. Ibid., I, 3 (PL volume 210, column 309). This same New-Testament saying is cited in the *Liber de duobus principiis*, at the beginning of the tract *De libero arbitrio*. Comparable philosophical reflections are found here as well.
35. Alanus ab Insulis, *De fide Catholica contra hæreticos*, I, 4 (PL volume 210, columns 309-310).

that imperishability is passed on from the Creator to the Creature is unfounded[36].

From the remaining teachings of the Cathars presented by Alanus, only two need be extracted here – the teaching on souls and the condemnation of the Old Testament Fathers. According to Alanus, the Cathars posit that human souls emerged through the fall of the angels, and could be bound into a body many times[37]. He does not however go directly into the writings of the Cathars, but rather assembles other records from the New Testament, to show the absence of any connection between the soul and fallen angels or demons[38]. Alanus tries to show that the Cathars' teaching on the soul is unfounded, as it can be based neither on Scripture nor on reason[39].

According to Alanus, the Cathars also held that the Old Testament Fathers were evil and damned – as in the case of Abraham, who was prepared to sacrifice his son. In this Abraham was acting against the law of nature; and such a command would not have been given by a good God but by an evil one[40]. With the attempt to refute the condemnation of Abraham, Alanus takes up the theme of temptation: God only wished to test Abraham's will and thus is not shown to be evil[41]. As appendix to this, Alanus introduces additional texts from the Bible to show that Abraham is not damned. The ethical problematic of the temptation-theme, however, is not raised again.

The Gnostic character of Catharism, as presented by Alanus, is above all seen in Dualism – in the attempt to relieve God of the burden of responsibility for evil in the world. The other characteristic of Gnosis – the special type of religious knowledge – is not as prominent in Alanus's report. And yet from his presentation of the teaching on the soul it emerges that the Cathar-system contains elements of a "narrated metaphysics" as well, which makes no distinction between belief and knowledge[42]. With regard to the difficulty of reconstructing a tradition-history between Late Antiquity and the Middle Ages, it cannot be said with certainty that Catharism is representative of a

---

36. Ibid., I, 5 (PL volume 210, columns 310-312).
37. Ibid., I, 9 (PL volume 210, column 316).
38. Ibid., I, 10 (PL volume 210, columns 316-17).
39. "Ubi autem nec adest auctoritas, nec ratio, non est probabilis opinio" (I, 12 [*PL* volume 210, column 317]).
40. Ibid., I, 37 (PL volume 210, columns 341-42).
41. Ibid., I, 38 (PL volume 210, column 343).
42. In Bonacursus these mythical elements appear more clearly (*Vita hæreticorum*, in PL volume 204, columns 775-76.).

"Gnostic current". In any case, Catharism may indeed be registered as belonging to the history of Western Gnosis, and as "Gnosis of the Middle Ages" – indeed, with regard to institutionalization, not as mere religiosity but also as a Gnostic religion that arose, in the context of the Western Christian tradition, as an alternative to the Roman Catholic Church. The kind of argument in Alanus's theological work shows the will and readiness to perceive the competing, Gnostic version of Christianity as a "shadow"; the Cathars' criticism of the Church is presented in detail, and opposed in a rational way – that is, it is not "repressed".

2.3 The Treatment of Gnostic Elements in the Poetical Works of Alanus ab Insulis

By the title of his allegorical poem, *Anticlaudianus de Antirufino*, Alanus makes a connection with the work of the late antique poet Claudianus. Claudianus's poem dealt with powers inimical to the divine that were trying with the help of a completely evil person to affect the progress of the world in a negative way. Alanus's poem has to do with good divine powers, particularly the goddess Nature, who for their part wish to create a wholly good person, in order to affect the world's progress in a positive way.

The goddess Nature regrets the errors she made in the creation of the world, and to correct this she shows her sisters a plan to make a perfect human being. Reason points out that only God can create a human soul. Prudence then undertakes the journey to the extra-worldly God, to request that the goddess Nature be given a perfect human soul. The Seven Arts build a chariot that can be drawn by the Five Senses as horses. Reason acts as charioteer. The ascent through the heavens and the spheres provides broad knowledge of the material world. In the highest heaven, Reason must however stay back, with four of the horses – only the fifth horse, Hearing, may accompany Prudence further along the way to God, led by Belief and Theology. Once there, Prudence recognizes the triune God and can relay Nature's request, which is for a soul fit for creating a perfect human being. The way back through the spheres to Earth passes some evil planets and is therefore not without difficulty. From the purest elements, the goddess Nature can create a body that binds with the new soul to make a perfect human being. The appearance of this new person, however, calls forth the counter-godly powers, and a battle ensues between Virtues and Evils. Virtue triumphs, along with the perfect human being, and the poem ends with the vision of a renewed Earth.

In this poem, Alanus strays far from Christian dogmatics[43]. Yet he avoids contradicting the belief-statements of the Church directly. For example, he alludes to Trinitarian and Christological dogma, and by doing so acknowledges their validity. Alanus's poetical work thus does not stand in direct contradiction to his theological tract, *De fide Catholica contra hæreticos*. The theological conception of his poetry, however, touches on Gnosis in some respects (Berner 2007).

The mention of counter-godly powers, demons of the underworld, and evil planets indeed recall something of Dualism insofar as evil is incorporated in powers beyond humans – not at all in the figure of Satan as known from the Bible. The introduction of Nature as a subordinate Creator-godhead partly constitutes a Dualistic approach, for in this way God is relieved of the responsibility for the world's imperfection. The idea that Nature as Creator-godhead is not in a position to create the human soul directly recalls the mythos of the Bogomils as rendered by Euthymios Zigabenos – except that in the latter it is the apostate, the fallen Satanael, whose powers are limited when it comes to creating a person. The goddess Nature, as *vicaria Dei*, can be seen directly as an orthodox equivalent to the Gnostic demiurge.

In the allegorical context, the distinction is made between Belief and Knowledge – Reason must remain on the edge of the world, and Prudence may only advance to the extra-worldly God if led by Belief. The encounter with the extra-worldly God, however, is then described in detail, which once more creates the impression of a higher knowledge of divine mysteries. The entire allegorical construction can be seen overall as an orthodox equivalent to the Bogomil and Cathar systems – a kind of "narrated metaphysics", in order to explain the imperfection of the earthly world. In a certain sense, the fact that Alanus's poetry has been of interest to anthroposophy proves that Alanus could be (mis)understood as one proclaiming a higher knowledge – that is, as a Gnostic[44].

A look at the theological work of Alanus has shown that in the twelfth century there was a reasoned dispute between Roman Catholic Christianity and Catharism, the Gnosis of the Middle Ages. The poetical work of Alanus

---

43. Curtius (1993, 131) pointed out the extra-Christian elements in Alanus's *Anticlaudianus*: "the salvific act of Christ does not appear to have helped […]".
44. On this point, see the introduction to Rath's 1966 translation.

shows the dispute reaching even deeper: Alanus accommodates Gnosis as far as possible, taking Gnostic elements and working them out in his poetry[45]. It would not make sense, however, to designate him as a Gnostic on this account. For such an extension of the concept would hide the differences between the systems and the acuteness of the dispute. On the other hand, it seems sensible to adopt Quispel's view, and to see the poetical work of Alanus as an attempt to integrate the "shadow" of Christianity. The allegorical poem provides space for such an attempt; and it was indeed in the artistic domain that the work of Alanus had the greatest effect[46]. In fact, this artistic "space" had already been exploited by another twelfth-century theologian – Bernardus Silvestris, who had worked with Gnostic elements in his *Cosmographia*[47]. In the theological field, such thoughts led to the risk of heresy-accusations. Thus Wilhelm von Conches, was accused of "showing himself to be a Manichaeist" in his natural-philosophical exposition of the Creation-story[48].

Quispel's theory could also serve to explain the fall of Catharism. For the Church's recognition of the thirteenth-century mendicant orders, which offered a way of living similar to Catharism, could also constitute an attempt at "integrating the shadow". In any case, this emergence and recognition of the mendicant orders must be included as one of several factors in this context.

3. Gnosis in the Religious History of Modern Europe. Conclusion and Outlook

The history of Gnosis in the West does not end with the decline of Catharism; for both elements of Gnosis – Dualism and the postulated special type of religious knowledge – are encountered in modern times as well. In each case one should first ask if both elements are present, or if there is only

---

45. See Fichtenau's interpretation of one of Alanus's sermons: "It was necessary to tell the believers a cosmic epos, in order to compete with the Cathars" (1992, 156).
46. On this, see Huber (1988).
47. See Bernardus Silvestris, *Cosmographia megacosmus* II, 2; and *Microcosmus* 1, 2. Curtius (1938, 196) had stressed the closeness to late antique paganism and spoken of the "Gnosis of Silvestris". See, however, on the other side, Silverstein (1948, 109).
48. "[...]; in altero manifestus Manichaeus est, dicens animam hominis a bono Deo creatam, corpus vero a principe tenebrarum" (Guillelmi Abbatis S. Theodirici Prope Remos, *De erroribus Guillelmi de Conchis ad Sanctum Bernardum,* in PL volume 180, column 340).

one. If the latter, we would not be looking at Gnosis but only at Gnostic elements; and one would have to cite the reasons on which the assessment is based – cosmology, epistemology, or perhaps mere *Daseinshaltung*[49]. In addition, one would need to determine whether the case merely resembles Gnosticism or if there is a historical connection with late antique Gnosis, at least in terms of reception-history – that is, whether the phenomenon belongs to the history of Gnosis or to the Gnostic current. One would have also to consider, on the one hand, that the meaning of Gnostic motifs in the history of reception can change completely; and on the other hand that congruity may occur without any historical connection at all.

In the following, we will present two examples of individual religiosity that could be classified, perhaps, within the Gnostic current of the European history of religions. Mere incidental reference is made here to religious movements that might be classified as Gnostic religions: the twentieth-century discovery of hitherto unknown sources from Late Antiquity had an effect on the emergence or institutionalization of religious movements who consciously use the Gnosis-concept as a self-designation. Thus, for example, the Ecclesia Gnostica and the Apostolic Johannite Church on their websites refer to the papyri of Nag Hammadi[50].

A possible example from the Gnostic current of Early Modern Time can be seen in the work of English painter and poet William Blake. Quispel has cited this work as an "influential Gnosis", without researching it closely. Enough comparable mythology seems at hand to suggest at least a Gnostic epistemology. According to the report of Henry Crabb Robinson, a contemporary, Blake did also maintain a Gnostic cosmology:

> "We spoke of the Devil, and I observed that when a child I thought the Manichaean doctrine, or that of two principles, a rational one. He assented to this and in confirmation asserted that he did not believe in the omnipotence of God ..."[51].

---

49. On the concept of *Daseinshaltung*, see Berner (1992, 58-59).
50. http://www.gnosis.org/ecclesia/ecclesia.htm (11.02.2009); http://www.johannite.org/johannite.html (11.02.2009). See Pearson (2007, 340-341).
51. See Morley (1967, 330). Cf. as well the remark of the same contemporary (Henry Crabb Robinson): "But I gained nothing by this[,] for I was triumphantly told that this God was

This report allows one to suppose hypothetically that Blake's well known poem, "The Tiger", expresses a Dualistic worldview – particularly in the penultimate stanza:

> When the stars threw down their spears
> And watered Heaven with their tears,
> Did he smile his work to see?
> Did he who made the Lamb make thee?[52]

It appears plausible to assume that the last line of this stanza plays on the distinction between the good God and the evil demiurge. In response to the poem's unresolved question, one would answer that it was the Demiurge, not God – in Blake's own mythology, perhaps Urizen – who had made the tiger[53].

Scholarship is divided, however, on Dualism in Blake's work[54]. Kathleen Raine views "The Tiger" as a direct reflection of Blake's occupation with the Gnostic tradition. And she suggests that Blake might have been reading the "Poimandres" and dealing with the hermetic tradition that is seen as part of late antique Gnosis[55]. And yet she is unwilling to ascribe to Blake a Gnostic cosmology:

> Instead of seeking to find a yes or a no, we will be nearest to the truth if we see the poem rather as an utterance of Blake's delight not in the solution but in the presentation of the problem of evil as he found it in the Hermetic and Gnostic tradition.[56]

W.D. Horn sees no problem in speaking of Blake's Gnosis[57]. For Horn, *The Book of Urizen* provides a particularly suitable entry-point for

---

not Jehovah, but the Elohim, and the doctrine of the Gnostics repeated with sufficient consistency to silence one so unlearned as myself" (Bentley 2004, 701).
52. See Blake (1989, 215).
53. See Damon (1988, 422-23).
54. On this, see Damrosch (1980, 165-243).
55. See Raine (2002, 16).
56. Ibid., 31. See also Damrosch (1980, 257), who stresses the distance from the Gnostics; also Curran (1973, 130).
57. See Horn (1987, 77 ["Blake's Gnosticism"] and 94 ["Blake's Gnosis"]).

interpretation, since it allows for a comparison with the Gnostic exegeses of Genesis[58]. P.J. Sorensen goes further than this, seeing in Blake's work a new creation of the Gnostic mythos overall. With respect to the Gnostic mythos, H. Jonas had earlier stated: "The tale has found no Michelangelo to retell it, no Dante and no Milton"[59]. This assertion is contradicted by Sorensen, who claims: "[…] that Gnostic myth has been retold, even recreated, in the works of the British poet-prophet William Blake (1757-1827), as a Gnostic reading of Blake's work will demonstrate"[60].

In discussing whether Blake's religiosity in the full sense – not just with reference to one or the other element – can be designated "Gnostic", one must first state that the work belongs probably to the reception-history of late antique Gnosis, and thus at least in this sense represents the modern Gnostic current. For it seems well founded that Blake had knowledge of the Gnostic tradition – likely mediated directly through contemporary historical representations, for their part based on the reports by the Church Fathers[61]. Thus the case could be made that Church-historical research has given the artist the possibility to recover a repressed interpretation of Christianity and to reformulate it for his own time.

Another artist associated with Gnosis is painter Paul Gauguin[62]. From his self-portraits it has been concluded that Gauguin viewed himself "in the role of the homeless, that is, Gnostic spiritualist"[63]. Once more, one must first ask if Gauguin's work belongs to the reception-history of late antique Gnosis. A point of departure for this supposition seems to be Gauguin's inscription to one of his paintings, which recalls a famous formulation from late antique Gnosis:[64]

---

58. Ibid., 77.
59. See Jonas (1963, XIII).
60. See Sorensen (1995, 1).
61. On the question of the sources available to Blake, see Raine (2002, 12-14), Curran (1973, 122), and Sorensen (1995, 7-8).
62. Buser (1968, 375) places Gauguin close to the realm of theosophy, but also speaks of "his own modern gnosis".
63. See Fischer (1977, 353).
64. On this, see Pauen (1994, 119) and Buser (1968, 380). On the painting itself, see Goldwater (1989, 115). The inscription in English: "Where do we come from? What are we? Where are we going?"

D'où venons-nous?
Que sommes-nous?
Où allons-nous?

One must first determine if this constitutes an intended connection to late antique Gnosis – that is, whether the work can be ascribed to the Gnostic current as above defined. And if so, one must then consider the meaning of the quotation in this new context – whether and how it is that "the central theme of the painting points to Gnostic thinking"[65], and if it really depicts Gnostic religiosity in the sense discussed here. In any event, it does not appear impossible that there is no connection between the painting, with its inscription, and Gnosis.

Blake's work as an example of an interweaving of science, art and religion shows especially clearly the complex situation of a European history of religions. Next to religious movements and institutions, with all of their confessional and theological differentiations, one must also consider individual religiosity as it may manifest itself – perhaps mediated by scientific research – in the arts. Quispel's attempt to pursue Gnosis-history right into modern-day science, literature and art seems thoroughly sensible; and his works on Hesse, Jung and Gnosis can serve as contributions to the European history of religions.

The Gnosis-concept, however, would lose its analytical value if it were applied indiscriminately to all documents of its reception-history, even if there are only singular themes or elements. Quispel himself had in fact pointed out that Jung and Hesse, in their reception of a Gnostic motif – the name of the god Abraxas – were engaged in reinterpretation. And yet he drew no conclusions from this as to the use of the Gnosis-concept[66].

The complexity of European religious history becomes more obvious when one introduces some terminological distinctions, for example the differences presented here between (Gnostic) religiosity and religion; between (Gnostic) elements and motifs; between a Gnostic current and the history of Gnosis; and between tradition-history and reception-history within the Gnostic current. To these must be added – not least – the difference between concepts (of Gnosis) in meta- and object-language: with respect to present-day religious movements and institutions that see themselves as "Gnostic", it

---

65. See Pauen (1994, 119); also Fischer (1977, 353).
66. See Quispel (1978, 499-500).

must again be asked whether their religion can be considered Gnostic as we have delineated the concept here.

THE FIRST SHALL BE LAST: THE GOSPEL OF MARK AFTER THE FIRST CENTURY

WILLI BRAUN

Nescit vox missa reverti[1].

Overheated and overdetermined by greater causes, parochial discords may escalate into fateful events[2].

Despite enormous labours over more than a century on pin-pointing the geographical and social setting for the originary composition of the Gospel of Mark[3], what John Donahue concluded twelve years ago is still the case: "there is no consensus on the *setting* of Mark, nor is there a method agreed upon for describing the social makeup of a [Markan] *community* on the basis of the text"[4]. Consensus is not lacking, of course, because scholars hate consensus, though they often do. Nor is the problem *fundamentally* due to disagreements on a method for describing a social entity on the basis of a text, though such

---

1. Horace, *Ars* 389-90 ("Once sent, the word never returns").
2. Marshall Sahlins, "Structural Work: How Microhistories Become Macrohistories and Vice Versa", *Anthropological Theory* 5 (2005): 5-30 (6).
3. In lieu of a long bibliographic note on quests for the setting and community of Mark, see Stephen C Barton, "The Communal Dimension of Earliest Christianity: A Critical Survey of the Field", *Journal of Theological Studies* 43 (1992): 399-427 (408-10 on the Markan community); John R. Donahue, "The Quest for the Community of Mark's Gospel", in *The Four Gospels, 1992: Festschrift Frans Neirynck* (ed. Frans van Segbroeck et al.; Leuven: Leuven University Press, 1992), vol. 2, 819-34; idem, "Windows and Mirrors: The Setting of Mark's Gospel", *Catholic Biblical Quarterly* 57 (1995): 1-26; Michael F. Bird, "The Markan Community, Myth or Maze? Bauckham's *The Gospel for All Christians* Revisited", *Journal of Theological Studies* 57 (2006): 474-86.
4. Donahue, "Windows and Mirrors", 1.

disagreements are serious in our field. No, the failure that Donahue so correctly notes is a failure because Mark's gospel actually *requires* it.

Why? As to geographical location, Mark's text gives us no indisputable address; even worse, there is nothing in Mark that prohibits us from plausibly locating its point of originary composition here, as opposed to there, or there as opposed to almost *anywhere* in the Roman empire.

As to a *community* that might have produced and used Mark in its own community formation (in the first century), the text is no *more* forthcoming. No matter how much we squeeze the text, we cannot extract from it enough specific information about a group of "[specific] historic agents, including the micro-sociology of their interaction" and the specifics of the groups interests vis-à-vis real-world problems, as Marshall Sahlins would define a "community"[5]. Any effort, thus, to get at what Mark was up to as a first-century "enunciation"[6] with reference to a precise locale and a describable community is something like trying to squeeze milk from a he-goat, so to quip with Diogenes the Cynic[7].

This does not mean that we can say *nothing* about what originary (first-century) Mark, whatever its precise textual form, was up to. The heat of the adversarial rhetoric, the almost shrieking tone of Mark's justification of the truth of his story, the massive and agitated preoccupation with the destruction of Jerusalem and the temple, for instance, suggest that at least in the mind of the author "Mark" we must reckon that there was a set of issues that presented interests and worries of crisis proportions for him – interests and worries that I have no trouble seeing as "specific" and "local" but that do not obligate me to

---

5. Marshall Sahlins, *Islands of History* (Chicago: University of Chicago Press, 1985), xiv.
6. The term condenses an agonistic understanding of speech/utterance generally and of "religious" utterances in particular; see Tim Murphy, "Toward a Semiotic Theory of Religion", *Method & Theory in the Study of Religion* 15 (2003): 48-67.
7. See William Arnal, "The Gospel of Mark as Reflection on Exile and Identity", in *Introducing Religion: Essays in Honor of Jonathan Z. Smith* (ed. W. Braun and R. T. McCutcheon; London: Equinox, 2007), 59: "The problem is not that Mark provides us with no clues about his context: it is that he provides us with so little data about the existence of a discrete "Christian" group – the omnipresent "community" – which is affected by this context and to which he is, more or less particularly and uniquely, directing his writing ... Indeed, Mark provides *so* little information about his audience that we cannot even be sure that he has *any* discrete Christian group in mind. Mark is simply not amenable to explanation in terms of precise intra-Christian developments".

locate "local" in a precise locale. An assist comes from a recent article by William Arnal, in which he argues that Mark is a narrative "reflection on exile and identity", possibly, and very plausibly, by a real-world historical author who is deeply troubled about what Arnal describes as a "double exile"[8]: once, by cause of a somehow tainted Jewish identity, a stranger in the Judean homeland; twice, from a destroyed, temple-less homeland from which he or she is now finally displaced and forced to make a home and identity in a strange land where homeland and temple do not, can not, function even as nostalgic treasures – because they were never granted to the author as treasures in the first place[9]. Arnal is able to make sense of Mark this way without needing a precise originary location or a community, much less one we should think of as "Christian"[10].

---

8. Arnal makes productive use of Benedict Anderson's story of and reflection on a certain Mary Rowlandson, an English colonist abducted in 1675 in Massachusetts, thus becoming a double exile, a displaced colonial and a kidnapping victim. See Benedict Anderson, "Exodus", *Critical Inquiry* 20 (1994): 314-27.
9. See Arnal, "The Gospel of Mark as Reflection on Exile and Identity", 61-66.
10. I am attracted to a "crisis scenario" as the occasion for the creation of Mark. That is, the magisterially elaborated occasion in Burton Mack's *A Myth of Innocence: Mark and Christian Origins* (Philadelphia: Fortress, 1988) strikes me as a muscular motive-set for the ultimately apocalyptic logic of Mark, so sharply focussed, as it is, on the devastation of Jerusalem and the temple and other fall-outs caused by the Jewish War. If I now prefer Arnal's view it is because it makes sense (in a way not fundamentally dissimilar to Mack's) of the genre, topical foci, and ideological agenda of Mark without needing to posit a specific locale or a Christian community. Noting that the Gospel of Mark "strenuously resists our usual procedure of positing a (usually "Christian") *community* and making inferences about the author's agenda in terms of interaction with that community" (Arnal, "The Gospel of Mark as Reflection on Exile and Identity", 59), he abandons the explanatory assist of a "Markan community" whose social interests and social-formational agenda are somehow encoded in the gospel-cum-myth-cum-social charter. Rather, he takes from Burton Mack (e.g., *Myth of Innocence*, 321) the point that Mark is the work of a scholar and suggests that the "what's he up to?" question posed by Mark's narrative might be answered more satisfactorily if we "focus on the intellectual problems solved by Mark, rather than the role of Mark in a distinct Christian group whose essential characteristics can be recovered by us" (Arnal, "The Gospel of Mark as Reflection on Exile and Identity", 59). The "occasion for thought" (Jonathan Z. Smith), Arnal argues on the basis of a persistent and multi-faceted preoccupation in Mark's narrative, is "the Jewish War and the fallout subsequent to the War" (60). The gospel is Mark's answer in narrative form "to the questions raised by the War, with its attendant dislocations, exiles, and opportunities for re-imagining identity, nation, and location. Mark's massive emphasis on the War, the

I move on to a second set of remarks as part of the set-up for the central point in the pages to come. First, I would like to claim readers' permission to suggest that NT-Mark is, in a complex way that is only opaquely discernible, a document of consequence only from the second century onward, when it was pressed into what would now be rather explicitly "Christian" duties, though largely of a political sort. What I will ask you to consider is that the second-century use of Mark either ignored or dismissed the authorial agenda of whoever first created the Markan narrative as an exercise in thought on matters about as presented by Mack in his *Myth of Innocence* or by Arnal.

Second, Mark's gospel had a literary history – a fact known well enough and not disputed, even if the precise stages of this history, and NT-version of Mark's placement in this history, are not clear and hence contested. Where NT-Mark should be placed on the timeline of this history does not matter much to me for my purpose; the *fact* of Mark's literary history *does*. What matters even more to me is that this history can *not* be understood on the model of a text that is changing, say, by means of friendly emendations or amendments at the hands of a single school or community over time, adapting or altering its own precious story of "the beginning of the gospel" (Mark 1:1) to suit changing sociological realities within the group and changing self-perceptions of the group as it interacts with its larger social environment over time. That is, the literary history of Mark appears to be very different from the composition history of, for example, Q, or the Gospel of Thomas, or the Johannine corpus. Rather than seeing the literary history (and reception history) of Mark as an organic unfolding of a "trajectory"[11], possibly in

---

destruction of the temple, and the peculiar movements made by Jesus between Gentile, semi-Jewish, and Jewish regions, and between Galilee and Judea, all point to the possibility that Mark is engaging in post-traumatic re-imagining of identity in his … Jesus-narrative" (60). What I like about this argument is that it correlates the form and content of Mark's narrative, an authorial agenda, a highly plausible historical "situational incongruity" that appears to be of "crisis" proportions to the author, and equally plausible real people (who need not be labelled "Christian") whom one can envision as thinking about the situation in about the way that Arnal proposes. And all this without having to postulate, contrary to what Mark allows us to do, a discrete community urgently given to its own formation with reference to a "social charter" encoded in a Jesus-*bios*.

11. Trajectorist-thinking has been influential in plotting early Christian texts along temporally unfolding developments of Christian thought. Although James M. Robinson and Helmut

coordination with the social history of a particular Christian group, I see it as a *history of confiscation*[12]. Two all-familiar examples in support of this claim: (1) Matthew and Luke purloined Mark's general literary structure as well as most of the discrete parts of his narrative, thus paying respect to Mark's genius as a literary inventor, but erasing or refracting Mark's argument about the import of Jesus for Mark's agenda. In short, Matthew and Luke confiscated

Mark's literary form and structure and erased it, by overwriting, his thought. (2) The critically reconstructed *editio princeps* of the original ending of Mark's gospel (at 16:8) is of course not how NT-Mark *really* ends. Mark 16:9-20 continues the narrative after what the text critics say was its "original" ending. This "longer ending" is a second-century addition by an unknown author who "made use of the four 'NT' Gospels in order to make his addition to Mark resemble documents that had attained at least some level of popularity in certain Christian communities"[13]. A case can also be made that the beginning of NT-Mark (1:1-3) has been tampered with[14]; it certainly was

---

Koester did not invent the trajectory model, their *Trajectories Through Early Christianity* (Philadelphia: Fortress, 1971) is the most programmatic demonstration of it.

12 "Confiscation" in Nietzsche's sense of "in Beschlag nehmen", grasping or claiming an item from the historical or cultural repertoire and redirecting its meaning or purpose, thereby erasing the item's originary meaning or purpose. See *On the Genealogy of Morals: A Polemic* (trans. W. Kaufmann and R. J. Hollingdale; New York: Vintage, 1967), 77, and Murphy's exegesis of Nietzsche concept of confiscation in "Semitioc Theory of Religion", 61: "The English word [confiscation] comes from the Latin, *com* (together) and *fiscus* (basket), or, by extension, treasury. The obvious implication is the taking of something into the treasury. The verb *nehmen* (to take) implies an active sense of taking, grasping, or even seizure. The preposition 'in' further suggests the figurative *relocation* of the something into some other kind of enframing structure. The phrase *in Beschlag nehmen* then suggests a violent seizing upon something, as in the violent transport of a thing from one place or setting into another. In military terms, the phrase is used for an order to requisition something; in nautical terms, it means to seize something off another ship, as in a naval blockade, or 'search and seizure' of contraband".

13. James A. Kelhoffer, "'How Soon a Book' Revisited: EUAGGELION as a Reference to 'Gospel' Materials in the First Half of the Second Century", *Zeitschrift für die neutestamentliche Wissenschaft* 95 (2004): 1-34 (10); see idem, *Miracle and Mission: The Authentication of Missionaries and Their Message in the Longer Ending of Mark* (WUNT 2/112; Tübingen: Mohr-Siebeck, 2000).

14. J. K. Elliott, "Mark 1.1-3 - A Later Addition to the Gospel?" *New Testament Studies* 46 (2000): 584-88.

prefaced later by means of the anti-Marcionite Prologue (ca. 160-180 at the disputed earliest). If so, both ending and beginning, that is, the two most crucial reading-bias storage sites in any literary work, show the work of secondary scribal/ authorial activity.

And so I dare to try out the thought that NT-Mark is a second-century "exaptation"[15] or a confiscation-by-redaction of some Mark, perhaps read by some second-century Christian groups in Alexandria. "Mark", I suggest, appears to have been a commandeered, wandering and variable "cultural operator"[16], not only in the temporal sense of moving from the first century to the second and back again, but also geographically and, importantly, as a kind of hapless child in intra-Christian custody battles. For instance, it is not too difficult to imagine, because the evidence allows it, that Mark's exultation about knowing "the secret [*mysterion*] of the kingdom of God" (4:11) could be exploited by some early Christian "mystery" group to serve its "research" and/or ritual interest[17]. It is just as easy to imagine that some other second- and third-century Christian intellectuals saw the chief value in Mark's gospel in its Passion story in support of an emerging theology of the cross and Jesus' death as "a ransom for many" (10:45). We know that a crucial battle line formed over the differences between the views of Jesus as a mystagogue and

---

15. Stanley K. Stowers, "Mythmaking, Social Formation, and Varieties of Social Theory", in *Redescribing Christian Origins* (ed. Ron Cameron and Merrill P. Miller; SBLSS 28; Atlanta: Society of Biblical Literature; Leiden: Brill, 2004), 493. The term "exaptation" (coined by Stephen Jay Gould) is used in neo-Darwinian evolutionary theory where it refers to a change of function of a trait in a species during evolution. For example, feathers on birds evolved for temperature regulation (adaption), but are then co-opted for flight (an exaptation). Stowers uses this term to refer to the "social equivalent". "A social exaptation would be a cultural artifact that in some sense originated in one social formation and environment but that came to serve a different use and function in another population, environment, and social formation" (493-94).

16. James A. Boon, "Further Operations of Culture in Anthropology: A Synthesis of and for Debate", *Social Science Quarterly* 52 (1972): 221-52; repr. in *The Idea of Culture in the Social Sciences* (ed. Louis Schneider and Charles M. Bonjean; Cambridge: Cambridge University Press, 1973).

17. Note Davies's argument for the *Gospel of Thomas*'s literary influence on Mark, notably visible in NT-Mark's interest in the "mystery" of its knowledge; see Stevan Davies, "Mark's Use of the Gospel of Thomas", *Neotestamentica* 30 (1996): 307-34. Although it is impossible to be sure exactly when and where literary crossings between Mark and Thomas took place, second-century Egypt is, as far as I know, the only place in which both gospels evidently were used in the second century.

Jesus as a suffering saviour in the second to the fourth centuries[18]. And I am suggesting that Mark was one text over which possession rights were fought in this larger battle, a battle in which both sides had long forgotten or chosen to ignore whatever agendas were so urgent for the first author of Mark.

* * *

Thus far the set-up. Let me now move to my main point by reconsidering the two best-attested data items about Mark in the second century. Both are actually well known and often remarked in scholarship, but *together* they pose a most interesting incongruity that, to my knowledge, has not received enough thought.

The first is the near-absence of evidence for use of Mark as a text of intrinsic interest for exegetical, apologetic, or liturgical purposes by the Christian *literati* in the second and early third centuries (and beyond) – in marked contrast to their extensive use of Matthew, John, and Luke (in that order)[19]. There is most emphatically not a single trace of evidence that there ever was anything like an "interpretive community" (Stanley Fish's term[20]) in which Mark enjoyed place, much less pride of place – with the exception, perhaps, of some Christian mystery association, perhaps in Alexandria. In lieu of a long recitation of a survey of the sources here, I piggy-back on the splendid work of Brenda Deen Schildgen on the reception history of the

---

18. See Graydon Snyder, *Ante Pacem: Archaeological Evidence of Church Life Before Constantine* (Macon: Mercer University Press, 1985; rev. ed., 2003).
19. Per multi: Helmut Koester, "History and Development of Mark's Gospel (From Mark to Secret *Mark* and 'Canonical' Mark)", in *Colloquy on New Testament Studies: A Time for Reappraisal and Fresh Approaches* (ed. B. C. Corley: Macon: Mercer University Press, 1983), 37-38; idem, *Ancient Christian Gospels: Their History and Development* (Philadelphia: Trinity Press International; London: SCM, 1990); Brenda Deen Schildgen, *Power and Prejudice: The Reception of the Gospel of Mark* (Detroit: Wayne State University Press, 1999); John Kitchen, "Variants, Arians and the Trace of Mark: Jerome and Ambrose on '*neque filius*' in Matthew 24:36", in *The Multiple Meaning of Scripture: The Role of Exegesis in Early Christian and Medieval Culture* (ed. I. van't Spijker; Commentaria 2: Leiden: Brill, 2009), 15-40.
20. Stanley Fish, *Is There a Text in This Class? The Authority of Interpretive Communities* (Cambridge: Harvard University Press, 1980).

Gospel of Mark[21]. I string together her bottom-line statements on what she calls Mark's "absent-presence"[22] in the early Christian documentary record:

> [T]he gospel was present in the canon, but essentially absent from attention ... [without] "intrinsic" merit ... The references or allusions to the gospel [of Mark] in citations and lectionary cycles in the patristic period point conclusively to the absence of Mark as a major text in the early Church ... The actual count of the citations ... shows that if there is a stepchild in the canon, Mark is the one about whom the Fathers spoke most infrequently[23].

Indeed, it appears that it was only when Mark said something outrageous enough to give the later Fathers serious theological heartburn that they paused at this gospel – for example, at Mark's inclusion of "the son" among those who did not know the arrival time of the apocalyptic Day of the Lord (οὐδὲ ὁ υἱός; Mark 13:32//Matthew 24:36), which was considered a bit of Arian nonsense at which the anti-Arian Christian intellectuals would take such umbrage[24]. But even here, they seemed willing to ignore the phrase in Mark, preferring instead to apply their philological and hermeneutical ingenuity to treat the polluting phrase in Matthew's gospel, which, after all, was the flag of the emerging centrist Christian clan and thus much more "assiduously read" than Mark[25]. All in all, Augustine's off-hand dismissal of Mark as *breviator*, in the context of proposing his two-source theory of gospel relationships, reflects the judgment about Mark in the centuries preceding Augustine:

---

21. Schildgen, *Power and Prejudice*.
22. "Absent-presence" is Schildgen's use of John Dominic Crossan's term in *Cliffs of Fall* (New York: Seabury, 1980).
23. Schildgen, *Power and Prejudice*, 37-41.
24. I am indebted to my colleague John Kitchen for drawing my attention to this in his essay, "Variants, Arians and the Trace of Mark". An Arian preference for Mark's gospel has been alleged (Kevin Madigan, *The Passions of Christ in High-Medieval Thought: An Essay in Christological Development* [Oxford Studies in Historical Theology; Oxford: Oxford University Press, 2007]) and is intuitively sensible, but a careful study of Arian uses of the gospels remains a desideratum.
25. A. Bludau, *Die Schriftfälschungen der Häretiker: Ein Beitrag zur Textkritik der Bibel* (Münster: Aschendorff, 1925), 53 ("fleissiger gelesen"); see also Kitchen, "Variants, Arians and the Trace of Mark", 29, 36.

"separately, he has little to record" (*De consensu evangeliorum* 1.2). Whatever ideational, ideological, social, or political work the gospels were made to perform in post-first-century Christian formation, Mark's narrative, and much more so its originary agenda, was a silent sideline presence.

Why then is Mark in the canon at all, to repeat an oft-asked question once more? The second datum concerning Mark in the second century, and the Patristic period in general, provides an answer[26]. The answer has to do with how Mark became a "prestige good" without intrinsic value[27]. This is what I make of the patristic tradition of insisting that what the author of Mark wrote derived from Peter. I am referring to the idea that Mark was "Peter's interpreter" (ἑρμηνευτὴς Πέτρου), first claimed by Papias in the middle third of the second century (in Eusebius, *HE* 3.39.15, citing Papias's *Exegesis of the Lord's Oracles* [ca. 140 CE]), then repeated with some variation in detail by Justin Martyr, Irenaeus, Tertullian, Origen, and on and on into the third and fourth and fifth centuries, becoming a fact by means of repeated recitation until the onset of modern (post-Enlightenment) biblical criticism[28]. (I note in passing that this attempt to link Mark to Peter leaves traces in the manuscript evidence for Mark, explicitly in the so-called shorter secondary ending. And, although there is no manuscript evidence for claiming that the

---

26. I pass by altogether the discussion, beginning in the latter part of the second century, of the relation between the Gospel (theological truth) and the gospels (literary entities) and the emerging preference to think of this relation in the terms of Irenaeus's famous τετράμορφον τό εὐαγγέλιον formulation ("the gospel in four forms"; *Adv. haer.* 3.11.8); for a splendid study of "gospel" in Irenaeus see Annette Yoshiko Reed, "ΕΥΑΓΓΕΛΙΟΝ: Orality, Textuality, and the Christian Truth in Irenaeus' *Adversus Haereses*", *Vigiliae Christianae* 56 (2002): 11-46. In this "one Gospel-four gospels" argument Mark merely serves a gratuitous place-holding function that is not tied to the merits of the narrative itself. One might think of it as analogous to the structural completion of the College of the Twelve by the enrolment of Matthias in this College to replace Judas (Acts 2:15-26).

27. See William Paden, "Connecting with Evolutionary Models: New Patterns in Comparative Religion", in Braun and McCutcheon, *Introducing Religion*, 406-17. Paden presents a Durkheim-influenced analysis of the process of turning goods into prestige goods: either turning into high-status goods things that have little or no inherent value (such as wood into baseball bats, or cloth into flags or "sacred" head covers) or turning objects with intrinsic value into prestige objects without intrinsic value (such as books into "collectibles").

28. See Hugh M. Humphrey, *From Q to "Secret" Mark: A Composition History of the Earliest Narrative Theology* (London: T & T Clark, 2006); and Schildgen, *Power and Prejudice*.

curious καὶ τῷ Πέτρῳ in Mark 16:7 is also secondary, it is an odd parenthetical specificity that may well be a later insertion.)

In terms of historical authenticity, the claim that Mark was the ghost writer of what is really Peter's gospel is surely bogus, but that is quite beside the point of my interest. What *is* of interest is that this claim is made, and then repeated so often that it seems to reach the status of taken-for-granted and undisputed fact[29]. Why? One recurring answer in the scholarly commentary tradition is that the Mark-Peter connection enhances the merit and authority of Mark; it serves "to uphold the integrity and worth of Mark," in Hugh Anderson's words[30]. But this "integrity and worth" suggestion is rendered dubious by the striking lack of evidence that anyone actually *read* Mark, a lack, moreover, that is not mitigated by what appears to be such certain knowledge that Mark's gospel is really Peter's. Hence, I would think that the Petrine connection is a shibboleth that does not add lustre to the "the integrity and worth" of Mark, at least not with reference to its intrinsic value, for Mark's Petrine imprimatur appears not to have placed Mark on the "must read" list of early Christian associations.

Let's amplify the incongruity. It is also difficult to explain Petrine "authorship" of Mark by supposing that the status ascendancy of Peter in the second century and beyond should be appropriately recognized by a Petrine gospel, which, though he did not actually write one, nonetheless would be his ἀναγραφή ("record"; playing on Clement of Alexandria's term; *Hypotyposes*, in Eusebius, *HE* 6.14.5-6). This would require us to believe that Peter was responsible for a "record" that, on the evidence from Mark's narrative, is most anti-Petrine, matched only by the anti-Petrinism in Paul's letter to the Galatians and in John 1-20. It is in this connection that I find most amusing some slippage in Clement's credulity in one of his rehearsals of the Mark-is-Peter's-amanuensis claim. Clement intimates that Mark's ἀναγραφή may have been a case of an "unauthorized memoir". I paraphrase what Clement says: "When Peter learned of [Mark's project of writing out the εὐαγγέλιον

---

29. To my knowledge, the Petrine provenance of the substantive content of Mark's narrative is never questioned by early Christian writers, though not all who remark on Mark make a *positive and explicit claim* for its derivation from Peter (e.g., Augustine).
30. Hugh Anderson, *The Gospel of Mark* (New Century Bible Commentary; Grand Rapids: Eerdmans, 1976), 28.

that Peter had been preaching in Rome], he said 'I won't stop him, but I sure as hell won't give him any encouragement either'"[31].

Here we have it exactly: the incongruity concerning NT-Mark (post-first-century Mark) on brilliant display. Mark, a prestigious narrative by virtue of its emplacement in the emerging canon; Mark, apparently without intrinsic value in the very canon that bestows prestige on it, hence as really absent, even though present; Mark, presented as Peter's *anagraphê*, but without any consequence for Mark's influence; Mark, presented as Peter's *anagraphê* despite the fact that Mark's story features Peter as a dense-headed character who would seem ill-suited to ascend to the status of the holder of the "keys to the kingdom of heaven" (the church), as Matthew's gospel states (16:19), announcing what would become the axiomatic view of centrist Christianity after the first century.

* * *

And so a different tack is needed. It is noteworthy recently to see an appreciative, even rehabilitating, reconsideration of the once "heretical" argument made by F. C. Baur long ago that Paul, and his theology of "Christ crucified" and his view that the Law was passé in the new *Christos*-era, represented not a wide-spread, much less central view, but a sectoral, and embattled one, and a rather lonely voice crying in the proverbial wilderness of early Christian social and ideological formations[32]. With respect to Mark, it is just as interesting to observe, as Joel Marcus and others have pointed out, a remarkable return to what he calls "the question of the relation between Mark and Paul"[33], a question that had been considered as answered in Martin Werner's 1923 refutation of Gustav Volkmar's 1857 thesis that Mark's gospel

---

31. The Greek text from Eusebius, quoting Clement (*Hypotyposes*, in Eusebius, *HE* 6.14.6): ὅσπερ ἐπιγνόντα τὸν Πέτρον προτρεπτικῶς μήτε κωλῦσαι μήτε προτρέψασθαι ("when Peter discovered this, he neither urgently put a stop to it nor urged it on").
32. Joel Marcus, "Mark – Interpreter of Paul", *New Testament Studies* 46 (2000): 473-87; Snyder, *Ante Pacem*. Note Marcus's comment on Baur's thesis: "If Paul was a lonely and contentious figure rather than a universally approved one, it is more remarkable than it would otherwise be that Mark frequently agrees with him. Mark, too, has been portrayed in post-war scholarship as a polemical writer, and it is natural that sooner or later the attempt would be made to compare and even to draw lines of influence between these two contentious theologians" (474-75).
33. Marcus, "Mark", 473.

is an allegory in which Jesus is really Paul[34]. Werner's refutation argued for no cross-contamination (no pun intended!) at all. In fact, there are a good "number of striking similarities between Paul and Mark", as Marcus demonstrates. For example, now quoting Marcus:

> Not everyone agreed with Paul that the Law was passé for Christians – but Mark did. And he even expressed this point in terms that are remarkably similar to those of Paul in Rom 14 (καθαρίζων πάντα τὰ βρώματα, Mark 7.19; compare πάντα μὲν καθαρά, Rom 14.20). Not everyone was as negative as Paul about Peter and Jesus' family – but Mark was. And only Mark among the NT writers gives to one of his stories, that of the Syrophoenician woman, an interpretation that echoes Paul's formula 'to the Jew first, but also to the Gentiles'. If these are coincidences, they are amazing coincidences. If not – and I think not – they provide … evidence of Pauline influence on Mark[35].

Marcus is correct about the evidence. And on the preponderance of this evidence NT-Mark clearly is *not a Petrine but a Pauline* ἀναγραφή. *When* and *how* this "influence" took place is difficult to pin-point, though we should not by mere default assume, as we are surely tempted to do, that it is the result of a first-century encounter or collaboration between Paul and the author of Mark. In fact, the vocabulary similarities are of the sort to suggest that the cross-over happened when a Pauline letter corpus and a version of the gospel of Mark were available for cross-referencing. Hence the "influence" might in fact reflect a later assimilative redaction. The "when" and "how" questions are not important to me for now in any case. What *is* remarkable is that the patently Pauline echoes in Mark did not *positively* purchase Mark's ticket of admission into the canon. More precisely, insofar as Mark was amenable to being read in a Pauline key, it could gain admission to the canon only by masking Mark's Paulinism and giving it a Petrine imprimatur – both moves as

---

34. Martin Werner, *Der Einfluss paulinischer Theologie im Markusevangelium: Eine Studie zur neutestamentlichen Theologie* (BZNW 1; Giessen: Töpelmann, 1923); Gustav Volkmar, *Die Religion Jesu* (Leipzig: Brockhaus, 1857); a brief synopsis of the issues is in Marcus, "Mark", 473 n. 1.
35. Marcus, "Mark", 486-87.

patently transparent as unsuccessful.

If this is so, why not try another move and seriously consider the possibility that Mark should be placed on the same side of what Joseph Tyson, in his consequential recent book on *Marcion and Luke-Acts*, calls "the defining struggle" over marking the Christian "centre" in the second century[36]. This is the side marked by a totem pole on which are carved the faces of Marcion and his intellectual hero, Paul, and – I am suggesting – Mark. That Mark was considered to be on the "wrong" side apparently was at least a presumed, if not a known fact in the late second century – if an early dating of the anti-Marcionite Prologue to Mark is sustainable. In this sense Mark represents a confiscation, just as the Pauline corpus is in the New Testament as a "good" confiscated for a post-first-century myth of Christian origins in which Paul is not the flag of the Marcionite clan, but thoroughly subordinated to the Peter-led single-and-unified Christian church, as we see it presented in Acts and elsewhere.

In this struggle over defining a myth of Christian origins that could work as foundation myth and epic for the project of creating a single Christian entity, for marking its orthodox centre, in part by combating the threat of different voices as peripheral or entirely beyond the pale, it was important to claim texts that the second-century mythmakers *really* wanted. Apparently it was just as important to claim texts they did *not* really want, because what they wanted even less was for the so-called heretics to have them.

Now a little story. I get it from Marshall Sahlins who found it in an archive of colonial stuff in Hawai'i:

> Early in 1841, the irascible British Consul in Honolulu, Richard Charlton, fired off one of his habitual letters of complaint to the Governor of O'ahu. 'Sir,' he wrote, 'I have the honor to inform you that some person or persons are building a wall near the end of the bowling alley belonging to Mrs Mary Dowsett, thereby *injuring her property and violating the treaty between Great Britain and the Sandwich Islands*'[37].

---

36. Joseph B. Tyson, *Marcion and Luke-Acts: A Defining Struggle* (Columbia: University of South Carolina Press, 2006); see also Richard Pervo, *Dating Acts: Between the Evangelists and the Apologists* (Sonoma, CA: Polebridge, 2006).
37. Sahlins, "Structural Work", 6, emphasis added.

Sahlins's interest in this micro-story is in "how small issues are turned into Big Events; or in somewhat more technical lingo, the structural-cum-symbolic amplification of minor differences" into large-scale animosities. He provides me with an instructive meditation "on the historical dynamics by which relatively trivial disputes over local matters ... get articulated with greater political and ideological differences" on a grand scale: how is it that a very local and specific problem (Mrs Dowsett wants to prevent some guy from building a wall on her property) gets implicated "in conflicts of world-historical significance" (Mr Charlton's desire to avoid a show-down on the treaty governing the relations between Britain and the Sandwich Islands)[38]?

At the structural level Sahlins's story is an analogy to what happened to Mark after the first century. Like Mrs Dowsett who was worried about some guy's wall encroaching on her property, originary Mark's worry appears to have been over rather local problems concerning home and identity (Arnal); like Mr Charlton's amplification of Mrs Dowsett's micro-problem and intercalation of it into higher-level conflicts involving international empire and colonies, Mark's originary local problems in all their poignancy, and his response to them with near-panic urgency, were transposed into, confiscated for, a struggle over defining later Christian centres and margins, a struggle that coincided with the move, in late antiquity, to shift authority and prestige from secret knowledge (inspired speech) to sacred books[39].

* * *

I end with just a few comments of a methodological and conceptual kind. For reasons partly due to Mark's eventual achievement of "first gospel" status in post-Enlightenment gospel criticism, and partly due to the conflation of this source-critical fact with the theological value placed on "first times", we have overstressed our expectation of Mark as a key witness for what happened in the first century. Mark is in motion across time, place, and social setting; and the shifting, contingent, and local historical realities through which Mark passed are not best thought of in terms of *continuities* and *trajectories*, terms that obscure precisely those *contingencies* of greatest interest to us about

---

38. Citations from Sahlins, "Structural Work", 6.
39. See Bruce Lincoln, "Epilogue", in *Religions of the Ancient World: A Guide* (ed. Sarah I. Johnson; Cambridge: The Belknap Press of Harvard University Press, 2004), 657-67.

Mark's historical work (or work in history). NT-Mark is not what set Mark in motion, but a stop in this story's whither, hither, and yon – a stop that effectively "centres" Mark, where, now standing shoulder to shoulder with Paul and John (1-20), for example, he is largely muzzled concerning whatever originary problem he tried to think about. Indeed, NT-Mark now obliterates not only the purposes of its originary composer, but it also repudiates the interests of those who likely were his most avid readers (the Alexandrian Christian "secrecy" group) in exchange for acting as a ceremonial guard of the Christian palace that was under construction in the face of threatening Christian outposts (in the minds of the palace constructors). I have in mind here Jonathan Z. Smith's proposition that *"canonization [is] a process that moves from periphery to center"*[40] and what might be entailed in understanding the canonizing process as a centre-periphery struggle, a struggle for which the Gospel of Mark stands as a hapless case in point.

And thus, perhaps counter-intuitively: for a historiographical stance that may help us I suggest that *the second century precedes the first*. Later centuries too for that matter. Let's see how we can make sense of the same evidence if we take the post-first centuries, not the first century, as the grand era of Christian mythmaking, which includes the creation of the first century as a "Christian" century. This is not to say, of course, that nothing happened in the first century, but it *is* to say few *Christian* things happened and that whatever happened in the first century is massively mediated to us by what happened in the second century and later. This includes the all-important determination of the facts that are given the weight of evidentiary data for the first century in second-century Christian mythmaking. Linkages, trajectories, successions, traditions, and the like, go not forward in time, but backward. They are categories made for manufacturing "origins" in the full mythic sense of the term. They are categories that are employed in a retrospective mode that is in the mood for "first times". I would suggest that these terms, to which one might add others, especially canon, canon-making and legacy-making, even the current fondness for the making and transmission of "cultural memory", might become subject to what Jonathan Z. Smith has called "the rectification of categories." The incongruous burdens placed on the Gospel of

---

40. Jonathan Z. Smith, "Canons, Catalogues and Classics", in *Canonization and Decanonization: Papers Presented to the International Conference of the Leiden Institute for the Study of Religions (LISOR), Held at Leiden 9-10 January 1997* (ed. A. Van der Kooij and K. Van der Toorn; Leiden: Brill, 1998), 296, emphasis added.

Mark – on the one hand, it has nothing to report; on the other, it is a grade-A witness to what Christian things happened in the first century – seems to invite, if not require, such activity on the part of students of the formation of early Christianity[41].

The result of such activity not only would demythologize the early Christian myths of origins, but also, and more importantly, demythologize the scholarly historiography of early Christian mythmaking. Such double demythologizing surely would be a fine tribute and con-tribute to the scholarly agendas that have occupied Luther Martin throughout his remarkable scholarly life.

---

41. In 1992, at a University of Toronto conference devoted to Wilfred Cantwell Smith's contribution to the academic study of religion, Jonathan Z. Smith presented a paper entitled "Scriptures and Histories" (subsequently published in *Method & Theory in the Study of Religion* 4 [1992]: 97-105) in which he rather laconically but evocatively provides further foundation for the statement above, but also strategies for thought on distinguishing "chronology" as a temporal sequence of happenings and "chronology" as a timeline "of when we became interested in them [happenings] ... [which] is a significantly different timeline than the one we are accustomed to — for example, [in the second timeline] the Sumerians would not appear until some 70 years ago" (100); and on excessive worry about recovery of "first times" as an operational credo in history of religion scholarship.

48

COMPARATIVE RELIGION SCHOLARS IN DEBATE: THEOLOGY VS. HISTORY IN LETTERS ADDRESSED TO UGO BIANCHI

GIOVANNI CASADIO

The central focus of the comparative-historical method is to develop comparisons between religious phenomena (comparative) while accounting for their developments within particular contexts through time and space (historical). As such, this method differs both from a purely historical approach because it is cross-cultural and from a phenomenological approach insofar as phenomenology deals with timeless patterns of religious life extracted from their flesh and blood reality. As stated by John P. Burris (2005: 1871), 'it is reasonable to say that Bianchi viewed the comparative-historical method as the dynamic fulcrum of the history of religions'. In fact, for the Italian historian of religions Ugo Bianchi (1922-1995), far from being a univocal concept, 'the notion of religion that emerges from the continued comparison of new and varied historical materials is an analogical notion' (Bianchi 1987: 401). Accordingly, religions can be described analogically insofar as they conform to each other in certain important respects but differ from one another in other equally important aspects. Consequently, the task of the historian of religions is that of 'discovering the degree and quality of the affinity that exists between religions and that warrants characterizing them as such.' Ultimately, the goal of the comparative-historical research is 'the establishment of specific sets of synchronic and diachronic continuities and discontinuities that apply to more than one religion and perhaps to an entire cultural area' (Bianchi 1987: 403). The patterns of cult or the types of religion involved in the procedure of comparison, far from resembling ahistorical universal models or ideal types fixed in an archetypical reality, are multifaceted and dynamic according to variations of time and place. In synthesis, for defining the historical-comparative approach, Bianchi stresses the necessity of three requisites: historical typology, analogy[1] and concrete (historical) universals.

If we consider some leading scholars who have made up the panorama of western studies of religion in recent times we realize that: Ninian Smart (1927-2001) belonged to the Episcopal Church and described himself as a Buddhist-Episcopalian; Mircea Eliade (1907-1986), who practiced the comparative study of religion within the perspective of his particular 'creative hermeneutics', was influenced by both Christian orthodoxy and Hinduism; Seyyed Hossein Nasr (1933-) is a Shii Muslim based on the doctrine and the viewpoints of the perennial philosophy[2]; and Ugo Bianchi has practiced the comparative study of religion from a Catholic outlook. This does not make Bianchi a historian of confessional style[3] but implies that his Catholic education and creed have influenced his epistemology[4] (in an Aristotelian-

---

1. 'Analogy' as a technical term has two convergent meanings. 1. Theoretical: 'Resemblance in some particulars between things otherwise unlike'. 2. Practical: 'Inference that if two or more things agree with one another in some respects they will probably agree in others'. (See Merriam-Webster 11th Collegiate Dictionary, s. v.). Both meanings have an operative function in Bianchi's strategy of comparison.
2. For Nasr as a representative of a peculiar scholarship on Islam in the perspective of Religionswissenschaft see Waardenburg 2007, 235-242.
3. On the thorny issue of the religious insider as scholar of religion (in certain milieus, religious insiders are admitted into the precinct of critical scholarly discourse only if they profess, at least, a kind of methodological agnosticism), I share the judicious remarks of Alles 2008, 7: 'What the study of religions requires is not that those who practice it be outsiders to religion but that they take the most rigorous, critical stance to what counts as knowledge that human beings are capable of taking'.
4. In particular with regard to the adoption of 'analogy' as a specific strategy of comparison. As aptly observed by McKeown 2005 'analogy ... is both the heart of this Catholic intellectual tradition and the proper basis for any good comparison'. McKeown refers to her mentor David Tracy, the Chicago Roman Catholic theologian (1939-) who has been hailed as one of the most original theologians for his work in hermeneutics in a pluralistic context. Even if they probably knew each other (through Bianchi's Chicago alliances), Bianchi and Tracy developed independently their analogical methodology against the background of their common Christian Catholic (Thomistic) matrix. Tracy's (theological) notion of 'Analogical Imagination' (1981), conceived as a stance in understanding that simultaneously grants the otherness of the other (difference) while recognizing the possibilities that the other presents for the tradition in which one finds oneself (similarity), characteristically parallels Bianchi's (historiographical) notion of analogy as described above (introduced firstly in 1971).

Thomistic direction) and his thematic preferences (viz. God, destiny and man; creation, fall and salvation).

It has been lucidly affirmed that the emergence of a serious history of scholarship 'in the place of panegyrics written by adoring students or intended for the grieving widow' (Calder 1994: 71) has resulted in a growing awareness of the importance of new sources. The main sources of this virtually quite rewarding sub-discipline in the field of a theoretically and methodologically motivated study of religion can be listed as follows:

1. Critical footnotes in scholarly literature containing discussion of other scholars' views.
2. Reviews of other scholars' works[5].
3. Oral interventions with comments on papers presented by other scholars during colloquia of various kinds. These need not be stored in the memory of the participants, annotated in their notebooks or taped, but can also be printed in the proceedings with the opportune editing[6].
4. Correspondence between two scholars (letters in the past, e-mails, prevalently, in present times).

---

5. To be really rewarding, and worth-reading, reviews should be just the opposite of that (referring to Smith 2004) recently contributed by a celebrated and overbusy scholar: 'This collection marks Smith as the premier theorist of religion of the generation that has now completed its work. He has been, and remains now, the generalist and the essayist best able to explain the entire academic enterprise of studying not only culture, society, history, politics and anthropology but also religion. He alone has had the power to validate the academic study of religion in such multiple contexts, anthropology, philosophy, the history of the West and the history of the world [!?], with which it intersects....' (Neusner 2005: 130). The field in general, and J. Z. Smith in particular, do not need this kind of panegyrics (involving also auto-celebration).
6. For matter-of-fact reasons this procedure is not much practiced in current scholarly publications, but it was Bianchi's characteristic objective in all the proceedings in the publication of which he was involved. He promoted eleven international conferences and edited the proceedings of nine of them. The prolegomena to these colloquia were always groundbreaking and the resulting epilegomena were often epoch-making (which does not mean that they were always convincing) in their own specific fields. In seven cases he afforded the publication of the 'Discussions', which frequently contains materials and arguments more stimulating than the papers themselves. This procedure, of course, implied a delicate work of editing and consequent discussions with contributors and respondents.

In this paper we plan to concentrate on item 4: examining the correspondence between Ugo Bianchi and two prominent north American scholars of religion with whom he was in contact for some years. This is the third (See Casadio 2007 and Casadio 2009) enquiry into fresh materials preserved in Bianchi's legacy: presently, around three hundred letters prevalently written by Bianchi's correspondents and only a few letters penned by himself, based on carbon copies, have been collected, classified and transcribed.

What contribution can documents which were apparently not destined for publication give to the history of the discipline 'history of religions', and of scholarship in general? In a pre-internet era sources which can cast light on the public and private motivations that govern the progress (or regress) of scholarship in any field are first and always letters. Collecting, editing and, in some cases, deciphering unpublished handwritten letters and diaries is certainly not an easy task. But the gain deserves the effort.

> One sees how things really worked. How were appointments made? Who decided? On what grounds? A crucial contribution may be inexplicably neglected by a later scholar or uncritically endorsed. Letters tell why. Many acknowledge receipt of books or articles by informed friends who would never review them. They are the best evidence for contemporary reception, often more revealing than any critical reviews. There are often revelations on the part of scholars concerning their method or their acknowledged prejudices. Often colleagues are praised or damned with a candour that would not be suitable for publication even 100 years ago (Calder 1994: 72).

When the correspondents are two great savants we expect that their letters reveal thoughts too intimate to be exhibited in their published works. Not just the biographical anecdote but rather the secret motivations of their oeuvre in relationship to the intellectual history of their age. 'Nous imaginons que les échanges privés nous dévoileront le visage qui s'est absenté et retiré loin en arrière du travail scientifique. Nous voulons savoir comment une vie s'est donné cohérence, unité, puissance, dans la multiplicité des jours qui, pour nous, est chaos ou évanouissement' (Jambet 1999 : 5).

Ugo Bianchi was not a brilliant letter-writer because of his reserved temperament and his intricate thought-style. In his letters, however, sometimes he exposes his theoretical cogitations or his scholarly projects in a more detailed and clear-cut way than in his published works. Not all the scholars with whom he was in correspondence were historians of religions. Some were theologians (Bianchi was also involved in theological discourse as a specialist on crucial theological issues, interreligious dialogue, and as a consultant of various pontifical councils), some were philosophers, sociologists, or anthropologists (since he taught ethnology in various stages of his life, in Messina and then in Rome at the Università Urbaniana Propaganda Fide). The great majority of his correspondents were philologists and historians specialized in one of the subjects in which he worked as a comparative historian of religions. Among the full-fledged historians of religions you find the top representatives of the discipline in the 20th century, viz. H. S. Nyberg, G. Widengren, C. J. Bleeker, S. G. F. Brandon, A. D. Nock, W. Cantwell Smith, G. Dumézil, C. Lévi-Strauss, V. Lanternari, M. Eliade, Åke Hultkrantz, L. Honko, J. Pentikäinen, N. G. Holm, J. Waardenburg, C. Colpe, K. Rudolph, N. Smart, E. J. Sharpe, M. Pye, M. Meslin, and, last but not least, his mentor R. Pettazzoni and his most gifted student, I. P. Culianu.

For this brief contribution to honor a scholar, Luther Martin, who has been in the forefront in advocating a critical stance in a study of religion theoretically based, including 'a responsibility to venture judgments about religions'[7] we have drawn on the very short but remarkable correspondence between Bianchi and Kenneth Morgan, a scholar with a notable reputation in North America but virtually unknown in continental Europe, and a top figure in the history of the academic study of religion, Wilfred Cantwell Smith, who share a very empathetic approach in the study of the religious objects.

Kenneth W. Morgan (1908-2011) was born in Montana, educated at Ohio Wesleyan and Harvard Universities, lived in Ramakrishna Mission Ashrams in 1935 and apparently became involved in that spiritual atmosphere. (By contrast, Eliade stayed at the Svarga Ashram at Rishikesh in 1930 for three months, but he was not particularly attracted by the theosophical and syncre-

---

7. Martin 2000, 145. In this paper Luther offers a succinct and dense formulation of his own theory and method in the study of religion, defined as 'a social system legitimated by claims to the authority of some superhuman power, ... to be tested against the empirical research of *cognitive* psychology, on the one hand, and against *comparative* and *historical* research, on the other' (141-142: my emphasis).

tistic perspectives of Ramakrishna and Vivekananda.) Morgan has taught history and comparative religions at Colgate University from 1946 until 1974, serving there also as University Chaplain and then Director of Chapel House and the Fund for the Study of World Religions. Morgan is, in a way, a typical representative of a particular phase of the North American academic study of religion – the same field that for *approximately* thirty years (1960-1990) decreed the triumph of the Eliadean approach to the study of religion. During those years he was an active participant in efforts to encourage the study of Asian religions through college courses and research projects, conferences, fellowships for travel, and providing materials for teaching about Asian religious ways. He had ensured his reputation as a specialist of the religious traditions of the Indian subcontinent with three collectanea, viz. The Religion of the Hindus. Interpreted by Hindus (New York 1953), The Path of Buddha. Buddhism interpreted by Buddhists (New York 1956); Islam. The Straight Path (New York 1958), and a textbook which is also available on line, Asian Religions. An Introduction to the Study of Hinduism, Buddhism, Islam, Confucianism, and Taoism (New York 1964)[8]. His final piece of writing in late age, Reaching for the Moon (New York 1990), was written 'with profound affection, respect and understanding' of the other religious traditions, as J. M. Kitagawa described it in a dust jacket blurb. The first three volumes consist of collections of essays by specialists who are also Hindu, Buddhist or Muslim practitioners respectively. His perspective is patently 'emic', inasmuch as he was convinced that only the word of scholars who are also insiders of a particular religious tradition is authoritative. In other words, the standard for valuing the accuracy of statements about any religion should be the experience and testimony of adherents. In this passionate plea for giving the final word to

---

8. Aims and content of this series are explicitly supported by Smith 1959, 39, n. 17. Morgan was also the promoter of a more specific series aimed to present Shi'ism to the Western world from the point of view of Shi'ism itself: Shi'ite Islam, The Quran in Islam, Hadith (London-New York 1975), all written by a famous Iranian ulema Allamah Sayyid Muhammad Husayn Tabataba'i' (1892-1981), who in his later years used to hold study sessions with Henry Corbin and Seyyed Hossein Nasr, in which not only the classical texts of divine wisdom and gnosis were discussed, but also a whole cycle of what Nasr calls comparative gnosis. Symptomatically, many of his students were among the ideological founders of the Islamic Republic of Iran (Morteza Motahhari, Dr. Muhammad Beheshti, and Dr. Mohammad Mofatteh), others, like Nasr, continued their studies in the non-political intellectual sphere and had finally to abandon their native country and to settle abroad.

the believers – and, consequently, refusing it to the non-believers – Morgan anticipates the celebrated (albeit highly disputed and disputable) argument raised by Wilfred Cantwell Smith (1916-2000) in an epoch-making article dated some years later: 'No statement about a religion is valid unless it can be acknowledged by that religion's believers' (Smith 1959: 42; reprinted 1976: 46)[9]. The perspective adopted by Bianchi, in the two articles mentioned by Smith, for example, in his two letters [B. 1 & 2], is much more nuanced (although quite distant from that advocated by Dario Sabbatucci in Italy or Bruce Lincoln and other stern reductionists in North America).

That said, the contents and the tone of the letter that Prof. Morgan [A. 1], wrote to the IAHR General Secretary C. J. Bleeker one month after the Claremont IAHR international congress (September 6-11, 1965) [The conference had been initially organized by H. W. Schneider (Claremont), but Morgan had subsequently become chairman of the congress as delegate of the American Society for the Study of Religion: see Proceedings of the XIth International Congress of the IAHR, vol. I, 160 and 164)] and three years before the Jerusalem special IAHR conference (1968: see Werblowsky C. 1) are revelatory of the situation of the field that after forty years has not substantially changed. The arguments he raises in order to justify his resignation[10] appear as a mixture of arrogance, dogmatism and naiveté. He is arrogant in stressing his American diversity, if in a self-mocking way ('the naive, bumbling, well-intentioned American'), in a Mark Twain's 'Innocents Abroad' style. He is really naïve when he complains about certain procedures of the academy ('I was disappointed when at Claremont the officers of the IAHR largely ignored the delegates from Asia and Africa, and steps were not taken to extend the IAHR to the new areas represented; my disappointment increased when I discovered that in the IAHR the officers make the decision concerning office

---

9. See most recently Kent 2005: 231-234. It is no coincidence that both Morgan and Smith were adherent to a broadly ecumenical Protestant vision – the former having been brought up in a Methodist, Wesleyan milieu, the latter in a Presbyterianism converging into the United Church of Canada (As observed by Waardenburg 2007, 229, Smith's Presbyterian mentality is recognizable through his attitude to focus on 'the person's mind, and the personal relation to God').

10. All the USA delegates resigned: W. Harrelson from the Executive Board, W. C. Smith from the International Committee (Proceedings, 150-152). They were subsequently replaced by R. M. Grant and J. M. Robinson, two scholars of strictly philological observance and not involved in any interfaith dialogue agenda.

holders and policies without prior consultation with the International Committee which has final authority'). Nominations of office holders in the IAHR take place, as in other learned societies, according to the criterion of co-optation (electoral procedures are in most cases a pure formality). The accusation of discrimination against the Arabs ('I cannot participate in an international organization which is so insensitive to the feelings of scholars outside the European area, and I will not be a member of any voluntary organization which discriminates against any of its members') can evidently be reversed; not convening a meeting in Israel, because of a predictable Arab boycott, is in fact equivalent to a discrimination toward Israel[11]. Lastly, Morgan's well-intentioned defence of the delegates from Asia and Africa (prevalently religious adepts rather than scholars of religion) makes clear that the suspicion about his attempt 'to make an American interfaith society out of a learned academic group of European scholars' was somewhat grounded.

In Bianchi's response to Morgan (replying on behalf of the IAHR International Committee) we find the dialogic, irenic[12] attitude that one can expect in an open-minded Catholic scholar after the 2nd Vatican Council. (In the following years Bianchi served as consultant to the Secretariat for Non-Christians, a dicastery of the Roman Curia instituted by Pope Paul VI on May 19, 1964, in order to promote mutual understanding, respect and collaboration between Catholics and the followers of others religious traditions and to encourage the study of religions). Bianchi's personal attempt to conciliate this basically theological project committed to interreligious dialogue with a philological-historical fundamentally non-confessional approach to the study of religion is displayed in several contributions, two of which are mentioned in the letter of W. C. Smith (B. 2), dated two years later.

In his first methodological intervention Bianchi proposes a third approach to the study of religion standing between the search for phenomenological understanding (represented at Marburg as the 'intuitive approach' of the 'ori-

---

11. Bleeker's prudent response to the objections raised by an Arabo-American delegate was that 'Jerusalem was chosen, because Israel was the only country in the East which could and would organize such a conference at present'. (Proceedings, 154).
12. Irenic, but uncompromising with any sort of crypto-theological intromission in the IAHR affiliation. In the meeting of the International Committee he was in the forefront to stress the scholarly character of the association: 'He did not wish that the IAHR-Congresses would become 'religious conferences' (Proceedings, 150 and 153).

ental trend') and historical explanation (including any psychological or sociological or historicist reductionism). Both history and phenomenology should renounce something to avoid a hermeneutic short-circuit. Phenomenology should give up the pretence that the phenomenological structures have an absolute value, independent from the historical context, and, vice versa, the history should give up the pretence that religious phenomena are explainable purely by material causes[13].

In the 1961 essay written for his colleagues in the academy Bianchi launches his own – very balanced – vision of the scientific study of religion. In his later contributions he speaks as a historian of religions who is also a Christian believer, not as a theologian to an audience of teachers and students of missiology. As a scholar, he exhorts the Catholic missionaries coping with new challenges in addressing the problem of evangelization in a renovated world to circumvent the dilemma between predication finalized to conversion and spontaneous convergence leading to syncretism and confusion by making recourse to a study of the various religious traditions relying on a firmly historical base [Bianchi 1964, 92-4]. As a believer in the universality of the Christian message he invites the Catholic missionaries to preach the Jesus' Gospel without yielding to the facile temptation of ecumenicism[14]. Smith's declared appreciation for Bianchi's attempt to re-assess the task of the history of religions in the education of new Catholic missionaries ('I particularly am in sympathy with your general thesis… and am particularly interested in the missionary implications of such an observation. In fact, the entire theory of missions is transposed to a new level once your orientation is recognized….') is explainable in light of his notorious concern for a 'religiously related scholarship of religious diversity' (Smith 1959: 45). In Smith's vision, 'comparative religion studies are something that might or should serve to promote mutual understanding and good relations between religious communities' (Smith 1959: 42), and the ideal comparative religionist should serve as a mediator or interpreter between diverse religious traditions, 'or at least as a kind of broker helping them to interpret themselves to each other' (Smith 1959: 51). This attitude is instead considered by Bianchi theologically dangerous and scientifically unreliable (Bianchi 1965).

---

13. Bianchi 1961, 238-246.
14. Bianchi 1964, 96, where a bold – but convincing – comparison with the (successful) enculturation of European Stalinist communism in China recurs.

The present exchange between the Italian historian of religions and the two North American comparative religion scholars must be evaluated against the background of the debate raised in the aftermath of the Xth IAHR Congress held in Marburg (September 11-17, 1960). The prime mover of that very crucial event in the history of the IAHR had been Friedrich Heiler (1892-1967), professor of Religionswissenschaft (RW) at the local university and vice-president of the German branch of the World Congress of Faiths (a Lutheran minister after his conversion from Catholicism, he used to celebrate the Mass as a Gallican bishop). His ecumenicalism was even more radical than that of his Canadian confrere: he believed in an ultimate and most profound unity of all high religions in spite of all differences in doctrines and cults. For him, the task of the science of religion was 'to bring to light this unity of all religions' (Heiler 1959: 155). Coherently within this dream, his main preoccupation in gathering scholars (and non-scholars: theologians and dilettantes) together at the Marburg venue had been to convince them of the necessity of paving the way for an era of 'true tolerance and co-operation in behalf of mankind' (Heiler 1959: 160), on the route opened by his mentors Nathan Söderblom and Rudolf Otto.

The official response to this surreptitious shift of the IAHR towards tasks (the promotion of certain ideals – political and theological – which may be good in themselves but are extraneous to the foundations of Religionswissenschaft) was given by Claas Jouco Bleeker (1898-1983), the energetic IAHR Secretary General (1950 to 1970: from 1960 to 1977 he was also the editor of Numen and its supplements), in an address to the General Assembly at Marburg, Sept. 17 1960. He distanced himself both from R. Pettazzoni's utterly historical perspective and W. C. Smith's dangerously 'dialogic' attitude[15]. As a Dutch Reformed Church minister, however, he could not refrain from formulating a list of practical applications of the results of scholarly research, a list including the establishment of the value of religion and the fostering of sympathy and tolerant understanding. At last he pleaded for a third-way ap-

---

15. 'It is questionable whether the believer always understands his religion better than the outsider. It may be that the outsider, being a scholar, has a broader outlook and is in some respects better informed about the religion in question' (Bleeker 1960, 232).

proach, based on 'an ideal combination of the western and the eastern approach to the study of the history of religions'[16].

Much stronger was the response by R. J. Zwi Werblowsky (1924-), an Israeli historian of religions who was to become a very active Secretary General in the following decade (1975-1985). In the article he contributed to Numen (a place which gave it an official character), stressing that some papers presented at Marburg 'were inadmissible by any standards' (Werblowsky 1960: 217), he called attention to the three principal dangers against which the IAHR should have to guard most at its future meetings: a) invasion by dilettantes, b) theological preoccupations, c) un-scholarly (even if morally elevated) concerns. What is historiographically and methodologically crucial, he reminded his Eastern colleagues (and indirectly the European representatives of a 'traditional' approach in the style of René Guénon) that only recently the study of religions had 'extricated and emancipated itself from its matrix in religious studies'[17]. Werblowsky's statement containing the 'basic minimum conditions (or presuppositions) for the study of the history of religions' make some points even more forcefully. In five theses, Werblowsky stresses certain basic prerequisites for the scientific study of religions: 1) avoidance of an opposition between an 'occidental' and an 'oriental' method; 2) an emphasis on the anthropological, non-theological character of the discipline; 3) an avoidance of any reference to a 'transcendent truth'; 4) an emphasis on the autonomy of scholarship, finding a justification in itself; 5) protecting the IAHR against enterprises of a heterogeneous and non-scholarly character. The 'Werblowsky platform' was signed by 16 scholars, including prominent figures (of the most diverse adherence: Catholic, Protestant, Orthodox, Jewish and atheist) like S. G. F. Brandon, J. Duchesne-Guillemin, E. R. Goodenough, J. M. Kitagawa, C. H. Long, M. Simon, R. C. Zaehner, and – remarkably – the Japanese representative Hideo Kishimoto (1903-1964)[18] and Mircea Eliade[19]. The

---

16. Bleeker 1960, 230. For him the 'eastern', 'oriental' approach is based on intuition and seeks for the essence of religion. See the founded objections by Werblowsky in Schimmel 1960, 236 and the pungent remarks by Brelich 1960, 125. See also Fujiwara 2008, 193, pointing out that the Japanese attendees were 'shocked' by this distinction.
17. Werblowsky 1960, 220. Ironically enough he uses 'religious studies' as synonym of 'theological studies'. This is telling of the ambiguity of this now fashionable denomination.
18. The leader of the post-war generation of Japanese scholars of religion, 'who sharply contrasted religious studies as a purely empirical science both with theology and philosophy of religion' (Fujiwara 2008, 202 and 205-6).

document in question was not printed in the official congress proceedings, but in a report by the congress secretary published in Numen[20]. It still has critical value, if we keep in mind that — after 45 years — similar, if not more chaotic, circumstances reproduced themselves at the Tokyo 2005 IAHR Congress, with startling actuality.

---

19. Expectedly, all the Italians (A. Brelich, P. Brezzi, V. Lanternari, and A. Pincherle) signed. All, except U. Bianchi, who penned his own manifesto to be published in the following issue of Numen. He probably had no particular sympathy for the radical options of some of the signers and disliked the starkly anti-theological tune. Brelich 1960, 126 and 128), with characteristic virulence, used the label of 'concilio ecumenico' and 'orgia di comunione interconfessionale'. In this respect Bianchi's position was more nuanced: 'La méthode historique gardera son autonomie, qui ne signifie pas dédain, et qui n'implique pas non plus l'adoption de présupposés philosophiques anti-théologiques, impliquant p. ex. l'impossibilité du surnaturel etc.' (Bianchi 1961, 242).
20. Schimmel 1960, 236-237. Cf. Edsman 1974, 69-70, for a perceptive reconstruction.

LETTERS:

A. 1. Morgan to Bleeker, Bianchi and others.

COLGATE UNIVERSITY HAMILTON, NEW YORK

For your information

K. W. Morgan

KENNETH W. MORGAN

October 12, 1965

Professor of Religion
Professor Dr. C. J. Bleeker
Churchill-laan 290[1]
Amsterdam, Netherlands

Dear Professor Bleeker:

Please remove my name from the list of members of the International Committee of the IAHR. I have notified the executive Council of the American Society for the Study of Religion that I will no longer be able to serve as an American representative. I am not sure whether there is such a thing as individual membership in the IAHR, but if so, please remove my name from the membership list.

It is a matter of some embarrassment to me that for the past eight years I have sought inappropriately to participate in the International Association for the History of Religions. I was under the impression that the 'International' meant 'World-wide' and did not realize until quite recently that it is interpreted to mean 'European' or 'European-derived'. I also thought that 'History of Religions' referred to any scholarly study of religion, including the disciplines of philosophy, the arts, anthropology, and psychology, as well as history and linguistics. When I first heard of the IAHR I was excited at what I thought was the discovery of an organization we greatly need, one which would bring together from all over the world the best scholars who are studying religion critically and objectively, but I have made the naive mistake of seeing the organization as I thought it should be; rather than as it is.

For the past sixteen months I have given the major portion of my time to the IAHR, have spent some $ 3,000 from my Colgate University budget, have administered $ 40,000 granted to us by the American Council of Learned Societies, all in order that we might bring together scholars from all over the world for the encouragement of research in religion and for the strengthening of the IAHR. I was disappointed when at Claremont the officers of the IAHR largely ignored the delegates from Asia and Africa, and steps were not taken to extend the IAHR to the new areas represented; my disappointment increased when I discovered that in the IAHR the officers make the decision concerning office holders and policies without prior consultation with the International Committee which has final authority; then, when it became clear that I was suspected of attempting to make an American interfaith society out of a learned academic group of European scholars, I realized that I was in error in my picture of the IAHR and could no longer serve any useful purpose in the organization.

Furthermore, it was clear that I could not remain in the IAHR when it was announced, without consultation with the International Committee, that there would be a conference in Israel in 1968. I would be shocked at the anti-Semitism of a group that would hold a meeting where the Jewish members could not attend; I am equally shocked that an international organization, supported by UNESCO, would hold a meeting where Arabs could not attend. I cannot participate in an international organization which is so insensitive to the feelings of scholars outside the European area, and I will not be a member of any voluntary organization which discriminates against any of its members.

I am embarrassed. I have been the embodiment of the caricature of the naive, bumbling, well-intentioned American. I withdraw from the role. I'm sorry, for I enjoyed the hospitality in Stockholm and Amsterdam, and I cherish the friends I have come to know through the IAHR. But I can no longer be identified with the organization.

Copies to: IAHR Exec. Com.; International Com.;
Sincerely yours, ASSR Exec. Council Kenneth W. Morgan

2. Bianchi to Morgan.

Rome, le 24 oct. 1965

Cher Collègue,

Je viens de recevoir le texte de votre lettre au prof. Bleeker.

Je vous dis avant tout qu'elle m'a donné du chagrin, et qu'elle a troublé mes souvenirs très agréables de mon séjours américain, que je dois à votre obligeance et à celle de vos institutions, et pour lesquelles je conserve toute ma reconnaissance et mon obligation.

En deuxième lieu, je crois, et continue de croire, qu'une confrontation sereine des différents points de vue est toujours à désirer, et qu'elle peut rapprocher des points de vue, sans dissimuler les différences qui dérivent avant tout de l'histoire et des expériences qui sont derrière chacun de nous, et, peut-être, intégrer ces points de vue sans secousses pour les traditions légitimes, que l'approche scientifiques des problèmes peut rendre toujours plus compréhensives.

C'est pourquoi mon opinion est qu'une séparation hâtive ne puisse que nuire à nos études, et qu'une discussion paisible puisse éliminer les malentendus et clarifier les différences.

Pour des raisons de collaboration scientifique, aussi bien que pour des raisons de coopération personnelle, j'exprime donc le souhait qu'il soit donné à continuer, sur les deux cotes de l'Atlantique (tout comme des autres océans), un dialogue scientifique dont rien ne me parait justifier l'interruption. Je serais honoré si une petite contribution dans ce sens pouvait venir du Colloque de Messine de l'avril prochain. En effet, bien que la gnose soit un phénomène que l'on étudie d'ordinaire en occident, et que les textes considérés soient surtout en grec, j'ai invité bien des orientalistes européens et asiatiques[21] (aussi de l'Extreme Orient et indologues), dans l'intérêt de l'étude historique mais aussi typologique de ce phénomène. Naturellement, je ne fais ici que vous répéter mon invitation à cette rencontre.

Veuillez agréer, cher Collègue, l'expression de mes sentiments distingués

---

21. Je me suis servi, pour les premières invitations, que j'ai adressées jusqu' à la Birmanie, au Japan etc., des vos listes. Je viens maintenant d'inviter deux autres savants japonais. Une section du colloque sera dédiée à la comparaison sur échelle mondiale, bien que cela puisse vraisemblablement soulever des critiques, de la part de certain philologues un peu étroits.

et de ma parfaite estimation.

Ugo Bianchi

B. 1 & 2, Smith to Bianchi.

1. WITH THE COMPLIMENTS OF   Wilfred Cantwell Smith

I greatly appreciated your study 'Lo studio delle religioni non cristiane'
    Many Thanks for it

THE CENTER FOR THE STUDY OF WORLD RELIGIONS
HARVARD UNIVERSITY
42 FRANCIS AVENUE
CAMBRIDGE, MASSACHUSETTS

2. CENTER FOR THE STUDY OF WORLD RELIGIONS
HARVARD UNIVERSITY

42 FRANCIS AVENUE
CAMBRIDGE, MASSACHUSETTS 02138
U.S.A.

AUGUST 22, 1967

Professor Ugo Bianchi
Facoltà di Lettere e Filosofia
Università di Messina
Messina, Italy

Dear Professor Bianchi,

Your note of a year ago today, which you sent me with your offprints, should have been answered long before this. You will feel sorry for me, however, rather than critical, when I say that last academic session with its combination of teaching (and my teaching duties were multiplied because certain colleagues at the last minute were unable to take courses) and administrative duties (this Center is proving an immense, though eventually perhaps rewarding, task) meant that I was left with no leisure whatever for anything that did not imperatively need to be done by the next morning. A hopeless existence, which I hope not to repeat.

Anyway, the result was that I had only glanced at your items. It is this summer that I have read and studied them with care. The one on Gnosticism and the History of Religions confirms my feeling that that movement, of which you are the master and I know next to nothing, is a *locus classicus* for the conceptualization problem in our discipline. It was the 'Après Marbourg' and especially the 'Studio delle Religione [*sic!* —i] non Cristiane' that engaged me most closely. Of the latter I have read several passages more than once, and the whole with great care. I am most grateful for your having sent them to me. In your note you indicate that a further article, on the continuity/discontinuity question, is to be expected. An offprint would certainly be welcome.

I should be interested to know how the theologians received your presentation. I particularly am in sympathy with your general thesis summarized in the second paragraph of p. 135 and am particularly interested in the missionary implications of such an observation. In fact, the entire theory of missions is transposed to a new level once your orientation is recognized; and much theology in this realm is rendered inappropriate, it seems to me, by a failure to recognize it. It is exciting that the Church is moving into a radically new reassessment of mission concerns at this time. I am also reading these days Schlette on a theology of religions, but his view lacks the historical dimension.

Anyway, again my thanks

Sincerely yours
Wilfred Cantwell Smith

P. S. Recently I gave instructions that the publishers should send you a copy of a recent work of mine, not too distantly related to this field. Through a slip, we used your Rome address (Via S. Agnese, 12) from which you last wrote. Let me know if that will not reach you. Enclosed, for what it may be worth, is another recent item.

C.1 Werblowski to Bianchi and others.

INTERNATIONAL ASSOCIATION FOR THE HISTORY OF RELIGIONS (I.A.H.R.)

To all members of the I.A.H.R.

As you know, the proposal to hold a congress or study conference of the I.A.H.R. in Jerusalem was first mooted at the Marburg Congress in 1960 and was discussed again at Claremont in 1965. The Executive and National Committees were all the more anxious to have the meeting take place in Jerusalem as this would be a fitting continuation to the tradition we hoped to found at the Tokyo Congress in 1958, especially as the plan to hold a congress in India in 1965 failed to materialize.

Now we are pleased to inform you that a study conference of the I.A.H.R. will be held in Jerusalem from Sunday to Friday 14 - 19 July 1968. The theme of the conference will be 'Types of Redemption'. In keeping with the character of the meeting as a genuine study conference, only 20 - 25 papers will be presented. All papers will be read and discussed in plenary session. There will be no sectional meetings. Other excursions and tours will be organized in conjunction with the conference after the end of the academic and business part.

In order to enable us to make, the necessary preparations in good time, you are urgently requested to notify the local organizing committee in Jerusalem as soon as possible - and not later than 30 November 1967 - of your intention to participate.
Communications should be addressed to:

Professor R.J.Z. Werblowsky,
The Israel Society for the Study of Religion,
c/o The Hebrew University,
Jerusalem.
Israel

WHY DID GREEKS AND ROMANS PRAY ALOUD? ANTHROPOMORPHISM, 'DUMB GODS' AND HUMAN COGNITION

ALEŠ CHALUPA

Ancient Greeks and Romans used to pray aloud. It seems that this mode of behaviour was generally observed and only exceptionally disobeyed (Versnel 1981: 25-28; Graf 1997: 82). One provision must be mentioned, though: we are informed about this feature of the Graeco-Roman religiosity especially from literary sources which need not necessarily reflect the actual practice with complete precision. There are, however, other indications which strongly support the general accuracy of this view. In figural art, we sometimes find gods and goddesses depicted with a pair of unrealistically huge earlobes which probably referred to their ability to *hear* prayers devoted to them (van Straten 1981: 83, fig. 10-11). A helping hand is also offered by ancient inscriptions preserving prayers or at least their fragments. They often state explicitly that these prayers are destined for the *ears* of the divinity (Versnel 1981: 36)[1]. Also the divine epitheta *epékoos*, *hypékoos* or *audiens* (listening) are widely attested epigraphically (Weinrich 1912; Versnel 1981: 34-35).

In Greek and Roman religion, prayers were also intrinsically connected with rituals, much more than, for example, in contemporary Christianity. A prayer without a proper ritual concomitant, even though very rudimentary, was usually bereft of any meaning (Burkert 1985: 73; Zaidman Bruit – Schmitt Pantel 1992: 41-42; Scheid 2003: 97-98). On the other hand, rituals without prayers were not considered effective either (Pliny the Elder 28.10). Moreover, words were understood performatively. A ritual, if performed incorrectly, could be repeated. A prayer, on the contrary, could not, because words once pro-

---

1. See also relevant inscriptions, e.g. AE 1983, 798; CIL 3.986, 5.759, 8.11269. This fact is also mentioned by Roman philosopher Seneca (*Epistulae* 41.1).

nounced could not be taken back. To prevent undesirable and ominous disruption of rituals by wrongly uttered invocations, prayers were led (*praeire in verbis*, *verba praeire*), at least in the public domain, by a priest who assisted a person - usually a magistrate - that sacrificed on the behalf of the whole community[2]. Other participants, practically relegated to the role of mere onlookers, were asked, before the ceremony started, to keep absolute silence (*favete linguis*), so their voices did not disturb the words of a prayer being recited. Intrusive voices or noises could also be neutralized by a musical accompaniment which often formed an important part of ritual activities (Pliny the Elder 28.11).

Whether these rules were also obeyed in the domestic cult, remains an open question that cannot be answered with absolute certainty. We know sorrowfully little about this important sphere of Roman religious life, compared to our knowledge of the public cult. Nonetheless, domestic cult was one of the most widespread, stable and persistent forms of ancient religiosity (Wachsmuth 1980: 34). Everything seems to support the conclusion that even here the bigger part of ritual activities had a communal character. Even though the rules were probably less severely applied and improvisations could play a more important role, pater familias, the father of the family under whose supervision Roman domestic rituals were performed, certainly had to use many prayers during these religious activities (Harmon 1978; Orr 1978; Turcan 2000: 14-50). Their wordings were largely conventional and every pater familias learnt them, as well as the knowledge of how to perform necessary rituals, from his father or grandfather[3]. Nonetheless, we can suppose that also in the domestic context the custom to pronounce prayers aloud was rather a norm than an exception.

Silent Prayer in Graeco-Roman Antiquity

---

2. Pliny the Elder 28.11. In Roman religion, this role was usually taken over by one of *pontifices*, priests who were, according to Roman tradition, in possession of so-called *indigitamenta*, precise verbal formulae necessary for successful ritual performance (cf. Szemler 1972: 22).

3. Because of the absence of any religious education, this appears to be the most probable scenario. This fact also means that religious knowledge acquired in such a way was rather performative: to understand the meaning of a ritual meant to know how to perform it (cf. Feeney 1998: 138-139).

In spite of the above-mentioned facts, we cannot say that a silent prayer was a completely unknown phenomenon in antiquity (van der Horst 1994: 1-2). Prayers whispered or uttered in a voice inaudible to other persons, only with moving lips, did indeed appear, but usually only in situations which were anomalous or special in many respects. In some cases the fact that a prayer was pronounced in a muffled voice or only in mind could be caused by inconvenient circumstances. Drowning Odysseus, thrown into the sea from his ship by a mighty wave, prayed *kata thymon* (in his heart), because the tempestuous storm obviously precluded his prayer from being pronounced in a traditional manner[4]. These cases are, however, only rarely attested. If we come across silent prayer in works of ancient authors, we can usually identify the four following motives for why people suspended generally observed rules and prayed in silence (van der Horst 1994: 2-12).

1. One motivation would be to prevent enemies from hearing prayers devoted to their destruction and neutralize or undo their influence by application of more powerful counter-prayers. The context is therefore usually military: a prayer said out aloud could divulge the intentions of one of the warring parties or even draw attention to military activity already in progress[5].

2. Also prayers offered to deities characterized by ambivalent or even plainly malevolent abilities, for example to the netherworld deities, are usually portrayed as silent or at least whispered (E.g. Aeschylus, *Eumenides* 1035-1039, *Choeforoi* 95-96; Sophokles, *Oedipus epi Kolono* 124-133, 489). It was thought that mentioning their names aloud could possibly bring destruction to those persons who uttered them (Burkert 1985: 73). The Erinyes, typical representatives of these ambivalent deities, are probably for this reason often addressed as 'Silent Ones' (Nilsson 1967: 709) (E.g. Euripides, *Orestes* 37-38, 409).

---

4. Homer, *Odysseia* 5, 444. This explanation is provided by some ancient commentaries to this passage.
5. Homer, *Ilias* 7.191-196: Ajax asks his fellow combatants to pray for his victory before his fight with Hector, but silently (*sige*). A gripping alternative to this situation can be found even in the Old Testament (1 Sam. 1:13): 'Judith, standing beside his [Holophernes's] bed, prayed in her heart' a moment before she cut off his head with a sword. The motive for a silent prayer is here determined by the caution not to wake up her enemy.

3. A praying person could also feel embarrassed by the subject of his/her wish which could be, in case it was overheard by an unwanted audience, considered inappropriate and banal. Prayers of this type are usually found in elegiac poetry: they contain pleadings for help in matters of love - very often unrequited - or erotic adventures (E.g. Catullus 64.103-104; Tibullus 2.1.84-86; Juvenal, *Saturae* 10.289-291). Also prayers for help in regaining lost or stolen possessions or asking for satisfaction in petty injustices are often described as silent (Versnel 1991). Roman philosopher Seneca mentioned this practice in one of his letters: 'How foolish people are! They whisper to gods their meanest wishes, but fall silent when someone approaches. The things no one is allowed to hear they tell god' (Seneca, *Epistulae* 10.5 [my translation]).

4. The most typical context where we can regularly encounter silent prayers is the world of Graeco-Roman magic. We can even say that a silent prayer gradually became a synonym of magical activity. The reason is readily discernible: the content of these prayers was often socially unacceptable or even criminal. Apuleius in his judicial speech, in which he defended himself against accusations of magical activity, criticizes some contemporary stereotypes, among them also: 'You prayed in silence in the temple: therefore you are a magician. Or what did you pray for'? (Apuleius, *Apologia* 54 [my translation]) Even though his argumentation could be influenced by rhetorical exaggeration, his claim is fully consistent with many passages in the works of ancient authors who persistently depicted magical prayers as silent or whispered. It is very difficult, however, to establish to what degree this literary *topos* reflects the actual magical procedures. Possible distortion of reality cannot be completely excluded, but many magical prayers we know from the so-called binding spells obviously could not be pronounced aloud and in the presence of other people (See some of them in Gager 1992, e.g. no. 20,22, 24). Nevertheless, concerning their formal structure magical prayers strictly followed standard rules used in communication with the Graeco-Roman gods (Graf 1991: 196). All differences appeared especially in the ritual context in which magical prayers were realized and which can be most successfully characterized by the absence of social dimension rather than by different assumptions underpinning their efficaciousness.

From the sociological point of view, there can be little doubt that the prevailing custom of praying aloud was mostly preserved by the need to control rituals performed and wows made in the public domain. The main goal of these religious activities was to promote good and undisturbed relations between the Romans and their gods. All activities which could possibly jeopardize this 'peace with gods' (*pax deorum*) thus had to be suppressed. Although in antiquity the practice of praying in an inaudible voice was found permissible in certain situations, silent prayers always remained a mere alternative which could never fully compensate for traditional prayers uttered aloud. If you prayed without words clearly spoken or in an otherwise inaudible manner, the best thing you could hope for was that people around you would think that you did not pray at all. In a worst case scenario, your behavior could be understood as an antisocial or even criminal act punishable under the statues of Roman law. In antiquity, the private sphere where we find this practice most often was simply never seen as the genuine place for the practice of religion (Kippenberg 1997: 163). This could be the reason for the sometimes puzzling fact that a great part of private religious practices took place in domains we would nowadays see as public space (About the cultural dependency of the terms *private* and *public* see von Moos 1998).

But why did Greeks and Romans persist in praying aloud when communicating with the divine? And can this fact supply evidence for a sometimes expressed opinion that the spirituality of ancient pagans was just different? And can we claim that it was only later when this custom was at first slowly eroded and later completely supplanted by a new type of spirituality, propagated especially by Christianity which accentuated emotional attitude and inner truthfulness in prayers to God? Even though these questions are sometimes answered positively, I do not think that this scenario is sufficiently persuasive. It is my opinion that these answers in reality describe effects rather than causes. Roman religion, although ritualistic and sometimes very conservative in persistent keeping with prescribed ceremonial procedures no matter how outmoded they could appear, cannot be seen as completely void of emotion, a restrained system of exercising empty rituals (For a criticism of these opinions see Linder – Scheid 1993). Roman religion simply looks unfamiliar to us because assumptions and premises on which it was based were so different compared to modern Christianity or other monotheistic traditions (Durand – Scheid 1994).

To tell the truth, even in ancient Judaism or Early Christianities opinions on silent prayer probably varied substantially. Only one passage in the Old Testament mentions silent prayer (1 Sam 1:12-13: the prayer of childless

Hanna). But the plain fact that the moral of this passage is positive - the prayer was answered by God - could later be used in favor of this type of practice. About the opinions of early Christians and their attitudes to silent prayers we know only little (cf. Hamman 1980; Klinghart 1999; Peronne 2003). We can probably conclude that even here the silent prayer gained its support only gradually. Again, one sentence ascribed to Jesus in the Gospel according to Matthew holds a silent prayer in high esteem, even higher than a prayer uttered aloud (Mt 6:5-8).

Some Cognitive Considerations

At this point I would like to venture an idea that, maybe, the perseverance of the custom of saying prayers aloud, so deeply rooted in the Graeco-Roman world, has something to do with the strongly anthropomorphic concept of the divinity and different set of abilities ascribed to it, compared with those prevalent in modern monotheistic traditions. It seems that Christians gradually created such a concept of divinity that was less anthropomorphic and more abstract than the one usually found among their pagan contemporaries[6].

Anthropomorphism, if understood in a broader sense as an attribution of human abilities and intentions to objects or phenomena which do not have them (and not only as a banal and uninteresting fact that gods in a given cultures are depicted in human forms) gained new credibility due to the progress in Cognitive Sciences. Some scholars even hold the pan-human tendency to anthropomorphize the environment to be a decisive stimulus that made possible the appearance of human religiosity (Guthrie 1993, 1996, 2002, 2007).

The assumption of ancient Greeks and Romans that gods needed to hear a prayer to know what they were asked for is thus nothing but natural and logical. A praying person automatically and involuntarily activates those brain systems which are responsible for interpersonal communication (Martin 2004: 120-121; cf. also Versnel 1981: 38; van der Horst 1994: 1). A prayer pronounced in an audible manner is therefore much more natural than a prayer said only in the mind. But is it not possible that the situation will change dramatically if the divinity addressed in prayer has some higher abilities, for example that it is able to read human minds and thus realize what it is asked for

---

6. This does not mean that this concept of divinity was totally unknown in the Graeco-Roman world. But the philosophical schools where it can be usually found were rather small and secluded. Their power to influence prevailing modes of religious behaviour was thus rather limited (van der Horst 1994: 9-12)].

even before this was pronounced? And can Cognitive Sciences help us to delineate this process in greater detail?

It seems that various experiments recently undertaken by cognitive scientists can provide us with two valuable pieces of information. Firstly, that there really is a connection between the procedural form of a prayer and its concomitant ritual and the god/goddess to whom they are devoted. Secondly, that this fact notwithstanding, people who consciously operate with the concept of an omniscient and omnipresent god are not able to transform these premises into their practical behaviour under certain types of conditions.

Justin L. Barrett (2001, 2002) has recently done some experiments in which he tried to asses the influence of human cognition on the procedural structure of petitionary prayers. He reached the conclusion that norms required from a prayer and concomitant ritual change dramatically depending on the abilities of an addressee. If the god/goddess being petitioned is *smart*, it means that he/she is able to read human minds, this fact will cause the relative unimportance of a concomitant ritual performance. On the other hand, if these deities are *dumb*, therefore unable to read human minds, the requirement of a concomitant ritual act being performed exactly according to a prescribed manner will become a matter of the utmost importance (Barrett 2002: 106). I think that the same can be said concerning a prayer as well. If a god is *dumb*, you have to say aloud what you want, otherwise the deity will not understand. From the perspective of average, philosophically uneducated Greeks and Romans gods really needed to hear explicitly what they were asked for, so they could bestow the required favors or at least be persuaded to try. Contrary to the opinion widespread in contemporary monotheistic religions, Graeco-Roman deities were not (usually) omnipotent or necessarily omniscient. They were powerful, immensely so, but sometimes they failed to do what people expected, for no evident reason. The custom to be as precise as possible in the act of formulating what you wanted, in prayer or concomitant ritual, could thus be at least partially motivated by the human tendency to minimize chances for possible misunderstandings on the part of gods.

Another series of experiments done by Justin L. Barrett, this time in cooperation with Frank Keil (Barrett – Keil 1996) proved that even contemporary Christians are not completely immune to the influence of spontaneous anthropomorphization. Their research showed that although Christian believers are perfectly capable of declaring God's omnipotence, omnipresence and omniscience in the form of propositions - generally they consider God an entity unconstrained by distance, simultaneity of events or circumstances - in solution to practical problems they intuitively attribute to God the same limits and constraints as other humans would have. For example, they suppose that God will not be able to listen to a pair of singing birds when a jet plane is landing nearby, in spite of the fact that such an idea is in contradiction with their previously held concept of God as an omnipotent and omniscient (and therefore by jet engines unconstrained) being (Barrett – Keil 1996: 239). Intuitive understanding of conditions enabling verbal communication evidently, under some circumstances, wins over abstract theological concepts. Also the use of God images (for example an icon) during prayer substantially increases human tendency to his anthropomorphization, because it invokes the feeling of presence and physical closeness relevant for interpersonal communication (Barrett – van Orman 1996).

The discrepancy between people's 'knowledge' of God's powers and their inability to apply them when solving some practical situations is caused by a phenomenon named by Justin Barrett (1999) as a theological correctness. Christians formulate their answers according to theological correctness, if these are in the form of declarative statements. In practical life they, however, abandon them easily (and very often unconsciously). This situation is caused by the fact that concept of God according to theological correctness is highly abstract, with relatively low inferential potential and from the perspective of human cognition also quite unnatural. That these theological concepts are relatively widespread among Christians is probably brought about by mechanisms which Christianity (and also some other religions) evolved for the purpose of their spread and retention. We can name priestly institutions controlling the correct thinking and behavior of believers, the indoctrination of articles of faith by their frequent repetition in sermons, the institution of religious education and formulation of systematically expounded dogmas, etc[7].

In Roman religion, however, the existence of the concept of theological correctness is to some extent meaningless. There was no coherent body of doctrine to which Romans could adhere and which they could profess, and, quite logically, also no authority that would police the normativity of their theologically correct thinking. Roman religion was only a vaguely connected body of rituals, festivals, customs and traditions without any official, theolog-

---

7. Boyer 2001: 313-322. The existence of these mechanisms is also critical for the so-called doctrinal mode of religiosity proposed by Harvey Whitehouse (2000, 2004).

ically expounded doctrine. In case of some rituals, there was even no explanation for acts done or prayers said. Moreover, some Roman prayers were apparently so archaic that they were virtually incomprehensible, but they were still persistently recited (Beard 1991: 57-58). Aetiology of many Roman festivals, even the most important ones, was evidently lost, but these festivals still provided an important mainframe for public religious activity (For the case of Lupercalia, one of the oldest and most important Roman festivals, see Scholz 1980). If we can find some ritual exegesis, in most cases these are *ad hoc* explanations or rationalizations not universally shared even by those who took part in these rituals.

Conclusions

The fact that Greeks and Romans pronounced their prayers aloud, at least under normal conditions and in the public sphere, could have two main reasons: 1) to exercise social control over things people wished for; 2) to minimize the possibility of misunderstanding, because the abilities of Graeco-Roman gods and goddesses were often considered less than perfect: they were powerful, but not necessarily omnipotent and omniscient. Attitudes of the Greeks and Romans in their commerce with gods were thus grounded in common intuitions activated in cases of interpersonal communication (Barrett 2001). It is not necessary to conceive silent prayers as a product of some special type of Graeco-Roman 'spirituality'. They were rather a natural consequence of a particular set of assumptions processed by ordinary human cognition than different mental abilities or needs. The form of religious behavior appearing in a particular cultural system can be only imperfectly explained by vague terms like 'spirituality', 'religiousness' or 'religious mentality' whose existence and causal effects remain very doubtful[8]. On the other hand, they can be better understood and explained if detailed knowledge of underlying mental and cognitive processes causing certain type of behavior, its cultural stabilization and successful transmission across generation are taken into account.

Christianity gradually acknowledged silent prayer as an integral part of Christian religious life, although problems with its acceptance are to be seen even here (Klinghardt 1999). The question why this type of prayer was gradu-

---

8. The fact that these concepts induce some notion of reality is caused by their relative currency among the members of a particular society (cf. Sperber 1985, 1996).

ally integrated remains difficult to answer. We can probably suggest that Christians consider the relationship between God and humans much more personal and intimate than ancient Greeks and Romans did, at least in the framework of public cults. Christians also attribute to God many abilities that were not entirely common in Graeco-Roman antiquity, even though they were not completely unknown. One reservation has to be made, though. It is quite possible that silent prayer could be a much more common phenomenon even in antiquity, but this information just did not come down to us due to the peculiar character of our historical sources that tend to overrepresent the public aspect of ancient religions. An especially promising facet of Graeco-Roman religious life seems to be ancient mystery cults that presupposed the possibility of close relations between a deity and his/her human counterparts and aspired to establish them[9]. In the future, therefore, we ought to reinvestigate possible traces left in our sources by personal religious motivations of this kind and try to assess them in the global context of Graeco-Roman religious cults (Bendlin 2000: 131). Nonetheless, this task will not be easy and a chance that we will gain a large amount of information which will substantially increase our knowledge of processes leading to the growing acceptance of silent prayer in antiquity is rather small. The exact outlines of this process will thus most likely always remain shrouded in mystery and we will depend on the formulation of feasible models singling out some possible causes. The Cognitive Science of Religion can probably be very helpful in this regard.

---

9. Nevertheless, nowadays already classical monograph of Walter Burkert (1987) warns us against unwarranted overemphasis on spiritual 'otherness' of so-called 'Oriental religions' or 'mystery cults'. Even here, the standard mode of religious behaviour was fully consistent with the rules of votive religion known from other cults of traditional Graeco-Roman deities.

REFLECTIONS ON THE ORIGINS OF
RELIGIOUS THOUGHT AND BEHAVIOR[1]

ARMIN W. GEERTZ

Luther H. Martin is one of the pioneers in establishing the cognitive science of religion at the center of the social, human and natural sciences. Currently Founding Member and President of the International Association for the Cognitive Science of Religion (IACSR), Luther Martin has consistently exemplified how the cognitive science of religion can be used by scholars in the comparative study of religion. In particular, he is interested in history and evolutionary theory. I have had the great pleasure of Luther's friendship and collegiality for many years, and as a tribute to him, I present below some of my reflections on the origins of religious thought and behavior.

A Conscious Species

The British fantasy fiction novelist Terry Pratchett wrote a hilarious novel called *The Amazing Maurice and His Educated Rodents* (2001) in which a pack of rats eat magically infested garbage in a dump behind the sorcerers' university of higher education and thereby accidentally achieve consciousness and the ability to speak and read. A cat named Maurice happened to eat one of the rats, thus becoming conscious as well. Maurice and the rats join company with a young boy who plays the flute, and upon the urging of Maurice, the whole troop make a living by a scam based on the fairy tale plot of the Pied Piper of Hamlin who relieves village after village of their mysterious plagues of rats by mesmerizing the rats with his flute music. People don't notice of course that the rats were bickering among themselves and the boy's feline

---

1. This contribution is a reprint of a paper titled "The origins of morality and religion", published in *Origins of Religion, Cognition and Culture*, edited by Armin W. Geertz, London: Equinox Press, 2010 (in press).

side-kick was the brains behind the whole scam. All was not well among the rats. They were having trouble dealing with this new consciousness business. They were getting all kinds of strange ideas about existence, morality, creativity, and, even ideas of a far-off land where animals and humans live in perfect harmony. They knew about this place because they had diligently read a children's book series which they often found in the bookstores where they sometimes hid themselves, called *Mr. Bunnsy Has An Adventure*. These books are innocent tales about animals who walk on their hind legs, wear human clothes and hats, and generally act like any human would. I want to quote a passage which is relevant for my talk today. The senior rat can't understand why the younger rats are afraid of the dark and especially of shadows. They even light candles in the passages and tunnels underground. The senior rat can't understand why good food like candles is wasted by lighting them. One of the characters replies:

> We have to be able to control the fire, sir.... With the flame we make a statement to the darkness. We say: we are separate. We say: we are not just rats. We say: we are The Clan. (p. 44)

Which of course produces the shadows that they are afraid of. As the old rat rightly points out, this is ridiculous because it's the essence of rats to be in the dark. So the wiser rat explains:

> Being afraid of shadows is all part of us becoming more intelligent, I think. Your mind is working out that there's a *you*, and there's also everything *outside you*. So now you're not just frightened of things that you can see and hear and smell, but also of things that you can . . . sort of . . . *see* inside your head. Learning to face the shadows outside helps us to fight the shadows inside. And you can control *all* the darkness. It's a big step forward. Well done. (p. 45)

Despite the absurdity of the situation, the description strikes me as being characteristic of what it means to be human. In ways that are still unclear to us, humans developed consciousness and symbolic competence through culture. One way of phrasing it, in tune with two well-known books, is that brains made up their minds and started thinking through culture[2]. Taming fire for heat, lighting and cooking, developing tools and utensils for hunting, processing food, producing clothes and making adornments, all happened during the phylogenetic process of what we have become, namely, the human clan.

The amazing thing about our development as a species is that our consciousness made it possible for us to be aware not only of ourselves, but also of others. We think not only *for* ourselves, but also for others. We think not only *of* ourselves, but also of others. We develop during the first two years of our lives the art of looking in on our own mental states, or inner mindsight, as evolutionary psychologist Nicholas Humphrey calls it. Almost immediately, Humphrey points out, we use this mindsight to make sense of other people[3]. This is the art of doing psychology, and it is achieved Humphrey argues through empathy. I will get back to ontogenetic development and the biological and social techniques used to achieve it. But first, let me sketch out for you the evolutionary scenario which I am betting on, after which I will discuss Pascal Boyer's and Scott Atran's discussions of the origins of religion.

Cognition and Culture

The scenario I will sketch out takes its point of departure in insights formulated by cultural anthropologist Clifford Geertz. He is not someone one would normally expect to be mentioned at a cognition conference. And I think he would probably be disturbed if he found out that people were talking about him at cognition conferences. But the fact is that he was way ahead of colleagues who later developed cognitive anthropology and psychological anthropology, or outdoor psychology, as Geertz called it. As he wrote in 1966:

> And out of such reformulations of the concept of culture and of the role of culture in human life comes, in turn, a definition of man stressing not so much the empirical com-

---

2. Walter J. Freeman, *How Brains Make Up Their Minds*, London: Weidenfeld & Nicolson, 1999, and Richard A. Shweder, *Thinking Through Cultures: Expeditions in Cultural Psychology*, Cambridge: Harvard University Press, 1991.
3. Nicholas Humphrey, *The Inner Eye*, London: Faber and Faber Ltd., 1986; reprint, Oxford: Oxford University Press, 2002, 94.

monalities in his behavior, from place to place and time to time, but rather the mechanisms by whose agency the breadth and indeterminateness of his inherent capacities are reduced to the narrowness and specificity of his actual accomplishments. One of the most significant facts about us may finally be that we all begin with the natural equipment to live a thousand kinds of life but end in the end having lived only one"[4].

There is, thus, a fundamental tension between our universal, natural equipment and the kinds of lives we end up living in specific contexts. But this situation, this basic human condition, should not blind us to the rhetoric of particularism or methodologies that deny comparativism. We must also not fall victim to the other extreme, promoted by some cognitivists, that culture is epiphenomenal in relation to the internal processes of the brain. This particular claim I do not share, and my colleagues and I in Aarhus promote the pivotal counterclaim that symbolic systems are not just important, they are in fact fundamentally *formative* to cognitive development. We make this claim because hardcore neuroscientists agree with us. Clifford Geertz was ahead of his times when he wrote that human history has shown a fundamental dialectical relation between our evolutionary development (more precisely, the tremendous expansion of our brain) and the development of culture. The two go hand in hand. As Geertz argued, in comparison with other animals where genetic information plays a much larger role in controlling behavior patterns, humans are born with much more general response capacities that allow far greater plasticity, but which leave behavior much less regulated. Culture is not an added ingredient to an already completed animal, Geertz claims, rather, it is "centrally ingredient in the production of that animal itself" (1973, p. 47):

> Undirected by culture patterns - organized systems of significant symbols - man's behavior would be virtually ungovernable, a mere chaos of pointless acts and exploding emotions, his experience virtually shapeless. Culture, the accumulated totality of such patterns, is not just an ornament of

---

4. Clifford Geertz, "The Impact of the Concept of Culture on the Concept of Man", [1966], reprinted in Geertz, *The Interpretation of Cultures: Selected Essays*, New York: Basic Books, Inc., 1973, 45.

human existence but - the principal basis of its specificity - an essential condition for it (p. 46).

There is, in other words, "no such thing as a human nature independent of culture" (p. 49). These claims have greater impact today because of insights gained through the development of advanced techniques during the past two decades in paleoanthropology, archaeology, cognitive archaeology, evolutionary psychology and genetic analysis.

Thus, the unfinished animal becomes completed not just through culture in general but through very particular cultures. As Geertz wrote:

Man's great capacity for learning, his plasticity, has often been remarked, but what is even more critical is his extreme dependence upon a certain sort of learning: the attainment of concepts, the apprehension and application of specific systems of symbolic meaning. (p. 49)

As compared to beavers, bees, bower birds, baboons and mice that do what they do and build what they build primarily driven by encoded genes, humans live with an 'information gap': "Between what our body tells us and what we have to know in order to function, there is a vacuum we must fill ourselves, and we fill it with information (or misinformation) provided by our culture". (p. 50)

Along similar lines, cognitive philosopher Daniel Dennett emphasized the role of narrative in the human condition:

> *Our* fundamental tactic of self-protection, self-control, and self-definition is not spinning webs or building dams, but telling stories, and more particularly concocting and controlling the story we tell others – and ourselves – about who we are.... [W]e (unlike *professional* human storytellers) do not consciously and deliberately figure out what narratives to tell and how to tell them. Our tales are spun, but for the most part we don't spin them; they spin us. Our human consciousness, and our narrative selfhood, is their product, not their source[5].

In assuming that selves are nothing other than narrative fictions, Dennett claims that what is behind the drive to narrative is that the brain must do

---

5. Daniel C. Dennett, *Consciousness Explained*, Boston, Toronto, London: Little, Brown and Company, 1991, 418.

whatever it takes "to *assuage epistemic hunger* – to satisfy 'curiosity' in all its forms" (p. 16).

Merlin Donald expressed it a little differently. He called it our "cerebral boxing match with the cultural matrix":

> ...symbols of all kinds are the playthings of a fantastically clever, irrational, manipulative, largely inarticulate beast that lives deep inside each of us, far below the polished cultural surface we have constructed. That passionate and devious intelligence...is isomorphic with our conscious experience of the world.... This is why the human brain cannot symbolize if it is isolated from a culture.... The tension between cultural symbolic systems and the underlying intelligences that use them determines the quality of our uniquely human modes of consciousness.... Culture shapes the vast undifferentiated semantic spaces of the individual brain. The brain takes on its self-identity in culture and is deeply affected in its actions by culturally formulated notions of selfhood[6].

Again, this fundamental tension between the universal and the particular can be seen, according to Donald, as the fundamental and continual confrontation between culture and consciousness. The engine of culture, Donald claims is found in metacognitive awareness, "but the patterns of culture, the mazes we must penetrate, are generated by the cultural matrix itself.... The patterns that emerge at the level of culture are not only real but dominate the cognitive universe that defines what 'reality' is". (p. 287)

Human consciousness is endlessly self-curious and epistemically starved. It is caught between the innate memory banks of the brain and the vastly complex, externally available memory banks of culture. The external memory field is, according to Donald, a mirror of consciousness, but it also changes the ways that consciousness deals with its representations:

> We can arrange ideas in the external memory field, where they can be examined and subjected to classification, comparison, and experimentation, just as physical objects can in

---

6. Donald, Merlin Donald, *A Mind So Rare: The Evolution of Human Consciousness*, New York & London: W. W. Norton & Company, 2001, 2002, 285-286.

a laboratory. In this way, externally displayed thoughts can be assembled into complex arguments much more easily than they can in biological memory. Images displayed in this field are vivid and enduring, unlike the fleeting ghosts of imagination. This enables us to see them clearly, play with them, and craft them into finished products, to a level of refinement that is impossible for an unaided brain. (p. 309)

The core of awareness, then, is caught in a dynamic interrelation between two powerful cognitive fields, the internal and external. The external field allows consciousness to reflect on thought itself and to develop thinking into formal procedures and greater abstractions, the goal of which is to improve and refine our ways of thinking.

Theory of Evolution

My approach to evolution shares the basic assumptions of recent insights from zoology. As evolutionary biologist Mary Jane West-Eberhard explains it, "The universal environmental responsiveness of organisms, alongside genes, influences individual development and organic evolution"[7]. Thus she conceives of evolution in terms of nature *and* nurture, genes *and* environment within the framework of a "fundamentally genetic theory of evolution". It is a theory that must include "the ontogeny of all aspects of the phenotype, at all levels of organization, and in all organisms". What is new here is the addition of a developmental approach to evolutionary biology. Evolutionary biology has generally restricted itself to adaption and has traditionally not included "proximate mechanisms" (p. viii). One of the advantages of this approach is very relevant for us:

A deep look at the evolutionary role of development reaches beyond the issue of nature and nurture to illuminate such themes as the patterns of adaptive radiation, the organization of societies, and the origin of intelligence. For it is undoubtedly the assessment and management of environmental and social contingencies that has led to the evolution of situation - appropriate regulation, with the eventual participation of the

---

7. Mary Jane West-Eberhard, *Developmental Plasticity and Evolution*, Oxford: Oxford University Press 2003, vii.

> sophisticated device we call "mind". Indeed, seeing judgment and intelligence among other mechanisms of adaptive flexibility helps explain why learned aspects of human behavior so closely mimic evolved traits. (p. 20)

"Learning", Eberhard argues, "is just one among many environmentally responsive regulatory mechanisms that coordinate trait expression and determine the circumstances in which they are exposed to selection" (p. 338). Learning involves four components: 1) Resource-specific motivation, whereby motivation is understood as being an "internal value system" concerning such things as hormones, homeostatic responses, etc. 2) A repertoire of motivation-specific exploratory behaviors, which Eberhard calls "the bag of tricks that have been successful in the past". 3) Reinforcement by context-specific sensations of reward or punishment, and 4) Discernment and memory, whereby solutions that were successful are discerned and remembered. (Eberhard 2003, 341-342).

Thus, in my view, human evolution must be considered in significantly broader terms than psychological reductionist and cultural eliminativist models allow. As the British biological anthropologist Robert Foley wrote:

> Two inferences about the evolution of human cognition can be made thus far. The first is that the way in which the brain operates is likely to be strongly linked to social processes, as these are the underlying selective pressures, and secondly that it will evolve in an additive way on existing capacities. The first of these in particular has been the basis for suggesting that a key stage in cognitive evolution is what is referred to as a theory of mind[8].

A theory of mind is a kind of mind reading process. It assumes that a person is not only conscious, but is also self-conscious. It assumes that self-consciousness also implies consciousness of others and that the primary use of thought is to simulate activities, actions and conesquences in relation to other people.

This all makes good sense. But the problem, as I mentioned in my

---

8. Robert Foley, *Humans Before Humanity*, Oxford & Cambridge: Blackwell Publishers Ltd., 1995, 203.

opening talk, is that this sequence of givens have been shown to be found in other animals, especially apes. In other words, what we assume as being uniquely human was most likely achieved prior to the evolution of hominids. And that realization makes the problem of the large size of our brain even more mysterious. Why did the brain evolve?

Foley shows in his book *Humans Before Humanity* (1995) that the relative humanness of the various hominid taxa raise important questions:

> In looking at brain size, technology and ecological behaviour it was clear that the australopithecines were ape-like in everything other than gait, and that neanderthals had modern-sized brains but lacked modern behaviour. Indeed it is even the case that early *Homo sapiens* did not show 'modern' technology for over 60,000 years. Brains evolved late in hominid evolution, and there is a marked technological contrast between archaic and modern hominids. (p. 204)

This leaves a relatively slow rate of technological change for the period from 2 million years ago to roughly 300,000 years ago. According to Foley, this implies the absence of "the characteristics of thought and language seen in modern humans". The critical change, with a subsequent rapid development of culture, happened rather late. Foley argues that language was the essential factor in brain development and that this occurred over the last 300,000 years. As he wrote, "The archaic hominids of the Pleistocene were probably adept mind readers, but they were also silent ones" (p. 205). He argues furthermore that selective processes for thought and for communication were different and separate.

The interesting thing about Foley's arguments is that they are based on the fossil records rather than on general theories of cognition. This implies that the analogies assumed by Boyer and Atran for instance as aptly formulated in the title of one of Atran's chapters: "Stone Age Minds for a Space Age World?" may need reconsideration. In applying the results of modern day studies of human cognition to perceived Pleistocene conditions must be done with extreme care. I think that we have very little chance in imagining what humans were like then. One thing I believe we safely can assume is that they were not much like us except on some very basic levels. At any rate, we need to pay equal attention to the evolutionary environment as well as to general knowledge about human cognition. As Foley argues, human evolution has a very fragmented history and we must therefore pay attention to each stage of evolution in terms of costs and selective advantage. We must consistently react critically to the default assumption that evolution is a process leading to humans (p. 207-210).

Going a step further, we could argue, in agreement with biologist Terrence Deacon, that we need to push our evolutionary antecedents and possible default mechanisms much further back than the Pleistocene. As Deacon and Ursula Goodenough wrote:

To say that our brains have undergone critical reconfigurations as they evolved their capabilities for symbolic (self-)representation is not to say that our common-ancestor brains were left in the dustbin.... [W]e share strong cognitive and emotional homologies with our primate cousins, and, to the extent that degradation or reconfiguration went into generating our capacity for language, it occurred in a primate brain that remains very much a primate brain. Any perspective on the human condition that brushes this fact aside is an incomplete - indeed, we would say impoverished - perspective.

A common response to this interface is to propose a de facto dualism. Yes, it is acknowledged that much of who we are has primate antecedents, but, given our emergent minds, our rationality, our spiritual yearnings, and our culturally encoded meaning systems, we somehow have the where - withal to transcend these antecedents and operate in a set-apart matrix of human - specific truths.

An alternative to such forms of dualism, and one that we find more germinative and satisfying, is the notion that one of the things that we do with our symbolic minds is experience our primate minds symbollically. Our primate minds have not gone away (although some phylotypic instincts have been lost and perhaps reconfigured), nor are they experienced as apes would experience them. They are experienced as things are experienced by human minds: symbolically[9].

Origins of Religion

In cognitive theories of the origins of religion, it is held that religious thought and behavior are "parasitic" in relation to more basic cognitive processes. For instance, Pascal Boyer and Scott Atran seem to agree that the

---

9. Ursula Goodenough, & Terrence W. Deacon, "From Biology to Consciousness to Morality", *Zygon: Journal of Religion and Science* 38 (4), 2003, 813-814.

age-old assumption that religion produces morals and values is the wrong way around the issue. Rather, humans are more or less born with, or at least quite early on have, default moral sensibilities. The origins of such sensibilities are claimed to be found in basic social cognition. One of the concepts used by Boyer and Atran is the "check for cheaters" mechanism by which moral attitudes can be explained.

In much of the literature on the neural and psychological constituents of social intelligence and moral cognition, it is assumed that humans are born with or quite early on exhibit moral capacities. To be human is to be moral. Such an assumption explains how people know something is wrong without knowing why.

Pascal Boyer and Scott Atran maintain similar assumptions, but their point of departure seems wrong to me. And their time scale is off too, as indicated in the quote from Deacon and Goodenough. Boyer and Atran present similar programmatic claims about what religion is not. Boyer summarizes them in four points: 1) Religion provides explanations, 2) religion provides comfort, 3) religion provides social order, 4) religion is a cognitive illusion[10]. Even though Boyer rejects these scenarios for the origin of religion, he claims that they are not that bad (p. 6). But they point to phenomena that need explaining. Atran argues that "religions are not adaptations and they have no evolutionary functions as such". He says that religion did not originate to 1) cope with death, 2) keep social and moral order, 3) recover the lost childhood security of father, mother, or family, 4) substitute for, or displace, sexual gratification, 5) provide causal explanations where none were readily apparent, and 6) provoke intellectual surprise and awe so as to retain incomplete, counterfactual, or counterintuitive information[11]. As he writes:

> It is not that these explanations of religion are all wrong. On the contrary, they are often deeply informative and insightful. It is only that, taken alone, each such account is not unique to, or necessary or sufficient for, explaining religion. Rather, there are multiple elements in the naturally selected landscape that channel socially interacting cognitions and emotions into the

---

10. Pascal Boyer, *Religion Explained: The Evolutionary Origins of Religious Thought*, New York: Basic Books, 2001, 5.
11. Scott Atran, *In Gods We Trust: The Evolutionary Landscape of Religion*, Oxford: Oxford University Press, 2002, 12-13.

> production of religions. These include evolved constraints on modularized conceptual and mnemonic processing, cooperative commitments, eruptive anxieties, communicative displays, and attentiveness to information about protective, predatory, and awe-inspiring agents. (p. 13)

The problem is not that I think that the various elements which Boyer and Atran reject are sufficient explanations or scenarios for the origin of religion. My point is that I don't think we are capable of sorting religious elements out of the strands of the evolution of human cultural traits. If the assumption that the brain and cognition developed in a dialectical relationship with culture holds, then issues such as morality, identity, solidarity, meaning, and death will be religious issues right from the beginning and not as secondary to other matters. Boyer argues that "having a normal human brain does not imply that you have religion. All it implies is that you can acquire it, which is very different" (p. 4). But if you admit as I do that having a normal human brain implies that you have culture, otherwise you can't use your brain – an assumption that Boyer by the way does not seem to share – then why should religion be filtered out of the discussion?

A crucial problem of course is, what do we mean by religion? I have argued in a number of publications on this issue. I have argued that religion is social and cultural. It is "a socially imposed hermeneutical device that draws on the cultural competence of individuals with reference to conceived transempirical powers or beings"[12]. It may very well be that this definition needs to be adjusted. How this phenomenon arose is probably impossible for anyone to imagine, let alone prove. But whatever the most reasonable scenario might be, religion cannot play a subsidiary role. The powerful social and cultural processes of which we have knowledge today could not have been developed or conceived outside of a religious framework. And the reason for this, again, recalling Deacon and Goodenough, is that what sets our ape brain off from the brains of apes is symbolic competence.

We would probably have to concede that the first activities and ideas that we might conceivably recognize as being religious or spiritual – and here I am not talking about theologically correct or coherent ideas or highly developed priesthoods, shamans or other religious specialists – were what might be

---

12. Armin W. Geertz, "Definition as Analytical Strategy in the Study of Religion", *Historical Reflections/Réflexions Historiques* 25 (3), 1999, 460.

called partially formulated superstitions or proverbial responses, feelings and intuitions. If we can depend on neurologist Michael Persinger's findings on so-called temporal lobe religiosity, despite the fact that they evidently have not been replicated, then we might imagine a kind of proto-religious experience that religions quickly fill up with semantic content. Or, if we can depend on geneticist Dean Hamer's claims about a religious instinct that is triggered by a gene associated with the self-transcendence scale of spirituality, then the biology is there waiting to be given meaning. By the way, I am somewhat skeptical both of Hamer and of psychiatrist Robert Cloninger's questionnaire self-transcendence scale. Both Hamer's so-called God Gene and Cloninger's scale build on tremendously loaded assumptions which could disqualify their projects even before they are put to the test. On the other hand, there is sufficient evidence that religious or religious-like experiences affect or trigger brain areas through certain endogenous chemicals. So Hamer may be right somehow. Other matters, such as trust[13], have been shown to be if not caused at least significantly influenced by specific chemicals.

Ontogeny of Religious Sensibilities

As I mentioned in the beginning of this paper, Nicholas Humphrey argues that mindsight, or the art of doing psychology, is achieved through empathy. Empathy is achieved through the complex interactions between the child and its mother and closest family members. Through nursing, the child gains chemical and somatic access to its mother's inner world, according to neurologist Daniel J. Siegel and others[14]. Already here, we find the formative nature of religion and culture, since each culture has its own nursing styles. There are interesting studies which show how nursing styles reflect models for ideal individuals in different cultures[15].

Other mechanisms which help develop the child's empathy are mirror neurons in the brain which automatically stimulate areas that would normally

---

13. Michael Kosfeld, Markus Heinrichs, Paul J. Zak, Urs Fischbacher & Ernest Fehr, "Oxytocin Increases Trust in Humans", *Nature* 435, 2 June 2005, 673-76.
14. Daniel J. Siegel, "Toward an Interpersonal Neurobiology of the Developing Mind: Attachment Relationships, 'Mindsight,' and Neural Integration", *Infant Mental Health Journal* 22 (1-2), 2001, 67-94; and Louis J. Cozolino, *The Neuroscience of Psychotherapy: Building and Rebuilding the Human Brain*, New York & London: W. W. Norton & Company, 2002.
15. Judy DeLoache & Alma Gottlieb, eds., *A World of Babies: Imagined Childcare Guides for Seven Societies*, Cambridge: Cambridge University Press, 2000.

be stimulated if the child itself were performing the perceived action[16]. Children are also equipped with the uncanny ability to imitate their mothers' facial expressions even though they could not possibly have a conception of their own facial expressions. This factor has been demonstrated through innumerable experiments. The point here is that facial and bodily expressions produce mental effects. If you keep smiling, you would eventually become happy, and so on.

Right from the beginning, the child is subjected to massive socialization techniques which help internalize the objectivized assumptions and representations of a particular culture. The kinds of ideal patterns of behavior are controlled innocently at first through parental approval and disapproval, but later become more systematically routinized. An important device is the use of narrative. Daniel Siegel has shown that parents' ability to relate their autobiographical narratives are crucial not only to the child's mental world, but also to its physical brain. A process of neural and bicameral integration occurs through the use of coherent narrative. Such narratives also allow the child to beginning formulating its own story. But for many years, the child's narrative is controlled by its parents, its siblings and other authority figures. For many years, it is a question of being in someone else's narrative. Such narratives, seen in the canvass of human history, are mostly religious ones. They are narratives about a child's struggle to become the ideal person.

As the child grows older, stricter instruments replace routines and imitation such as persuasion, force, even punishment and violence in some cases. Systematic education is introduced and when the individual becomes a legal person, social instruments of power are employed: law enforcement, religious and political rituals and public displays, communal activities, initiation rituals, etc.. The individual is not simply a private person. He or she takes on a personality model and a social role. He or she becomes the values of a culture and a religion. Everything in an individual's life functions to that end.

As mentioned, we are born with certain capacities which allow the above-mentioned instruments and techniques to function. We are born with social intelligence and moral sensibility. We have not only the ability, but the very

---

16. Giacomo Rizzolatti, Luciano Fadiga, Leonardo Fogassi & Vittorio Gallese, "From Mirror Neurons to Imitation: Facts and Speculations", In *The Imitative Mind: Development, Evolution, and Brain Bases*, eds. Andrew N. Meltzoff & Wolfgang Prinz, Cambridge & New York: Cambridge University Press 2002, 247-66.

need, to take on worldviews. We have the capacities to deal with imaginary worlds and make them real, with imaginary beings and make them personal, with imaginary narratives and make them embodied.

If religion is something that was with us from the beginning, what would the advantages be? Here is where I take issue with Boyer and Atran. All the elements in their catalogues of things that are supposedly not the origins of religion should be found here: meaning, dealing with uncertainty and death, developing and maintaining individual and group identities, group mobilization and so on. Especially the latter would be a highly significant factor. Religion is a terrifyingly effective instrument for the mobilization of group violence. Almost all wars and terrorism thrive on religiously motivated group identity, no matter whether other factors lurk behind the scenes. Both Boyer and Atran present, I feel, solid arguments concerning fundamentalism and violence. Here the argument is that religious behavior must be demonstratively costly in order for individuals to prove their commitment. In fact, Atran defines religion in such terms:

> Roughly, religion is (1) a community's costly and hard-to-fake commitment (2) to a counterfactual and counterintuitive world of supernatural agents (3) who master people's existential anxieties, such as death and deception. (p. 4)

"Emotionally motivated self-sacrifice to the supernatural", Atran argues, "stabilizes in-group moral order, inspiring competition with out-groups and so creating new religious forms" (p. 268). Rituals serve to produce attunement in a group. The ritualized use of the body stimulates particular mental states. Public rituals help groups of individuals attune themselves to a common mental state, whether it is terror, ecstasy or violence.

Back to Morality

The above-mentioned activities have their ultimate source in social intelligence, especially what both Atran and Boyer call the "check for cheaters rule". The latter is part of a special inference system for social exchange. Even though Boyer does not show how the neurological evidence might support his theory of default inference systems, it does rest on psychological experiments. By all accounts, we seem to have social intelligence mechanisms and assumptions, if not in-born, then at least acquired quite early on. Such matters as gossip, social exchange, trust, coalitional dynamics, etc., all stem from hominid social cognition which is highly dependent on strategic information from and about others (pp. 122-129). Boyer argues that a social mind and concomitant inference systems that regulate social interaction are also basic to notions of supernatural beings (pp. 150ff.).

Basing his work on the neural constituents of moral cognition, neurologist William D. Casebeer argues that moral cognition "might not be a tightly defined 'natural kind'" as in the case for instance of visual modality[17]. Others, such as Jorge Moll et al. have shown that inferring cognitive and neural mechanisms from behaviors can be misleading. As they note:

> Westerners and East Asians differ in categorization strategies when making causal attributions and predictions, and moral values and social preferences are shaped by cultural codification. The PFC [prefrontal cortex] has a central role in the internalization of moral values and norms through the integration of cultural and contextual information during development. Assessing the relationships between culturally shaped values and preferences in social interactions will therefore be a logical next step in designing experiments with which to study moral cognition[18].

Whether we are born with it or acquire it, there is no doubt that we become aware of our moral experience and motivations, reflect on them, share them with others. The major texts and doctrines of religious traditions produce ready-made scenarios, codes, answers to moral dilemmas, etc. As Deacon and Goodenough argue, culture complements our moral sensibilities by leading us to moral experience and insight (p. 814).

There is much more that could be said on this topic. I haven't had time to mention the results of studies of psychopaths, brain damage, social deprivation, evil, abuse, dehumanization, rampant self-interest and so on. But these matters illuminate the backside, one might say, of our social intelligence

---

17. William D. Casebeer, "Moral Cognition and Its Neural Constituents", *Nature Reviews – Neuroscience* 4, 2003, 842.
18. Jorge Moll, Roland Zahn, Ricardo de Oliveira-Souza, Frank Krueger & Jordan Grafman, "The Neural Basis of Human Moral Cognition", *Nature Reviews – Neuroscience* 6, 2005, 804.

and attitudes. Again, the negative aspects are shared with our primate cousins. And we would do well to take this important fact into consideration.

Despite the great advances introduced by pioneers in the cognitive science of religion, such as Boyer and Atran, there is still a lot of work to be done. We need to increase our time scales and pay close attention to the growing field of moral cognitive neuroscience. We must also pay close attention to the interdependent causal links between culture and cognition, and the role that religious symbolic systems play in it.

WHY IS IT BETTER TO BE A PLANT THAN AN ANIMAL? COGNITIVE POETICS AND ASCETIC IDEALS IN THE BOOK OF THOMAS THE CONTENDER (NHC II,7)

INGVILD SÆLID GILHUS

Introduction

The *Book of Thomas the Contender* is the last tractate in Codex II from Nag Hammadi[1]. Through questions and answers and eventually a monologue, the resurrected Christ teaches Thomas his secret message. Like several of the Nag Hammadi texts, the *Book of Thomas the Contender* probably developed in stages[2]. The ancient readers of this particular version of the text, however, must have seen it as a unity. So shall we.

When Thomas, disciple and brother of Jesus, is called 'Contender' (*athletes*), the term implies that he struggles against the passions of the body. The title thus signals what the tractate is about. It is an encratic text, transmitting a radical form of ascetic Christianity. Its main themes are the abominations of sexual desire and sexual reproduction and the punishment after death. It is also an example of the 'masculine patterns of transcendence' which Luther H. Martin has pointed out emerged in ancient religion and especially in Christianity (Martin 1987: 150). In honor of him I will analyze cognitive poetics and ascetic ideals in the *Book of Thomas the Contender*.

---

1. The *Book of Thomas the Contender*, (or rather earlier versions or parts of it), is usually seen as deriving from Eastern Syria and being a translation from Greek to Coptic. Several translations of the tractate, three of them with Coptic text and extensive commentaries have been published (Turner 1975, Kuntzmann 1986, Schenke 1989). The translation used in this article is mainly Turner 1975 and 1989.
2. John D. Turner launched a two-text hypothesis. According to him the tractate is put together by a revelation dialogue between Jesus and Thomas, and a monologue by Jesus (Turner 1975 and Turner 1989).

The focus is the metaphorical language of the tractate, and especially four key-metaphors, two negative and two positive. Negative key-metaphors are *animal* and *fire*. Animal describes sexual desire and bodily appetites in general. Fire describes sexuality, as well as the punishment of hell. It lurks in the bodies of humans and makes them behave like animals. After death, humans who have been sexually active and, according to the vocabulary of the text, have behaved like animals will be punished by means of fire.

Plant and *twin* are positive metaphors, used to describe spiritual relations and processes. Ideally humans should be like plants, giving fruits in due time. The twin concept is found in the introduction where Jesus describes Thomas as 'my twin', 'my true companion' (138: 7-8) and 'my brother' (138: 10, 19). Thomas is a prototype for those who read or listen to the text, a paradigmatic figure, seen, for instance, when the text apparently without effort changes between 2nd person singular, implying Thomas, and 2nd person plural, speaking directly to ancient as well as late modern readers.

What does the metaphorical language of the *Book of Thomas the Contender* say about bodily and soteriological processes? How does this language manipulate emotions? Why is it better to be a plant than an animal?

Context

The tractate suffers from an acute lack of context which it shares with all Nag Hammadi texts. One initial challenge is to establish relevant contexts which are plausible and make the tractate into a meaningful object of enquiry. Interpretation of it therefore includes suggesting possible relevant contexts and re-inscribing it in a wider universe of meaning. Three obvious contexts for the *Book of Thomas the Contender* are Codex II[3] and the other Nag Hammadi codices; ascetic life in Upper Egypt in the fourth century; and Christianity and other religions, especially Manichaeism. In this essay I will focus on the tractate's relation to asceticism, keep the other tractates in Codex II as a background, but not dive further into the fascinating, but thorny question of the tractate's specific religious and social milieu in fourth century Egypt.

The *Book of Thomas the Contender* takes asceticism to a higher level. It discerns between 'deeds of the truth that are visible' and 'deeds of the truth

---

3. Codex II consists of the *Secret Book of John*, the *Gospel of Thomas*, the *Gospel of Philip*, the *Nature of the Rulers*, the *Origin of the World*, the *Interpretation on the Soul*, and the *Book of Thomas the Contender*.

that are invisible and pertain to the exalted greatness and the fullness' (138: 31-34). I will argue that this is a division between sexual acts on the one hand and sexual thoughts and desires on the other. Both are forbidden, but the text concentrates on the second category. It is not enough to spurn sexual acts, but necessary to get rid of sexual thoughts and desire as well, to reach spiritual perfection. Thus the tractate is clearly advocating asceticism on a higher level, and it makes derogatory remarks about those who do not live an encratic ascetic life (cf. Turner 2007: 604 f.). This may explain why it is the last text in a codex which contains several ascetic tractates, some of them encratic, but where none of the other texts take asceticism to quite the same compromiseless level as the *Book of Thomas the Contender*. The reason why *Thomas* was placed at the end of the codex, could be that it represented the high point of radical ascetic teaching within the codex and constituted its grand finale[4]. That the codex was seen as a unity is implied in a colophon placed just after *Thomas*, because this colophon seems to include all the texts in the codex: 'Remember me also, my brethren, [in] your prayers: Peace to the saints and those who are spiritual' (145: 20-23). However, an alternative explanation to why *Thomas* was placed at the end could also have been that it was regarded as extreme as and less important than other texts in the codex. The rationale behind the construction of ancient codices is difficult to determine and the statistical material is not large. In the case of *Thomas* I will opt for the first solution and treat this seventh tractate of the codex as its concluding text.

The *Book of Thomas the Contender* uses several persuasive techniques to raise its audience to action. These are rhetorical strategies, intended to work in a specific context, and especially in a radical ascetic setting. Basic among these persuasive techniques is the play on the audience's knowledge/lack of knowledge. This ambiguity is first seen in the introduction, when the Savior says to Thomas (138: 7-21):

> Now since it has been said that you are my twin and true companion, examine yourself and learn who you are, in what way you exist, and how you will come to be. Since you will be called my brother, it is not fitting for you to be ignorant of yourself. And

---
4. According to Michael A. Williams, the dialogue between Christ and Thomas is 'hammering home a lesson of ascetic discipline that could easily have been seen as the implications of the doctrine and myths in the earlier tractates'. (Williams 1996: 225).

> I know that you have understood because you have already understood that I am the knowledge of the truth. So while you are walking with me, although you are uncomprehending, you have already come to know, and you will be called 'the one who knows himself'. For one who has not known himself has known nothing, but one who has known himself has at the same time already achieved knowledge about the depth of everything. So then you my brother Thomas have seen what is hidden from human beings, that is, what they stumble against, being ignorant.

Has Thomas 'understood' or is he 'ignorant'? Has he 'come to know' or is he 'uncomprehending'? Or everything at the same time, verging on confusion? Modern commentators have pointed out that this passage is self-contradictory and has been compiled from other sources in a rather incoherent way[5]. This could be the case, but this does not necessarily mean that what the text says is impossible *per se*. Technically, those who are attracted to the message of the *Book of Thomas the Contender* can be described as those who are potentially perfect. They are perfect when they realize their perfection (and behave accordingly). However, as long as they do not realize it, they are not perfect.

By treating them simultaneously as ignorant children and as knowing and perfect, the audience is kept alert. They are described as aiming at perfection, but are later in the text characterized as 'apprentices' (*shouei*) and 'small ones' (*kouei*) (138: 34-36, 139: 11-12). One function of this language may have been to make its audience meek and receptive.

There is generally a strong emotional component in the *Book of Thomas the Contender* which is part of the tractate's rhetorical strategies and persuasive power. Because this tractate is associative, with several gaps, and because it makes use of a striking imagery, it could be read more like religious poetry than a theological treatise. It combines passion with damnation and develops these themes through myths, metaphors, and metonymies. By means of such devices the tractate speaks rather loudly to human emotions.

---

5. Turner says that 'the passage is full of inconsistencies'. (1975: 122). Schenke sees 'eine Diskrepanz zwischen Rahmen und Inhalt'. (Schenke 1985: 270).

Animals and the cyclical character of bodily life

In the second interchange between Thomas and Jesus, Jesus is about to reveal profound truths, but does he? Thomas asks: 'Tell us about these things that you say are not revealed, (but are) hidden from us'. (138: 36-39). In what follows it seems like a profound question does not really get a profound answer. Hans-Martin Schenke says that Jesus 'spricht nicht über Dinge, die verborgen sind, sondern gerade über etwas, das allen sichtbar ist'. (1989: 77). What the text reveals is the bestial nature of procreation (139: 2-12):

> But these visible bodies consume creatures similar to them with the result that the bodies change. Now that which changes will decay and perish, and has no hope of life from then on, since that body is bestial. So just as the body of the beasts perishes, so these modeled forms will also perish. Do they not derive from sexual intercourse like that of the beasts? If the body too derives from sexual intercourse, how will it beget anything different from them? So, therefore you are children until you become perfect.

What sort of disclosure is that? That beastiality is glued to sexuality and procreation is a traditional Christian insight. But revelatory knowledge and religious truths are not necessarily new and not necessarily impressive *per se*. Insights, religious or non-religious are established through persuasive techniques and as part of social situations. They offer alternative models of the world, which, when they are internalized, radically change peoples' lives. Accordingly, it is not necessarily the novelty of the message, but its potential effects on people when they succumb to it, which really constitutes its newness.

However, the beastiality of humans *is* more radically expressed in encratic circles than anywhere else and might for that reason be called a revelation. To know that bodies are made through sexual generation is visible to all, but it is not visible that sexual generation is an evil mechanism that could and should be turned off completely, which is the view of the tractate. According to Julius Cassianus, a well-known radical encratite: 'Man is like a beast when he marries and like a wild beast when he wants to mount a strange woman for intercourse'. (Clement of Alexandria, *Strom*. 18.102; 9.67). The *Book of Thomas the Contender* seems to share similar views on *encrateia*: Sexual intercourse is always sinful, also within marriage.

To know oneself according to the *Book of Thomas the Contender* involves several types of knowledge (138: 7-21). Gillian Clark has pointed out that in Delphi to know oneself probably meant 'know yourself to be mortal'; in most philosophical traditions it meant 'know your true self to be your immortal rational soul', while for Christian authors it may have meant, 'know your desires'. (Clark 2004: 76, cf. Turner 1975: 120-121). In Hermetic texts the relationship between self-knowledge and knowledge of the All is a major theme (Turner 1975: 120-121, DeConick 2007: 219).

The *Book of Thomas the Contender* is heir to several types of knowledge. It presupposes the mortality of the body and the immortality of the soul and elaborates on both these themes. However, the tractate is especially preoccupied with self-knowledge as knowledge about sexual desires. It's most important insight is to show that a human being is a sexual being and, as a consequence of this insight, the tractate struggles to persuade its audience that they ought to change completely and root out all sexual thoughts and desires. The text seems to presuppose that males and females are equally eager to give vent to their sexual lust, though the passage where it is described is rather damaged (139: 38-41). The *Book of Thomas the Contender* presupposes a male audience. On its penultimate page the text says: 'Woe to you who love the customary usage (*sunetheia*) of womankind and polluted cohabitation with it'. (144: 8-10). Though, it is, of course, a question if this ideological and transcendent maleness also included women, who had made themselves malelike (Elm 1994: 341). The woe over those who have intercourse with women is followed by three other woes: to those who are in the grip of the powers of their body, in the grip of the forces of the evil demons, and who beguile their limbs with fire. The focus is here on the effects on sexual feelings and thoughts on their subject, more than on the object of these desires and feelings.

What does it mean that the body is a beast and begets beasts? The term for animal (*tbne*) usually means 'domesticated animal' (cf. Layton 1987). But in the *Book of Thomas the Contender*, it is probably better to interpret *tbne* as a generic term for all sorts of animals, especially because it is explicitly said that animals (*tbne*) eat other animals (141: 27). The expression 'the fruits of the evil trees, beings punished, being slain in the mouths of beast and men' (142: 14-17) also refers to eating living beings. The metaphorical connection between animals and humans implies that the body derives from intercourse and perishes, and that animals devour each other. Schenke says that of all the characteristic features that were connected to animals in antiquity this text is 'geradezu neurotisch fixiert auf diese einzigen Punkt, die sexuelle Lust'. (1989: 87). He is right, but it must be added that sexual desire should not be seen in isolation from other types of human desires.

In the quotation, 'visible bodies consume creatures similar to themselves with the result that the bodies change' (139: 2-4), there is a reference to the diet of carnivores and a connection between this diet and bodily change. Later, however, the text refers to those who accept human procreation and says that 'do not esteem them as human beings, but regard them (as) beasts, for just as beasts eat one another, so also human beings of this sort eat one another'. (141: 26-29). Taking into consideration that men do not usually devour each other, eating should in the last case be interpreted as a metaphor (cf. Turner 1975: 157-158). Generally speaking eating is one important source domain for a metaphorical expression of sexuality. Thus eating, as a bestial trait, seems rather easily to be interpreted as sexual conduct. In other words, the most important characteristic of beasts in this text is their sexual drive and it is especially this trait that is mapped on humans when humans are likened to beasts.

That does not mean, however, that the tractate is only preoccupied with sexuality. One thing is that meat-eating is mentioned as well. Another thing is that sexual desire should rather be regarded as the most acute instance of human appetites, the prototypical appetite so to speak. It is the key to the whole field of appetites as well as to the cycle of bodily life, consisting of procreation, birth, eating, drinking, growth, decay and death. In this way there is a clear rejection of material creation going on in this tractate.

Sexual desire is the dynamic force in the cycle of life and death, because it points out over individual lives and assures that life will continue. Because it assures that life goes on from generation to generation, it is a more dynamic force than eating and drinking. Still more basically in this text, *epithumia* is 'desire for those visible things that will perish, change and dissolve' (140: 32-34). It implies the clinging to life in the material world, and is an expression of the intense grasping of bodily existence.

Fire of passion and fire of damnation

An even more prominent metaphor than animal (*tbne*) is fire (*koht*). The implications of clinging to this life are clearly seen in the use of this metaphor (140: 33-37):

'And it (the fire) has fettered them (human beings) with its chains and bound all their limbs with the bitterness of the bondage of lust (*epithumia*) for those visible things that will perish and change and dissolve. They have always been attracted downwards: as they are killed, they are assimilated to all the beasts of the perishable realm'.

When fire is used for *epithumia*, 'appetite', a poetry of passion, though definitely with negative connotations, is developed. In the passage quoted below, all parts of human beings are involved (139: 33-38):

> 'Oh bitterness of the fire! You blaze in the bodies of humans and in their marrow, blazing in them night and day, and burning the limbs of humans and [making] their hearts drunk and their souls disturbed....'

In this quotation, bodies, marrow, (which was seen as the seat of the production of semen), limbs, heart and soul are mentioned. The conceptual metaphor LUST IS HEAT/LOVE IS FIRE, relies on embodied experience including human biology and physiology. Cognitive linguists have investigated the metaphorical language of love and lust. Thanks to them we know that in English romantic fiction the most highly 'metaphorized' emotion concept is love. According to George Lakoff and Zoltan Kövecses, the most prominent conceptual metaphors for lust are fire/heat and hunger/eating. Then come animal behavior/wildness, war, insanity (Lakoff 1987: 380-415, Kövecses 2003: 29-31). Most of these conceptual metaphors are found in the *Book of Thomas the Contender*. In the passage quoted above the conceptual metaphor LOVE IS TO BE DRUNK and LOVE IS INSANITY were referred to. The metaphor of fire may also blend with the other metaphors, for instance 'you are drunk (*tahe*) with the fire' (143: 27). Generally, emotions are seen as a natural physical force, and as occurrences inside the body, therefore they are something that take control over and divide the self, EMOTION IS FORCE is also an important conceptual metaphor in this text (Kövecses 2003: 37).

In antiquity there were a range of conceptual metaphors for love and lust. Prominent among them was obviously fire. A few random examples are Paul, according to whom it is better to marry than to burn (1 Cor 7: 9); magical papyri where the magician wishes to set the object of his desire on fire

(Winkler 1990: 86); and more close to our text, the *Origin of the World*, the fifth tractate in Codex II, where Eros kindles everyone he comes into contact with as from a single lamp (109: 1-22). The metaphorical domains for love/lust had, of course, specific cultural formations and interpretations in Egypt in the fourth century.

It should also be added that fire is not only a metaphor, but a physical substratum and a vital connection as well, which is clearly seen in the *Book of Thomas the Contender* where it is said that the fire of lust blazes in the bodies of human beings and in their marrows (139: 33-35). According to this thinking fire can be used as a metonymically based metaphor for sexual desire.

The metaphor of fire also implies the possibility that the fire/desire might be extinguished. In a similar way as it is possible to completely extinguish fire, it is possible to completely quench sexual desire. At the penultimate page of the text, it is said: 'Woe to you (pl.) who beguile your limbs with fire! Who is it that will rain a relieving (*mton*) dew on you to extinguish the mass of fire from you along with your burning' (144: 14-17)? Correspondingly, salvation is described as *mton*, 'to be relieved' or 'to rest'.

Either the attempt at controlling the fire and, ultimately, extinguishing it will succeed or, if the control is lost, the burning fire will lead to death and punishment in hell (143: 2-7):

> 'If he flees west, he finds the fire. If he turns south, he finds it there as well. If he turns north, the threat of seething fire meets him again. Nor does he find the way to the east so as to flee there and be saved, for he did not find it in the day he was in the body, so that he might find it in the day of judgment'.

Fire as an element of punishment in the afterlife is traditional. It is present in Egypt as well as in Mesopotamia, Palestine and Greece. The New Testament and later Christian apocalyptic literature reckon on postmortem punishment, including both physical and mental factors. The fire of *Thomas* shares the retributive quality of hell-fire with other Christian texts. According to Origen, (who relies on Isaiah 50: 11: 'Walk by the light of your fire and by the brands you have kindled.'), the 'fire by which each person is punished is proper to oneself'. (*Principes* 2.10.4). (Bernstein 1996: 311). There are some distinctions between different types of punishment in the *Book of Thomas the*

Contender (cf. Turner 2007: 620). But while the fire in Origen is beneficial and curative, it is fatal in the *Book of Thomas the Contender*. What is most striking in *Thomas*, however, is the associative interaction between the fire of sexual desire and the fire of hell that makes these two types of fire close to each other. It could be noted in passing that fire is, in fact, a prominent conceptual domain for both anger and lust – generally, and in Christian antiquity (especially Lakoff 1987: 409-415). The closeness of the fire of sexual desire and the fire of hell furnishes the tractate with strong metaphorical and emotional groundings.

To be like a plant
In a rather damaged passage, the *Book of Thomas the Contender* suggests that man should strive to become like a plant and not remain with the animals. The text speaks about 'those that are above' and who 'live from their own root and it is their fruit that nourish them' (138: 42-139: 2). This is in contrast to visible bodies that copulate like beasts.

In relation to abstract concepts like life, death and salvation, plants are rather simple, and therefore good to think with when it comes to guiding and structuring thought. What sort of thoughts do plants help the tractate to think? When people are described as plants, it usually refers to the human life cycle and mortality. In the *Book of Thomas the Contender* the plant metaphors are especially focused on growth and fruitfulness. The imagery has a platonic reference in *Timaeus*, where it is said that 'we are not an earthly, but a heavenly plant' (*Timaeus* 90a), and a biblical resonance in the Septuagint as well as in the New Testament, especially in images and similes that focus on the producing of good fruits. Different from traditional plant metaphors, however, the *Book of Thomas the Contender* employs the striking and rather surrealistic image of a plant which is nourished by its own fruit.

There are three positive uses of plant metaphors in this text. In addition to the one already mentioned, there is the wise man who 'will be nourished by the truth and will be like a tree growing by the torrent (*mou nsorm*)'. (140: 16-18). Here the *Book of Thomas the Contender* alludes to a quotation from Psalm 1.3: 'And he shall be like a tree planted by the rivers of waters, that bring forth his fruit in his season'. The fruit is not explicitly mentioned in the *Book of Thomas the Contender*, but is presumed in the 'truth' from which 'the wise man is nourished'. Again it is the strange image of the plant which lives on its own fruit that is implied. We note that there is a difference between eating/devouring (*ouom*), which is connected to sexuality and is always used negatively, while nourishing (*saanch*) is a positive concept, used once in connection with 'fruit' (139: 2), and once in connection with 'truth.' Bodies devour other creatures, while souls are nourished by spiritual fruits. We also note that the rivers have become more destructive than in the *Septuagint*. Psalm 1.3 uses 'streams of water' (*dieksodous ton hydaton*), while Thomas uses 'torrent,' literally 'erring waters' (*mou nsorm*) (140: 18). Such a phrasing is in accordance with the general outlook of this text: The world is a place where people throw themselves into dangerous maelstroms of bodily desires and emotions.

The last positive use of plant imagery is developed into an extended story. It is a composite simile which includes and condenses several biblical motifs and allusions to biblical texts (Turner 1975: 185). It is about the grapevine which, when the sun shines on it, will shade the weeds besides it so that they wither and die and the grapevine dominates the land on which it stands and is bountiful for its master (144: 19- 36) (cf. Schoenborn 1997). Some of the details of this simile are difficult to understand, but a sensible interpretation is that it refers to the goal of developing spirituality at the cost of bodily existence, and not giving in to the body and its appetites so that they destroy the spiritual growth of the soul. In consonance with the radical ascetic message of the text, the weeds/appetites should be completely destroyed. Also in this case the *raison d'être* of the metaphorical plant is to produce good fruit, though here the fruit is made for the master of the field, *i.e.* Christ. Taking in consideration that the goal of Thomas and therefore of the perfect Christians is to be *like* Christ, also in this case the plant's production of fruit can be seen as a self-contained act.

In connection with the grapevine the fragrance, which the sun and moon will give 'to you' (pl.), implying the audience, is mentioned. The fragrance can be read as a parallel to the developing of fruit. It is a positive result produced by the plant. Fragrance refers to something which is immaterial and often used as a metaphor for salvation.

What these images of plants and fruits do not include, is the sowing of the seed. The reason is probably that the process of sowing was generally thought of in sexual terms. When Plato describes the 'sowing upon the womb as onto a field' (*Timaeus* 91 d), he alludes to the male seed and the female earth. It is striking that the *Book of Thomas the Contender*, in contrast to most of the tractates in Codex II and other ancient Christian texts, deliberately evades all sexual allusions when it speaks about spiritual matters. When sowing is twice mentioned, the tractate refers to the sexual organs and birth, describing them as 'that which sows and that which is sown'. (142: 11), and connecting them to evil trees (142: 9-18):

"The savior replied: Listen to what I am going to tell you (sg.) and believe in the truth. That which sows and that which is sown will dissolve in the fire – within their fire and the water – and they will hide in tombs of darkness. And after a long time they shall show forth the fruit of the evil trees, being punished, being slain in the mouth of beasts and humans by means of the rains and winds and air and the light that shines above".

In line with the interpretation of sowing, referring to sexual generation, 'hiding in the tombs of darkness' probably refers to the gestation of the foetus in the womb, while 'the fruits of the evil trees' is a metaphorical expression which refers to the product of the sexual organs which is the new born child. The rest of the passage is more difficult to understand, but it may refer to the complete life cycle and the death of the body and to what happens to bodies when they are killed or in other ways die, and to what happens to these bodies after death.

Consequently, when the plant metaphor is used to describe the soul and its development, the key feature that is mapped from plants to people is fruit; the producing of fruit; fragrance; having a root and being nourished by the sun. It is the process of internal growth of the plant, not the sowing of the seed, which is mapped on the target domain, namely humans, to describe spiritual growth and relate humans to the world above.

One more thing must be stressed in relation to plants. In antiquity their sexual life was unknown. Exactly because plants were regarded as being without sexual life, they were perfect models for the spiritual growth and fruitgiving of the elect (cf. Laquer 1992: 172-3).

Twins and the lack of spiritual marriages

Some of the tractates in codex II promote an ascetic outlook and condemn sexuality at the same time as they indulge in a marital poetics with its source and references in the world above. By means of conceptual metaphors of procreation, relations in the world above and unions between souls and heavenly counterpart are described. In these texts sexuality has clearly moved into the spiritual world as well as into the upper parts of the body (Gilhus 2008).

This is definitely not the case with the *Book of Thomas the Contender*. There are no sexual relations in the world above, sexual metaphors are not employed to describe spiritual matters, and salvation is not conceptualized as a heterosexual reunion with a heavenly double. On the contrary, such metaphors are deliberately restricted to the sinful and damned material creation.

The most positive human relation in the tractate is the relation between Jesus and Thomas, described in the introduction. Thomas is described as 'twin' (*soeisj*), 'true companion' (*sjbrmme*) and 'brother' (*son*) of Jesus[6]. The tractate draws on a twin-motif which is well-known from the Thomas tradition as well as from Manichaean sources. When it is said at the end of the tractate that 'you (pl.) will reign with the king and be joined with him and he with you' (142: 14-15), it may suppose a male union based on imitation and similarity.

There are similar mechanisms at work in the twin-model as in the rest of the tractate, sexual relations and opposite sex models are deliberately avoided. This is most likely part of the radical encratic strategy which this text employs. All uses of sexual metaphors with positive references are banned and anything to do with the flesh is condemned.

Models, metaphors and reality

In the *Book of Thomas the Contender* there is a more direct connection between some of the metaphors and the reality they describe. When, for instance, the sun is described as a good helper to the world above, it seems to be the real sun that is alluded to (cf. Schenke 1989: 91). At the same time Jesus is the light, and light is used metaphorically for spiritual enlightenment. Fire burns in the marrows of men, but fire is also used metonymically and metaphorically for sexual desire and other appetites. One could ask if the moving back and forth between material references and 'metaphoricity' gives the text persuasive power, because a perceived factual basis eases the acceptance of metaphorical meanings. Also in the case of the concept of 'twin' and 'brother', the sibling relationship between Jesus and Thomas, which is a direct non-metaphorical reference, is the basis for a metaphorical relation between Christ and the perfect Christian.

---

6. The word for 'twin', *soeisj* in Coptic, has, according to J. Osing, who is referred to by Schenke, (cf. Schenke 1989: 65) the meanings 'Doppelganger', 'Paargenosse', 'Paar' and may render both Greek *suzugos* and *didymos*. (cf. also Uro 2003: 10-15).

One important example of the change between metaphorical and non-metaphorical concepts in this text is reflected in its use of animals. Animals and humans exist in a common field. In many ways humans are not only like animals, they *are* animals. The connections between animals (mammals) and humans are based on perceived identity and vital relations, which include that both animals and humans have four limbs and a head, eat, drink, multiply by means of sexual reproduction and finally die. However, all these similarities are not mapped when animals are used to describe humans. It is primarily sexual desire and reproduction that are singled out as the defining common traits that make humans into beasts. Eating meat is mentioned, but this characteristic is not so striking as sexual behavior. This means that even if humans still have four limbs and a head, and continue eating and drinking – just like animals – when they put an end to sexual activities and reproduction, (and probably also to eating meat), they have removed themselves from the animal world.

The animals on which humans are modeled are carnivores. Carnivores are animals that attack, kill and eat other animals. The fact that most carnivores copulate only rarely does not stop them from being described/constructed as sexually active in this tractate. The mapping between humans and animals are usually based on perceived identity and some rash conclusions. In Christian interpretation carnivores got a demonic quality that made them extremely fitting to be used in a negative construction of sexuality (Gilhus 2006: 205-226).

The metaphorical relation between animals and humans is different from the metaphorical relation between plants and humans. To put it briefly, humans are animals, not plants (cf. also Schenke 1989: 84). When humans are metaphorically described as plants, these descriptions are constructed in other ways than when humans are conceived of as animals. The plant metaphor describes aspects of human life which relate to spiritual matters. In fact it describes a human being who lives the perfect spiritual life in this world as an anticipation of the life to come.

The different metaphors used in the *Book of Thomas the Contender* are not totally in consistency with each other, which is common when metaphors are employed. But these metaphors have a strong consistency in some points, which makes one overlook inconsistency in other points, and contributes to make the tractate at large convincing. There is, for instance, a consistency in what we could call the superior symbolic co-ordinate system in the text: the dual system of light/darkness and visible/invisible, and how connections are established between metaphors in this co-ordinate system. A few examples must suffice: *LOVE IS FIRE* presupposes that the destructive heat of the fire is mapped and not its illuminating light. Another option, when it comes to keeping fire on the destructive side, is that when the light of the fire is mentioned, it is implicitly compared to the tremendously stronger light of the sun and in relation to the light of the sun, fire is described as a false light.

In addition to *LOVE IS FIRE*, the other prominent conceptual metaphor for sexual desire is *LOVE IS AN ANIMAL*. Also in this case there is a certain consistency and connection between the two conceptual metaphors. Bodies, bestial and human, are driven by internal fire, while plants live by means of the light from above. Sexual fire is opposed to spiritual light. The body is like an animal, the spirit like a plant. The opposition between animals and plants can also be interpreted as an opposition between common people and perfect Christians.

There is further a consistency in the use of procreation metaphors to describe material creation and the use of the plant model and the brother/twin model to describe a spiritual ideal. Metaphors of sexual procreation are never used for developing knowledge and for spiritual relations in this text. On the contrary, it seems to be a conscious avoidance of this type of language, which is otherwise rather common in Christian texts, not least in most of the other texts in Codex II.

Conclusion

The *Book of Thomas the Contender* does not present asceticism for beginners, but for advanced students, and it does it by means of a metaphorical language that is remarkably consistent. In line with its avoidance of using heterosexual models for spiritual matters, its paradigmatic soteriological model is a male-male relationship. According to the twin-model and its prototypical ideal, Jesus and Thomas, the goal of the perfect Christian is to become Christ-like. Thus the *Book of Thomas the Contender* reflects a special and extreme version of the Christian 'masculine pattern of transcendence', described by Luther H. Martin.

The tractate further works with an opposition between a *phytomorphic* and a *theriomorphic* model of reproduction, and launches the plant as the proper and the animal as the inappropriate model. Why is it better to be a plant than an animal?

According to the ancient conceptions of plants, they lack a sexual life, do not copulate and are not nourished by eating. Plants are nourished by their root and get light from the sun. When the beastly body is contrasted by the plant-like soul, it implies that the ascetic goal is to produce fruit as a solitary and self-sufficient activity. In this model one produces one, this implies that a human being grows into a perfect spiritual entity. Opposite to the plant model and the twin model, the theriomorphic model describes sexual generation which implies that two become three. While carnivores eat other animals and produce more animals through sexual intercourse, the plant is nourished by its fruit which it produces itself. Sexual desire is further prototypical in describing a more general connection between the body and the world. It covers the grasping of all material things that change and lack a connection to the world above.

We do not know who were the readers and users of the tractates from Nag Hammadi, and we do not know their specific religious belonging. So even if asceticism is a likely social milieu for the *Book of Thomas the Contender*, its ascetic context must be constructed rather than re-constructed. What we know about the diversity of asceticism in Upper Egypt is fragmentary, but we do know that different versions of radical encratism were widely spread in Egypt in the fourth century (Elm 1994: 331-372, Hunter 2007: 131-146, DeConick 2005: 175-194). The *Book of Thomas the Contender* presents a consistent and persuasive version of such a rigorous Christian asceticism.

MIRACLES, MEMORY AND MEANING:
A COGNITIVE APPROACH TO ROMAN MYTHS

ALISON B. GRIFFITH

Not long ago Ilkka Pyysiäinen argued that miracles, whether real or imagined, are counter-intuitive events that we represent to ourselves using the same cognitive processes as we do for other counter-intuitive phenomena (Pyysiäinen 2002 [2004]). Furthermore, he hypothesized that 'such cognitive processing [about unexplainable events] occurs even in cultures in which no *explicit* concept of miracle exists, because miracles are exceptions to panhuman intuitions' (2002, 738, original emphasis). He defined a miracle in the 'weak sense' as an event that occurs counter to our intuitive expectations and that we cannot explain, and in the 'strong sense' as 'those weakly miraculous events and phenomena that are given an explanation or interpretation employing the counterintuitive representations present in one's database (other than scientific ones)' (Pyysiäinen 2002, 735-36). Following Pascal Boyer (2000, 200 and Boyer and Ramble 2001), he also asserted that miracles, like all other counter-intuitive phenomena, must have relevance to existing knowledge and a potential for inference in order to be culturally successful (Pyysiäinen 2002, 734).

It is the aim of this paper to investigate the way in which Roman myths[1] contextualized miraculous events and, without necessarily explaining them, presented them in a cultural framework that promoted recollection and facili-

---

1. This paper focuses on the concepts of 'miracle' and 'miraculous' rather than the definition of 'myth' per se. On C. Scott Littleton's useful graph showing the distinction and inter-relationship between the different types of fabulous stories, Roman myths would be plotted in 'history', with elements causing them to overlap with legend and myth (Littleton, 1965). Classicists must cope with the fact that such fabulous occurrences appear in what are otherwise historical sources. Although some would call them legends, a good case has been made recently for identifying them as proper 'myths' (Wiseman, 2005).

tated inference. My data consist of the multiple versions of some of the best known ancient Roman myths about miraculous events. The Romans had an explicit concept of 'miracle' and 'miraculous', such that this study does not constitute a test of Pyysiäinen's hypothesis (above). However, by examining the variations in the versions of the stories I will show how these meet the requirements he outlined for the conception and transmission of miracles as counter-intuitive phenomena, and how the Romans were able to derive meaning from counter-intuitive events and reconcile them with 'reality.'

The Romans recognized a range of miraculous occurrences and had a rich nomenclature to distinguish them; no single Latin word translates the English 'miracle.' In *De Natura Deorum* (2.7; 2.12) Cicero lists *ostentum* (prodigy, wonder), *monstrum* (evil omen), *portentum* (sign, omen or portent), *prodigium* (prophetic sign, omen) and *signum* (sign) as words that indicate future events and that are understood as communications from the gods. These events, which evoked a range of formal religious responses, form the basis of Anders Lisdorf's insightful study addressing their memorability and transmission (Lisdorf 2004). The Latin noun *miraculum* (a wonderful, strange or marvelous thing, a miracle) from which we derive the English word 'miracle' comes from the verb *mirari* (to wonder or be amazed at, to admire). It encompasses the specific English sense of a 'miracle', but can be used in a more general sense as well. Interestingly, ancient authors did not always use the words *miraculum* and *mirari* in connection with a miraculous event. It is clear, however, that the Romans recognized a number of events as miraculous, and that such events formed the basis of myths from which the Romans inferred that a divine entity was at work. Moreover, in these myths the wondrous occurrence is integral to the narrative (Boyer and Ramble 2001, 539), and the stories are told and retold in different ancient sources over several generations. In this paper I focus on a select body of myths that recount miraculous occurrences on which the plot of the narrative turns but which were *not* specifically recognized as omens or signs from the gods and did not inspire or require a religious response. It is outside the scope of this paper to delve into the complicated distinction between myth and legend, which is meaningless if one is only interested in how counter-intuitive events are reconciled with 'reality'. Most of the stories I have chosen contain a kernel or two of historical fact, which makes their counter-intuitive elements all the more interesting to study. I do not pretend that the myths presented here form a representative sample by any scientific definition. My aim is to show that cognitive studies help answer a persistent question among Classicists: did the Romans believe their myths?

Among the many, recorded Roman myths it is possible to identify those that conform to Pyysiäinen's distinction between miracles in the strong sense and miracles in the weak sense. The stories of Horatius Cocles and Attus Navius, which I discuss below, might be said to belong to the 'weak sense' category because the central counter-intuitive event is not directly attributed to a god, but is merely noted as miraculous and left unexplained. In several other myths that might be identified as miracles in the strong sense, the stories run almost according to a template in which the central character is at the outset a person of low status, socially peripheral, or of tarnished reputation. The miraculous occurrence focuses on that individual and often consists of his or her unique exercise of extraordinary (i.e. counter-intuitive) powers. After the performance of a 'miracle' the individual is either elevated in status or restored to a previous high status (as with women whose doubtful chastity is proven to be intact). To this group belong the stories of Romulus and Remus (the founders of Rome), the miraculous conception and childhood of Servius Tullius (a king of Rome), and several stories about the proven virtue of Vestal Virgins. The Romans also told a number of stories in which the timing of an ordinary occurrence was so fortuitous as to be attributable to a divine cause - for example, a heavy rainstorm in battle that helped the Romans and debilitated the enemy despite its larger numbers. Unfortunately, I do not have the space to discuss such occurrences in detail.

It is now well established by experimental data that counter-intuitive concepts are easier to remember and, subsequently, to transmit to others. In a series of four experiments, Justin Barrett and Melanie Nyhof presented unfamiliar stories to a group of undergraduates (Barrett and Nyhof 2001). One story was adapted from Native American folk tales and the other, about a diplomatic mission to another planet, was especially constructed for the purpose of the experiment. Both stories contained a specific number of counter-intuitive, bizarre and commonplace elements. Students read the stories, performed a brief distraction task, and then wrote down or retold the stories in as much detail as they could remember. These new versions were then told to a second generation of participants, who then rewrote or retold the stories to a third generation, who wrote down the stories as they recalled them. The results indicated that counter-intuitive concepts are remembered better than those that are merely bizarre, or those that are downright commonplace, even after a delay of three months. Put in terms of Roman myth, these experiments predict that:

Counter-intuitive: Servius Tullius, on whose head a flame burned without hurting him, became a king of Rome when he grew up[2].

Will be better recalled immediately (and after a delay) than:

Bizarre: Servius Tullius, who was a captured slave but was raised in a palace, became king of Rome when he grew up.

Will be better recalled immediately (and after a delay) than:

Commonplace: Servius Tullius was born in Corniculum and later became king of Rome.

In a later study, experiments focused on the effects of context (Gonce et al. 2006) and the ability to visualize or 'image' a particular concept (Slone et al. 2007). The results of this later study indicated that: 1) The more easily and quickly a person can visualize or 'image' a concept, the better it is recalled if the items are presented in a list and without a context, including maximally counter-intuitive concepts with high imagery and 2) The ability to visualize a minimally counterintuitive concept had no effect at all on the rate of its recall. Barrett and Nyhof's study results indicated that minimally counter-intuitive concepts are better recalled when presented in a context, for example a narrative. These results reinforced earlier findings that commonplace and bizarre items are recalled better than counter-intuitive items when they are presented as lists (Atran and Norenzayan 2005). Thus the story (used in the experiment) about horses that giggle and admire each other had a higher rate of recall when the counter-intuitive concept of 'giggling admiring horses' was presented in a simple context than when 'giggling admiring horses' appeared on a list of other three-word concepts (Gonce et al 2006).

Through their experiments involving serial retelling Barrett and Nyhof documented not just which concepts were successfully recalled, but the gradual distortion of concepts into a form more easily recalled. Thus after the third retelling of a story (Barrett and Nyhof 2001, 82):

---

2. Cic. *Div.* 1.53.121; Livy 1.39.1-4; Dion. Hal. 4.2.4; V. Max. 1.6.1; 1.8.11; Ovid, *Fast.* 6.635-36; Plin. *Nat.* 2.241; Plut. *Mor.* 323c-d (*Fort. Rom.* 10); *Vir. Ill.* 7; Flor. *Epit.* 1.1.6.1; Serv., *A.* 2.683; Zonar. 7.9.

- 12.5% of bizarre items became common, but 22.2% of common items became bizarre (i.e. easier to recall later)
- 7.2% of counter-intuitive items became bizarre, but 37.5% of bizarre items became counter-intuitive

Although we cannot prove it for Roman miracles, such results help explain why counter-intuitive events crept into and became so structurally integral to myths. An additional finding by Barrett and Nyhof further illuminates this process. When a story is retold any of the following might occur:

- Transformation: an unfamiliar or confusing element may be made more familiar
- Rationalization: a counter-intuitive idea may be expressed in simpler terms
- Omission of details altogether

The difference in detail in the various accounts of a miraculous occurrence during the removal of Juno's statue from the city of Veii to Rome illustrates these processes. The story is told within the context of an historical event - the sack of Veii in 396 BCE after a 10-year siege. It is said that the Roman commander, Camillus, called on the Veientine goddess Juno to aid Rome in defeating Veii and vowed to build a temple for her if she would do so. After the Romans successfully captured the city they wanted to remove Juno's statue to Rome in order to install it in the promised temple. Just before lifting the statue one of the Roman soldiers, either in jest or in earnest, was said to have asked Juno if she wished to go. Her response, a miraculous event (though not referred to specifically as such), is recorded in four sources as follows:

Livy 5.22.5-6[3]	Dion. Hal. 12.13.3	V. Max. 1.8.3	Plut. *Cam.* 6.1-2

---

3. The chronological span of the sources is noteworthy. Livy, an Italian, lived around 59 BCE-17 CE and wrote this section of *Ab Urbe Conditum* (*A History of Rome from its Foundation*) by 27 or 25 BCE. Dionysius was born in the Greek colony of Halicarnassus but taught rhetoric at Rome in the early 20s BCE. His *Roman Antiquities* (written in Greek) presented a history of Rome for a Greek audience. Valerius Maximus was a slightly later contemporary of these two, and wrote *Factorum ac Dictorum Memorabilium Libri* IX (*Memorable Deeds and Sayings in Nine Books*) in the early 1st century CE. Plutarch, a

Young soldiers perform the task:	Knights perform the task:	Soldiers perform the task:	Workmen are assigned to perform the task:
1) The goddess nods her assent. 'It is added that...' 2) The goddess also speaks her assent. 3) The heavy statue is moved with little effort.	1) The goddess speaks her assent (loudly). 2) The goddess repeats her assent.	1) The goddess speaks her assent.	1) The goddess (softly) voices her assent while Camillus is sacrificing and praying, without being directly asked.

Across the accounts the young soldiers who are addressed by the goddess are transformed to knights by Dionysius of Halicarnassus, and to Camillus himself in Plutarch's account, as if a miraculous event of such importance could not occur to mere soldiers. The statue's nod and easy removal are omitted in all accounts after Livy's, and the double agreement (first the nod, then the vocal response) is rationalized to repeated assent, and then a single instance of assent. Moreover, Plutarch introduces an error by wrongly attributing Camillus' presence at the event to Livy (in whose version Camillus was not actually present).

To summarize thus far, then, successful cultural concepts are those things that are memorable, and are successfully recalled and transmitted. However, experimental data indicate that minimally counter-intuitive phenomena have a greater chance of recollection and transmission; these grab our attention. Despite having intuitive knowledge about the way things are (reality), the human brain is able to form 'counter-intuitive representations' that contradict this

---

Greek born in Chaeronea around 50 CE, wrote numerous biographies of famous Greeks and Romans, among many other surviving works.

knowledge and that may be religious, fictional, magical or miraculous. Miraculous events are those that violate our intuitive ontological expectations, as with the nodding, speaking statue of Juno or Servius Tullius' flaming head, and I have described the way in which context assists recall and transmission of these myths. A moving, speaking, opining statue has little value as a cultural concept until we learn that it is a statue of a goddess who has been called upon to assist the Romans and whose statue was removed to Rome after a tremendous victory. A slave-boy with a flaming head is merely a curiosity without a context, but the event becomes more meaningful when we learn that the miracle prognosticated his future as a king of Rome. Although the narrative framework makes these central miracles easier to recall and transmit, narrative alone does not fully account for their success as a memorable tales.

Counter-intuitive phenomena can only be successful if they have some significance or meaning within a given culture. Myths and other stories put counter-intuitive events, including miracles, into context by iterating the circumstances of the miracle and the outcome after its occurrence. Such stories do not explain *how* the miracle happened, or *why*, since miracles are in effect a representative category we have created for events whose causes we cannot fully explain or understand. In addition to attributing a counter-intuitive cause to a miraculous event, we often embed it in a narrative context with detail sufficient to stimulate later recollection but sparse enough to invite reflection on and infer meaning from the episode. This opportunity for inference is an additional prerequisite for successful transmission (Boyer 2000; Pyysiäinen 2002). The multiple versions of the most popular Roman myths show this process of reflection and inference at work. To return to the story of Juno of Veii, one can observe that despite slight changes from one version to another, the central miracle of a speaking statue that expresses her preference is retained in all. Juno's desire to be removed to Rome was interpreted in one of several ways. Most specifically, it affirmed the Romans as victors, but in a more general sense this is a story about Roman *pietas* (religious devotion). Although the young soldier's question might have been a joke, the ensuing miracle validated his action: Juno wanted to be asked. Thus generations of later Romans interpreted her assent as an affirmation of their *pietas* and this, in turn, substantiated the Romans' view that their imperial success was a sign of the support of their gods.

Another myth about Rome's most famous hero shows how variation facilitated inference. Among the multiple versions of the story of Horatius Cocles, all but one recount the miracle of his being able to swim to shore while fully armed. The story runs as follows. After the Tarquin kings were ejected from Rome in the late 6th century BCE, Lars Porsenna took the Janiculum Hill and prepared to attack Rome using the Pons Sublicius (Pile Bridge) over the Tiber River. Horatius Cocles and two other men defended the bridge, but seeing that they could not hold out forever, Horatius begged his companions to retreat to safety and ordered the other Romans to cut the wooden bridge out from underneath him. When the bridge was cut, Horatius:

Polybius 6.55.1	Livy 2.10.11	Dion. Hal. 5.24	Val. Max. 3.2.1	*Vir. Ill.* 11

1) Plunged into the Tiber river, fully armed.	1) After a prayer to Father Tiber, leapt into the Tiber fully armed.	1) Leapt into the Tiber fully armed.	1) Threw himself into the Tiber fully armed.	1) Fell into the Tiber.
2) Deliberately sacrificed his own life for Rome.	2) Swam while avoiding Etruscan weapons.	2) Swam with difficulty against eddies and a strong current.	2) Swam to the bank despite the fall, his armour, eddies, and Etruscan weapons.	2) Swam to the bank fully armed.
3) Became an example to Roman youth.	3) Reached the bank safely.	3) Reached the bank safely and still fully armed.	3) Reached the bank safely.	3) Reached the bank safely.
	4) Was honored with land and a statue in the Comitium (i.e. the Roman Forum).	4) Was honored with a statue.	4) Was honored with glory, land, a statue in the Forum, food during famine.	4) Was honored with land and a statue in the Volcanal (i.e. the Roman Forum)

Polybius' version (the oldest) is the only one that logically reports that Horatius drowned under the weight of his armor. But Polybius was a Greek attempting to explain to a Greek audience the Romans' meteoric rise to power in the 2nd century BCE, and thus this variation relates to his purpose in recording the story. His aim was not to report a miraculous occurrence, but to underscore that Romans esteemed most those who were willing to sacrifice their own lives for the welfare of Rome. Horatius, he says, was the foremost example of this trait and was revered for it. One might quibble that the ability to swim the Tiber fully armed and under fire is merely bizarre rather than miraculous. However, the Romans' reward to Horatius - a statue in the Forum and free land - and the tenor of the stories in the sources, as demonstrated by the increasing number of impediments to his safe return to the riverbank across the different versions, attest their amazement at this outcome. Significantly, Horatius Cocles is cited as the foremost example of bravery by Valerius Maximus, who wrote *Factorum ac Dictorum Memorabilium Libri* IX (*Nine Books of Memorable Deeds and Sayings*) in the early 1st century CE, and he was still described as such in the anonymous *De Viris Illustribus (Concerning Distinguished Men)* written around the 4th century CE.

The many versions of the story of Romulus reflect a similar tendency to adapt details to suit an authorial agenda. Romulus experienced two miracles in his life. The first and most famous, that he and his twin brother Remus were saved from drowning (by a fortuitous ebbing of the flooded Tiber) and were then suckled by a she-wolf until they were rescued, is widely known in Western culture[4]. Romulus' death is a lesser-known but no less miraculous occurrence.

Cicero *Rep.* 1.16.25; 2.10.18	Livy 1.16	Dion. Hal. 2.56.4; 2.63.3	Plut. *Rom.* 27.6-7; 28.1
1) Romulus was	1) Romulus disap-	1) The senators tore	1) Romulus

---

4. Liv. 1.4.6-7; Dion. Hal. 1.79.6-8; Ov. *Fast.* 3.37-38; Prop. 2.6.20; 4.1.32-38; 4.4.53; Plut. *Rom.* 4; *Fort. Rom.* 8; *Quaest. Rom.* 21; Just. 43.2.5; and Serv. *A.* 1.273

raised to heaven during an eclipse on account of his own virtue (*Rep.* 1.25). -and- 2) Romulus disappeared during an eclipse and was apotheosized (*Rep.* 2.10.18).	peared in fog during a storm and was hailed as a god by the Roman people. -or- 2) He was torn apart by senators. -and- Julius Proculus claimed that an apparition of Romulus appeared to him, predicted Rome as capital of the world, and disappeared into heaven.	Romulus to pieces in the senate-house (during a full eclipse) and these were smuggled out individually under their togas. -or- 2) He was murdered in the dark by the Roman people (during a full eclipse). -and- His successor, Titus Tatius, ordered that he be honoured as the god Quirinus (because of doubts about the cause of his sudden disappearance).	disappeared suddenly and no portion of his body was recovered. -or- 2) The senators murdered him in the Temple of Vulcan, cut his body to pieces, and smuggled these out individually under their togas. -or- 3) He disappeared during an eclipse while addressing the people in the Campus Martius. -and- Julius Proculus claimed that Romulus appeared to him and said he was now the deity Quirinus.

The major difference in the versions comes between the earliest - Cicero's, from around 54 or 51 BCE - and all later versions, which were written after the assassination of Julius Caesar in 44 BCE and the Senate's decision to worship him as divine (a *divus*). Cicero's account conspicuously lacks the variation in which Romulus was murdered by the senators. Writing after Caesar's death, Livy (if he is in fact the first) and later authors appear to have embellished the story in order to draw a direct comparison between Caesar, who was accused of aspiring to become a king of Rome, and Romulus, who was the (mythical) first king of Rome. Though indirect, the analogy undoubtedly made both stories more memorable by the reduplication of the bizarre and conflicting accounts of Romulus' murder and by the similar outcome - apotheosis and divinization - for Caesar and Romulus[5].

Several other miracle myths focus on authority, either of an institution or of an individual of humble status or questionable character. The authority of augury is confirmed by the myth of a miraculous act by Attus Navius, who, when challenged by the Roman king Tarquin the Elder to perform the task he had in mind, took the auspices and declared that the task could indeed by accomplished. On learning that Tarquin had in mind to cut a whetstone in half with a razor, Attus Navius promptly completed the task. Both items were said to have been buried in the Comitium (the meeting place of the plebs in the Roman Forum) because of this extraordinary feat[6]. One must note that the miracle does not relate to Attus Navius' ability to read minds, otherwise he would surely have told Tarquin that what he had in mind was impossible. Rather, the miracle affirms the Roman religious observance of augury, the taking the auspices, by proving their veracity and authority over and above that of the king. This is underscored by a variation in the account by Dionysius of Halicarnassus in which it is Tarquin himself who performs the miracle and successfully cuts the whetstone in two.

Myths of divine conception were aetiologies for the later high status and authority of particular individuals. Rhea Silvia's conception of Romulus and Remus after being raped by the god Mars is a well-known example. It is said that her (illegitimate) children were in line for the throne of the city of Alba and were deprived of their birthright by a jealous uncle. After being saved by the she-wolf the twins grew up as mere shepherds, until through a series of coincidences their true identities were revealed and they reclaimed their royal birthright. Later, in a slightly better documented period, the sixth king of

---

5. I am grateful to my colleague Dr. Peter Keegan of Macquarie University for bringing this detail to my attention.
6. Cic. *Div.* 1.17.31-32; Livy 1.36.3-5; Dion. Hal. 3.71.2-5; V. Max 1.4.1.

Rome, Servius Tullius, not only experienced a portentous but harmless flaming head as a child (predicting his future greatness), but several Roman sources also said he was the son of Ocrisia, a captive slave impregnated by a divinity (Vulcan) in the form of an erect phallus that appeared in the ashes of the sacrifice to the household god, the Lar Familiaris[7]. So well established were these myths of divine conception that later, completely historical figures were able to promote themselves by claiming divine birth. In this way Livy tells us that Scipio Africanus, the conqueror of Carthage in 201 BCE, did nothing to contradict a rumor that he was the son of a snake that used to appear in his mother's room (Livy 29.19).

Many Roman miracle myths concern aspects of religious tradition, as has already been noted with Attus Navius, but the Romans seem not to have treated these stories as sacred narratives. While some of the most popular myths, such as the story of Romulus and Remus, were almost certainly presented as dramatic performances, no evidence substantiates that this was ritual drama *per se*. The story of the arrival of the statue of the goddess Magna Mater in Rome was re-enacted both as a formal religious procession and as a play performed at the Megalensia, the annual festival in her honour. In this myth the miraculous occurrence centers on Claudia Quinta, a woman, and in some versions a Vestal Virgin of tarnished reputation, who was said to be among the prominent and chaste women who welcomed the arrival of the statue of Magna Mater in 204 BCE. Livy does not stipulate why her participation in this event restores her good reputation (29.14.12), but the more famous accounts by Ovid, Propertius, and Silius Italicus relate that the goddess' barge miraculously shifted from a sandbar when Claudia Quinta touched the tow rope - in some accounts after a prayer beseeching the goddess' help. This miracle acted as a proof of her chastity[8]. The theme of the Vestal-proved-innocent is a topos in Roman myth. Similar miraculous acts that prove the chastity of Vestals are also recorded for Tuccia, who proved her accuser false by carrying water in a sieve from the Tiber to the Forum, and for Aemilia, who rekindled

---

7. Ov. *Fast.* 6.627-34; Plin. *Nat.* 36.24; Plut. *Fort. Rom.* 10 (323a-c); and Zonar. 7.9.
8. Ovid, *Fast.* 4.291-328; Prop. 4.11.51-52; Sil. It. 17.23-45; Appian, *Hann.* 9.56 and Herodian 1.11.4. The story is also briefly confirmed by Pliny, *Nat.* 7.35.120; Suet. *Tib.* 2.3, and Stat. *Silv.* 1.2.245-46.

the cold embers of Vesta's extinguished fire with a linen cloth[9].

It will by now be clear that Roman myths with miraculous occurrences have counter-intuitive features in the same way that other religious phenomena do. Indeed, even though such stories were never regarded as sacred narratives, they reveal another way in which interactions with the gods permeated Roman life. However, subsequent interpretation may be traced through variations and consistencies in the way the myths are retold and seems, at least to modern sensibilities, to straddle a sacred/secular divide. Miracles might be caused by divine agency, but their memorability and transmission arose as much from their social or political significance as from any specifically religious significance ascribed to them. In the foregoing discussion, I have argued that Romans, like all other humans, accepted the reality of miracles on some level. The multiple versions of some of the most oft-told myths - those with a miraculous occurrence - attest that these were successful by virtue of being both remembered and transmitted to a new generation of listeners using counter-intuitive elements. More importantly, these myths had the potential of inference because they contextualized a miraculous event within Romans' cultural framework in such a way as to encourage reflection on the moral or meaning of the story. Some of the most famous miracle myths had considerable socio-cultural relevance, which in some cases is revealed by a single variant in the story. The stories of Horatius Cocles and Claudia Quinta extol them as *exempla*. Other miracles, such as those involving Romulus and Remus and Servius Tullius, had aetiological relevance on a number of levels. Although these myths are not treated as sacred, many had religious relevance by demonstrating a god at work through the miraculous action of an individual. In sum, I believe that the question 'On what level did the Romans believe their myths?' is more appropriate and more fruitful than 'Did they believe their myths at all?'

---

9. Tuccia: Dion. Hal. 2.69; Val. Max. 8.1; Pliny, *Nat.* 28.3.12-13; Tert. *Apol.* 22.12; August. *C.D.* 10.16 and 22.11. Aemilia: Dion. Hal. 2.68; Val. Max. 1.1.7; Prop. 4.11.53-54 (vestal not named).

ABBREVIATIONS

All abbreviations for Classical authors and works follow those used in the *Oxford Greek Dictionary* and the *Oxford Latin Dictionary*.

'WHATEVER STORY SINGS, THE ARENA DISPLAYS FOR YOU'

PERFORMANCE, NARRATIVE AND MYTH IN GRAECO-ROMAN DISCOURSE

[Martial, *Liber spectaculorum*, Kenneth Scott, *Imperial Cult under the Flavians* (Mythology and Religion; New York, N.Y.: Arno Press, 1975, reprint of Stuttgart/Berlin: Kohlhammer, 1936), 56]

GERHARD VAN DEN HEEVER

In the course of a long career in the study of religion, Luther Martin has made his influence felt in such diverse areas of the general field of the study of religion as history of religion (particularly mystery religions), with very substantial contributions to the study of Mithraism; changing trajectories of ancient religions and their mutual interactions (attendant to the issue of religious change); to theory of religion, and lately prominent in the promotion of cognitive study of religion (and applied to theorising ancient religions and mystery cults including Mithraism); as well as contributing to the general field of the contemporary study of religion in America and Eastern Europe as scholarly practice and the study of discourse. It is a privilege to celebrate the honorand with this reflection on one aspect of the living context of mythic narrative and representation in the Graeco-Roman world, especially given his long-enduring influence on my own work in this field.

1. A Culture of Public Spectacle – To Textualise the Visual and to Visualise the Text

The title of the essay derives from the celebratory epigrammes Martial wrote for the occasion of the inauguration of the Flavian amphitheatre in Rome, the Colosseum, to wit: *De Spectaculis* (the citation is from no. 5). It reads:

> Pasiphae really was mated to that Cretan bull:
> believe it: we've seen it, the old story's true.
> old antiquity needn't pride itself so, Caesar:
> whatever legend sings, the arena offers you[1]

Even allowing for the language of flattery and sycophancy, the epigramme plays on the vividness of the theatrical performance – its *enargeia* giving it the quality of reality ('we've seen it, the story is true') – and then drives home the point: even better than residing in the display cabinet of ancient mythical tale-telling, the performance in the arena makes the myth come alive and by giving it a visual presence weaves a world of cultural references that simultaneously creates a vision of the world, and acts as vehicle for the discursive strategies and embedded ideologies that constructed Roman 'impereality'.

It has been pointed out in recent scholarship on the making of Roman 'impereality' that imperial power and authority was constructed in the performance of different kinds of spectacle – the more 'spectacular' events like staged wild beast hunts, executions, gladiatorial games, staged battles, and triumphs; as well as the slightly less over the top ritual, pomp, and circumstance[2]. The workshop of imperial discourse pro-

---

1. Martial, *De Spectaculis* (trans. A. S. Kline; Http://www.tonykline.co.uk/ PITBR/ Latin/ Martial.htm#_Toc123798956, accessed 11 November 2007). On Martial, see Marion Lausberg and Franz Tinnefeld, "Martialis", in *Der Neue Pauly* (ed. Hubert Cancik, et al.; Brill Online. http://0-www.brillonline.nl.oasis.unisa.ac.za:80/ subscriber/entry?entry= dnp_e725280, accessed 11 November 2007).
2. A wealth of literature exists, but see the classics in this regard: S. R. F. Price, *Rituals and Power. The Roman Imperial Cult in Asia Minor* (Cambridge: Cambridge University Press, 1984); Paul Zanker, *The Power of Images in the Age of Augustus* (trans. Alan Shapiro; Ann Arbor, Mich.: University of Michigan Press, 1988); and also, for an application to early Christian 'textual spectacles' in anti-Roman literature, Christopher A. Frilingos,

duction is richly furnished with other tools as well, and one should, in this regard, also pay attention to the visual world of the plastic and graphic arts, as well as novelistic literature[3]. My interest here, however, is not so much the spectacles themselves, but the 'inter-space' between spectacle and text, or, the making of textualized spectacle, as well as the wider encompassing effects of this textualization of the visual[4].

The stylistic quality of vividness (*enargeia* in Greek, *demonstratio* in Latin), had long been part of the arsenal of rhetorical theory (*Rhetorica ad Herrenium*; Cicero *Brut.* 261, describing the rhetorical ornamentation in Caesar's *Commentarii*: 'It seems, as if he had placed a well-painted picture in a good light')[5], but it is noteworthy that the Roman historian Livy was particularly known for his vivid style (e.g., Tacitus, *Ann.* 4.34), and in fact, this was the way in which he derived authority for his historical narrative[6]. Thus, when Livy stated in the preface to the *History of Rome* that:

> [t]his in particular is healthy and profitable in the knowledge of history, to behold specimens of every sort of example set forth in a conspicuous monument [*in inlustri posita monumenta intueri*]; thence you may choose which models to imitate for yourself and

---

Spectacles of Empire. Monsters, Martyrs, and the Book of Revelation (Divinations: Rereading Late Ancient Religion; Philadelphia: University of Pennsylvania Press, 2004).

3. This was the point of my Ph. D. dissertation, Gerhard A. van den Heever, "'Loose Fictions and Frivolous Fabrications'. Ancient Fiction and the Mystery Religions of the Early Imperial Era" (Ph. D. diss., Pretoria: University of South Africa, 2005).
4. On the interplay between such cultural scripts or frames and the making of Christianity, see among others V. Henry T. Nguyen, "The Identification of Paul's Spectacle of Death Metaphor in 1 Corinthians 4.9", *New Testament Studies* 53, no. 4 (2007): 489–501; and J. Cilliers Breytenbach, "Paul's Proclamation and God's *Thriambos*. (Notes on 2 Corinthians 2:14–16b.)", *Neotestamentica* 24 (1990): 257–71.
5. Andrew Feldherr, *Spectacle and Society in Livy's* History (Berkeley, Calif./London: University of California Press, 1998), 5.
6. Feldherr, *Spectacle and Society*, 4.

your *res publica*, and which corrupt in their beginnings and corrupt in their outcomes, to avoid (*Praef.* 10),

he imagined his work to be a 'visual artifact subject to the gaze of the er[7]'. In contrast to Thucidides's promise of his work presenting a 'clear view of the events', meaning that the readers will be able to 'see through' the text to the events described in a certain immediacy of representation, Livy holds up his work as an *inlustre monumentum*, the text itself an *exemplum*, eliding the boundaries between the representation and the event described[8]. His descriptions in the *History* of proto-imperial spectacle, as asserted in the preface that the entire history acts on its audience through being gazed upon, signify not only 'a parallel between his text and the public spectacles of state, but makes his narrative the medium through which these spectacles reach a new audience'[9]. The text, as it were, becomes a site for the reprise of the event itself. In a sense, then, the spectacles of Livy's historical narrative are the effects produced by his text. The increasing interest in *ekphrasis* in late Roman panegyrists due to the convergence of effect produced by oratory and the visual elements of the ceremonial context in which it was performed, was characteristic of the 'splendid theatre' of Late Antique ceremonial, itself the crucible for the making of society and political reality[10]. Signalling this later development, Livy made his own narrative into a spectacle in which the imperial state is constructed through the 'experiences it makes available to the audience'[11].

---

7. Feldherr, *Spectacle and Society*, 1.
8. Feldherr, *Spectacle and Society*, 12.
9. Feldherr, *Spectacle and Society*, 17.
10. A reference to Sabine MacCormack, Feldherr, *Spectacle and Society*, 17. See also Ann Vasaly's study of Cicero's use of *enargeia*, which 'demonstrates how even purely literary representations of visual scenes can approximate the effect produced by direct visual contact', Feldherr, *Spectacle and Society*, 18.
11. Feldherr, *Spectacle and Society*, 19.

2. Picturing a Novel: A *Mise en Abyme* in *Leucippe and Clitophon*

If *ekphrasis* as the confluence of oratory and visualized spectacle constitutes a kind of 'verbal theatre', then the double *mise en abyme* in the second century romance of Achilles Tatius, *Leucippe and Clitophon*, deserve attention as visual artifacts of quite a special kind, for the romance narrative is itself a simulated *ekphrasis* on two works of graphic, representative art (themselves narrative representations). The first occurs at the beginning of the story, at the outset of book 1, when the narrator recounts the votive painting depicting the rape of Europa in the temple of Astarte in Sidon. In a highly graphic description that can only be properly labelled a visual celebration the narrator describes in flowering detail the abduction of the beautiful maiden, Europa, on the back of the bull, led by Eros towards Crete[12]. At this point, telling about the Loves 'cavorting in the sea', Eros leading the bull ('a tiny child, with wings spread, quiver dangling, torch in hand. He had turned to look at Zeus with a sly smile, as if in mockery that he had, for Love's sake, become a bull') leads into the story proper, a story about Eros and the suffering caused by passion:

> Though the entire painting was worthy of attention, I devoted my special attention to this figure of Eros leading the bull, for I have long been fascinated by passion, and I exclaimed, 'To think that a child can have such power over heaven and earth and sea'.
>
> At this point a young man standing nearby said, 'How well I know it — for all the indignities Love has made *me* suffer'.
>
> 'And what have you suffered, my friend? You have the look, I know it well, of one who has progressed far in his initiation into Love's mysteries'.

---

12. B. P. Reardon, ed., *Collected Ancient Greek Novels* (Berkeley/Los Angeles/ London: University of California Press, 1989), 175–77.

'You are poking up a wasps' nest of narrative. My life has been very storied'[13].

And then the narrative sets off recounting the stranger's tale of lovers' travels and travails, a long exposé of exactly what was depicted in the painting. At mid-point in the romance narrative, at the start of book 5, the narrator finds himself, in company with the two lovers Leucippe and Clitophon, in Alexandria in the holy month of Zeus/Sarapis. An evil omen, in the shape of a hawk pursuing a swallow which in swooping collides with Leucippe, foretells new trouble. Immediately, as the narrator turns around, a painting is introduced into the narrative – the rape of Philomela: 'the plot of the drama was there in every detail', says the narrator[14]. Again the described painting emplots the plot of the following narrative:

> Menelaos then remarked: 'I suggest we put off our trip to Pharos. You see these two unfavourable signs: the bird's aggressive wing and the threat implicit in this painting. Interpeters of signs tell us to consider the story of any painting we chance to see as we set out on business, and to plot the outcome of our action by analogy with that story's plot. Well, just look at the disasters proliferating in this scene: lawless sex, adultery without shame, women degraded! I therefore advise that we go no further'[15].

Leucippe asks for an explanation of the painting, provided by the narrator. And immediately following upon the '*ekphrasis* within the *ekphrasis*' the plot sets off, the story indeed a mirror of the explanation of the painting as if in predetermined imitation: 'So for the time being we escaped the snares of this plot. But our margin of grace was only a single day ...'.

Just as on the narrative level the story is an *ekphrasis* of the two paintings, so the romance story is on another level also an *ekphrasis* of the myth of Isis and Osiris. There is an uncanny, but very real similarity between the plot of

---

13. Reardon, *Collected Ancient Greek Novels*, 177.
14. Reardon, *Collected Ancient Greek Novels*, 234.
15. Reardon, *Collected Ancient Greek Novels*, 234.

Achilles Tatius's *Leucippe and Clitophon* and the portrayal of the myth of Isis and Osiris in Plutarch's *De Iside et Osiride*. I would like to contend that *Leucippe and Clitophon*, as a whole but especially the core book 3, is a 'refictionalized' version of the myth of Isis and Osiris. According to the myth Osiris reigned over Egypt and 'delivered them from their destitute and brutish manner of living'. He later travelled through the world spreading civilization. Then he is killed by Typhon 'by a treacherous plot'. Osiris is tricked and locked into a chest specially made to measure. The chest is dropped in the river and so sent off to sea. The chest washes up, is 'shipwrecked', at Byblos where Isis eventually finds it and brings it back to Pelusium in the delta area of Egypt. Here Typhon stumbled across the chest, recognizes who is inside and dismembers Osiris's body and scatters the parts in different places. Isis dutifully searches the swampy delta area, finds the parts (save the phallus) and buries them. Isis's son, Horus, eventually avenges Osiris's death and dismemberment by defeating Typhon in battle. To deepen the intrigue of the text, the narrated space of the novel coincides with the geographical settings of the Isis-Osiris myth.

What is important in this context is the interpretation given to this myth by Plutarch:

> Stories akin to these and to other like them they say are related about Typhon: how that, prompted by jealousy and hostility, he wrought terrible deeds and, by bringing utter confusion upon all things filled the whole Earth, and the ocean as well, with ills, and later paid penalty therefore. But the avenger, the sister and wife of Osiris, after she had quenched and suppressed the madness and fury of Typhon, was not indifferent to the contests and struggles which she had endured, nor to her own wanderings nor to her manifold deeds of wisdom and many feats of bravery, nor would she accept oblivion and silence for them, but she intermingled in the most holy rites portrayals and suggestions and representations of her experiences at that time, and sanctified them, both as a *lesson in godliness and an encouragement for men and women who find themselves in the clutch of like calamities.* She herself and Osiris, translated for the virtues from good demigods (*daemones*) into gods, as were Heracles and Dionysos later, not incongruously enjoy double honours, both those of gods and those of demigods, and their powers extend everywhere ... (*De Iside et Osiride* 27) (my emphasis, GvdH).

Later in the same commentary Plutarch, after exploring the etymology of 'Osiris' as joy and fructifying and regenerating moisture, claims that this whole narrative 'is an image of the perceptible world' (54).

If 'encouragement for those in the clutch of like calamities' is the point of the Isis and Osiris myth/narrative, then Achilles Tatius's *Leucippe and Clitophon* is a treasure-house of ready references for 'being in the clutch of like calamities.' In fact the whole novel is not only a long *ekphrasis* on the opening scene of the interpretation of the painting of the rape of Europa (1.1), and the rape of Philomela in book 5, but the many accounts of dangers, quasi-deaths, and quasi-resurrections are a play on a syncretized divine power (i.e., Isis>Aphrodite/Artemis> Tyche) so that one should understand the image of Isis Tyche (as in the Cumaean aretalogy: 'fate hearkens to me' – a typical depiction of Isis) implicit in the narrative.

On one level then, *Leucippe and Clitophon* is the story of two young lovers buffeted by the machinations of fate/Fate, that is on the story level, as romance, as novel. On another level, the story is a Sophistic declamation through the repeated *ekphraseis* on the pictures of mythic themes that portray mythic rapes. Mythic rapes and murders are central to the myths of Demeter and Persephone, Dionysus, and Isis and Osiris. If the mysteries, then, are enactments of mythic narratives, this novel can, in parallel, be understood to be a declamation on a myth of rape similar to the myths purportedly underlying the mysteries.

According to the conventional viewpoint, the myths associated with the mysteries concern the sufferings of the gods, but this needs to be recast. The phrase 'suffering of the gods' allows too quickly a shortcut into theology (like fate, the 'gods' *can* lead astray), whereas what is the core of the 'myth' is a tale of rape, abduction, murder, and good ending.

It is my contention that it is the 'gods' of the conventional view that cause the misdirection (and therefore it does not matter whether this novel is an Isis novel, or *Daphnis and Chloe* a Dionysus novel, or *An Ephesian Tale* an Artemis/Isis novel, or *An Ethiopian Story* a Helios novel)[16] – the Sophistic char-

---

16. As was argued famously by Reinhold Merkelbach, *Roman und Mysterium in der Antike* (München/Berlin: Verlag C. H. Beck, 1962); idem, "Novel and Aretalogy" in *The Search*

acter of these novels with their *ekphrases* and declamations should direct our attention to the central fact of these novels, and that is the plot – vicissitudes and good endings. And this is what one should see – how narratives function as the stuff of mime and dramatic enactment.

Furthermore, the myth of Isis and Osiris and the romance of *Leucippe and Clitophon* create the same narrative world. And this can be seen in a number of illuminating points of contact between the two narratives. Both are stories of shipwrecks, the hints of adultery, the dismemberment, the parallel between the bandits and Typhon (= forces of uncivilization), geographical locales (between Phoenicia and Egypt; Typhon and bandits in swamps; Pelusium, the burial place of Osiris and landing place of Leucippe and Clitophon after the shipwreck).

Even in antiquity it was realized that the story line embedded in the romance of *Leucippe and Clitophon* is a refictionalized version of a mythical narrative, and Plutarch was himself aware of the intersections between myth and fiction. After recounting the myth to his interlocutor, Clea, he turns his nose up, ever the snob, at the popular fictions feeding off the myth or repackaging the myth in novelistic fashion:

> That these accounts [that is, the recounted myth – GvdH] do not, in the least, resemble the sort of loose fictions (*mytheumasin ararois*) and frivolous fabrications (*diakenois plasmasin*) which poets and writers of prose (*logografoi*) evolve from themselves, after the manner of spiders, interweaving and extending their unestablished first thoughts, but that these contain narrations of certain puzzling events and experiences, you will of yourself understand'. (*De Iside et Osiride* 20).

By arguing for a qualitative difference between his recounted version of the myth and the other, fictional, versions of it (calling them by the derogatory term 'empty/idle concoctions'), Plutarch actually witnesses to the existence of story plots that encapsulate or imitate the mythic narrative. If the *plasmata* of the Plutarch citation indeed refers to what we now call novelistic fiction (as I think it does), then these words attest to the existence side by side of Greek

---

for the Ancient Novel (ed. James Tatum; Baltimore/London: John Hopkins University Press, 1994), 283–95.

novel and the myth/mythic narrative (or the *logoumena*) of mystery cults. When he later interprets the plastic representation of Osiris as an erect male member to be the signification of his creative power (*De Iside et Osiride* 51) and the representation of Isis by means of lunar symbols (the moon governing love affairs) as signifying her presiding role in love affairs (*De Iside et Osiride* 52)[17], Plutarch himself establishes the link between the myth of Isis and Osiris and adventure love romances.

3. A World of Representations

The Greek novels cannot be taken to be a popular mass medium in our modern sense of the term (despite the existence of a number of complete novels and quite a number of papyrus fragments, the novels did not enjoy mass readership, a conclusion borne out when comparing the remaining texts and text fragments of novels with other works of 'high culture')[18], nevertheless, the fact is that they floated in a sea of other representations of versions of the same narratives. Lucian recounts, in *De saltatione* 2 and *Pseudologista* 25, how scenes from novels or scenes similar to these, were sometimes mimed by street performers, and although these were by no means performances of the novels themselves, they did circulate stock episodes and scenes that also occurred in novels and so kept them alive in the public mind and eye[19]. For

---

17. See for instance the best known of the Isis aretalogies, the one from Cyme, Asia Minor: '... I brought together woman and man (l. 17) ... I compelled women to be loved by men (l. 27) ... I devised marriage contracts (l. 30)', Marvin W. Meyer, ed., *The Ancient Mysteries. A Sourcebook. Sacred Texts of the Mystery Religions of the Ancient Mediterranean World* (New York: Harper San Francisco, 1987), 173.
18. Cf. Berber Wesseling, "The Audience of the Ancient Novels", in *Groningen Colloquia on the Novel* [ed. H Hofmann; Groningen: Egbert Forsten, 1988], 67–79, and Tomas Hägg, "Orality, Literacy, and the 'Readership' of the Early Greek Novel", in *Contexts of Pre-Novel Narrative. The European Tradition.* [ed. Roy Eriksen; Approaches to Semiotics 114; Berlin/New York: Mouton De Gruyter, 1994], 47–81 and Susan A. Stephens, "Who Read Ancient Novels?" in *The Search for the Ancient Novel* [ed. James Tatum; Baltimore/London: Johns Hopkins University Press, 1994], 405–18 – the novels seem not to have been popular at all [p. 414]; a similar argument in Ewen Bowie, "The Readership of Greek Novels in the Ancient World", in *In Search of the Ancient Novel* [ed. James Tatum; Baltimore/London: Johns Hopkins University Press, 1994], 440–41).
19. Cf. Stephens, "Who Read Ancient Novels?" 409.

instance, as David Balch has shown, the mythic narrative of Io's peregrinations to the land of Isis and eventual restoration to humanity, occurs in various preserved frescoes in Pompeii, Rome and elsewhere, and it also reflects a similar plot to Achilles Tatius's *Leucippe and Clitophon* and Plutarch's version of the myth in *De Iside et Osiride*. What is more, their occurrence in the dining halls and other public spaces of patrician homes inspired many verbal interpretations of the allegorical meaning of the image portrayed[20]. But consider, too, Bowersock's exposé on the Late Antique Dionysus in the eponymous epic by Nonnos (cf. chapter 4: 'Dionysus and his world') for the portrayal of scenes from Nonnos's *Dionysiaka* in various floor mosaics from the Levant[21]. Nonnos's *Dionysiaka* is in this context particularly interesting, for the fifth century epicist and mythographer has produced in this text an almost complete compendium of Greek myth, but given the prominence of Dionysus in the period, tailored to a representation of Dionysus as conquering emperor, as participant-actor in every mythical scene available. Not only is the depiction of mythical scenes in Nonnos characterized by dramatic liveliness (and all the mythical plots occurring in ancient fiction, as well as those from mysteries themselves, are found in Nonnos's epic poem), the mythic cycles of Nonnos's myth-compendium reoccur outside of the epic in graphic depictions. So, if one just pays attention to the vivid description of the war between Dionysus and Poseidon for the hand of the beautiful maiden, Beroë (Nonnos *Dionysiaka* 53), it is clear we move in the world of mime-acting and theatrical performance. The epic text itself has become a textualized mime-spectacle.

Next to these one may also mention the Metiochus and Parthenope mosaics evoking the eponymous novel (known otherwise only from a papyrus fragment) from Daphne, a suburb of Syrian Antioch, datable to the Severan period, as well as two Ninus mosaics, one from Daphne and the other from Alexandreia ad Issum, of which the former show Ninus contemplating a por-

---

20. David L. Balch, "The Suffering of Isis/Io and Paul's Portrait of Christ Crucified (Gal. 3:1): Frescoes in Pompeian and Roman Houses and in the Temple of Isis in Pompeii", *Journal of Religion* 83, no. 1 (2003): 26–28.
21. G.W. Bowersock, *Hellenism in Late Antiquity* (Cambridge: Cambridge University Press, 1990), 41–53.

trait of Semiramis[22]. Apart from settings in the dining halls of patrician homes, paintings of mythical narrative scenes could also be admired (and declaimed on) in temples, as depicted in *Leucippe and Clitophon* and *Daphnis and Chloe*. In the case of these two novels it could be argued that the novels themselves are *ekphraseis* of a graphic portrayal of mythic narrative. Putting it like this one might, with proper reservation, indeed liken the function and effect of the Greek novels to that of modern mass media. Although relatively few had access to the novel as text for the purposes of (private) reading, whatever form that took, the narratives themselves lived in the public domain and could be seen and 'read' by a far wider public.

As Glen Bowersock argued in *Mosaics as History*[23], the rich tapestry of mosaic decoration in Levantine Roman houses evidences a sense of urban consciousness and civic pride illustrative of imperial class consciousness and comportment. In effect, then, the densely present Dionysus in the represented Nonnos scenes in urban environments, upholds a graphic imperial spectacle constructing, undergirding, and maintaining the romanitas of the inhabitants of these environments.

4. The Visual World of Spectacle: The Context for the Adventure Novel

One should not underestimate the effect of processions, spectacles, and triumphs in the 'paradoxification' of the empire, the point being that the over-the-top nature of the spectacle was a vehicle for announcing, advertising, and promoting Roman hegemony in the Greek East[24]. While the art of spectacle

---

22. Even if one wants to follow Ewen Bowie's interpretation that these rather suggest mimed roles, following Lucian, cf. Bowie, "Readership of Greek Novels", 448–49.
23. Glen W. Bowersock, *Mosaics as History. The Near East from Late Antiquity to Islam* (Revealing Antiquity 16; Cambridge, Mass./London: Harvard University Press, 2006).
24. See Jonathan C. Edmondson, "The Cultural Politics of Public Spectacle in Rome and the Greek East, 167–166 B.C.E", in *The Art of Ancient Spectacle* (ed. Bettina Bergmann and Christine Kondoleon; Studies in the History of Art 56; New Haven/London: Yale University Press, 1999), 77–95 for a description of the interplay between Roman and Greek forms of spectacle at the crucial period of Roman ascendancy in the Greek world, that is after the defeat of Macedon and the battle of Pydna.

and the triumphal procession were not Roman inventions[25], they gained a particular significance in the context of the empire. Spectacle and triumph, far from sedate and sober occasions[26], constituted extreme and overwhelming experiences: apart from the noise, vociferous response and shouting from both audience and soldiers, there were also pageants in which paraded, of course, the triumphator made up in the image of and impersonating Jupiter Capitolinus, but also defeated generals and royalty, captured images of their gods, placard bearers, troupes of musicians and large brass bands of trumpets and horns, paintings depicting battle scenes, models of destroyed cities, moving mechanical set-pieces portraying important campaign events or mythic episodes and animated statues[27], captured and looted trophies carried on portable

---

25. The visually and theatrically extravagant procession combined with a festival that ran over several days, was known from Philip II of Macedon (348 B.C.E.), and we have an example of Alexander the Great's lavish procession plus contests on his return to Phoenicia from Egypt in 331 B.C.E., also the famous and paradigmatic procession of Ptolemy II Philadelphus in the 270s B.C.E. As the Roman republic made inroads into the eastern Mediterranean world, triumphant field commanders had their triumphal processions, both abroad and repeated in Rome itself. For a discussion of the confluence of Roman and Greek culture in the art of spectacle, see Edmondson, "Cultural Politics". The triumphal processions following on the conquest of Macedon (those of L. Aemillius Paullus at Amphipolis, 167 B.C.E., L. Anicius Gallus in the Circus Maximus, Rome, 166 B.C.E.), and on the other side, the victory celebrations of Antiochus IV 'Epiphanes' of Syria at Daphne, September – October 166 B.C.E., demonstrate how Romans started to borrow Greek cultural expressions (choruses and musicians), and the Greeks borrowing Roman elements (gladiatorial displays), albeit with due adaptation for the cultural context – all in the service of promoting the imperial ideal and the incomparable sovereignty of Rome (Antiochus' victory celebrations came after his aborted campaign in Egypt, where he was prevented by the Roman commander from advancing into Egypt, nevertheless the celebration was put on to assert and consolidate his power internally within the Seleucid empire). Edmondson, "Cultural Politics", 85.
26. If the earliest Roman triumphs were more sober, by the time of the late Republic and the advent of the Principate they had become noisy and boisterous extravaganzas, costly, carefully scripted theatrical events ('ostentatious display', 'visual splendor'). Richard Brilliant, "'Let the Trumpets Roar!' The Roman Triumph", in *The Art of Ancient Spectacle* (ed. Bettina Bergmann and Christine Kondoleon; Studies in the History of Art 56; New Haven/London: Yale University Press, 1999), 224–25.
27. '... similar fabulous machinery marked Hellenistic stage effects and was taken to a high pitch in Rome', Ann Kuttner, "Hellenistic Images of Spectacle, from Alexander to

platforms – treasures on display, herds of exotic animals like tigers, lions, and especially elephants, and tableaux-vivants in which mythical and historical scenes were enacted in a kind of allegorical commentary on the present celebrated event[28]. These assaults on the senses not only grew more elaborate (each new staged procession aiming to surpass the previous), but also preserved, and consciously evoked, the *pompa triumphalis* of Dionysus (the god's 'raucous epiphany', so Brilliant) as descrybed by Callixenos and preserved in Athenaeus' *Deipnosophistae* 5, 196a–203b)[29]. Add to these the enacted military campaigns, naval battles (with ships in flooded amphitheatres), performed violence (gladiatorial fights, killings of various kinds of undesirables), and wild beast fights and displays in theatres and amphitheatres, and one finds oneself within the broad sweep of narrative scenery encountered in the novels, but which existed everywhere for public consumption[30]. Apart from the Isis/Io mythic narrative discussed by Balch, one should think of such examples as the exotic Nilotic scene preserved in mosaic (ca. 125 B.C.E.) from a nymphaeum-like hall on the forum of Praeneste, depicting Egyptian scenery (which include Egyptian architecture, priests, and peasants, and black hunters chasing exotic animals), with a romanized cuirassed imperator enjoying a spectacle victory banquet (with an automaton – a moving statue of Victory – pouring the wine); as well as the famous Dionysiac marriage scenes

---

Augustus", in *The Art of Ancient Spectacle* (ed. Bettina Bergmann and Christine Kondoleon; Studies in the History of Art 56; New Haven/London: Yale University Press, 1999), 99.

28. Compare also Claudius' spectacle as part of a triumph in the Circus Maximus in Rome, in which he presided, dressed in military cloak, over the enacted storming and sacking of a town and the subsequent surrender of the British kings: a reconquest of Britain to justify his claim to the title Claudius Imperator Brittanicus (Suetonius *Claudius* 21.6), Brilliant, "Roman Triumph", 228.

29. Brilliant, "Roman Triumph", 223. Dionysus' triumphal procession from India to Greece through Asia Minor is not a reflection of the 'original' myth of Dionysus, but a Hellenistic invention, one which became very important in the maintenance of imperial ideology, the figure of Dionysus crafted as an imperator himself, cf. Brian Bosworth, "Augustus, the *Res Gestae* and Hellenistic Theories of Apotheosis", *Journal of Roman Studies* 89, no. 1 (1999): 2–3.

30. Kuttner, "Hellenistic Images of Spectacle", 100.

from the Villa of the Mysteries in Pompeii which prefigured by about half a century the enactment of the marriage
(and banquet revel) of Aphrodite/Cleopatra and Dionysus/Antony, when Cleopatra came to meet Antony on barge, imitating Aprodite 'in the manner of a painting', surrounded by a costumed crew of 'Nymphs', 'Graces,' and 'Erotes'[31]. In a very real sense the enacted and performed spectacle, as well as the textualized and the graphic and plastic portrayals of spectacle, forms the diorama of the discursive world in which the ancient novels had their home, and which created the world of novelistic references in a kind of intertextuality on a very grand scale. Moreover, the novels stood at the confluence of narrative scenes, mythic 'portraiture', ritual as habituated action and scripted performance – a world of images and narratives of which the novels constituted but one element.

5. A World of Narrative and Performance: Rethinking the Settings of the Mysteries

In this essay I explored the inter-place between ancient fiction, mythical discourse, and theatrical spectacle. Taking as my point of departure Achilles Tatius's *Leucippe and Clitophon*, I argued how on the most immediate level, this fictional narrative is a textualized spectacle of a mythical discourse. On the next level, the fictional narrative intersects with large-scale public performances of mythic events, relating this as well to the theatricality of Roman culture in general. I would venture that a new conceptualization of what is generally regarded as one of the defining features of the religious scene of the early imperial era up to the onset of the early Middle Ages – mysteries, or mystery cults – is called for, namely that the mysteries should be understood as large-scale *son et lumière tableaux vivants*, that themselves operate in the same discursive space as other spectacle 'performances' such as triumphal processions. Inversely, given the re-registering of Egyptian mythic discourse since the through-set of Hellenistic Ptolemaic dynastic ideology with its identification of Isis-Osiris/Sarapis and Dionysus, the 'spectacleization' of the double myth of conquering Isis-Dionysus, especially as performed in the famed procession of Ptolemy II Philadelphus (which established itself as

---

31. Kuttner, "Hellenistic Images of Spectacle", 101.

paradigmatic processional performance), traces a trajectory from public spectacle to myth (Sanchuniathon; Philo of Byblos), to myth-fiction as textualized spectacle – Achilles Tatius's *Leucippe and Clitophon*, and Nonnos's *Dionysiaka*.

In this, spectacle, narrative, myth all were vehicles for the enactment of imperial dispositions. It is in this sense that one can indeed maintain that the mysteries flower exactly in the imperial era, as do the ancient novels, because they were subsets of the same vast repertoire of representations that characterized the theatrical culture of the Graeco-Roman world, driven to especially high levels in the Roman Empire.

With spectacle everywhere, performed and plastically installed, the Roman Empire was everywhere too, in the vividness of the reality resulting, in part, from its graphic evocation.

RELIGION AND VIOLENCE: A PSYCHOANALYTIC INQUIRY

Respect for minds generates respect for self,
respect for other,
and ultimately respect for the human community.
— Peter Fonagy

Religion remains of course the one topic we are enjoined
to treat with kid gloves; indeed,
the one area where critique is verboten.
— Walter A. Davis

MARSHA AILEEN HEWITT

It is hardly contested among a number of scholars of religion that the object of their study contains tendencies to extremism, militancy, and violence. Walter Burkert, commenting on ancient sacrificial religious practices in Judaism, Christianity and Greek religion observes that 'blood and violence lurk fascinatingly at the very heart of religion', where the 'worshipper experiences the god most powerfully not just in pious conduct or in prayer, song and dance, but in the deadly blow of the axe, the gush of blood and the burning of thigh-pieces' (1983, 2). In his empirical studies of recent forms of global religious militancy, Mark Juergensmeyer concludes that violence 'has always been endemic to religion (1992, 1) while David Rapoport observes that 'all major religions have enormous potentialities for creating and directing violence' (1993, 446). According to Bruce Lincoln, 'all religions sanction, even enjoin the use of violence under certain circumstances, the definitions of which have proven conveniently elastic' (2003, 73). Jonathan Z. Smith summarizes these shared conclusions when he writes that 'Religion has rarely been a positive, liberal force. Religion is not nice; it has been responsible for more death and suffering than any other human activity' (1982, 110). For these authors, the historical evidence is quite clear that violence has always been a key feature in religious beliefs and actions that is expressed through ritual forms, myths and cosmologies organized around the necessity for sacrifice and sustained struggle between the forces of good and evil, order and chaos, purity and danger. 'Disturbing as it may be to those accustomed to the anti-sacrificial attitude that seems to characterize late modernity religion', observe the editors of *Numen*, 'engaging in violence, sacrificing oneself and others, far from being a betrayal of religion, constitutes its core' (2005, 4). At its most lethal, the images and narratives of violence that are integral features of religions become mobilized in contexts of political and social crises. When these crises are experienced and interpreted in religious terms, they focus upon a threatening other or object that must be contained, dominated or even destroyed in order to protect and/or restore the believing community. 'God's world is pure, not pluralist', and the enemies of God and his faithful communities must be abolished one way or another (Almond, Appleby, Sivan 2003, 148).

My own research is directed toward offering psychoanalytic explanations that may illuminate the kind of mind, or more precisely those mental states that are present in conceptions of and enactments of religious violence. In considering these mental states, it is important to consider both the internal world of the individual as well as the cultural, social and political contexts in which she/he is located. Minds do not exist in isolation from culture, and interactions with other minds. As will become clear throughout this essay, my focus is not on empirical studies of individual religious militants, although I will draw upon some empirical studies of those who engage in religious violence in order to provide illustrations of some of my major theoretical points. However, this essay will focus upon those psychoanalytic writers, particularly the work of Peter Fonagy and his colleagues, who offer interesting avenues of theorizing what sorts of mind states might well be operative in acts of religious violence and how those minds are structured and organized. This line of inquiry does not pretend to make absolutist claims about the nature of religions per se[1], and certainly not that religions solely promote violence. It cannot be denied, for example, that Western moral norms and values derive from Jewish and Christian religious traditions (Habermas 2002, 149). Nor do I claim that religious persons who engage in violent acts are necessarily psychopaths, borderline personalities or insane[2]. In fact, I am not convinced this line of inquiry is useful in any event, since it cannot possibly be empirically established. Further, there is no single, universally agreed upon definition of psychic normality, and this has been the case in the field of psychoanalysis since the time of Freud.

Again, the focus of this essay is to theorize about those mental states that most likely underlie acts of religious violence and how religious ideas and feelings facilitate and provide external legitimation, stability and expression to those mental states. It must also be noted that violence is not restricted to physical actions, but also includes worldviews, social practices and relationships and social institutions that are structured in binary terms of good vs bad, sacred vs profane, believer vs unbeliever, true religion vs idolatry and insider vs outsider whose exclusionary and often dehumanizing effects are harmful to others in multiple ways. The space between dehumanizing modes of thinking about others along with hierarchical practices of social, political and economic domination and marginalization and their physical annihilation is

---

1. Although too lengthy to go into here, suffice to say that the very category 'religion' is controversial and contested, with little widespread agreement about its definition or even relevance as a conceptual construct beyond Western scholarship. See, for example, Benson Saler, Conceptualizing Religion: Immanent Anthropologists, Transcendent Natives, and Unbounded Categories, Leiden: E.J. Brill, 1993.
2. However, as one psychologist interested in terrorism observes, 'suicide terrorists...are not mad but on the other hand they are not normal either' (Anne Speckhard, 2006). I take this to mean that the mind states of individuals who engage in violent acts involve some degree of pathology.

alarmingly narrow. In this sense religion and politics are inseparable. In what follows my point is not that persons who engage in religious acts of violence are necessarily insane, psychotic, or irrational. As shall be observed below, these militant practitioners are usually very clear about their motives, their goals, and how to achieve them (Victoroff 2005, 12). Viewed from within the perspective of their own religious ideology and social and cultural worldview, they may be considered as entirely rational, perhaps even 'terrifyingly normal', to borrow a most controversial phrase from Hannah Arendt.

This essay will utilize certain concepts and categories derived from psychoanalysis which provide useful theoretical tools that offer insight, understanding and partial/provisional explanations of religious violence. A psychoanalytic approach to the issue of religious violence has no direct interest in moral evaluations of religious experiences, beliefs, or actions and has no interest in establishing whether the object of religions, be they gods or superhuman entities exist or not. Rather, the point is to shed some light on the possible modes and structures of thought about experiences of self and other that involve internal structures of emotional organization and development that are both deeply compatible with and reflected in aspects of religious beliefs and cosmologies. For example, religious attitudes structured along the lines of antagonistic dualisms such as good vs. evil, pure vs. defiled and believer vs. infidel provide the external or conscious conceptual and explanatory architecture whereby varying degrees of impaired patterns of emotional development that involve psychic processes of splitting (good versus evil, endangered self versus persecuting other, etc.), projection (sense of inner badness evacuated and deposited onto another), disavowal (disruptive, disturbing feelings not-mine), denial (it's not happening), and dissociation (sequestered feelings vertically split off from conscious awareness) are both 'normalized' and unconsciously operationalized within acceptable religious and moral attitudes. Viewed from this perspective, religion offers a protective haven from threatening internal and external forces that stabilizes, consoles and regulates internal (unconscious) experiences of endangerment, turmoil and disorder (Freud 1927; Hewitt 2008).

The capacity for intersubjectivity, reciprocal recognition and emotional distance where one can differentiate between self and other, internal self-states and the external world, one's own beliefs and those of others, between 'doer and done-to' is less developed in such mind states, which involve not only cognitive and intellectual faculties. Such mind states are profoundly affective as well, as neurologist Antonio Damasio (1994) and neuropsychoanalyst Allan Schore (2003) have shown. The inability to regulate and contain affective states compromises the reflective function with the result that the world and people in it are as one feels them to be. There is only right and wrong, black and white and no shades of grey, which is often a particularly relevant description of many extreme fundamentalist religious believers worldwide (Davis 2006; Stein 2006; Juergensmeyer 2003; Stern 2003; Strozier 2002). Reality is experienced as conforming with unconscious phantasies and expectations to the point that there is little sense, if any, of considering alternative perspectives (Fonagy 2002: 280) and where there is, those perspectives are rejected as dangerous and threatening. Being a true believer, devoting oneself to doing God's will, living one's life in conformity to the strict dictates of one's religious beliefs and traditions necessarily exclude taking seriously alternative perspectives since that would compromise one's relationship with God and the believing community. Cognition and self-awareness become restricted since belief and knowledge heavily overlap, or may be even indistinguishable at times. The conflation of belief with knowledge is based on a further lack of a sense of separateness between the mental, internal world, and the external, material world (Britton 1998, 14ff).

MENTALIZATION, AFFECT REGULATION AND PROBLEMS OF SELF/OTHER DISTINCTIONS

The work of Peter Fonagy and his colleagues is particularly useful in offering ways of theorizing mental/mind states that underlie and are operative in ideologies and acts of religious militancy and violence. Not only is his work psychoanalytic, it is also rooted in developmental psychology and attachment theory which combines both theoretical, clinical and empirical (laboratory experimental) methodologies. Fonagy's interrelated notions of 'reflective function' and 'mentalization' (2002) in his theory of mind are particularly promising in contributing to psychoanalytic theorizing about those mental states that may be present in and facilitate religious violence. Although too lengthy to pursue his theory in depth in the space of this essay, a few key highlights of his work will provide a sense of the potential importance of his contribution to the study of religious violence. Fonagy describes reflective function as 'ideally [providing] the individual with a well-developed capacity to distinguish inner from outer reality, pretend from 'real' modes of functioning and intrapersonal mental and emotional processes from interpersonal communications' (25). In the earliest experiences of the infant with his primary caregivers (usually the mother) the child learns to recognize and regulate his own internal affective states through a complex set of multiple interactions with his mother where he looks for and is able to 'find' or locate himself in her mind. As the child experiences feeling, mother recognizes the feeling and 'mirrors' it back to him accurately, but not precisely, so that the child is able to recognize his feelings not only as what they are–anger, fear, hurt, confusion–but recognize that they are his. If things go well, if mother is able to parent in a way that provides 'good-enough' attunement and recognition of her child's feeling and mental states, then the intersubjective attachment dynamics that develop between them will result in a securely attached child-adolescent-adult. The child will have been able to successfully internalize his mother's mental representation of him and map those representations into his own mind. While these representations are intersubjectively generated, they are experienced as and become felt as his. The child's capacity for and experience of affect-laden and cognitive structures that organize self-other relationships is part of a developmental process that will structure or mediate all his relationships throughout his life span. Affectively, this means that the child will have learned to regulate himself, recognize and distinguish his own affects from that of others, and most important, develop the capacity for mentalization, that is, the ability to mentalize his own self-affect states, fantasies and motives and the mind states of others. For all this to happen, the child must develop within a stable, secure family where the caregivers are mentally and emotionally healthy, are themselves able to mentalize the child's mind states and those of others reasonably well, and are not distracted by sustained poverty, war, depression, violence, feelings of shame, helplessness or terror. Thus social context is crucial to psychological and mental development. Reflective function then, operationalizes the 'psychological processes underlying the capacity to mentalize (24). This involves the capacity to make sense of and regulate affects and behavior, and is an 'automatic' or implicit procedure 'unconsciously invoked in interpreting human action' (27). The quality of reflective function and the capacity for mentalizing that arises from it depends upon the nature of the attachment relationships between the infant and his caregiver. Distracted, anxious, depressed caregivers cannot provide the conditions necessary for the fostering of secure attachments.

Mature forms of mentalization allow individuals to distinguish between internal emotional and mental states, the minds of others and interpersonal events, which is necessary for more psychologically complex forms of thought that do not collapse into the concrete and the literal . The experience of having a mind capable of mentalizing one's own internal affective states and those of others is not a 'genetic given' but a structure that evolves through emotional development sustained by caregivers capable of mirroring, containing, and holding the child's affective states so that the child learns to regulate and understand them, as previously explained Fonagy emphasizes that mentalization is both a cognitive and affective developmental process that involves 'the experiential understanding of one's feelings in a way that extends well beyond the intellectual understanding. It is in this realm that we encounter resistances and defenses ... against entire modes of psychological functioning; not just distortions of mental representations standing in the way of therapeutic progress but also inhibitions of mental functioning ... the inability to imagine psychological and psychosocial causation may be the result of the pervasive inhibition and/or developmental malformation of the psychological processes that underpin these capacities' (Fonagy and Target 2003, 271).

Such resistances and defenses would include the inability/refusal/ fear of critically questioning one's beliefs and explanations of reality and would seriously inhibit any motivation to consider reality from different perspectives and thereby consider negotiating those perspectives with others. Fundamentalist forms of religious belief that foster a tight relationship between internal and external reality, such as experiencing oneself as god's 'instrument' or whose self worth and lovability lay in pleasing god, may strengthen a dissociative mental structure that requires absolute views of reality in order to maintain its own stability. Charles Strozier, in his psychological study of American Christian fundamentalists observed that, 'the move toward literalism and away from metaphor defines the religious experience of the fundamentalist. Literalism means control over sin and badness and ultimately control over death' (2002, 95). Perhaps this is why certain fundamentalist mind sets reject negotiation with different perspectives; to do so would undermine one's sacred faithful connection to God, thereby endangering internal stability. The unconditional love and acceptance of the believer by God may be the only constant experience of such love the individual has ever had (Kirkpatrick 2005). In her interview with Yoel Lerner, who was involved in a plan to destroy the Dome of the Rock in order to rebuild the Third Temple in Jerusalem, Jessica Stern quotes him as declaring that 'the highest value for a Jew is not the preservation of human or even of Jewish life. The highest value is doing '*what God wants you to do*' (90). The locus of individual agency is displaced and subordinated to a higher divine agency that absolves the religious militant of having to critically reflect on the moral implications of violent acts or consider that the 'enemy' might have a point of view at least worth thinking about. In fact, one is spared the effort of having to mentalize one's own internal states and those of others; it is neither required nor desirable to see or feel what it is like to be another person, or consider their motivations, needs or desires. In such cases the will of God provides the locus of agency and compensates for these developmental failures of mentalization.

Fonagy writes that in earlier developmental stages in the process of mentalization, there is the 'psychic equivalence' mode, where 'mental events are equivalent in terms of power, causality and implications, to events in the physical world' (2003, 274). Appearance and reality are identical, and internal thoughts and feelings which are seriously distorted by fantasy are projected onto external reality without conscious awareness that this projective distortion is happening. Gradually the developing child learns that the world does not correspond with appearance – psychic equivalence – but that others may have feelings, thoughts and motivations that are different from him/her while not necessarily posing any threat or endangerment. Later stages of mentalization involve the ability to adjust one's affect states and regulate one's self both internally and interpersonally (270-71). The individual learns that fantasy is not reality and belief is not necessarily knowledge. Mature forms of mentalization transcend and negate, for example, those particular modes of religious thought that rely on and promote simplified worldviews that reduce complexity and uncertainty to binary oppositions. At the same time, the awareness that comes with developments in the capacity for mentalizing about self and other, that one cannot always know what is in the mind of the other (individual others, the society or even the world at large) and that there are real limits to one's ability to know and understand and that there are no absolute truths of meaning can arouse intense affects of anxiety, helplessness, vulnerability and shame. In minds for whom such affects are intolerable there must be a resort to psychological processes to relieve these unbearable feelings by locating them outside the self where they can be identified, controlled and potentially destroyed. A psychoanalytic discussion of religious violence helps illuminate ways in which literalist religious ideas that divide the world into opposing forces of good and evil, sacred and profane, male and female etc. facilitate psychic mechanisms such as splitting and projective identification, dissociation, denial and disavowal. Such concepts offer at least partial explanations about religious mind states that underlie, motivate and inform religious beliefs and practices on conscious and unconscious levels that can result in violence in the name of God. It is important to remember that mentalization as a developmental concept takes place in the context of attachment relationships and that the quality of those attachments is crucial in the child's growing capacity for mature forms of mentalization. Children who have been subject to sustained trauma/neglect or other forms of abuse (which doesn't necessarily include trauma) or chronic misattunement are susceptible to *impairred mentalization* capacities which results in a variety of forms of relationship dysfunction. Religious faith may offer a positive 'replacement' of problematic 'real' attachments by substituting an imagined idealized sacred attachment with a loving father that calms internal conflicts resulting from impaired attachments in earlier and current life.

DYING AND KILLING FOR GOD

The application of Fonagy's ideas of reflective function and mentalization offers deeper psychological understanding into the motivations and mental states of some of those people who engage in religiously inspired violence. Nasra Hassan's widely cited article 'An Arsenal of Believers', on suicide bombers, their families and their communities in the Gaza strip, provides a valuable illustration of some of the mental states of a number of people who support violence in the name of religion. While there are a number of motivations that inspire people to become suicide or 'human' bombers (Strenski 2003), an element common to all of them was religion (Hassan 2001, 38; Pedahzur, Perliger, Weinberg 2003, 417). The 'sacred explosions' were understood in terms of spiritual purification (Hassan, 39), a desire for the love of God and to be pleasing to him (39), immediate access to Paradise and honour for the bomber, his family and community. These actions are profoundly relational in that they are viewed not only in terms of a relationship with God, but also in terms of a relationship with the community on whose behalf it is enacted (Strenski 2003). One of Hassan's informants described his state of mind in these terms: 'The power of the spirit pulls us upward, while the power of material things pulls us downward ... We were floating, swimming, in the feeling that we were about to enter eternity ... All martyrdom operations, if done for Allah's sake, hurt less than a gnat's bite!' (37). Debates about whether such statements indicate insanity miss the point of what is being expressed in these descriptions. The intensely religious state of mind described here is absolute, unquestionable and any possible feelings of anxiety concerning death or pain are dissociated from conscious awareness; such intense religious experience may be understood psychoanalytically as a dissociated mind state. 'Dissociation', writes Philip Bromberg, 'is not only a mental process but also a mental structure ...that hypnoidally sequesters certain self-states and limits their communication with one another' (2006, 6). Intense religious beliefs may facilitate and help sustain dissociative structures because they mesh with what is perceived as objective reality. Hassan notes that the human bombers she researched and those who desired to emulate them were 'normal' and 'deeply religious' (38).

Mark Juergensmeyer writes that 'violent images are endemic to religious ways of thinking because of the nature of religion', (1992, 114) whose aim (among others) is to impose order on the world for the purpose of containing chaos. Enactments of religious violence arise out of 'specific attempts to impose perceptions of order on disorder', and these violent actions derive in part from cosmologies and fantasies of cosmic warfare that provide a 'grand context in which all of life's struggles, including political ones, make sense. Cosmic war and real war become one' (114). Elaborating on Juergensmeyer's line of reasoning from a psychoanalytic perspective, it may be interpreted that the need to impose external order is an attempt to regulate and contain internal chaos, where the psychically endangered self is locked in a perceived life and death struggle against emotional forces that threaten to overwhelm the self unless rigidly controlled through a process of evacuation (projection) and located in the external world (projective identification) where they are more easily identified, contained, and potentially destroyed. Religion can provide the internal/external regulating and containing structures that support and legitimate this psychological process. It is significant that the preparations for a 'sacred explosion' involve isolating the bomber from his family and friends, focusing his attention 'on Paradise, on being in the presence of Allah', which is perceived to be just 'on the other side of the detonator' (40).

Although people who embrace physical violence in the name of religion may have a variety of differing conscious motivations for specific acts, one element they share includes a strong aversion to certain aspects of cultural modernity associated especially with values of gender egalitarianism, toleration of difference and diversity, and secularism, all of which is experienced as giving rise to widespread moral corruption and loss of social control in its discrediting of traditional world views, belief systems, and moral cultures based on kinship and family ties[3]. In response to a question by Jessica Stern (who traveled the world interviewing militants representing different religious traditions) about his views of the West, the late Dr. Abdel Aziz Rantissi, a founder of Hamas, replied that 'Technological advances, democracy, the information revolution, the industrial revolution, and elections – these are things that should be absorbed by Islam ... But dancing, drinking, seductive behavior – these are forbidden by Islam. There should be no inappropriate mixing of the sexes' (2003, 56). In similar vein, Susan Rose's essay on the Christian education movement in the U.S. explains that parents who place their children in Christian schools 'desire to restore religious authority in American society, to reinforce parental authority, and to provide quality education for their children while protecting them from drugs, sex, violence, and the lack of discipline in the public schools' (1993, 455).

Juergensmeyer agrees that the modern secular world is experienced as 'dangerous and chaotic', and that religious militants report a feeling that their lives are out of control and their self and collective identity are endangered. (2003, 227). 'Indeed, the resistance to modern forms of secularization is a defining common feature of religious fundamentalisms', write Almond,

---

3. The reaction against certain aspects of modernity is one of the elements that constitute what the editors of The Fundamentalism Project refer to as a 'family resemblance' that is shared by a variety of disparate religiously inspired movements which confront certain processes of modernization and secularization in the 20th century. (Volume 2, Introduction, p. 2).

Appleby and Sivan (2003, 20). Jewish, Muslim and Christian fundamentalist ideologies emphasize dichotomies that structure reality in terms of light vs darkness, truth vs falsehood, the party of God vs the forces of Satan, purity vs impurity, and so on (36). In certain circumstances no sacrifice of human life is considered excessive in this attempt to do 'God's work'. Such religious world views and the people who espouse them cannot be dismissed as mere extremists. 'Religious enemies', writes David Rapoport, 'are important' (1993, 431). I suggest that the kind of religious militancy that requires external enemies as part of a need for clarity and conviction against affective, internally disruptive experiences of ambiguity and confusion is connected in part to weaknesses in the developmental capacity of mentalization.

In his work on terrorist organizations and individuals, psychiatrist and political psychologist Jerrold Post also observes that the 'psycho-logic' of violence involves splitting and externalization (projective identification) where individuals have not been able to integrate fully the good and bad parts of themselves into a psychic whole. What the internal self experiences as badness can be terrifying and therefore must be evacuated and displaced, or projected onto another, which in turn helps, for a time, the subject of this process to achieve internal equilibrium. The existence of an external enemy is a necessary construct for one's own disavowed feelings of weakness, shame and inadequacy (Reich 1990, 27). Religiously inspired hate transforms shame into pride, weakness into strength while by-passing a reflective process that opens up the possibility of considering the moral implications of hate and its potentially violent consequences. According to Post, '[F]or the militant Islamist movement, we have a situation where children from infancy on in some cases, are led onto the path of terrorism, where their parents are proud of my son the martyr and where hatred is really bred in the bone. And it isn't an aberration isolated from the mainstream, quite to the contrary' (2006).

In describing internal conflicts that require an 'evil' other in order to feel resolved psychoanalyst Ruth Stein writes that 'the two warring parts of the psyche that have been entwined in an insoluble conflict become increasingly divided and externalized'. The seductive power of the secular world with all its vices and freedoms arouses both desire and loathing, with the result that internal longings are in fact the 'infidels,' 'whereas the part that originated from one's religious tradition, the God-fearing part, is reinforced and becomes the one and only 'God'. Through the processes of projection and projective identification, the two parts now confront each other in the external world at war' (2003, 300). The need to counteract the devastating effects of shame requires further study as a motivating factor in current religious violence. In writing of the mind states of religious militants that he interviewed, Juergensmeyer observed that the modern, secular world poses a serious, destabilizing threat, where religion provides 'an anchor in a harbor of calm' (2003, 227) that functions as an internal soothing experience against feelings that one's personal self is imperiled. Stein (2003) speculates that such intense feelings of helplessness and shame may become transformed into an identificatory vertical love and devotion toward the Father/God whose blessing is earned by acts of (often self-sacrificing) violence, religious devotion's 'magical solution' that 'bypasses physical reality and psychic reality' that is 'otherworldly and manic' (300). The desire to win God's love is a powerful motivating factor in acts of religious violence and terror, as illustrated in a letter Stern received from a young woman who worked closely with the Laskar Jihad group: 'Don't ever think that we're afraid of death in defending our religion. Even death is our goal to reach the true glory... This is our call ... we're just seeking for bigger love from Him' (83).

While acknowledgment of shame is an important part of psychoanalytic insight and explanations of the motivations of those who commit acts of religious violence (Hassan, 38), it must be placed in a more nuanced and critical perspective lest we mistake moral indignation for what is more likely, that is, moral embitterment. The reflections of South African writer Njabulo Ndebele (1994) are relevant here. Basically, Ndebele's view that South Africans must transcend the rhetorical limits of 'protest literature' highlights and applies to the necessity of critically engaging and challenging the collapsed reasoning processes of religious militancy. Ndebele argues that for those of the oppressed whose actions remain at the level of protest, the sense that 'evil abounds' leads to and remains an unquestioning identification of social and political evil which is claimed as equivalent to moral and intellectual insight. In the context of religious violence, the world is full of enemies and oppressors, be they in the form of individuals, cultures, secularism, moral degeneracy, disrespect for religious tradition and male authority–the list is extensive. Thus, further expanding and applying Ndebele's ideas, the 'Zionist Occupied Government' (ZOG) of the United States is experienced by some groups as the source of all the humiliation and suffering of the Muslim people, and must be eradicated. Such 'entrapment of resistance in an unreflective rhetoric of protest' (64) becomes a source of reactionary politics of the oppressed, and, I would add, of moral blindness concerning the meaning of deliberately killing innocent people in the struggle for political freedom and religious purity. Oppressed and oppressor become locked in a lethal embrace of distorted representation which only has room for doer and done-to perceptions, foreclosing on a more genuinely liberating emotional and political space where each may consider the needs and motivations of the other without necessarily condoning them (72). There needs to be a 'freeing of the oppressed social imagination' (68) where possibilities for transformation of both oppressor and oppressed may be envisaged, as opposed to the simplistic belief that if the enemy is destroyed, liberation is achieved. For this to be at all possible requires minds that are capable of entering into intersubjectively organized relationships, without which negotiation and peace are impossible goals. Reciprocal recognition breaks open the frozen epistemological structures of oppressor and victim, 'structures which can severely compromise resistance by dominating thinking itself' (67). For Ndebele, what is required for new forms of intersubjective relatedness and the experience of individual agency is a shift in discourse from the 'rhetoric of oppression to that of process and exploration' (73). However, analyses such as Ndebele's are only partial, and fail to take into account the psychological realities of people who have been damaged by sustained political, economic and social trauma as was most certainly the case in Apartheid South Africa. People who have been subject to sustained forms of trauma and brutal neglect will be impaired in a number of ways in their capacity for mentalizing their own affective states, motives and worldviews as well as those of others. This is not because they are insane or incapable of reason, but because their minds have not had the opportunity to develop in psychologically optimal ways, as described earlier.

Many of these ideas of course resonate with Freud, who understood the consoling effects of religion on feelings of vulnerability and helplessness in a world revealed by science as staggeringly incomprehensible as well as increasingly in flux. One of the great emotional protections of religion in his view shields the individual from thought and the strenuous efforts involved in the 'education to reality' (1927, 81). Religion encourages belief over knowledge, where belief gives 'the force of reality to that which is psychic, just as perception does to that which is physical' (Britton 1998, 11). Beliefs are saturated with conscious and unconscious phantasies that arouse powerful feelings, influence perceptions and promote actions (11). Beliefs can be invested with the qualities of a psychic object so that 'believing is a form of object relating' (11-12). The resistance that arises when beliefs are challenged is similar to the refusal to accept that one has lost an object, or other person; the acceptance of the loss of another and the loss of a belief involves mourning, and this can only come about when one accepts that a belief is not knowledge, and that what one believed to be true no longer exists. 'Only through psychic development', writes Britton, 'do we recognize that we actively believe something and that we are not simply in the presence of facts' (12). Cognitive, scientific and cultural development challenge existing beliefs where the individual is confronted with bringing together his/her subjective experience with an 'objective self-awareness' that results in one's ability to move beyond the stage of psychic equivalence and experience oneself *'in the act of believing something'* (13;). For this to happen requires a further level of developmental achievement in the creation of a 'triangular psychic space–a third position in mental space ... from which the subjective self can be observed as having a relationship with an idea' (13). This 'third' is a perspective that is part of self-experience where the perspectives of others, the larger society and social institutions foster critical self-questioning in encounters with difference and diversity which in turn help the developing subject to distinguish between material and mental reality. One realizes this distinction and becomes able to tolerate it with the result that one can maintain the ability to differentiate between belief and knowledge.

In personalities where these distinctions are compromised, the experience of internal stability and coherence is constantly endangered. If we accept a relational psychoanalytic premise that aggression is a derivative of a sense of endangerment (Mitchell 1998), it is then understood as a response to a threatening environment. The violent militancy of fundamentalists 'displays a paramount concern to defend their sacred symbols and to desanctify those of secular communities and other religious ones' (Rapoport 456). If religious feelings and beliefs are motivated by a quest for order (Juergensmeyer 1992) and meaning and intelligibility (J. Z. Smith 1982) then environmental instability, episodic chaos, complexity, pluralism and the claims to truth of secularism are real threats that become terrifying in their perceived ability to overwhelm and destroy. Religion provides a sense of security in contexts of rapid change, dislocation, poverty, hunger and warfare that are often accompanied by intense feelings of humiliation and vulnerability. When aggrieved individuals express their anger and shame in 'religious or spiritual terms...contestants [have] the feeling they are fighting over eternal, spiritual values rather than fleeting, material ones' (Stern 84). Thus an important attraction for those who embrace religious violence is the inner transformation of these feelings, which is largely facilitated by powerful religious feelings. 'Rage turns to conviction ... they enter a kind of spiritual trance, where the world is divided neatly between good and evil, victim and oppressor. Uncertainty and ambivalence...are banished. There is no room for the other side's point of view ... They believe that God is on their side' (282). This is an excellent description of a serious breakdown of the mentalizing capacity where inside and outside experience collapse into undifferentiated rage, with no nuances or shades of grey. Stern is describing an emotionally dissociated state ('trance') where all experience concentrates and distills into a perceived alliance with God. Hassan's informants exhibited a similar sense of uncritical conviction in 'the rightness of their cause and their methods' (38).

Although most religions provide individuals and groups with a sense of internal and external control, a still point in a turning world, the intensity and degree is much more extreme in the forms of religious militancy described here. Psychoanalyst Sudhir Kakar, writing about Hindu/Muslim violence in India, observes that for many communities, religion provides a sense of identity, a sense of 'we are' as opposed to a 'we-ness', either with oneself or with others (1996, 191). Identificatory constructs of 'we are' relieves the individual of the burden of thinking through for himself the implications for moral responsibility for his actions and beliefs, relinquishing individual agency in extreme forms of accommodation to the demands of an external force imagined as God's will and mediated and transmitted through sacred scripture, revered teachers, ancestors and traditions. It asserts exclusionary boundaries of identity where adherence to heteronomously established dictates of belief and practice lend a feeling of group coherence; to be part of a 'we are' identity, one must conform and submit. On the other hand, 'We-ness' involves relational, reciprocal identity organization where individual selves are conscious of their agency and active participation in constructing their communities through communication and negotiation. This involves states of mind that are receptive to new ideas, engaging in risks associated with change and the emotional ability to tolerate them. Here, relationships are experienced in terms of mutual recognition and respect and the willingness to examine beliefs and possibly relinquish them to what Juergen Habermas would call 'the force of the better argument' (1984). In its fundamentalist forms (Almond et al), religious attitudes rely on frozen interpretive frameworks that are resistant to critical transformation of their foundational, most sacred concepts. The boundary between what is religious and what is secular must be vigilantly maintained and guarded.

SHAME THAT ANNIHILATES/ANNIHILATING SHAME

In her psychoanalytic examination of the instructions to the 9/11 hijackers found in the luggage of Mohamed Atta, Ruth Stein remarks upon a tone in the document that suggests a 'severance of the outer world from human meaning, made possible by a long-held and cultivated contempt for that world [which] enables terrorists both to focus on monitoring the instrumental tasks at hand, and to remain immersed in an intensely religious state of mind, which, by its acuteness, screens out all undesirable affects and thoughts' (2003, 296-297). In such mental states, violence is religiously sanctioned so that the perpetrator is disconnected from how another might feel or experience that violence. One passage in the letter instructs the highjackers that, 'If God grants any one of you a slaughter, you should perform it as an offering on behalf of your father and mother, for they are owed by you. Do not disagree amongst yourselves, but listen and obey' (cited in Mneimneh & Makiya 2002). Hassan Mneimneh and Kanan Makiya point out that the deaths of the victims on the planes were 'no longer real but part of a sacred drama' which must be interpreted in terms of 'ritualistic motions'. However degraded or distorted the interpretation of Islam expressed in the letter, it is important to remember that 'the sense throughout [the letter] is that the would-be martyr is engaged in his action solely to please God'. Mneimneh and Makiya further observe that 'the idea that martyrdom is a pure act of worship, pleasing to God, irrespective of God's specific command, is a terrifying new kind of nihilism'. The intense state of spiritual and religious feeling suggests a dissociated mental state wherein all connection with other human beings, including the ability to identify with their terror and suffering dissolves in a 'vertical desire' for God's love and approval. All 'undesirable affects and thoughts' that may weaken the hijackers resolve has been psychically sequestered and split off from conscious awareness (Stein 2003, 297). Psychoanalytically, religious faith in this sense functions as 'a 'splitter' of affect...a degenerative process of literalization [that] allows religious men...to experience heights of exalted love for God *and at the same time* be coldly determined to destroy lives' (307). The terror and humiliation experienced by the victim of violence represents the disavowed, displaced feelings of the murderer who temporarily annihilates those dissociated feelings within him/her self and replaces them with feelings of power and strength (Fonagy 2001).

Hans Kippenberg's observations on the religious character of the 9/11 attacks are relevant here: 'The power of the faithful consists in his ability to overcome the natural fear of death and to make the arrogant Western civilization tremble. The battle cry, which also resounded in the plane, expresses that aspiration. Just as the Prophet, with only a few followers, was able to defeat the superior armies of the jahiliyya, so the faithful Muslim today is able to humiliate Western civilization by means of its own panic. By fearing God more than all other powers and acting accordingly, a faithful Muslim can spread terror among the unbelievers' (2005, 48). A clinical vignette of an individual woman, 'Henrietta', who murdered her boyfriend can provide some insight that psychologically illustrates Kippenberg's observation. It also illustrates some of the internal processes of dissociation, splitting and projective identification that are operative in violent acts, and which cross gender lines. In the moments leading up to her stabbing him, Henrietta experienced an intense shame in his ridiculing, contemptuous attitude. 'She resorted to violence to destroy a mental state ... that was hers yet not hers at the same time. She wished to disown it, but felt no control over it and a murderous act was the only solution to the problem. At the moment of the murder, her object turned from feeling her shame to becoming shaming, a feeling she was desperate to disown ... Perceiving the terror in the eyes of her victim, she felt strangely reassured' (Fonagy 2001, 7-8). During the moment of inflicting violence, the person may feel enlivened and powerful, 'coherent and real, out of reach of deadly rejections, insults and taunts, momentarily once again feeling vital self-respect' (Fonagy 2001, 8). The feelings and thoughts of the one who commits violence are changeable and dissociative, since the victim represents those disavowed 'impure', 'infidel', 'godless' parts of themselves (Stein, 305). In such mental states, it is impossible for the violent individual to adequately distinguish his/her own internal emotional turmoil from the external world or regulate his/her affects, so that feelings of intense shame, powerlessness and vulnerability are attributed to the persecuting 'other' who threatens to overwhelm and psychically destroy the individual. Such limitations of mentalization capacities result in a self-serving capacity to engage in hurtful actions that the perpetrator is not fully involved with at the time. As psychologist Albert Bandura observes, violent militancy is sustained in 'shared, fervent beliefs' where dehumanizing processes (disassociation) leads to 'self-exonerative patterns of thought' (1990, 180). An example of this can be seen in rationale of the militant anti-abortion group, Army of God, which advocates killing doctors who provide abortions as morally 'justifiable homicide' (Stern, 150). Paul Hill, a former Presbyterian minister who murdered Dr. John Britton in 1994 and his driver, James Barrett, believes that '[k]illing fetuses is the moral equivalent of Hitler's killing of Jews in gas chambers' (167). Such moral and religious justifications for violence affectively distance the killer from his victim, obliterating the possibility of empathic responsiveness to the victim's suffering.

Psychoanalyst D. W. Winnicott suggests that notions of a personal God are rooted in the infant's earliest attachment relationships, and that one's image of god is related to the quality of those attachments: 'The child who is not having good enough experiences in the early stages cannot be given the idea of a personal God as a substitute for infant-care ... moral [religious] education is no substitute for love' (1965, 97). More current, empirical attachment research that focuses on religious images and experiences of God confirms Winnicott's insight (Kirkpatrick, 2005; Kirkpatrick & Shaver, 1990). Identities that are organized in terms of conformity to the demands of an external authority result in compliance, not growth, and the demands of the other–God, parent(s), religious institutions–may become internalized as frightful, evil, alien others that need to be split off, disavowed or disowned and located in a series of external others experienced as persecuting, humiliating and threatening. Winnicott writes that '[r]eligion ... has stolen the good from the developing individual child, and has then set up an artificial scheme for injecting this ... back into the child, and has called it moral [religious] education' (94). Fonagy would agree that individuals whose caregivers (parents, teachers, religious leaders) prize obedience over independence, and identifications over maturation are not able to develop fully a capacity for mentalization that will allow them to rely on their own individual powers of reason and critical discernment relationally mediated by real others where differences in values and world views can be resolved in terms of rational justifications in sustained discursive settings. Self-critical, reasoned reflection is impossible when an individual experiences severe internal disruption, such as anxiety, shame or rage. Susan Rose in her study of Christian education movements in the United States points out that in many of the fundamentalist and evangelical Christian schools she examined, discipline is the 'primary concern' in education. The curriculum prepares children for 'ultimate submission to God' by being taught to obey parents, teachers, pastors and civil authorities but not 'non-believers' (461). Educational instruction minimizes 'critical thinking' and a diversity of perspectives by presenting knowledge in the form of a 'fixed body of facts', originating with God and confirmed in Holy Scripture (463, 476). Evangelical and fundamentalist Christian educators and parents are 'united in their opposition to secular humanism and values clarification in public education' (456). Rose further observes that graduates of these educational institutions are highly valued employees by corporations and as members of the military (477, 478).

'Modernity', wrote Baudelaire, in commenting on modernism in art 'is [also] the transient, the fleeting, the contingent; it is the one half of art, the other being the eternal and the immutable' (Harvey 1990, 10). Religion understood as a meaning-making and world-constructing activity is in part an attempt to master this very transient, fleeting and contingent nature of human experience in modernity which undermines the individual's sense of ontological security and emotional stability. From a psychoanalytic perspective, religion offers consolation for anxieties aroused by the limits of one's knowledge and feelings of helplessness and vulnerability. Freud writes, 'As we already know, the terrifying impression of helplessness in childhood aroused the need for protection... which was provided by the father; and the recognition that this helplessness lasts throughout life made it necessary to cling to the existence of a father, but this time a more powerful one' (1927, 30). From Freud's perspective, religion limits the possibility of freedom of thought, insisting that believers adhere to traditions, beliefs and practices because they are dictated by revelation and are the way of the ancestors. In this sense, religion can function as a psychic defense against the more threatening values of cultural modernity and secularism.

Recent psychoanalytic theories have much to contribute to ways of thinking about religiously motivated violence. As suggested here, rigid adherence to religious beliefs is not ultimately compatible with a well-developed critical self-awareness that understands itself as part of a multiplicity of diverse selves who hold a wide variety of perspectives about the world different from and perhaps as valid as one's own, and that what one feels to be so is a representation of reality and not reality itself. The February 1998 Fatwa issued by bin Laden and others is an example of the psychic equivalence mode, where the distinction between belief and knowledge, representation and reality break down, where not only America but all Americans are enemies of his version of Islam and Allah without exception: 'Nothing is more sacred than belief except repulsing an enemy who is attacking religion and life ... The ruling to kill the Americans and their allies – civilian and military – is an individual duty for every Muslim'. All Americans are a threat to the Muslim world in part because their 'crimes and sins' are 'a clear declaration of war on Allah, his messenger, and Muslims'. Meir Kahane's attitude to all non-Jews was similar, where 'any anti-Gentile violence' was 'kiddush hashem', sanctification of the name of God (Sprinzak 1991, 67)[4]. The division of the world into people of god and enemies of god generates a sense of strength and moral superiority in those religious militants who need the external world to contain their own disowned, disavowed sense of fear, humiliation and rage that if not evacuated into the other, threatens to overwhelm and destroy the threatened communal and individual self. In individuals and groups whose sense of identity is fragile or not well established, such religious fantasies easily fill the void. This kind of psychoanalytic approach is echoed in David Rapaport's observation that religious conflict involves questions of 'self-definition' and identity; 'and struggles involving questions of identity, notoriously, are the most difficult to compromise because they release our greatest passions' (446).

Religious fantasies aid in displacing humiliation and release narcissistic and grandiose needs to impose homogenizing order on the entire society.

---

4. Many scholars would rightly contest that neither Bin Laden nor Kahane represent either 'true' Islam or Judaism, or the prevailing views of the majority of Muslims and Jews. My interest here is not to inquire into these issues, but rather to provide examples of the kind of mentality or thinking expressed in extreme or fundamentalist religious views, and how such views can be offered as 'legitimate' interpretations of religious traditions by those who hold them.

'With their calls for prayer in the schools, evangelical revival, and a return to the Bible, [Christian fundamentalists and the religious right] seek to reverse the Enlightenment's restructuring of culture', writes Bruce Lincoln. Their aim, like that of religious fundamentalists everywhere, is to restore 'traditional, God-given values' on the entire society (Lincoln 2003, 61). 'While the understanding of, and reactions against, secularization may vary ... fundamentalists across religious traditions and regions of the world share an animus against political cultures that would deny religion what they feel to be its central place in ordering society' (Almond et al, 20-21). The evidence is mounting that these attitudes are spreading, not abating, throughout the world.

This paper is a revised version of an earlier paper delivered at the International Association for Relational Psychoanalysis and Psychotherapy, Boston, January 28, 2006. My thanks to Donald Wiebe and the late Gary Lease for their helpful comments on earlier drafts of this paper.

DISCIPLINARY CLANS

STEVEN M. HROTIC

Abstract: Academic disciplines are commonsensically defined as a 'data set' – a subset of the facts, theories, and methods that constitute academic knowledge. However, this view cannot alone account for the boundaries between disciplines, nor disciplines' characters. I suggest that disciplines are quasi-social networks. Interactions within academic alliances result in discipline-specific cultural schema, which are used to identify 'social' group members. This plays a beneficial functional role by organizing academia into coherent 'clans'. Disciplines are not imposed by academic leadership, but are an emergent property of academia as a self-organizing system. The Religion department at the University of Vermont is presented as an example.

Defining a Discipline

Some years ago, I had the good fortune of being a Teacher's Assistant for Prof. Luther Martin in the Religion department of the University of Vermont (UVM). It was interesting to see how students' perceptions of Religion classes change. For example, in the first meeting of introductory classes, Martin would ask students one at a time what religion was. (Note: henceforth 'religion' will refer to the subject of study, 'Religion' [capitalized] to the academic discipline.) After a few vaguely existential answers, Martin would theatrically roar, 'Don't tell me what you *believe*; tell me what you *think*!'

This declaration was a catalyst. Astonished students gradually came to understand that religion could be studied by social science. There was a certain amount of soul-searching; a few shared with me that they found the evolutionary viewpoint used by Martin discomforting, or even immoral. (Also during this period, I tutored for a Physics class. Though misunderstandings of the content of this class were also common, none worried about the morality of the subject.) But though introductory students did not initially understand the approach, relatively few dropped the class. Many of the students I talked to in more advanced classes related that it often took several semesters before they 'got' the academic study of religion.

Mistaken assumptions like these are also common in Anthropology. From the Harvard *Crimson*:

> Beginning concentrators in most fields have some idea of what they are getting into. If they don't know much about the methods of their discipline, they are at least somewhat familiar with the nature of the subject matter. In Anthropology, on the other hand, there are always a few concentrators who choose the field for the wrong reasons and regret it later. There are also several who subject [sic] Anthropology because, likewise they aren't well informed ('Anthropology' 1966).

The most commonsensical definition of academia is that it consists of a number of facts, theories, and methods; disciplines are equivalent groups of professionals who have learned and teach a subset of the data. This view is, at best, incomplete. First, it implies that disciplines were determined 'top down' as if the administration conducts a rational epistemological taxonomy; however, as I argue below, administrative recognition merely acknowledges divisions which already exist – the catalyst for divisions lies elsewhere. Second, there is reason to believe that disciplines do not neatly map onto a data subset; students' misunderstandings stem not from the data subset they assume, but from the assumption there *is* a standard data set. Consider: (1) as one progresses in a disciplne, the data set *changes*. Advanced students may never again work with some of topics they encountered as neophytes, while professionals may work exclusively with problems of which students are entirely unaware. (2) Academic success within a discipline results seems to be as much about the flexibility necessary to adapt to new research paradigms as the right initial choice (e.g., of a personality-compatible subject [*sensu* Holland 1997]). (3) Academic departments (one assumes) train students in a discipline. Yet departments may teach disparate and distinct data sets; for example, Anthropology's 'four fields' approach. (4) Some specific data sets fall into multiple disciplines, and some classes are 'cross-listed' in multiple departments.

In this essay, I will focus on the disciplines with which I am most familiar: Anthropology and Religion. Both of these are remarkably heterogeneous. Many U.S. Anthropology departments (including UVM's) define themselves as studying humanity through the combination of archaeology, biological anthropology, linguistics, and cultural anthropology.

This variety sometimes leads to tensions within departments. Biological anthropologists at Harvard University contemplated establishing a separate department (Troianovski 2005). Stanford University did split anthropology into 'anthropological sciences' and 'cultural and social anthropology' (Leslie 2000). Stanford's action may be interpreted as suggesting that Anthropology is not a discipline, but an unstable collection of two science disciplines (biological anthropology and archaeology) a social science discipline (linguistics), and a humanities discipline (cultural anthropology). However, Stanford anthropologists did *not* follow the expected paths:

> One of the ironies of this saga is that the Stanford anthropologists didn't split up neatly by disciplines. Anthropological sciences is a four-field department that includes an evolutionist, an archeologist, a geneticist, two cultural anthropologists, a paleoanthropologist and a linguist. Cultural and social anthropology houses seven cultural anthropologists and a linguist, plus two recent hires - an archeologist and a medical anthropologist. The rather hodgepodge nature of these final lineups suggests that the events at Stanford, while influenced by philosophical differences, had as much to do with personality conflicts, intra-departmental politics and festering disappointments (Leslie 2000).

Religion departments can be even more pluralistic. Harvard's academic concentration 'Folklore and Mythology' seems fairly specific, but even this is not situated in a single discipline. As described by its Chair, Folklore and Mythology is: 'a discipline in and of itself but it also stands at the crossroads of many other disciplines'. [A previous Chair] agreed. 'Folklore is sometimes referred to as the bastard child that English begot on anthropology ... I wear it as a badge of honor. What they're really trying to say is that we bridge the social sciences and humanities' (D'Gama 2008).

UVM's Religion department includes a much broader range of topics. Over the last years, Religion faculty members would be at home in several of the 'four fields', as well as fields like Economics, History, and Theology. Kevin Trainor, the current Chair of Religion at UVM, pointed out that there is no collective label (i.e., 'religionist') for academics in his department (personal communication). Nevertheless, UVM's Religion faculty operates with mutual support and a shared departmental identity – an identity that explicitly includes the belief that there is no discipline called 'Religion'.

So, 'disciplines', if defined epistemologically, only weakly corresponds to academic 'departments.' Note that above Leslie described issues at Stanford as involving personalities and politics; perhaps a more useful definition of disciplines can be based on academics' *social* bonds – not categories of research, but *clans* of individuals (cf. Becher and Trowler 2001).

A Cognitive Definition

The germ of my social view of academic disciplines came from fieldwork I conducted in 2005 primarily at the national meetings of the American Academy of Religion and the American Anthropological Association. My goal had been to discern disciplinary patterns in the way academics critiqued presentations, both in the official 'question and answer' sessions and in informal hallway conversations. To my (perhaps naïve) surprise, presentation content was only superficially examined after the sessions.

Instead, academics tended to discuss common acquaintances and introduce colleagues. I was particularly intrigued by the number of times professional genealogies were described: 'Smith supervised your Ph.D.? I worked with Smith at State'. I could only conclude that a primary goal of these conferences were to create and maintain social connections - potential collaborators were approached as if long-lost cousins, whose value was determined by pedigree. (I would mention that my anthropology mentor at UVM, Jim Petersen, studied under UVM's first Chair of Anthropology Bill Haviland. Haviland studied under Meade, and Meade under Boas. Therefore, I am Franz Boas's [intellectual] great-great -grandson.)

Disciplines ultimately take on many 'cultural' aspects. Some may seem trivial, (modes of dress, musical preferences,) but even these may be expression of social identity (e.g., Bryson 1996), and thus used to distinguish 'insider' from 'outsider'. But in the next two sections I will focus on features that would more directly impact important behaviors, such as peer evaluations. First, I must describe why and how such features may accumulate in the collective mind of a discipline.

Mr. Weasley of the *Harry Potter* books said, 'Never trust anything that can think for itself if you can't see where it keeps its brain'. A culture has no mind, nor a brain in which to put it, yet one notes consistencies as one moves from context to context within a culture, as if *someone* has been maintaining some ineffable constant. Sperber (1996) suggested that culture *does* have a brain – many brains, in fact: one in each member's skull. The ideas that shape behavior exhibit consistency because members continually transmit them back and forth. He calls this an 'epidemiology of representations', using as his metaphor the spread of a disease – including both endemic traditions and epidemic fashions (Sperber 1996: 56-61).

These representations or *schemas* are fundamental to the way we see the world. Schemas are heuristic tools which explain how we can process especially complex situations. They tell us the significance of behaviors, symbols, and objects – and declare what *ought* to be. Importantly, they allow members to identify and judge each other (Strauss & Quinn 1997). These generalizations are not ideal forms, or prototypes from a single experience, but are fictionalized simplifications of memories. Even if factually incorrect, they allow us to negotiate novel situations without painstaking analyses (Matlin 2002: 264-279).

One advantage of schema-theory approaches is that they can account for both universality and variation. In other words, schemas as construed by connectionist models are well-learned but also flexibly adaptive rather than rigidly repetitive. They can adapt to new or ambiguous situations with what Bourdieu called 'regulated improvisation', (Strauss & Quinn 1997: 53).

Individuals who share similar prior experiences share schemas; those who share schemas may react to new situations similarly: for example, they may use similar criteria in interpersonal judgments. But importantly '[s]chemas tend to remain out of awareness, with only the responses a network settles into arising to consciousness' (Strauss & Quinn 1997: 57). That is, individuals may not be consciously aware of judgment criteria, nor why particular schemas exist.

Pilot Study

A pilot explored whether academics more positively evaluate individuals who conform to (explicit and implicit) 'fieldwork' schemas of their own discipline. Though academics from both Anthropology and Religion departments are familiar with on-site data collection, fieldwork is more central to anthropologists' self-image. Additionally, anthropologists seem to prefer certain *kinds* of fieldwork.

Ten Religion professors and nine Anthropology professors from several New England universities were asked to read eight brief autobiographical statements from fictional research candidates. Five of the eight statements related topics of value to both Anthropology and Religion: personal growth, pride in learning a language, an interest in practical applications of academic work, etc. The other three described fieldwork experiences. Participants were asked to evaluate the potential of each candidate by rating them from 1 (least promising) to 10 (most promising).

Religion professors judged candidates who did and did not discuss fieldwork roughly equally (5.8 vs. 5.9), suggesting fieldwork experience had little impact on their evaluations. Anthropology professors assessed the candidates who mention fieldwork higher (6.4 vs. 5.0). For each professor, the average score given to statements who did *not* mention fieldwork was subtracted from the average score given to candidates who did. A one-way ANOVA on this figure found anthropologists' preference for fieldwork experience to be significant (Religion, $M = -.10$, $SD = .83$; Anthropology $M = 1.33$, $SD = 1.73$; $F(1,17) = 5.42$, $p < .05$).

From observations of anthropologists in informal settings, I hypothesized that anthropologists specifically valued *dangerous* fieldwork. To test this, each of the fieldwork statements appeared in two forms, describing high and low emotional, social, and physical danger, by making small changes to key sentences. For example: 'I became [extremely/a little] depressed, but after a few days started feeling at home'; '[My parents] [never/finally] understood my interest in Buddhism'; 'I [ran/walked] every day to get in shape, and even got a [malaria/flu] shot just in case'.

Religion professors *slightly* preferred the candidate whose statements indicated low danger (high emotional danger $M = 4.8$ vs. low emotional danger $M = 5.3$, high physical $M = 4.3$ vs. $M = 4.5$, high social $M = 5.3$ vs. $M = 6.5$). In contrast, Anthropology professors preferred the candidate who reported more emotional and social danger (high emotional $M = 7.0$ vs. low $M = 5.6$, high social $M = 7.8$ vs. low $M = 7.0$) but *less* physical danger (high physical $M = 4.3$ vs. $M = 6.4$). Though the size of the effect of danger on anthropologists' evaluations was similar to anthropologists' preference for fieldwork-oriented candidates, the differences were not significant because of the small number of participants. However, the trend is intriguing.

It appeared as if anthropologists more positively evaluated emotional and social danger, but were prejudiced against candidates who reported *physical* danger. When I discussed results with participants, two features recurred. The first was that while they hadn't consciously thought about danger as a positive part of fieldwork, the pattern matched their intuitions. While some Religion professors felt that reporting danger implied a lack of preparation, Anthropologists felt fieldwork was a 'rite of passage', and was *supposed* to be somewhat traumatic. A candidate who reported preparations for mere physical danger was not prudent but complaining or simply paranoid. A third feature was *not* present: no participants reported being instructed or instructing students in this normative view of the ideal anthropologist. These results suggest that disciplines may implicitly transmit complex evaluatory schema.

Meta-schemas

If discipline-defining schemas are transmitted implicitly, they must be simpler than, for example, the explicit information students in class work so hard to acquire. Also, though one may expect the most influential schemas to have historical connotations, these connotations are not necessary to the efficacy of schemas. For example, the history of Anthropology and Religion are presented in detail only to advanced students (e.g., at UVM only to majors and only in their Senior year) – *after* students have likely begun to absorb specific schema.

Their efficacy, then, is not based on strengths of past rhetoric, but on decontextualized tradition. Merton (1972) discussed a similar process for academic norms. 'These imperatives, transmitted by precept and example and reinforced by sanctions are in varying degrees internalized by the scientist, thus fashioning his scientific conscience' (p. 66). And ultimately, the legitimacy of academia rests on the authority of its traditions (Polanyi 1962). Above I describe a quite specific 'fieldwork' schema related to fieldwork, but do not explain how this schema reveals anything about the central character of Religion vis-à-vis Anthropology. Broad, foundational 'meta-schemas' may link disciplines' histories to specific schema.

One possible example relates to the differences in the way religion is approached in Religion and Anthropology departments. Respective data sets appear to overlap, for example in the authors frequently taught as theoretically important (e.g., E. B. Tylor, Evans-Pritchard, etc.). When (as part of the above study) I asked what the difference was, the usual response from Anthropology was that they didn't *know* what 'went on' in the Religion department. Participants from Religion resisted the idea that Religion *had* a collective character, beyond perhaps Religion's greater reliance on written texts, and perhaps more political motivation in Anthropology. Though expressed cautiously, I believe these statements are key to explicating a crucial meta-schematic difference between these departments.

Religion as the object of a secular study in U.S. institutions is modeled on the German tradition of a 'science of religion' (Sharpe 1986). This tradition argued that whatever the underlying reality, religion was a thing amenable to scientific study, and that such a perspective would enrich both religious and scientific endeavors (e.g., Müller 1872). As channeled through the University of Chicago, theologians' zeal for the texts of one tradition merged with a passion for finding commonalties between traditions (e.g., Eliade 1959).

Anthropology, on the other hand, attempted not to find similarities between cultures, but to account for cultural proliferation. Once the biological inferiority of certain races is disproven, the equality of humankind must be the rational and ethical *precondition* to the study of man (e.g., Boas 1938). Anthropology's *raison d'être* developed into an attempt to give voice and legitimacy to those (cultures, individuals) *not* of the elite. In getting to know a culture, one recognizes that differences between cultures reflect differences in historical circumstances. Many anthropologists in a now globalized world focus on critiquing the relationship between 'core' elites and those on the periphery (e.g., Appadurai 1996; Tsing 1993).

This is only a cursory sketch of decades of development, but even so a historical pattern is clear. Religion's emphasis on texts is not the *cause* of distinctiveness, but an effect of a subtly distinct interpretive framework. The German word for the 'science of religion' is *die Religionswissen-schaft*, incorporating the verb *wissen*: to know, as one knows a fact. For Religion, religion consists of relatively static, knowable data – *therefore*, one should look but in learned texts written by elite religious specialists. However, the historical circumstances of Anthropology's genesis challenged the assumed authority of the elite, seeking not to understand a central theme but to legitimize peripheral *variations* on the theme – and in extreme cases, question whether central themes even exists. They seek not 'to know' a set of static religious facts, but to *come* to know the people as agents. In German, this relates to a different word for 'to know'. I would argue that if Religion has as a meta-schema *die Religionswissenschaft*, Anthropology has *die Religionskennen-*

schaft.

The Function of a Discipline
There may be disadvantages to implicitly transmitted evaluative schema. As they are unexamined, they seem to undermine academia's central norm of objectivity (cf. Merton 1973: 267-278). However, I believe this process also has a distinct benefit.

Anthropologist Mary Douglas (1986) wrote that many anthropological quandaries can be solved by positing functional solutions via a collective mind; her prototypical examples being Durkheim's sociological epistemology (e.g., Durkheim 1912/1954) and Fleck's *denkkollectiv* (Fleck 1935/1979). Similar to the schemas described above, Fleck argued that '[t]he individual within the [thought] collective is never, or hardly ever, conscious of the prevailing thought style which almost always exerts an absolutely compulsive force upon his thinking, and with which it is not possible to be at variance' (Fleck, quoted in Douglas 1986: 13).

However, Douglas objected to 'loose' functional explanations. Functional explanations, for Douglas, are not necessarily false, but, referencing Elster (1983), she insisted they must both have a mechanism and be judged according to strict criteria. In addition, one must consider rational choice: if your model focuses on the good of the group at the expense of the individual, why would individuals act *against* their own interests?

Douglas suspected no such mechanism could exist, as the human psyche does not exist in the same causal domain as more mechanistic operations. 'No general theory equivalent to biological evolution applies to human behaviour', she wrote (Douglas 1986: 33). But cognitive science had started solving this problem before Douglas's book was published. Boyd and Richerson (1985) suggested that the same principles that apply to biological selection (trait variation, selective advantage of some traits over others, resulting in the proliferation of advantageous traits) can metaphorically explain 'cultural evolution', and Tooby and Cosmides (1992) have advocated for an Integrated Causal Model, which denies Douglas's separation of mental and biological causalities. (See also Alcock 2001, Laland & Brown 2002). Within a cognitive paradigm, Sperberian schemas are Douglas's requisite mechanism.

Regarding strict criteria
Jon Elster has declared provocatively that in sociology it is almost impos-

sible to find examples of functional analyses which include all the features a functional explanation would logically require:
1. The function is an effect of individuals' behaviour.
2. The function is beneficial to the group.
3. The function is not the intent of the individuals' behaviour.
4. Individuals do not recognize the function of the behaviour.
5. The function maintains the individuals by a causal feedback loop passing through the group (Douglas 1986: 32-33).

I submit that the purpose of academics' quasi-social activity is to acquire allies within the large, anonymous body of academia. But the *function* of their behaviour is to adjust disciplinary boundaries, splitting disciplines or creating new ones when necessary. They seek out unfulfilled potentialities and viable splinter groups, while isolating redundancies. Though academics act rationally for their own good, because of their actions academia benefits from a taxonomy of epidemiology more robust than could be constructed purposefully, and that, in response to academic progress, continually adjusts to changes in the political environment, etc.

I believe Elster's criteria are met. This re-organization (1) is a direct effect of individuals' behavior, even though (3) the intent of the behavior was to seek allies, not to subdivide academia. (4) I doubt they typically recognize the connection between their behavior and disciplinary divisions. Rather, academia *self-organises*; like flocks of birds following extremely simple rules, academics cluster themselves around academic topics (cf. Reynolds 1987). Established disciplines contain too many people for the individual to maintain ties with all members (Dunbar 2003), so this process benefits the individuals; rational choice is not violated. But established disciplines are also too large to be conceptually unified; (2) in subdividing disciplines into comprehensible 'chunks' academic progress becomes possible, and academia benefits. Finally, (5) there is a group-level feedback mechanism in place. Once an expanding group of allies (and their research paradigm) reaches a certain threshold, it may gain the support of institutional recognition – that is, they form new departments. Official approval gives the rebels an estate of their own, in which they can grow – and if successful, begin the process anew.

Creating a New Department

In the late 1960s and early 1970s, universities in the U.S. experienced unprecedented growth, resulting in the establishment of many new departments. At UVM in this period, Religion became a department independent from Philosophy, and Anthropology split from Sociology. A comprehensive history of these splits would be a valuable addition to the literature, particularly as UVM's Religion department was either the first or the second secular comparative religion program (depending on whether one asks alumni of UVM or the other candidate, Western Michigan), and as such became the prototype for many of the rest. Unfortunately, only points directly relevant to this essay can be included here.

Religion from a *religious* perspective had long been topic of study at the University of Vermont: the curriculum from 1831 ended with a course on 'Evidences of Natural and Revealed Religion'. But by the 1960s, the trend in American universities had been increasingly toward religion as a secular topic as Eliade's students from the University of Chicago graduated and began teaching. By 1965, a secular Religion major had already been established at UVM within the Philosophy and Religion Department. The third and fourth secular Religion instructors, William Paden and Luther Martin, were hired respectively in 1965 and 1967. Religion became independent in 1973 (Paden 2005). A few items from this period (i.e., 1965-1973) are especially noteworthy.

The first is that in 1970, Paden and Martin presented a paper on the emerging discipline at an American Academy of Religion conference:

> For departments and programs have a protean life indeed. Unlike the second person of the Trinity, there was a time when religion studies was not, and though it may be subversive to our professionalistic ears at a meeting dedicated to annual renewal rites, there most likely will be a time when it is not. Though spontaneous in its mutations, unpredictable in the way it will re-crystallize around changing social ideals, religion studies seems bound up with the particular dialect of its cultural and educational forms.
>
> Thus, whenever the value of abstract philosophical considerations about the nature of religion – as though it were an ontological problem – we find that primarily what religion studies *is*, is what it *does*, and what it *does* is a function of *where* it is situated, and that *what* it becomes is shaped by questions that arise from that environment (emphases in original, Martin & Paden 1970).

Note that the ebb and flow of topics of study is clearly in their minds, and that the *context* of the study of religion (to which I would add *people*) influences the study itself – the approach developing at UVM was not a top-down imposition, but rather a localized response to the academic environment.

Despite sharing a department, by 1973 Religion and Philosophy had separate identities.

1973 was simply the chronological time when institutionally they became separate departments. However, we very much had an identity before that ... The institutional split specifically came because Philosophy wanted its own identity. We were OK with that, each cohorts was big enough. They perhaps had nine faculty, we had perhaps six. They were clearly different fields; we shared the same building and the same department title, the Department of Philosophy and Religion, for years, but we were functioning completely separate. Religion in particular had a very strong identity (Paden 2005).

Interactions between Religion professors extended beyond faculty meetings:

> We would go every year to rent a place in Stowe. It started as two overnights – three days – but was at least one overnight. These were our retreats where we would get together to discuss our vision for the identity of the Religion Department which was linked to what was then the new academic phenomenological approach to the study of religion, compared to the sectarian, theological approach. That was already going on before the split (Paden 2005).

When the split formally occurred in 1973, the concerns of the administration of UVM were only that the two new departments were large enough (in terms of students and faculty) both to be viable and to warrant the additional administrative costs. Academic justification was not an issue. Students were similarly unconcerned; there had been Religion majors for years, and the curriculum was little affected.

At about the same time, a split in 'Sociology and Anthropology' was preceded by many of the same features: a long-standing practical division within the dual department, the accumulation of faculty and students who focused exclusively on one or the other, regular extra-curricular social outings, and eventual administrative acknowledgement. And when the division occurred, neither students nor administration were particularly involved or affected. Again, divisions developed organically. Explicit data (e.g., newly emerging methods) merely formed the focus of cooperative and social groupings, and were not the primary impetus.

Between them Professors Paden and Martin chaired the department for over three decades, as the UVM Religion department developed a unique character and curriculum. Much of the character reflected its charismatic Chairs. Decades after their arrival, students related the origin stories of the department: how the fetal department sought out Paden as a young graduate student to replace a retiring professor, how Martin drove from California to Vermont in a mail truck painted with peace signs and flowers to join his friend. Social ties seem to be both reality and symbol for the department.

Through much of these decades, UVM's curriculum was ahead of the curve. 'As new programs appear, they're looking to see what a program should be like. It's no longer the seminaries' model [...] It turned out that what we had done all along now, well I don't want to say everybody does it, but lots do it' (Paden 2005). In a sense, UVM's Religion department is only now entering its second generation. The majority of universities have now caught up along the trail blazed by UVM. It will be interesting to see what UVM will do next.

On Choosing a Discipline

I began this essay by describing students who choose classes without understanding what the disciplinary categories really mean. I went on to argue that a definition of a discipline that was limited to explicit theories, methods, or data was incomplete; that disciplines operate as pseudo-cultural networks of professional allies – more social group than professional category. I described the kinds of cultural schemas that may influence important judgments of academic work and the academics themselves, and how, despite the potential lack of objectivity, such a process may ultimately benefit academia. As an example, I briefly sketched the formation of the Religion department at the University of Vermont.

I focused primarily on the disciplines with which I am most familiar: Religion and Anthropology. One may object that these examples 'stack the deck' favor of a non-epistemological definition of disciplines. This may well be; I leave it to those in other disciplines to decide whether what I discuss here applies. Though other disciplines (e.g., 'hard' sciences) may be methodologically more homogenous, I suspect that even disciplines with narrowly-defined data sets exhibit the patterns I describe here.

I graduated from UVM having majored in Anthropology and Religion – not one but *two* disparate fields. I'm currently in a Biology department, though in my more frustrated moments I sometimes wish I had stuck with Music History. This may look like the career of an intellectual dilettante, but I've come to believe that I'm not much more varied than the average academic; it's just that my focus happens to fall between disciplines.

Prof. Martin's history is, I think, similarly diverse. He used to describe his specialty as 'dead religions'. His philosophical inspiration moved from Jung, to Foucault, to Gauchet (e.g., Martin 1985, 1998, 2002). He has the distinction of teaching UVM's only course on alchemy, and threatens to open a Mithraic temple in the Religion department's basement. Today, he continues to pursue his interests with utter disregard for academic boundaries. He has also importantly contributed to academia by encouraging collaborations between experimental psychologists, anthropologists, and others in the development of an interdepartmental discipline, the 'cognitive science of religion' – a discipline I fully expect will someday create its own undergraduate departments. I don't know if a university full of academics like him would be advisable – but I strongly believe the role he plays is vital to the health of academia.

The holy Qur'an says 'if ye mix their affairs with yours, they are your brethren'. As a post-graduate, I currently face a crossroads similar to that faced by incoming freshmen – with whom should I mix my affairs? Who shall be *my* clan? In writing this essay, the answer revealed itself: the University of Vermont, Department of Religion … circa 1970.

THE SOCIAL CAPITAL
OF RELIGIOUS COMMUNITIES
IN THE AGE OF GLOBALIZATION

HANS G. KIPPENBERG

Belonging and Believing

For years the situation of religion in Europe was conceived of as 'believing without belonging'. With regard to the issue of religious communities in the age of globalization, this would imply that in Europe religion has lost its power to establish social bonds. For that reason it is worthwhile to note that Grace Davie, author of that formula, has abandoned it. As early as 1994 Grace Davie showed in a study of Great Britain that a low level of church commitment can coexist with a large diffusion of faith convictions; she called this 'believing without belonging'[1]. This observation makes good sense of two discrepancies that can be observed in the religious statistics of almost every European country[2]. The first concerns the numerical relationship between individual religious convictions and the participation of these individuals in church activities. In Sweden, for example, only 6% of Christians go to church once a month; but personal religious views (such as faith in God, or belief in life after death) are more widespread and are held by more than half the population. At once, however, we see in this case a second discrepancy. Although half of the population tends towards atheism, 70% of Swedes have remained members of their church. In the case of Sweden, we must take into account the reverse: 'belonging without believing' In a later study in 2006, Grace Davie addressing that discrepancy brought another significant perspective into the picture: *de facto*, in today's Europe the churches have become voluntary

---

1. Grace Davie, *Religion in Britain since 1945: Believing without Belonging.* Oxford: Blackwell, 1994.
2. On this see José Casanova, 'Die religiöse Lage in Europa', in: Hans Joas & Klaus Wiegandt (eds.), *Säkularisierung und die Weltreligionen.* Frankfurt/M.: Fischer, 2007, pp. 322–357.

associations, are actors in civil society, and are able to mobilize considerable support though their sympathizers do not translate their approval into an active participation in the life of the local parish. Grace Davie called this 'vicarious religion', a concept that better should replace 'believing without belonging'[3].

The forms in which religious communities are formed in today's Europe are more diverse than at any time in the past. The synagogue in Judaism, the church in Christianity, or the mosque in Islam were never in fact the only form in which religious communities took social shape; *a fortiori*, however, if we investigate the forms of religious community today we cannot limit our researches to these institutions, since they no longer offer a sufficient criterion for measuring the degree of belonging. A great variety of organizations operate in the sphere between the state, economic life, and the private realm today. All of these are characterized by a high degree of societal self-organization, and legal forms that regulate their existence. Religious communities, too, profit from this situation, as Gunnar Folke Schuppert writes[4]:

'Religious communities are among the actors in the public realm of civil society, provided they perform as competitors and not as established church'.

Thanks to these new legal forms, private religiosity can emerge in the public sphere. For many years, sociologists of religion took privatization as the connecting thread of their investigations of religion in modern society, until José Casanova reversed the argument and demonstrated that the supposedly privatized religions are no longer restricted to the private sphere[5]. Rather, religions articulate in the public realm experiences and claims that do indeed have their origin in personal experiences and evaluations, but are shared by others - and hence are presented in a communal form. Indignation at wars, violations of human rights, protests against the destruction of the environment, and so on, bring believers onto the streets and into NGOs. Casanova

---

3. 'Vicarious Religion: A Methodological Challenge', in: Nancy Ammerman (ed.), *Everyday Religion: Observing Modern Religious Lives*. New York: Oxford University Press 2006, pp. 21–37.

4. Gunnar Folke Schuppert, 'Skala der Rechtsformen für Religion: vom privaten Zirkel zur Körperschaft des öffentlichen Rechts. Überlegungen zur angemessenen Organisationsform für Religionsgemeinschaften', in: Hans G. Kippenberg/Gunnar Folke Schuppert (eds.), *Die verrechtlichte Religion: Der Öffentlichkeitsstatus von Religionsgemeinschaften*. Tübingen: Mohr Siebeck 2005, pp. 11–35 on p. 21.

5. José Casanova, *Public Religions in the Modern World*. Chicago: University of Chicago Press, 1994.

sees this 'de-privatization' of religiosity as basic to the civil-societal constitution of religions today. This social form of religion is linked to a new kind of public presence, which is fundamentally different from the traditional state religion. Due to the fact, that detachment of church activities cannot function in Europe as a measurement of the weakness of religions, eminent scholars today are considering religion as a source of cultural integration of the new Europe[6].

The Social Capital of Private Associations as Public Concern in the U.S.

A similar turn to the social power of religions beyond the local parish has occurred in the US. Here in particular Robert Putnam, Professor of Public Policy at Harvard University, has taken the lead. Putnam had examined the efficacy of twenty regional governments in Italy which had been installed in 1970 and given the responsibility for a wide range of public tasks. While studying the results of this reform he encountered considerable differences. In the north of the country it had been a success, in the south, however, a failure. The cause he found to be in a differential development which can be traced back at least to the 14th century. In the towns of northern Italy citizens' associations actively took matters concerning their town into their own hands. In contrast the townspeople of the south, dependent upon landlords, expected matters to be dealt with by 'them up there'[7]. Putnam conceived of these differences in terms of transmitted patterns of social relations. In the former case they were based on principles of reciprocity and trust among the townspeople, in the latter case on dependency and hierarchy. Putnam conceived of his observations as a case of path dependency. 'Where you can get to depends on where you're coming from, and some destinations you simply cannot get to from here' (Putnam, *Making Democracy Work*, p. 179). The civic engagement of the north that can be traced back to a longstanding history of civic organizations constitutes a 'social capital' which was decisive for the economic and political success of the newly formed regional governments[8]. The differences

---

6. Kryzstof Michalski (ed.), *Religion in the New Europe*. Budapest: Central European University Press 2006 with contributions by Charles Taylor, José Casanova, Danièle Hervieu-Léger, David Martin, Peter L. Berger a.o.
7. Robert D. Putnam, *Making Democracy Work. Civic Traditions in Modern Italy*. Princeton: UP 1993, pp. 130-131.
8. Putnam, *Making Democracy Work*, chapter 6 ('Social Capital and Institutional Success'), pp.163-185.

still had an impact on the success or failure of politics in the 20th century.

Immediately after publishing his study, Putnam applied its findings to the U.S.A. and published his results under an inspiring title: *Bowling Alone*. Americans have increasingly been bowling alone for the past two decades - a development that brought Putnam to make an alarming prediction for the U.S.A.[9]:

'American social capital in the form of civic associations has significantly eroded over the last generation'.

Putnam presented this disturbing finding against the background of Alexis de Tocqueville's classic work democracy in America. According to the French observer of the early 19th century, American democracy was thriving on the readiness of citizens to join all kinds of associations. And Putnam began his article with a preface which made every politician sit up and take notice.

When Tocqueville visited the United States in the 1830s, it was the Americans' propensity for civic association that most impressed him as the key to their unprecedented ability to make democracy work. Recently, American social scientists ... have unearthed a wide range of empirical evidence that the quality of public life and the performance of social institutions (and not only in America) are indeed powerfully influenced by norms and networks of civic engagement. Researchers in such fields as education, urban poverty, unemployment, the control of crime and drug abuse, and even health have discovered that successful outcomes are more likely in civically engaged communities.... Social scientist in several fields have recently suggested a common framework for understanding these phenomena, a framework that rests on the concept of *social capital*. By analogy with notions of physical capital and human capital – tools and training that enhance individual productivity – 'social capital' refers to features of social organization such as networks, norms, and social trust that facilitate coordination and cooperation for mutual 'benefit'.

The discovery of a new kind of capital created a sensation. Bill Clinton invited Putnam to Camp David for a consultation lasting several days and worked his ideas into two 'State of the Union' addresses. Tony Blair received him in London. It almost appeared as if a political scientist had discovered the

---

9. Robert D. Putnam, 'Bowling Alone: America's Declining Social Capital', in: *Journal of Democracy* 6 (1995) pp. 65-78, quotation p. 73.

magic independent variable, on which both the success of political action and the well-being of the society depend and which is moreover low-cost[10].

Putnam assumed a declining social capital in the US – with one notable exception. The voluntary associations of private citizens are losing members. If there was hope, it could only be found in religious communities[11]. However, even in America the statistics indicate a decline in participation in church-activities, even though membership in religious communities was not affected. Since Putnam believed that 'Religious people are exceptionally active social capitalists'[12] he prepared the ground for arguments that legislation should be used to make that social capital work[13].

The American Welfare Reform of 1996

The discovery of the social capital of private associations happened at a time, when doubt was voiced about the efficacy of public spending for social welfare. Marvin Olasky argued in his book *The Tragedy of American Compassion* (1992) that public welfare establishes dependency, instead of making people independent. The welfare system reproduces the misery it wants to fight[14]. His argument: If an unmarried mother gets a contribution to the rent of her apartment, she is encouraged to stay unmarried, otherwise she loses it. Big organizations are unable to stimulate people to change their life style. Only communities of dedicated Christians can do so. The ecclesiastical welfare organizations likewise are too big and bear likeness to state agencies. In order to

---

10. Nicholas Leman, 'Kicking in Groups', in: *The Atlantic Monthly* 277 (1996) pp. 22-26.
11. There is an unresolved tension between Putnam's thesis of a general decline of associations and his praise of religious communities. See Friedrich-Wilhelm Graf: ' 'In God we Trust'. Über mögliche Zusammenhänge von Sozialkapital und kapitalistischer Wohlfahrtsökonomie', in: Friedrich Wilhelm Graf/Andreas Platthaus/ Stephan Schleissing (eds.), *Soziales Kapital in der Bürgergesellschaft*, Stuttgart: Kohlhammer 1999, pp. 93-130. Putnam sees church-going as decisive and ignores membership in affiliated religious associations.
12. Robert D. Putnam, , *Bowling Alone. The Collapse and Revival of American Community* New York: Simon & Schuster 2000, pp. 65-79 'Religious Participation' on p. 67. According Putnam religious communities contribute 15-20 billion $ and extensive social activities to welfare in the US (pp. 67-68).
13. For an analysis of the link between 'Religious Organizations and Social Capital' distinguishing social network and cultural norms or values see Pippa Norris/ Ronald Inglehart, *Sacred and Secular Religion and Politics Worldwide*. Cambridge: UP 2004, Chapter 8 pp. 180-195.
14. Marvin Olasky, *The Tragedy of American Compassion*. Washington D.C.: Regnery 1992 p. 18.

be genuine any assistance must be personal and face-to-face; the donor must look into the eyes of the needy. And the people in need deserve personal attention, not a pay check by an anonymous authority. Not only the society must be blamed, for their misery, but also their own morality[15].

Both Republicans and Democrats in Congress found their own views confirmed by Putnam's diagnosis: the former in their belief of the superiority of self-initiative and the market above legislation, the latter in their belief in the state to pass laws to contribute to an activation of a worthy common good. Together they passed the Welfare Reform Act in 1996, with its integration of religious communities into the welfare system and the obligation for unemployed people, to accept every kind of work. A time restriction was added to the existing federal support programs for unmarried mothers, for those unfit for work, for the ill, the homeless, and those dependent on drugs[16]. Payments from these programs were limited to sixty months/five years over one's life span[17]. Welfare offices were to become 'Job Placement Centers', and welfare to be replaced by 'workfare'. When signing the rules into law President Clinton remarked (August 22: 1996): 'this legislation provides an historic opportunity to end welfare as we know it and transform our broken welfare system by promoting the fundamental values of work, responsibility, and family'.

Section 104 (b) of the new law 'allows the States to contract with religious organizations, or to allow religious organizations to accept certificates, vouchers, or other forms of disbursement under any program described in subsection(a)(2), on the same basis as any other nongovernmental provider without impairing the religious character of such organizations, and without diminishing the religious freedom of beneficiaries of assistance funded under such program[18]. This legal provision should make it possible to local congre-

---

15. Ronald J. Sider/Heidi Rolland Unruh, 'Evangelism and Church-State Partnerships', in: *Journal of Church and State* 43 (2001), pp. 267-296; Rainer Prätorius, *In God we Trust. Religion und Politik in den USA*. München: C.H. Beck 2003, pp. 173-4.
16. Franz-Xaver Kaufmann, *Varianten des Wohlfahrtstaates*. Frankfurt: Suhrkamp, 2003 pp. 118-121.
17. Elmar Rieger/Stephan Leibfried, *Grundlagen der Globalisierung. Perspektiven des Wohlfahrtsstaates*. Frankfurt: Suhrkamp 2001, p. 205.
18. The law was published with comments by the „Center for Public Justice' : *A Guide to Charitable Choice. The Rules of Section 104 of the 1996 Federal Welfare Law Governing State Cooperation with Faith-based Social-Service Providers*. Washinton: Center for Public Justice 1997. URL: http://www.cpjustice.org/files/CC_Guide_0._pdf. The center is a Protestant 'Think -Tank', that rejected the established welfare system and exerted some in-

gations (Churches, Synagogues, Mosques) to supply services under the federal programs, although persons in need could decide turn instead to non-religious organizations for assistance. The legislators were interested in including local congregations in the relief systems, as is indicated by the degree of freedom religious communities were given in their participation in the program. For example, they were not obliged to hide their religious commitment and were allowed to choose their employees according religious criteria independent of the prohibition of discrimination (Section 104 f). But they are obliged to help everyone whether religious or not, in the same (Section 104 g) manner as the welfare structures.

Immediately after George W. Bush became President, in the White House an office of 'Faith-Based and Community Initiatives' was established and put this legal provision into practice[19].

Religious charities had been receiving public funds long before 1996. Catholics, Protestants, and the Salvation Army had established charities of their own - 'religious affiliated entities' - that became part of the new public welfare system. But the new law also accepted into the program 'pervasively sectarian organizations' that had previously been excluded[20].

Legal developments had facilitated this inclusion of religious congregations into the public welfare system. The 'First Amendment' to the constitution of the US (1791) had determined: 'Congress shall make no law respecting an establishment of religion, or prohibiting the free exercise thereof'. Religions should be protected from any interference by the federal government. The provision was restricted to Congress, and did not apply to the individual states where regimes of established churches existed without being challenged by the Supreme Court. By the mid-twentieth century the federal Court began

---

fluence on the new legislation (Amy E. Black/Douglas L. Koopman /David K. Ryden, *Of Little Faith. The Politics of George W. Bush's Faith-Based Initiatives*. Washington, D.C: Georgetown UP 2004, pp. 46-48).

19. For details see Black/ Koopman/Ryden, *Of Little Faith*, S. 218-222; a good analysis also by Birgit Oldopp/ Rainer Prätorius, Faith Based Initiative': Ein Neuansatz in der U.S. – Sozialpolitik und seine Hintergründe', in: *Zeitschrift für Sozialreform* 48 (2002) pp. 28-52.

20. John S. Coleman, S.J., 'American Catholicism, Catholic Charities, and Welfare Reform', in: Hugh Heclo /Wilfred M. McClay (Hg.), *Religion Returns to the Public Square. Faith and Policy in America*. Washington, D.C.: Woodrow Wilson Center Press 2003, pp. 239-267; Stanley W. Carlson-Thies, 'Charitable Choice: Bringing Religion Back into American Welfare', in: H. Heclo/ W.M. McClay (Hg.), *Religion Returns to the Public Square*, pp. 269-297.

to respond to religious matters at the state level. In the beginning attempts prevailed to erect a 'wall of separation' between civil authorities and religious bodies; later the principle changed and the jurisdiction stopped discriminating between religious and secular groups and introduced principle of equal treatment. Religion should not be placed at a disadvantage, since it was regarded as a good that deserved legal protection[21]. This legal change was favorable for a legislation that allowed public authorities to include religious congregations in their fight against well defined social ills. The costs of transactions were expected to be reduced, the people in need better served. But a tension remained: church-organizations do not understand themselves as supplier of state services that people can ask for when they are legally entitled. Christian charity, rather, is in essence personal[22].

Institutionalization of religious ethics of solidarity

This reform in the welfare system, then, transformed a principle of religious ethics into a legal institution and made it work for a public good. The Christian text fundamental to that ethic derives from the gospel of Matthew (25, 31-46):

But when the Son of Man comes in his glory, and all the holy angels with him, then he will sit on the throne of his glory. Before him all the nations will be gathered, and he will separate them one from another, as a shepherd separates the sheep from the goats. He will set the sheep on his right hand, but the goats on the left. Then the King will tell those on his right hand, 'Come, blessed of my Father, inherit the Kingdom prepared for you from the foundation of the world; for I was hungry, and you gave me food to eat. I was thirsty, and you gave me drink. I was a stranger, and you took me in. I was naked, and you clothed me. I was sick, and you visited me. I was in prison, and you came to me'. Then the righteous will answer him, saying, 'Lord, when did we see you hungry, and feed you; or thirsty, and give you a drink? When did we see you as a stranger, and take you in; or naked, and clothe you? When did we see you sick, or in prison, and come to you?' The King will answer them, 'Most

---

21. Winfried Brugger, *Einführung in das öffentliche Recht der USA*. 2. A. München: C. H. Beck 2001, pp. 187-188.
22. Nancy T. Ammerman, 'Connecting Mainline Protestant Churches with Public Life', in: Robert Wuthnow /John H. Evans, (Hg.), *The Quiet Hand of God. Faith-Based Activism and the Public Role of Mainlaine Protestantism*. Berkeley: University of California Press 2002, pp. 129-158 on p. 150.

certainly I tell you, inasmuch as you did it to one of the least of these my brothers, you did it to me'. Then he will say also to those on the left hand, 'Depart from me, you cursed, into the eternal fire which is prepared for the devil and his angels; for I was hungry, and you didn't give me food to eat; I was thirsty, and you gave me no drink; I was a stranger, and you didn't take me in; naked, and you didn't clothe me; sick, and in prison, and you didn't visit me'. Then they will also answer, saying, 'Lord, when did we see you hungry, or thirsty, or a stranger, or naked, or sick, or in prison, and didn't help you?' Then he will answer them, saying, 'Most certainly I tell you, inasmuch as you didn't do it to one of the least of these, you didn't do it to me'. These will go away into eternal punishment, but the righteous into eternal life'.

Those who are saved and those who are condemned in the final judgment hear, as criterion for the verdict, 'What you did to one of the least of these my brothers, you did it to me'. The brother cannot be defined in terms of belonging to Israel but in terms of needs: anybody who is hungry, thirsty, a stranger, naked, in prison. To recognize him is the true proof of faith in Jesus Christ. This parable is regularly invoked in sermons, when ministers address the situation of the destitute, social inequality and emphasize the value of caring[23].

In agreement with that powerful parable Christian congregations are sensitive to all kinds of social needs. They assist unmarried mothers, serve food, advise people in legal issues, help the homeless to find shelter, arrange for a visit with a doctor, support unemployed in additional training and serve in many other causes[24]. Solidarity with the needy is a Christian value; yet need is no timeless category; it changes with time and social conditions and has to be permanently newly redefined. This explains the broad and open scale of activities in which Christian congregations in the US are engaged in. Most important among them is providing connections. Since one and the same local congregation is unable to cope with all urgent cases, its strength depends on

---

23. Robert Wuthnow, *Saving America? Faith-Based Services and the Future of Civil Society*. Princeton: UP 2004, pp. 66-69. Between 11% and 15% of the American population earn less than $ 9000 annually, the official limit of poverty; 5% earn less than half of that amount (pp. 177-181).
24. A detailed study of these services, based on a national survey, was done by Mark Chaves, *Congregations in America*. Cambridge (Mass.): Harvard UP 2004, 'Social Services' chapter 3, pp. 44-93; pp. 213-221 deal with the 'National Congregations Study' as data basis. The same data together with an own sample are the base to Nancy Tatom Ammerman, *Pillars of Faith. American Congregations and Their Partners*. Berkeley: UCP 2005, 'Extending the Community: Serving the Needy, Saving Souls', chapter 5, pp. 115-157.

its relations with other relief agencies, private and public. While a local congregation is strong in performing direct immediate assistance, for longer lasting solutions charitable organizations and public offices are more qualified[25]. This model of religious solidarity was not confined to Christian congregations and spread across all other religious communities in the US. Immigrants belonging to Judaism, Islam, Buddhism, Hinduism and other religions emulated the model of the 'congregation'. Though differences between these congregations should not be ignored, the social services they offer for their members, and strangers, are similar[26].

These services are a major contribution of religious congregations to the social bond in society. But negative consequences are also imaginable. In the beginning Putnam assumed that the social capital inherent in this kind of congregation would automatically serve the well-being of society. In contrast to that view Alejandro Portes pointed out that also destructive effects upon civil order can also emanate from networks and from norms. In reaction to this view, Putnam made subtle changes to his concept and differentiated between an outward-looking social capital that builds bridges between different people who are unlike one another, from an inward-looking social capital which only establishes a bond between equals. These closed communities can become a source of tensions in a society[27].

The affinity of religious ethics of solidarity with modern society: a thesis of Max Weber

According to Max Weber there exists an inherent link between the rise of modern society and a religious ethic of solidarity. This is the essence of his argument in his *Economy and Society* in which he analyzes the interaction and relationships between the modern economical system and the coexisting social orders of family, household, neighborhood, nation, religion, market, domination and law. In that sequence he dealt with these orders and studied them

---

25. For s survey of services supplied by four different types of relief organisations (public, secular, faith-based and local congregation) see Wuthnow, *Saving America?*, p. 203.
26. Wendy Cadge, 'De Facto Congregationalism and the Religious Organizations of Post-1965 Immigrants to the United States: A Revised Approach', in: JAAR 76 (2008) pp. 344-374.
27. Alejandro Portes/Patricia Landolt, 'The Downside of Social Capital', in: *The American Prospect* 26 (1996), pp. 18-21; Alejandro Portes, 'Social Capital: Its Origins and Applications in Modern Sociology', in: *Annual Review of Sociology* 24 (1998), pp. 1-24; Robert Putnam, *Gesellschaft und Gemeinsinn. Sozialkapital im internationalen Vergleich*. Gütersloh: Bertelsmann 2001, pp. 27-31.

as types of communal action. What normally is regarded as an autonomous institution, disconnected from the individual, Weber turned into meaningful actions of individuals. He also did this with the neighborhood: the household, he argues, is self-sufficient, but in times of crisis it depends on its neighbors. Therefore neighborhood is not only physical proximity but also a community of interests. 'The neighbor is the typical helper in need, and hence neighborhood is brotherhood, albeit in an unpathetic, primarily economic sense'. That kind of neighborhood is not restricted to agrarian societies, but exists also in the apartment houses and slums of modern cities. Here too 'neighborhood is the typical locus of brotherhood'[28]. That does not mean that the neighbors necessarily maintain brotherly relations. Most often conflicts and enmity dominate their relations. 'But the essence of neighborly social action is merely that sober economic brotherhood practiced in case of need'[29].

According to Weber this popular morality was adopted and integrated into the ethics of religious communities. Historically, most religions have known only occasional associations of their adherents. Only a few developed a full-fledged congregational religiosity: Buddhism, Judaism, Christianity, and Islam. When this occurred the popular ethics of brotherliness changed its character and became the centre of a communal religiosity. As Weber puts it[30]:

> Congregational religion set the co-religionist in the place of the clansman. 'Whoever does not leave his own father and mother cannot become a follower of Jesus' This is also the general sense and context of Jesus' remark that he came not to bring peace, but the sword. Out of all this grows the injunction of brotherly love, which is especially characteristic of congregational religion in most cases because it contributes very effectively to

---

28. Max Weber, *Economy and Society*. Ed. by Guenther Roth and Claus Wittich. 2 vols. Berkeley: University of California Press 1978, vol. 1, pp. 361-363. Critical German edition: Max Weber, *Wirtschaft und Gesellschaft*. Volume 1: *Gemeinschaften*, ed. by Wolfgang Mommsen together with Michael Meyer. Max Weber Gesamtausgabe I/22-1. Tübingen: Mohr Siebeck 2001, pp. 121-122.
29. Max Weber, *Economy and Society*, vol. 1, p. 363. Critical German edition: Max Weber, *Wirtschaft und Gesellschaft*. Volume 1: *Gemeinschaften*, p. 125-126.
30. Max Weber, *Economy and Society*, Vol. 1, p. 580; cf. Max Weber, *Wirtschaft und Gesellschaft*. Vol. 2. *Religiöse Gemeinschaften*. Ed. by Hans G. Kippenberg together with Petra Schism and Jutta Niemeier. Max Weber Gesamtausgabe I/22-2. Tübingen: Mohr Siebeck, p. 372.

the emancipation from political organization. ... The obligation to bring assistance to one's fellow was derived ... from the neighborhood group.

Religious communities cannot be regarded as serving economic or political interests. When becoming part of modern society they develop an ethic that is not only independent of these new powers, but is antagonistic to them. The roots of this ethic are retaliation against offenders and fraternal assistance to neighbors[31]. The ethic of brotherliness differentiates itself principally from the imperatives for action of the political organization and evolves a dynamic independent of the state. Such congregations faced a major challenge, however, when the religion took the direction of world-rejection as the means to salvation. For, in Weber's account, the more a religion of salvation developed and the more its adherents experienced 'tensions' with the world and these tensions became systematized, the more an *ethic of commitment and responsibility* replaced an ethic of compliance with laws. In their turn, these tensions elicited new forms of religiosity. Weber sketched these dynamics of religious communities for the first time in chapter 11 of 'Religious Communities' and later revised and expanded it in the 'Zwischenbetrachtung' of 1915. At the core of his argument is the claim that with increasing tensions religious communities with respect to the spheres of economics, politics, sexuality, and art, the believers develop a culture of their own, independent of the constraints of these spheres. If one accepts that reasoning, one easily can understand that American congregations with their extensive social assistance networks are at the same time deeply involved in the wars about the official culture of the U.S.[32].

Islamic Community Building

In Islam too prayer alone does not make a genuine Muslim. Here the basic text is Surah 2:177:

It is not righteousness that ye turn your faces to the East and the West; but righteous is he who believeth in God and the Last Day and the angels and the Scripture and the prophets; and giveth wealth, for love of Him, to kinsfolk and

---

31. Max Weber, *Religiöse Gemeinschaften*, p. 371; Max Weber, *Economy and Society*, Vol. 1, pp. 579-580.
32. James Davison Hunter, *Culture Wars. The Struggle to Define America. Making Sense of the Battles over the Family, Art, Education, Law and Politics*. New York: Basic Books 1991.

to orphans and the needy and the wayfarer and to those who ask, and to set slaves free; and observeth proper worship and payeth the poor-due' (zakat).

Unlike Christianity, in Islam particular classes of community members have a documented right to be supported by their fellow believers. These are the poor and needy, those who maintain the alms tax, slaves (for their redemption), those in debt, wayfarers, participants in Jihad (Surah 9:60), and proselytes[33].

In Islam justice characterizes an admittedly material asymmetric yet socially reciprocal relationship among unequals[34]. The wealthy Muslims are obligated to support the cause of justice and common good (*maslaha*) of the community, pay the official alms tax (*zakat*) and support the needy with voluntary gifts (*zadaqa*)[35]; systems of state welfare have only very marginally taken their place. Help for the needy remained predominantly a matter for the religious community. Nowadays the Islamic Mission (*dawa*) includes, even more emphatically than was the case in previous times, the setting up of social institutions [Dale F. Eickelman/ James Piscator, *Muslim Politics*. Princeton: UP 1996, pp. 35-36]. Alongside the powerful and the rich it is private Islamic organizations that are taking over welfare functions. Some of these welfare organizations have demonstratively proceeded to newly assign the boundaries between the public and private sector and to embrace public functions in their religious activities[36].

These activities were enabled by two institutions. Firstly, private persons are legally permitted to form associations in the whole of the Middle East. In the 1990s in Egypt, where this situation has been particularly well researched,

---

33. For the actual practice of 'Financial Worship' see Jonathan Benthall/Jérôme Bellion-Jourdan, *The Charitable Crescent. Politics of Aid in the Muslim World*. London: Tauris 2003, pp. 7-28.
34. See for this social model James Scott, 'Patronage or Exploitation?', in: Ernest Gellner/John Waterbury (Hg.), *Patrons and Clients in Mediterranean Societies*. London: Duckworth 1977, pp. 21-39.
35. Jacqueline S. Ismael/Tareq Y. Ismael, 'Cultural Perspectives on Social Welfare in the Emergence of Modern Arab Social Thought', in: *The Muslim Word* 85 (1995), pp. 82-106]. Even the rulers emerge as benefactors to the needy [Michael Bonner/Mine Ener/Amy Singer (Hg.), *Poverty and Charity in Middle Eastern Contexts*. New York: SUNY 2003, Part III: 'The State as Benefactor'.
36. Diane Singerman, 'The Networked World of Islamist Social Movements', in: Quintan Wiktorowicz (ed.), *Islamic Activism. A Social Movement Theory Approach*. Bloomington: Indiana UP 2004, pp. 143-163.

there were approximately 14,000 private associations (*jamaʿiyya*) registered[37]. They are voluntary, small, local, and they operate predominantly in the areas of health and education. The Muslim Brothers have contributed to numerous founding of organizations without being officially recognized as an organization themselves. These private organizations which lie between the individual and the state generate their own public sphere in modern Islamic societies of the Middle East[38].

The Islamic donation system also operates in this public sphere. It constituted a further condition for common welfare. Wealthy Muslims, including the sovereign, are expected to donate from their personal wealth to the Islamic community. By means of *waqf*, the term for this institution, income from land, rent, business or financial assets will irrevocably at some point benefit Islamic institutions and groups[39]. The donation systems in these countries contribute to the existence of their own public realms[40].

The associations of Palestinians Muslim Brothers in the Conflict with Israel

The Near-East conflict illustrates how these opportunities have shaped the course of events. In the centre are the Muslim Brothers. When the Gaza Strip

---

37. Denis J. Sullivan, *Private Voluntary Organizations in Egypt. Islamic Development, Private Initiative, and State Control*. Gainesville: University Press of Florida 1994.; idem/ Sana Abed-Kotob, *Islam in Contemporary Egypt. Civil Society vs. The State*. Boulder (Color.): Lynne Rienner 1999.
38. See for the applicability of the concept of 'public sphere' regarding Islamic societies some recent studies: Dale F. Eickelman/Jon W. Anderson (ed.) *New Media in the Muslim World. The Emerging Public Sphere*. Bloomington: Indiana UP 20032; Miriam Hoexter/Shmuel N. Eisenstadt/Nehemia Levtzion (ed.), *The Public Sphere in Muslim Societies*. Albany: Suny 2002; Armando Salvatore/Dale F. Eickelman (eds.), *Public Islam and the Common Good*. Leiden: Brill 2004.
39. Henry Cattan, 'The Law of Waqf', in: Majid Khadduri/Herbert Liebesny (eds.), *Law in the Middle East*. Bd.1. *Origin and Development of Islamic Law*. Washington: Middle East Institute 1955, pp. 203-222; for the revival in the 20th century see Bethall/ Bellion-Jourdan, *The Charitable Crescent*, pp. 29-84 ('Waqf and Islamic Finance. Two Resources for Charity').
40. One must add the restriction that the terms 'public' and 'Körperschaft' should be used with care. See: Miriam Hoexter, 'The 'Waqf' and the Public Sphere'; Jan-Peter Hartung, 'Die fromme Stiftung [*waqf*]. Eine islamische Analogie zur Körperschaft?,' in: Kippenberg/ Schuppert (Hg.), *Die verrechtlichte Religion*, pp. 287-314.

passed from Egyptian to Israeli rule in 1967, the new occupying power gave the Muslim Brothers a free hand, since it regarded them as a welcome counterweight to the nationalistic liberation movements. The Muslim Brothers on the other hand were convinced, that in view of the decline which Islam had undergone since the dissolution of the Ottoman empire, the time for armed struggle against Israeli occupation had not yet come. For years they pursued the idea of building an Islamic order from below. The driving force of the Islamization of Gaza in the 1970s was Sheikh Ahmad Yasin, who had risen to the status of dominant spiritual leader among the Muslim Brothers[41]. In 1973, he founded the Islamic Centre (*Mujamma' al-Islami*) as a bulwark against the unbelievers; Israel officially recognized the institution in 1979. The obligation to engage in Jihad, to 'strive' for the establishment of an Islamic order, consisted not only in a readiness to carry out warlike acts; of equal importance was doing one's best to promote justice and the common good (*maslaha*) of the community. Driven by a religious ethics of solidarity, Muslims created social institutions and networks. By the mid-1980s, the Centre had developed into the most powerful institution in Gaza, featuring mosques, libraries, nursery schools, businesses, schools, clinics and a university.

Only the Intifada compelled the Muslim Brothers to change their position. Since they did not want the coordination of the uprising left to secular national organisations, they founded a Supreme Command of their own called 'Hamas' (literally 'zeal'), an acronym for 'Islamic resistance movement' (*harakat al-muqawama al-Islamiyya*)[42]. Years later, in 1991, in the course of the conflict with Israel Hamas added to its social institutions another one: their own militia, the al-Qassam brigade.

The societal network which the Muslim Brothers had created, actively promoted the willingness of those who belonged to it to defend its territory against attack. At the same time, however, it was precisely this milieu that required a certain flexibility in the struggle against Israel. Unlike the Islamic

---

41. On the life and work of Sheikh Ahmad Yasin, see Ziad Abu-Amr, 'Shaykh Ahmad Yasin and the Origins of Hamas', in: R. Scott Appleby (ed.), *Spokesmen for the Despised,. Fundamentalist Leaders of the Middle East*. Chicago/London: University of Chicago Press 1997, pp. 225-256.

42. The story of the rise of Hamas has been told by a couple of recent studies. Helga Baumgarten, *Hamas. Der politische Islam in Palästina*. München: Diederichs 2006; Zaki Chehab, *Inside Hamas. The Untold Story of Militants, Martyrs and Spies*. London: Tauris 2007; Joseph Croitoru, *Hamas. Der islamische Kampf um Palästina*. München: C. H. Beck 2007.

Jihad, which practiced a martial ethic of conviction, Hamas did not view the struggle exclusively from the perspective of a martial testing of the faith. The social Islam of Hamas had achieved an intensive penetration of civil society, but this success had also given the Islamists new tasks and forced them to accept new rules. The institutionalization in civil society was accompanied by limitations on militancy. We see this attitude in the leaflets of Hamas, which never preached only violence, but always spoke of patience as well[43]. Hamas cannot imagine a genuine peace with Israel, but a truce is by now conceivable [Mishal and Sela, *The Palestinian Hamas.*, p. 86]. This option is indignantly rejected by those who bear political responsibility in Israel and the US. However, historical studies show that Islamic states have practiced an international law and that there is a tradition in their relationships with non-Islamic states of making treaties which ensured their security[44]. Even more significant is the contemporary case of the Egyptian Jama'a al-Islamiyya, that in 1997 reversed its strategy of violence and made a truce with the institutions of the Egyptian state which it had previously abhorred[45].

Conclusion

Since the seventies of the 20th century religious associations have been disseminating on a global scale. Beside the local congregations new social forms and new kinds of communication contribute to this dynamic. As an explanation for this phenomenon I would like to refer to the transformation of the state as a crucial condition. Substituting the state's public functions by market relations has effects on the structure of modern societies. Services in the fields of education, of welfare, of medical care, and even security are turned over to the private sector. As the observer of this trend, religious communities are expanding their activities into these fields, some of them claiming to do so in the public interest. Based on the religious ethics of solidarity,

---

43. Shaul Mishal/Avraham Sela, *The Palestinian Hamas. Vision, Violence, and Coexistence*. New York: Columbia UP 2000, p. 63.
44. Rüdiger Lohlker, *Islamisches Völkerrecht. Studien am Beispiel Granada*. Bremen: Kleio Humanities 2006, pp. 33-34, 41-43.
45. For documentation, see I. Fazwi and I. Lübben, *Die ägyptische Jama'a al-Islamiya und die Revision der Gewaltstrategie.*. Berlin: Deutsches Orient-Institut 2004; on this, see also Gudrun Krämer, 'Aus Erfahrung lernen? Die islamische Bewegung in Ägypten', in Clemens Six, Martin Riesebrodt, Siegfried Haas (eds.), *Religiöser Fundamentalismus. Vom Kolonialismus zur Globalisierung*. Innsbruck: Studien Verlag 2005, pp. 185-200.

religious associations are becoming the source for trust, networks and institutions. But their new social power is not always socially productive; it involves the danger of closed powerful militant brotherhoods[46].

---

46. I elaborate on this side effect in my book *Gewalt als Gottesdienst. Religionskriege im Zeitalter der Globalisierung*. München: C. H. Beck 2008.

TOWARDS A COGNITIVE HISTORIOGRAPHY – FREQUENTLY POSED OBJECTIONS

ANDERS LISDORF

Few have with a force and vigor equal to Luther H. Martin argued for the value of cognitive theories in historical research. Not only has he provided general arguments for the utility of the cognitive approach to history (Martin 2004a; 2004b), but also by his own example demonstrated how cognitive theorizing can help advance our knowledge of ancient religions (Martin 2004c; 2005). His work on ancient mystery cults in general and the Mithras cult in particular has significantly advanced the history of religion in those areas. This has left its marks in the work of a number of scholars ranging from the most established ancient historians (Beck 2006), to a number of younger scholars (Lisdorf 2005; Gragg 2004). Like ripples in the water the field is beginning to move after Professor Martin.

This essay is my humble attempt to enhance the amplitude of these ripples in the sea of historiography. A cognitive historiography has the potential not only to solve problems in history, but also to reflect back on problems in cognitive science. Strangely, though, not everybody seems to concur that a cognitive historiography is necessary, worthwhile or even possible. One possible reason for this opposition is that history and cognitive science are polar opposites: whereas cognitive science is interested in phenomena that are synchronic, isolated and general, history is interested in diachronic, composite and particular phenomena. The objections to and arguments for a cognitive historiography are many and have been dealt with in the work of Professor Martin. The purpose of this paper is merely to collect them in one place and deal with them for people who are interested in cognitive historiography. Just as most larger corporations have a section on their website with Frequently Asked Questions to deal with the most common problems that people encounter with their products, this small essay is intended as a list of

'Frequently Posed Objections' to cognitive historiography that I have encountered using this new cognitive historiography. First we will consider the typical objections made by historians then those by cognitive scientists.

1. **'We cannot access the minds of historical persons'**. That is obviously true, but that does not mean that the minds of historical people were not important to them. Their actions were guided by their mind just as the actions of contemporary people are. It does not, therefore, follow that we should not care about minds at all. Second we also have to consider what we mean by 'accessing the minds' of anyone. Do we really have more direct access to contemporary persons mind? It is true that being face to face with someone may give us more information: we can hear the intonation of their voices, see their facial expressions, observe their gesticulations and so forth, but that does not give us any more direct access to their minds. I think the objection is hinting at the historian not having access to the same amount of information. Phrased like this, it is perhaps easier to see that the objection is hollow: just because we don't have a lot of information it does not mean that we cannot say anything. So, the response would be 'no, but we cannot access the minds of contemporary persons either'. Thirdly, the point of cognitive historiography is not to get into the minds of historical persons. That project may very well be more prominent in the hermeneutical big-man type of historiography that spews out endless amounts of biographies of Hitler, Napoleon, Caesar etc. Perhaps the objection is really more apposite in connection with mainstream blockbuster historiography.

2. **'Knowledge of modern minds will not help us in understanding historical minds'**. This objection assumes that there is a fundamental difference between the minds of modern and historical people. It is, however, almost unanimously agreed that no significant mutations have occurred in the human brain since around 100.000 years ago (Mithen 1996; Tremlin 2006: 24). If there is any difference between the minds of historical and modern people it cannot be attributed to any biological

difference in the brain. This leads to two other options: a) the mind is not constrained by the brain. This is a dualist assumption implying a fundamental difference between spirit and matter. This removes the discussion to a meta-theoretical plane, where we should be discussing whether historiography should be based on a dualist or monist ontology. In this context we can merely state that cognitive historiography is based on monist ontology. All modern science is. It may of course also be the case that you don't believe in science or the productiveness of scientific method or that history is not accessible to scientific inquiry (more on that below). That is not the view taken by cognitive historiography. b) A weaker version of this admits that the brain to some degree in principle is important. In practice, though, it is the local culture and socialization which determines the mind of a person. This is a version of the so called nurture view in the nature/ nurture debate of the human mind. This view assumes that the mind is a blank slate just waiting to be filled out with the contents of the nurturing culture. Compelling arguments against this assumption have been made by cognitive scientists and anthropologists alike (cf. Brown 1991; Pinker 2002). This is not to say that culture is not important, just that the human mind of modern and historical people are sufficiently similar in general cognitive function to warrant a meaningful comparison. To put it differently: historical minds are not more different from modern minds than those of other cultures in our contemporary world. A counter argument would also be: if the psychic chasm is so great that the minds of modern people are not in principle as those of historical people, then how is it possible at all to say something about historical people's actions, motivations or thoughts? There has to be a basic similarity between historical and modern minds in order for us to understand them at all.

3. **'We cannot understand historical particulars by the aid of the universal'**. This is historiography's version of another classic debate in scientific theory: the one between idiographic and nomothetic sciences. We may actually turn the basic premise on its head and state that we cannot understand anything

particular without the general. Just as Wittgenstein showed there can be no private language (Wittgenstein & Anscombe 2003), there cannot be any 'private history' consisting only of particulars. Just using language, and historians tend to use language, sometimes even a lot of it, is making use of the general (Jensen 2003: 141). In this sense, the general is a precondition for understanding the historical particulars. Maybe some would feel that there is a difference between the general and the universal. Let us therefore look at an example of how the universal aids our understanding of the particular. Imagine a geologist who wishes to understand the magnitude and character of a volcanic eruption thousands of years ago. In order to understand this particular historical eruption, he will make use of the universal laws of chemistry and physics. From soil samples in different regions he will measure the chemical contents to find out where the soil has been contaminated with the contents of the eruption. Once he has established a pattern of distribution he may proceed to calculate by using the laws of physics to find out how much force it would have taken to produce this distribution. By using the universal laws of chemistry and physics the geologist has solved a historical problem and found out about the magnitude and character of a particular volcanic eruption. There is consequently no reason to assume *prima facie* that universal characteristics of the human mind will not help us understand historical particulars. Rather we should expect the opposite, as the example of the volcanic eruption shows, that the knowledge of universal characteristics (I will refrain from using the word law until I have discussed it further below) of the human mind.

4. '**Theories will not help in history. It merely distorts the sources**'. This is a typical objection in ancient history (Morley 2004: 1). It is, however, built on a number of controversial assumptions, first being that the sources may speak for themselves. This is not a tenable assumption. Historical sources are always actively used and interpreted: historiography is not the same as publishing all available evidence. Any historical exposition is based on a theory. What may differ, however, is

only whether it is explicitly stated or not. In historiographical research, theories tend to be implicit and the historian bases his analysis on his own assumptions about human motivations, social forces etc. (cf. Morley 2004). These are not shaped by explicit, tested theories, but by folk notions and the historians own subjective experience of the world. That is perhaps the reason why every generation of historians has to write its own history. In this light, what actually distorts the material is a lack of explicit theorizing.

5. '**There is an essential difference between history and science. Science works by reducing phenomena to deterministic laws. This cannot account for history since human behavior cannot be reduced to any such law**'. This objection is founded on two misunderstandings of the nature of science. a) It seems that the historian fears that scientific reduction will render the historical discipline obsolete, because a lower level of explanation will replace the higher. There are however no examples of such a usurpation of a higher level science by a lower level one (McCauley 1986: 197). b) The conceptualization of law as a deterministic statement on reality is a dated remnant of a 19th century logical positivist conception of law which apparently thrives in classical history (Morley 2004: 15). This kind of law applies in every case and can be falsified if a situation arises in which it does not accurately predict an occurrence. It is correct that we do not find this kind of law in history, but that is not sufficient to disqualify it as a science. Contemporary views in scientific theory prefer to speak of invariance instead of laws (e.g. Woodward 2000: 205-209). Invariance is a question of degree and may apply locally not necessarily universally. Counterexamples are therefore not falsifications that render the search for regularities futile. There is thus no reason to assume a basic divide between history and science. So far so good, these were the most common objections from the historians. However, we still have to deal with a number of possible objections to a cognitive approach to history from cognitive scientists:

6. '**The results do not live up to scientific standards of verification. The evidence is too sparse to prove anything**'. To this, one could point out that not all psychological research lives up to these standards either. Evolutionary psychology, the branch of psychology investigating the evolutionary origins of human cognition, typically does not adduce any evidence of the historical processes stipulated. As for the history, it is pure (informed) guesses without any direct evidence. If one would want to dismiss a cognitive approach to history (which is actually based on evidence), on the objection that the evidence is not good enough, one would also have to dismiss evolutionary psychology altogether. One could also point to other scientific areas working on even more tenuous evidential basis, such as human palaeontology or a recent favourite of mine: astrobiology. The objection that there is not enough evidence is therefore not a valid criticism that precludes the investigation from being scientific.

7. '**Since there is no possibility for controlled experiment, there is no possibility for falsification, and therefore it does not qualify as science**'. While the falsification criterion is an important lever for scientific quality, few today would agree that it is the only necessary and sufficient criterion for science (see above). Once again let us consider what other sciences we would have to exclude from the academy if such a criterion was stringently applied. We would have to deem unscientific all of climatology apart from the past few decades; all of geology apart from contemporary research; all of astronomy; all palaeontology; basically all sciences with a historical objective would be unscientific. The possibility of controlled experimentation is consequently not a defining characteristic of science. Other meta-theoretic criteria should be given equal priority, such as consilience and parsimony.

8. **'When you cannot do controlled experiments, you cannot isolate variables and therefore it is not scientific'**. This may be a slight variation of the former. It is a mistake to believe that the fact that you cannot undertake controlled experiments means that you cannot isolate variables. Variables may be measured as well as in controlled experiments. Most of economics, sociology and parts of medicine operate under the same conditions. It is true that historical research is constrained by what evidence there is. It is not possible to produce new evidence and the historian sometimes has to live with the fact that some things will never be known for sure. What is of importance for scientific explanation, however, is when changes in one variable will invariably result in changes in another variable (Woodward 2000). In historical investigations you may not be able to see all possible combinations of manipulations of variables, but it does not follow that the results are therefore unscientific.

I do not presume to have convinced either historians or cognitive scientists that a cognitive historiography is necessary, but I do hope to have succeeded in providing arguments that would at least warrant the attempt feasible to both sides. It should also be pointed out that both camps may actually learn something from the integration of cognitive science and history. Historians may gain a better model of the human mind than the implicit common sense psychology typically employed by the historian. This can help the historian to make better and more accurate hypotheses about the historical reality he tries to understand. The cognitive scientist, on the other hand, may learn that factors that are important in the lab are not always very important in the real world, while others may turn out to be so.

HOW SCIENCE AND RELIGION ARE MORE LIKE THEOLOGY AND COMMONSENSE EXPLANATIONS THAN THEY ARE LIKE EACH OTHER: A COGNITIVE ACCOUNT

ROBERT N. McCAULEY

No one has explored the implications of cognitive theories and findings about religion for understanding its history with any more enthusiasm or insight than Luther Martin. Although my focus here is not historical, I assume that I will be employing cognitive tools in ways that he finds congenial. In the paper's first section, I will make some general comments about standard comparisons of science and religion and criticize one strategy for making peace between them. In the second section of the paper, I will delineate two cognitive criteria for comparing science, religion, theology, and commonsense explanations. Finally, in the third section, I will suggest that such a comparison supplies grounds for thinking that our longstanding interest in the comparison of science and religion is, oddly, somewhat misbegotten from a cognitive perspective.

1. Some Comments on Traditional Comparisons of Science and Religion

Standard comparisons of science and religion have not generally waltzed to cognitive tunes. Traditionally, most scholars (whether philosophers, scientists, or theologians) have focused on science and religion's comparative epistemological and metaphysical merits. Their focus has been either on how each activity does or does not contribute to our knowledge or on what each discloses about reality. Two trends have emerged. Generally, the champions of science have tended to headline its epistemological merits. They tout the fact that science stands unmatched in its ability to increase and improve our knowledge. By contrast, defenders of the faiths, in the face of what they see as the metaphysical severity of science, usually commend religions' metaphysical liberality. Those defenders concur that assumptions about invisible sources of agency, both in us and in other kinds of beings, help to make sense of human experience, to undergird what they see as proper moral and social arrangements, and to frame the most daunting questions humans face, concerning their own mortality, in particular.

An examination of these enterprises' cognitive foundations not only provides new views of science and religion, it also explains these trends. From the standpoint of popular conceptions of the world, science can appear metaphysically hobbled. Hawking *radically* counter-intuitive representations (that the earth is spinning at one thousand miles per hour, that solid objects are mostly empty space, that all current organisms are descendants of other organisms who would not have qualified as members of their species, etc.) whose appreciation requires painstaking cognitive processing that takes years, if not decades, to master, science carries some substantial liabilities into the marketplace of ideas, let alone into everyday marketplaces. Cognitively awkward representations that are often inconsistent with the representations of things that human minds most readily deploy are never a quick or easy sell. In particular, science's abandonment of agent causality across its history in a progressively wider set of domains inevitably leaves human minds, with regard to at least some of those domains, floundering and incredulous. Over the past fifty years the sciences of the mind/brain have even begun to constrain appeals to invisible sources of agency *within us*. It is the undoing of agency in the biological realm that has been the principal political flashpoint in contemporary American public life and that is at the crux of those battles over Darwinian evolution.

In the short run, science, just like anything else, most effectively grabs human attention when it seems wondrous. For the first fifty years of television in America, the best known purveyor of scientific insights was, not coincidentally, called Mr. *Wizard*. (Alas, American television has had no comparably sustained or well known purveyors of science since.) For most of the public, science's only major selling points are connected with those occasions when its effectiveness at explanation, prediction, or control are timely or when related technologies either thrill or fascinate. When the work of scientists develops effective vaccines for deadly diseases or successfully transplants organs or predicts celestial events or explains the mechanisms of inheritance or inspires the latest advance in computing, the public is less inclined to challenge science's epistemic authority, even if people find it's shifting verdicts and its underlying metaphysical commitments utterly perplexing.

By contrast, the recurrent ontological commitments of religions are far easier to swallow cognitively. They square almost perfectly with the deliverances of humans' maturationally natural cognitive systems, and they capitalize, especially, on the penchant of human minds to presume that noteworthy events are the results of the actions of mindful agents. Proliferating agents poses no special cognitive problems for human minds in standard operating mode. That mode relies on perception, cognition, and dispositions to act that are automatic and unreflective. Across a vast range of physical, cultural, and historical circumstances, human beings routinely develop intuitions about a variety of domains. On the basis of a paucity of cues in those domains, humans can, in an instant, draw elaborate inferences and act effectively. From such things as their command of the basic physics of solid objects, to such things as the recognition of agents, the comprehension and production of complex utterances, the knowledge of how to deal with environmental contaminants, and the discernment of emotional and intentional states on the basis of facial expressions, bodily postures, and tones of voice, humans at the onset of middle childhood have developed skills of perception, cognition, and action that enable them to manage a host of mechanical, biological, and social problems. Those intuitions and dispositions rarely result from any explicit instruction, yet most of them are normally in place by the time children reach the age of seven. These maturationally natural capacities concern matters and result in actions that are so fundamental to human life that their appearance in development helps to define what counts as 'normal'. Such capacities count as maturationally natural on the basis of their spontaneity, their ubiquity, their early onset (for the most part), and their independence both from explicit instruction and from other forms of culturally distinctive support.

Pascal Boyer (1994 and 2001) has argued that religious representations violate humans' maturationally natural presumptions only modestly. These modestly counter-intuitive representations that dominate popular religion are *easy to use*. (Tweney et al. 2006) By nearly always presuming, in any particular context, but one or, very occasionnally, two violations of intuitive knowledge, the representations of popular religions permit participants to utilize a huge range of *default inferences* that accompany our maturationally natural ontological knowledge. Consequently, these modestly counter-intuitive representations possess an abundant inferential potential. Knowing that a something is an *artifact* allows us to infer that it has a determinate size, shape, and weight that human beings have had some influence on its current state, but also that it does not indulge in respiration, contemplation, or copulation. On the other hand, knowing that something is an *agent* allows us to infer that it has goals, desires, and preferences, that it finds some attitudes and behaviors offensive, and that it is disinclined to help anyone who manifests such. That some agent has biologically counter-intuitive origins (a breach of folk biology) does not block our ability to draw all of the standard inferences about that agent's mental states, aims, interests, values, and likely behaviors that we can draw about any other agent (Tremlin 2006, pp. 112-113).

Boyer holds that representations that conflict so modestly with humans' ontological intuitions, while simultaneously drawing on all of their associated default inferences, approximate cognitively optimal arrangements from the perspective of making sales within the marketplace of culture (Boyer and Ramble 2001). Such representations approach the best available balance among the multiple ends of simultaneously attracting human attention, enhancing human memory, and increasing inferential potential. That is another way of saying that standard religious wares sell comparatively easily. It is also a way of saying that religious representations probably never completely lose their natural attractiveness, regardless of intellectual training. The most valuable evidence here is not the steadfast denials of the non-religious about their conscious mental lives but, rather, indirect tests that tap cognitive influence and activity that operate below the level of consciousness. Unshakeable, subterranean forces are the more interesting marks of some representation's natural cognitive allure.

The downside, though, is that ease of swallowing from a cognitive standpoint does not guarantee ease of digestion from an intellectual standpoint. Enduring texts afford systematic assessments of the truth of their claims. Because religious representations typically wear their violations of ontological intuitions on their sleeves, many of the logical problems they engender are transparent in literate contexts where methodical reflection is prized. Such conditions spawn *theological* reflection and proposals, which can end up appearing nearly as convoluted as the most puzzling claims of science. Because theological and scientific claims part so substantially from our maturationally natural knowledge, people often find them baffling. Generations of Calvinists have been bewildered by Calvin's notion of predestination (Slone 2004, chapter 5). Once the claims of popular religion undergo inspection in a literate culture, though, the conundrums they generate can become uncomfortably clear to thoughtful participants and, often, laughable to outsiders (I have yet to meet a scholar of religion or a religious person who has not admitted to finding some belief, practice, or artifact of *someone else's religion* nothing short of hilarious). The claims of popular religion, especially those in behalf of religious experience, cannot easily bear the unencumbered scrutiny of a literate public and the rigorous application of methods employed to study other areas of human conduct (Dennett 2006; Silk 2006). In these precincts the religious and, all too often, even scholars of religion break into special pleading, which is not a script for creating durable epistemological credentials[1].

Although the link is hardly deductive, these two trends among conventional comparisons of science and religion spur on a popular strategy for parceling the pertinent intellectual territory out between them. The best known advocate of that strategy recently has been Stephen Jay Gould in his book *Rocks of Ages*. There Gould assigns science and religion to two different 'magisteria'. He asserts that 'the ... magisterium, of science covers the empirical realm' ... while 'the magisterium of religion extends over questions of ultimate meaning and moral value'. This strategy for dividing up the turf is popular, because it promises intellectual peace. Gould stresses that '*these two magisteria do not overlap....*' (Gould 1999, p. 6, emphasis added). No overlap eliminates any possibilities for conflict. In this two state strategy, each activity, according to Gould, rules in its own realm.

This strategy for achieving peace faces problems, though, on at least two counts. First, it is not obvious that these ventures are the sole authorities in the respective magisteria Gould assigns them. For example, what specific religions have to say about meaning and morality always ends up turning, sooner or later, on their particular contents, commitments, and practices. The problem, if these religious systems' recommendations are to be persuasive to anyone other than their current subscribers, is that these distinctive features of religious traditions carry little, if any, authority precisely where they need to here, viz., *beyond* the confines of that particular religious system's followers.

---

1. See Lawson and McCauley (1990, chapter 1) for illustrations of the latter. Drees (1996) constitutes a welcome corrective to such special pleading.

These contents, commitments, and practices must retain their credibility in a diverse world, if they are to prove any basis for either general, morally obligatory prescriptions or what people, across cultures, take to be meaningful arrangements (Remember all of that laughing about other people's religions that I mentioned above). Arguably, a *particular* religion is exactly what any grounds for binding moral authority *cannot* depend upon, if rational and psychological purchase *across* religious systems and cultures is the aim. This is just one of those areas where it is difficult to underestimate the influence that culture exerts on *conviction*, even if we are inclined to overestimate its influence on *contents* (Hinde 1999, chapters 12-14 and Boyer 2001, chapter 5).

On the other hand, although science is second to none in the empirical realm that is not the same thing as claiming that it is the exclusive authority on empirical matters. Science is young, it operates with limited resources, it is difficult to learn, our lives are short, and the world is huge and complex. We have only just *begun* to question the world scientifically. Moreover, science is a never-ending process. As we do better science, we learn that much more about what we do not know and, as noted above, some of the conclusions invariably change as science progresses. Over the last few decades larger numbers of people have had sufficient time and material support to learn some science, and, occasionally, the particularly diligent get the opportunity of consulting informed, up-to-the-moment scientific judgment, but we should not be embarrassed about the fact that *most of the time* we are stuck with relying on little more than our maturationally natural intuition in our dealings with the world. It is the inevitable consequence, in the face of the practical necessity of getting about from day to day, of the immense variety of the problems that we face, of our limited resources, of the fallibility of our inquiries, and of the substantial intellectual challenges attached to comprehending the sciences.

The second reason why purchasing peace between science and religion on the basis of claims about their non-overlapping magisteria may prove too dear is that it involves some normative sleight of hand. I will only mention two related examples. First, one of the easiest ways of minimizing the tensions between science and religion is simply to deny that the religious people, who remain especially exercised about the apparent conflicts, deserve to be designated as 'religious' in the first place. Make no mistake about it, such dismissive legislation lurks behind all gentle and, apparently, conciliatory talk of 'true' religion among the faithful, among the theologians, and among many academics. This includes, for example, claims by members of each of those groups that the terrorists who attacked New York, Madrid, and London were not *true* representatives of religion or, more specifically, of Islam (Sullivan 2001). But the pressing questions are (1) who gets to say whose religiosity is or is not *true* or whose version of Islam (or any other religion) is the right one? and (2) on what rationally convincing basis do they get to say it? Or consider Gould's (1999, p. 148) declaration that 'creationists do not represent the magisterium of religion'. Gould proceeds as if the religious, let alone the logical, sensibilities of literally hundreds of millions of people should not count when sorting these matters out.

Gould and his allies here invent prejudicial norms where norms of the sort they desire, i.e., non-prejudicial ones, are not to be had. The second trend in conventional comparisons of science and religion, which accentuates religions' metaphysical liberalities while downplaying or even ignoring their epistemological liabilities, amounts to a tacit recognition of that fact. This asymmetry between religion and science is not coincidental. I shall argue in the next section that, as reflective activities, science and theology have different relations to the maturationally natural moorings from which they are born. Theology, like Lot's wife, cannot avoid the persistent temptation to look back – in the case of theology to look back to popular religious forms. By contrast, the radically counter-intuitive commitments at which the sciences inevitably seem to arrive commonly produce unbridgeable gaps with the intuitive assumptions underlying commonsense explanations. The sciences fairly quickly get to a point where they can no longer look back to our maturationally natural predilections, even if scientists wanted to. Theology is largely devoted to making sense *of* and bringing some logical order *to* the claims of popular religion. Science, by contrast, follows wherever its inquiries lead and across *all* of the sciences, *that* has reliably been *away* from the automatic deliverances of our maturationally natural mental systems that inform our commonsense understandings of the world.

As the *Hebrew Bible* amply documents, peoples have routinely construed their own conflicts as conflicts between their gods. The invention of literacy not only made proselytizing religions possible, it also created the possibility for *reflection* on conflicts about religions' comparative intellectual and moral merits. It is not from any lack of effort that advocates for any particular religious view have yet to come up with anything remotely close to the sort of case for their preferred versions of religiosity that comparatively disinterested observers from around the world would collectively find at all persuasive. This contrasts with the way that overwhelming majorities of the world's professional scientists *do* find the resolutions of so many of the controversies in their fields of study convincing, at least for the time being. Scientists regularly arrive at such views on the basis of relevant evidence and without epistemologically troublesome coercion. That, of course, is not to say that they *always* do so without epistemologically troublesome coercion or to say that they *ever* do so completely independently of extra-scientific social influences. The difference here between science and theology is not trivial, but, on the other hand, it should not be overplayed. That is because the sciences' verdicts, even their most fundamental ones, are constantly eligible for reconsideration and because, as noted, evidence sometimes emerges that the influence of scientifically arbitrary forces are *not* negligible.

These considerations lead to a second, related illustration of how designating non-overlapping magisteria for religion and science carries problematic normative consequences. Gould (1999, p. 211) urges both science and religion 'to stay on their own turf'. On his account science is concerned with empirical explanation while religion's magisterium covers morals and meaning. Religions certainly do try to make sense of our lives and of the world in which we find ourselves. The problem, though, is that that process of making sense of things inevitably involves appeals to explanations about the origins, the make-up, and the behavior of things generally and about *our* origins, make-up, and behavior in particular. Religious meaning making, indeed *all* meaning making, *always* makes explanatory assumptions. Some of those assumptions, such as those creationists proffer, are explicit. Many more, connected with such maturationally natural cognitive systems as theory of mind, are usually implicit (Lanman 2007). But in either case making meanings depends on, among other things, explanatory accounts of how things hang together, of how events are connected, of how the world works, and of how we operate. Whether advocates of exclusive magisteria like it or not, all religions explicitly traffic in explanations some of the time, and all religious meaning making makes explanatory presumptions all of the time (Lawson and McCauley 1990, chapter 1). Much of the time those explanations are superfluous from the standpoint of scientific accounts, if they are not downright inconsistent with the claims of science. The attempt to buy peace by designating exclusive magisteria requires either (a) ignoring the place of explanations, whether religious or scientific, in the processes of finding or assembling meaning or (b) ignoring the logical tensions between the explanations that science and religions favor or (c) ignoring both. Gould's conception of the relation of science and religion is not exactly peace at any price, but it does seem, in light of these normative problems, to be peace at too high an intellectual price.

2. Two Criteria for a Cognitive Comparison

Comparisons of science and religion have been so numerous over the last century as to constitute a cottage industry. The tensions surrounding the relations between science and religion concerning metaphysical and epistemological matters that I sketched in the previous section have been clear even to the casual observer. Contributors have been anxious either to dissolve those tensions or to emphasize them as grounds for extolling one or (like Gould) for extolling both of these enterprises. Such epistemological and metaphysical preoccupations are perfectly legitimate concerns and perfectly understandable philosophically. However legitimate and however understandable they are, though, while they clarify some things, these preoccupations blur others. In a recent volume (Harper 2005) surveying perspectives on science and religion that covered everything from quantum mechanics to the contemplation of the virtues, cognitive approaches received no attention. That oversight is unfortunate, since the exploration of the cognitive foundations of science and religion suggests that these traditional comparisons of science and religion are, from a cognitive standpoint, misbegotten in two related respects.

Of course, anything can be compared with anything. Still, science and popular religion diverge on two kindred cognitive criteria that expose reasons for thinking that their conventional comparisons are less revealing than is typically presumed. Those criteria permit science and religion, along with theology and commonsense explanations of the world, to be distinctively situated in a two by two table. (See figure 1) The first criterion, represented vertically at the left of figure 1, is a distinction between the relative prominence of two types of cognitive processing in any of these activities.

	appeals to agent explanation/causality	
preferred type of cognitive processing	**unrestricted**	**restricted**
reflective	1 *theology*	2 *science*
maturationally natural	3 *popular religion*	4 commonsense explanation and understandings of the non social world

figure 1. Cognitive asymmetries

Reflective processing is conscious, deliberate, and comparatively slow. By contrast, the operations of maturationally natural cognitive systems are typically unconscious, intuitive, and fast. Cognitive undertakings that tilt toward reflective, off-line, cognitive processing and away from maturationally natural cognition are across the top (represented by cells 1 and 2), whereas those that rely more prominently on maturationally natural, on-line, cognitive processing that tends to preempt conscious, off-line reflection are in the bottom row (represented by cells 3 and 4). Reflective, off-line cognition is the most plausible candidate available for thought that is under conscious control. Literacy has played a pivotal role in its enrichment, since the external representation of such thought in publically available texts permits conscious minds to produce and contemplate the elaborate ideas and extended arguments that the most sophisticated forms of reflection involve.

Maturational naturalness is not the only form of cognitive naturalness. Perception, cognition, and action can become intuitive and automatic in domains in which people have invested considerable effort over time to master something. Given that literacy is not much more than five thousand years old, there is no reason to think that human brains have evolved to learn how to read. On the face of it, the widespread incidence of reading disabilities like dyslexia is further evidence that this is so. Human brains were not built by nature to learn to read and write. Those are cultural accomplishments. Thus, most humans must acquire those skills laboriously. The only naturalness that can ever accrue to these forms of cognition is *practiced* naturalness. Their practiced naturalness is best illustrated by how automatic so much of reading becomes for experienced readers. That it is *practiced* naturalness (as opposed to maturational naturalness) is best illustrated by how effortful reading is for inexperienced readers, regardless of their age. How much time humans devote to explicit teaching and structured learning of literate skills is a further indication that any naturalness arising here is a function of extensive practice. Prolonged exercise at reflective activity in some field can yield a practiced naturalness on various intellectual fronts. With considerable experience, experts obtain developed intuitions about their areas of expertise. Research on lapses in deductive and probabilistic reasoning and in the application of scientific theories and concepts indicate that such practiced naturalness in intellectual matters is both hard won and, often, surprisingly inflexible. Small shifts in an otherwise familiar scenario can cause even experts' performance to crash sometimes (Piatelli-Palmerini 1994; Gilovich et al. 2002).

The second criterion represented horizontally at the top of figure 1 concerns the explanatory prominence accorded *agent causation*, in particular. This distinction arises from my earlier observation about the increasing restrictions that, over its history, science has imposed on the legitimacy of appeals to agent causality. Over the past four centuries science has progressively curbed the use of such explanations – in the physical sciences first, then in the biological sciences, and now increasingly so in the psychological and socio-cultural sciences. Scientific abstemiousness concerning intentional agents and their putative actions is to be contrasted with religions' pervasive recruitment of theory of mind and appeals to agent explanations.

Theory of mind concerns humans' intuitive knowledge about others' minds and what goes on in them and the enriched social world that that knowledge sustains. Humans' ability to draw inferences about others' mental states explains the scope, diversity, and complexity of human social arrangements and plays a pivotal role not only in individual (Dunbar 1996, p. 87) and collective survival but in individual and collective accomplishment.

Acquiring theory of mind involves a series of attainments (Leslie 1994), and it apparently takes some years just to get the maturational basics down. Infants are keenly aware of the fact that people are numbered among a restricted set of things in the world that are prominent because they are *animate*. People are the most conspicuous members of the subset of animate things that qualify as *agents*. Agents not only move about in irregular ways; their movements constitute *actions* with specific *goals*. Philippe Rochat and his colleagues (1997) have provided evidence that infants are sensitive to goal-directed actions at three months of age. No theorist questions that such a capacity is adaptive. Detecting agents, their goals, and their actions is a prerequisite for managing complex social relations in human communities, but it pertains to far more basic matters as well, such as detecting predators and prey.

By the time they are six or seven years old, children not only come to adopt what Dan Dennett (1987; 2006, pp. 109-111) calls 'the intentional stance' toward other agents, they come to regard them as fully qualified 'intentional systems', i.e., they regard them as if they possess not only goals but mental lives and mental representations of their own (Tomasello 1999, pp. 53 and 174). Once they reach early school age humans know about a world filled with other humans and have already acquired the *basic* skills and knowledge necessary for handling the problems such a world presents. Gaining social experience and ingesting the voluminous narrative materials (stories, myths, dramas, novels, etc.) that saturate cultural spaces provide humans ample bases for elaborating, extending, and embellishing their theory of mind (Frith 2005, p. 48).

The speed, facility, and sophistication with which human beings deploy the intentional stance to make sense of their social world contrasts starkly, though, with their liberality and frequent lack of insight about what qualifies as an intentional system. Deborah Kelemen (1999a; 1999b) has documented pre-school age children's 'promiscuous teleology'. This refers to children's penchant for over-attributing functions to things as a result of their new ability and growing experience with purposeful agents pursuing goal-directed actions. Unlike most adults, most children this age are willing to attribute functions to biological wholes (e.g., tigers) and to parts of natural objects (e.g., a mountain protuberance) as well as to the natural objects themselves (e.g., icebergs).

Adults as well as children are remarkably profligate in their ascriptions of agency, yet any individual who fails to take the intentional stance toward effectively structured systems of much complexity will be at a distinct disadvantage when it comes to predicting their behavior. One of the benefits of employing Dennett's technical terminology to discuss these matters is that it readily accommodates the fact that humans are so often indiscriminate in their attributions of intentionality. If not, upon reflection, in their assignments of minds to things in the world, then, at least, often in their *treatment* of many things, humans proceed not only as if inanimate things are agents but as if they are agents who understand what we say. This proclivity of the human mind manifests itself in everything from children's play to adults talking to, coaxing, even begging for cooperation from machines like cars and computers. The point is not so much that humans, even children, take the intentional stance toward inanimate things (though *that* is certainly noteworthy too) as much as it is that we so often feel compelled to do so (Mithen 1996, p. 55) and that we so often derive some comfort from doing so. In some ways, the adults' behaviors are more revealing than the children's. Children generally know when they are pretending, however steadfastly they may keep up the pretense for a time. Their own on-line, i.e., unreflective, episodes of taking the intentional stance toward inanimate things, though, regularly seem *unremarkable* to adults.

Evolutionary psychologists have a ready account for these extravagances (Atran 2002; by contrast, see Harris 1994, p. 308). So long as the costs of false-positive signals are not too high, it pays to have an agent detection system that is easily cued. In a hostile, competitive world that is red in tooth and claw, the costs of false-negative signals are *prohibitively* high. All else being equal, the creature that is inattentive to the movement in the periphery, the shadow passing overhead, or the rustling in the leaves (let alone the sound in the basement) is less prepared to protect itself from predators, competitors, and foes. A mechanism with a low activation threshold for spotting agents may leave a critter a little jumpy, but, again, so long as the costs are not exorbitant, a hyper-sensitive agent detection device (HADD) is also more likely to leave it alive to be cautious another day (Barrett 2000 and 2004; Baron-Cohen 1995, p. 35; Buss 1999, p. 88).

Supplementing this basic equipment with a rich theory of mind equips an individual to manage in a complex social universe, where, among other things, people make alliances, have conflicts, cooperate, compete, joke, threaten, ameliorate, inform, trust, and deceive. Among social animals, human beings are unmatched in their appreciation of an entire social world fashioned by individual agents' actions. A HADD disposes them to look for agents and, thus, to deploy the categories of agent causality when things go bump in the night (an intruder?) or when an unexpected event occurs amidst complex social arrangements (a conspiracy?). This maturationally natural proclivity steers human minds away from inventing or investigating other causal conceptions (cf. Tomasello 1999, pp. 22-25) concerning *things* going bump (at any time of day) and, especially, concerning *human* affairs, where the detection of intentional agents is as unproblematic as it can *possibly* be.

The next section examines the implications for science and religion and for theology and commonsense explanations of the physical and biological world of people having minds that naturally mature in the ways that human minds do. Their maturationally natural systems equip human minds to readily generate, retain, deploy, and transmit religious representations. By contrast, the prominence of those maturationally natural systems is, usually sooner but always later, mostly an obstacle to the invention and the investigation of alternative causal conceptions. Broadly speaking, this is why science is so hard to learn and why it is so hard to do.

3. Traditional Comparisons of Science and Religion Are Cognitively Misbegotten

Although discrete cells seem to imply differences in kind, both criteria that define the table in fig. 1 only gauge differences in degree. The table captures the *comparative* priority each venture places on these cognitive variables. The resulting array situates religion and science relative to theological reflection and commonsense understandings of the (non-social) world and illustrates two telling asymmetries.

Concerning cell 2: science is a reflective activity involving forms of thought and types of representation that depart radically from the pronouncements of our maturationally natural cognitive systems. Consequently, they also substantially constrain reliance on agent causality for the purposes of explanation, prediction, or control. The progress of science has gradually but steadily whittled down the range of areas in which the most accurate and comprehensive explanations for phenomena involve taking the intentional stance. The prohibition of agent causality from physical and biological science has, in effect, become a tacit methodological maxim. (McCauley 1988) Still, this contrast should not be exaggerated. The success of mechanistic modeling in the cognitive sciences notwithstanding, the psychological and socio-cultural sciences continue to call upon agents, their mental states, and their resulting actions in many of their explanatory theories. In fields such as social psychology, classical economics, and cultural anthropology, theories about intentional agents, their preferences, and their actions remain the standard mode of analysis and explanation. Thus, even in science the use of agent causality is unlikely to wither away completely, at least for the foreseeable future.

Concerning cell 4: not all of the verdicts of maturationally natural cognitive systems involve summoning agent causation or theory of mind. In fact, most do not (Frith (2003, e.g., p. 109) argues that even many social accomplishments may not rely on theory of mind). By school age, human beings seem to possess all sorts of detailed dispositions about matters as various as the basic physics of solid objects, grammatical form, fair distributions of resources, and the avoidance of contaminants. What makes many of our commonsense understandings and explanations common is precisely that they arise, in part, from maturationally natural dispositions of mind that human beings share. Certainly, humans are not incapable of reflection about such matters, though it rarely occurs to them to undertake such musings. But in many situations, especially those that call for quick judgment or fast action, these intuitive systems and the accompanying emotions they often involve kick into gear before opportunities for conscious deliberation even arise. For example, when people feel cheated, it dominates their awareness and drives their actions. Sometimes such dramatic circumstances cue these cognitive systems' automatic operations, but far more mundane matters can trigger dispositions that also have nothing to do with theory of mind either. Michael McCloskey (1983) showed that large numbers of naive subjects attempted actions that were aimed at producing physically impossible motions in order to carry out a task that did not require such a wondrous accomplishment. So, for example, when asked to roll a ball in such a way that its path crossed both the entrance and the exit of a curved passage drawn on a flat surface, many of McCloskey's subjects tried to do so by attempting to impart a curving motion to the ball that would follow the arc of the curved passage.

Concerning cell 1: *nothing* I have said rules out off-line, reflective activity in domains that have no inherent restrictions on appeals to the intentional, like the restrictions that now reign in the physical and biological sciences. By no means is theology the only kind of intellectual project that falls within this cell. It also contains traditional moral philosophy and somewhat more rarefied areas of contemporary philosophy such as action theory. With respect to matters religious, though, such reflection is principally the occupation of theologians. In the literate cultures where they arise, theologians regularly carry out the same forms of inference (deductive, chiefly, but probabilistic too) that philosophers and scientists do, and they brandish representations that can sometimes be as counter-intuitive as those that scientists use. Boyer diagnoses the underlying cognitive bases for how and why the violations of intuitive ontology that dominate the representations of popular religion turn out to be quite limited. By contrast, theologians have, by now, been generating radically counter-intuitive representations for millennia. Attributing esoteric abstract properties such as omniscience, omnipotence, and omnipresence to some gods are the sorts of examples that leap to mind, however, the conceptual recalibrations required, for example, of Christians to accommodate what are far more fundamental notions, historically, are plenty challenging enough. Understanding God as a triune entity (each person of which is alleged to have had temporary, divergent physical manifestations) presents all of the conceptual adjustments that the modern psychological account of multiple personality disorder demands and a good deal more.

Concerning cell 3: religion enlists humans' maturationally natural cognition and it engages theory of mind especially. Thus, it falls in cell 3. Folklore, fairy tales, and fantasy literature fall into this cell as well, but religion is the interesting case for present purposes. Popular religious forms, including icons, sacred spaces, rituals, priestly status, glossolalia, CI-agents with full access to people's thoughts, and more, variously activate mental systems that develop early on in human minds. Those mental capacities do not operate as they do in order to manage religious inputs, but, instead, arise in human cognitive development to handle problems of perception, cognition, and action that are far more basic to human survival. Particularly central to making our way in religious worlds are the automatic inferences and intuitive calculations about agents, their intentional states, and their actions that also happen to be particularly central to making our way in the everyday social world as well. These mental tools that humans routinely use are what make religious materials captivating for human minds. They are also what, by school age, equip human beings to grasp religious forms and enable them to acquire religion.

My aim here is not to restate my entire case for the cognitive naturalness of religion and the cognitive unnaturalness of science (McCauley 2000 and [in progress]), but rather to underscore how this analysis suggests that traditional comparisons of science and religion on epistemological and metaphysical grounds disclose little about the underlying cognitive factors that give them their shape. From the standpoint of *cognition*, science and religion are asymmetric on two crucial counts that correspond to the two cognitive criteria that define the table in fig. 1. First, they operate at wholly different cognitive levels. One, popular religion, is thoroughly dependent on the natural proclivities of human minds and, hence, recurs in every human culture, whereas the other, science, is a function of comparatively rare social arrangements that require familiarity with both norms of reasoning and radically counter-intuitive conceptions and the public availability of the pertinent processes, products, and evidence. The second asymmetry hinges on their critically different default assumptions about the way the world works. Religions presume that the most penetrating accounts of the world will always, ultimately, look to agent causality. Science does not.

Nor, in all domains, do our commonsense understandings of the world. That observation hints at how the disinterest of conventional comparisons of science and religion in these cognitive and cultural considerations can obscure some revealing connections. For example, *both* science and popular religion are more similar cognitively to *both* theology and commonsense explanations of the non-social world *than they are to one another*. Consider science first. As I just noted, neither scientific nor commonsense approaches to accounting for the non-social world assume that agent causality, finally, provides the most telling explanations. On the other hand, both science and theology are reflective activities that are mostly pursued by highly trained specialists and that are most credibly pursued by highly trained specialists. Popular religion, by contrast, shares neither of these properties with science. On both of the cognitive considerations just reviewed, it too is more like both commonsense explanations and theology than it is like science - though, of course, in exactly opposite ways. It is their mutual emphasis on maturationally natural cognitive capacities that link religion and commonsense understandings of the world, while it is the priority they set on agent causality in their explanations of things that religion shares with theology. There is a respect, then, in which the longstanding interest in the comparison of science and religion is, from the perspective of reflection on human cognition, somewhat misbegotten. Without systematic attention to these cognitive questions and explicit discussion of the place of theological reflection and commonsense views of the world as well, conventional comparisons of the metaphysical and epistemological statuses of science and religion seem a bit contrived.

A footnote: in his book *Inevitable Illusions*, Massimo Piatelli-Palmerini discusses findings from experimental psychology indicating humans' penchant for relying on the deliverances of their maturationally natural cognitive systems even when those deliverances are thoroughly contrary to the norms of deductive and probabilistic inference. 'We have come to see that our minds spontaneously follow a sort of quick and easy shortcut, and that this shortcut does not lead us to the same place to which the highway of rationality would bring us'. A few pages later he adds that 'our spontaneous psyche is not a kind of 'little' or lesser reason, nor is it an approximate form of rationality' (Piatelli-Palmerini 1994, pp. 142 and 159). I stand by my comments in the previous paragraph about the greater similarity between either science and commonsense explanation, on the one hand, or theology and popular religion, on the other, than between science and popular religion. Still, Piatelli-Palmerini's observations counsel that the cognitive affinities between science and maturationally natural commonsense explanations should not be overestimated. Behind these two approaches to the world lurk differences that make a difference cognitively. The symmetries that fig. 1 displays suggest that if that is true, then neither should the cognitive affinities between theology and popular religion be overstated. Systematic reflection seems to generate intellectual working space beyond that which our maturationally natural tendencies supply. Karl Barth's famous and much revered rendition of 'Jesus Loves Me' notwithstanding, the maturationally natural cognitive processes and inferences that prevail in popular religion are no more a 'little' or lesser version of systematic theological reasoning than are the intuitive shortcuts of our commonsense explanations a 'little' or lesser form of scientific reasoning.

THE TRANSMISSION OF HISTORICAL COGNITION

WILLIAM W. MCCORKLE JR.

> Similar to the case of culture, we might conclude that there are no historical contexts apart from 'structurally enduring relationships' among the mental states and behaviors of individual historical and historiographical agents.
>
> - Luther Martin (Whitehouse & Martin, 2004: 9)

The historical sciences have played an important role in the evolution of humans for over three millennia. With the advent of writing some five thousand years ago, the rise and introduction of literacy via professional guilds, and the promotion of *history* rather than myth, recent, modern humans have been able to carve out their own niche for technological and cultural advancement. This is not only a distinct cultural feature in one or two localized milieu. In fact, humans appear to have used history to preserve important cultural representations much in the same way that folklore, ritual, music, dance, and language were also used.

Twenty five hundred years ago, Kongfuzi (Confucius) via a written text known as *The Analects* described the earlier Zhou dynasty in China. Kongfuzi argued that the time he lived in had fallen out of harmony with Heaven's will (*T'ien-ming*). Only by returning to the values, ceremonies, vertical segmentation by filial piety, and virtuousness could humankind return to a *time* of historical promise (the earlier Zhou dynasty). Kongfuzi argued that man and heaven were bound together in a web of historical (and cultural) significance.

Ancient Greek and Roman historians carefully detailed in historical chronicles the advancement of their respective civilizations. Suetonius,

Tacitus, and Pliny all carefully described the rule of the Caesars, their histories constructed by a top down (individual) approach where great rulers created history and the masses followed. Alexander, on the other hand, commissioned coins, made with his likeness, throughout the Macedonian empire chronicling his military conquests and political exploits throughout the *Occidental* world, and in lands far beyond (South Asia).

Modern History (or the historical sciences) may appear to be a fairly recent academic enterprise. From Hegel's philosophy of historical change to Marxist dialectical materialism, and from Max Muller's 'disease of language' to Michel Foucault's 'archaeology of knowledge', humans (especially scholars) are (and have been) interested in the construction of history as a science.

In many ways the history of the study of history reveals a much larger theoretical commitment to science in several ways. The modern field of history was fortunate that it co-evolved with Charles Darwin's theory of natural selection. While evolution and history are not interchangeable concepts, both of these important scientific discoveries provided a peculiar theory (Dennett, 1995) for the study of humans and the production of culture. Many in the field of history are keenly aware that dates and facts are many times meaningless without theoretical and methodological commitments on the historian's part. Nevertheless, historians are very much aware that saying 'in 1492 that X occurred' is an extremely naïve view of the nature of physical and theoretical time and space in history.

I will argue that history is in fact a dialectical process that takes place both in vertical (generation to generation) time and horizontal (peer to peer, culture to culture) space. Historical knowledge is best understood in this vertical and horizontal relationship. The vertical transmission of history is always meaningless. In fact, most (if not all) representations (including public representations like ritual) are semantically meaningless in vertical transmission, yet those historical representations that survive the natural selection of memetic transmission may contain (and probably do) structure (McCorkle Jr., 2007; Staal, 1979). Moreover, historical memes that replicate and survive (via humans and their artifacts) are given meaning by horizontal contexts. Let me turn to an example.

I remember reading J.R.R. Tolkien's work *The Hobbit* when I was twelve years old. I found it to be an incredible source of knowledge that expanded my ever growing imagination. Later at the age of twenty five, I enrolled in a class where I had to read *The Hobbit* again. The meanings that were extracted by me at the age of twelve and again at the age of twenty five were tremendously different; however, the book had not changed. Its main structure remained intact. Nevertheless, I changed and, therefore, the meaning of the book changed. Todd Lewis calls this particular feature of horizontal transmission – 'the domestication of the text' (Lewis, 2000: 2-7). Texts are always given meaning in the present, even if they purport to give a meaning that is in the past (or in some cases the future).

Historians spend a great deal of energy trying to (re)construct what happened at a particular time in history. Many times scholars try to imagine, via comparative historical analysis, what an individual or society thought and how this thought influenced behavior and belief. However, taking Lewis' concept seriously, almost all (if not all) historical reconstructions are always informed by the present world of the historian. So, the historian constructs the past into a horizontal material portrait that speaks volumes of the present, rather than the past. Those scholars interested in the hermeneutic endeavor have always known this. However, hermeneutics are not the entire picture of historical knowledge and the transmission of history.

Historical knowledge has also played an important role in the academic study of religion (*Religionswissenshaft*). History has provided one of the best methodologies for the study of religion over the last hundred years or more. From linguistics to textual and area studies, many scholars in the field of Religious Studies pay particular attention to the hermeneutic stance of the construction of history; however, very little attention (until recently) is paid to how history is transmitted vertically. Dan Sperber (1996) argues that representations are spread epidemiologically. Like viruses, representations replicate, transmit, and survive based upon natural selection pressures related to their host (humans) and their environment. For Sperber, humans are hospitable for some memes because human beings are good at making 'meaning out of almost anything' at will (McCorkle Jr., 2007: 89). Sperber coins this theory *relevance*. Relevance is the theory that humans make meaning using a complex cognitive process that connects representations into symbols (Sperber, 1975) and therefore, *post hoc* meanings are attached to these meta-representations (Sperber, 1996: 71-72, 146-150, 2000). Therefore, representations stand a good probability of replication, but the down side is that they often mutate (Dawkins, 2006: 191-201). This is obviously the opposite to what happens in genetic transmission. Sometimes, however, according to biologist Richard Dawkins, some representations (memes) become 'selfish' and spread without any concern for their host (Dawkins, 1976). Dan Dennett argues that this feature, known as the *selfish meme* (in this case religious memes), spreads throughout a given host population at will, (Dennett, 2008). It has survived selective pressures and become an epidemic in the evolutionary pool of representations. Selfish memes become epidemics until their hosts become immune to them. However, sometimes certain types of memes re-occur throughout human history and become epidemics again. Many times this epidemic is due to the hospitable world of the host, the human mind.

History operates in much the same way. Historical facts (if we can call them that) are analogous to memes (or memeplexes) (Blackmore, 1999: 19-20). Out of the billions and billions of possible representations, historians pick and chose the most select, the more fit of the group. Historians in some ways become part of the environmental selective pressures because they pick the fit memes from the weak memes in the pool of ideas. Moreover, the memes that are selected for by historians are not just facts, but more relevant historical phenotypes that *win out* over weaker, less plausible ones.

Meme theory has generally worked at the level of *the extended phenotype* (Dawkins, 1982); meaning memes, like genes, are *selfish*. Contrary to human individuality and social theory (vis-à-vis Durkheimian sociology) selfish memes and genes utilize human hosts to travel the variable paths of evolution to survive, transmit, and replicate without regard for human survival. It just so happens that these genes and memes have allowed complex organisms - in this case humans - to co-evolve because they simple need us to travel the evolutionary superhighway. Every once in awhile, memes go so selfish as to cause harm to their host (like genes that mutate and cause cancer), eventually destroying their vehicle for survival.

History is dependent upon this meme theory because all historians have modern human brains and the individuals that they study typically are humans. The transmission of historical cognition is simply possible because of this epidemiological feature. It would be very difficult to try to tease out what ants think, much less try to figure out what ants were thinking three thousand years ago when they built anthills that utilized trees and animal bones in their construction. Human historians and their subjects typically share one major feature in common, the evolved mind.

Cognitive Archaeologist, Steven Mithen, has traced the material remains of humans over the last several hundred thousand years to argue that humans have developed an evolved cognitive fluidity out of specialized systems of the brain that were designed to perform certain specific functions (Mithen, 1996). Mithen, using the theory of John Tooby and Leda Cosmides (Barkow *et al.*, 1992), argues that the mind is analogous to a 'Swiss army knife': where like the knife, it has a multi of 'tools' dedicated to specific functions. The knife, for example, is designed to cut things. One could try to use the knife as a hammer, but without much success. The knife, that is, is a very simple tool dedicated to a very specific function. However, the magnifying glass on a Swiss army knife is quite a different story. The magnifying glass is designed to magnify small objects. Nevertheless, it can also magnify sunlight to help start a fire. To use this tool involves at least two steps to get the desired result. First the tool is made to magnify things. Second, together with the sun, this tool can be used to create another tool - a fire. So, the knife like our minds evolved out of simple and complex tools to handle certain functions.

The magnifying glass tool is analogous to a certain human mental system called the *Intuitive Psychological System*, or more commonly known as *Theory of Mind*. Theory of Mind is a social intelligence cognitive system that informs individuals about other agents' intentions, emotions, and psychological states. In other words, I know that Suzie is angry with me. I could be completely wrong about my inference that Suzie is angry with me; however, this psychological inference is an important step for human social interaction and cooperation. Contrary to some social theories, individuals have evolved a social intelligence to bond together in groups; human minds, that is, are not blank slates at birth but have specific innate capacities (Pinker, 2002). Theory of Mind is also an exciting mental tool because in humans it allows us to hold strategic social information about individuals in the group. This is known as *The Gossip Hypothesis* (Dunbar *et al.*, 2005). Gossip exists in every human culture on Earth. By making the inference that Susie is angry with me because John told her that I said something bad about her because he is jealous is an important piece of socially strategic information. Imagine if I needed John's help for hunting prey in the ancestral environment. John might just assume that he is better off letting me be eaten by the animal, rather than cooperating with me in killing it. He is after all (hypothetically) in competition with me for Susie's affection.

Understanding Theory of Mind is critically important to understanding the transmission of knowledge for it presupposes that the historian can 'get into' the mind of the subjects s/he studies to make certain inferences about *why* the historical actions took place. As Dawkins has pointed out, why questions are important because humans may be natural teleologists (Dawkins, 2006: 181). If we have a design, we want to know the designer and we want to know the intention behind the design. Here historical change is the design; a subject *caused* something to happen. The historian analyzes all the relevant data on the period, the biography of the subjects involved, and the modern theoretical insights on the relevant topics of interest. So, the historian is not only interested in when the Roman emperor Constantine converted to Christianity (if he really did at all), but *why* he converted to Christianity. Did he have a revelation? Did he do it for political purposes? What were the social, cultural, and environmental factors that shaped his decision? Or was this a later invention by biographical redactors of Constantine? All of these inferences by the historian are made possible by the evolved architecture of human minds. Without Theory of Mind humans wouldn't be able to create history, much less study it.

This point brings up a very interesting argument by Harvey Whitehouse. Whitehouse, in a dialogue with Jack Goody, argues that certain mental systems may have 'kick-started' specific types of technology (i.e. writing) (Whitehouse, 2004)[1]. Whitehouse states:

---

1. In fact the archaeological and historical data from early China may provide evidence of Whitehouse's position. I have argued elsewhere (McCorkle Jr., 2007, 2008, forthcoming) that dead bodies, their relics, and remains/cremains stimulate several mental systems that activate 'hazard-precaution' (Boyer & Lienard, 2006a; Boyer & Liénard, 2006b) and hyperactivity of Theory of Mind. Early religious guilds and experts may have used corpse remains as a pre-literate 'placeholder' for representations of the dead person in ritual scripts prior to literacy. The Shang oracle bones are some the earliest known writings in China. Corpses and their remains are physically hard to destroy, but also mentally hard to ignore; thus, ensuring ritualized compulsions to them again and again (McCorkle Jr., 2007: 183). Ritual experts may have found these compulsions useful in the early adoption of ritual scripts, as they connected these compulsions into texts (here a text can mean an action, artifact, or utterance). I argue that the Shang data provides evidence that early guilds (writing and religious) were most likely using the bones as placeholders for powerful representations and therefore, writing on them would not only last a very long time but also connect these material anchors with newly 'jumpstarted' literacy.

The homogenization of a regional tradition required, first and foremost, a method of transmitting (acquiring, remembering, and passing on) complex standardized teachings. Although literacy may have come to the aid of such projects, the latter's initial appearance and flowering need not have depended upon that. Indeed the doctrinal mode may have been a major stimulus for the development of writing systems, rather than the other way around (228).

Whitehouse's claims lend weight to the argument that the production of history (and the transmission of it) is correlated to the evolution of the modern, human mind[2]. Therefore, any responsible scholarship concerning history might utilize the more recent developments in the cognitive sciences.

The cognitive sciences were a breakthrough of the twentieth century. By combining sophisticated modeling techniques with computers, linguistic theories, and evolutionary, developmental, and experimental psychology, scholars interested in modeling the mind (directly or indirectly) triggered a paradigm shift in the study of humans and the production of culture. This cognitive revolution has produced multiple cross-disciplinary research especially in cognition, culture, and religion, resulting in the creation of multiple centers, institutes, and research programs around the world (Whitehouse & Laidlaw, 2007: 13).

The cognitive sciences may provide historians with an important methodological breakthrough. Historians should use any methods they can beg, borrow, or steal from other disciplines if it makes their own field a stronger one[3]. Some critics of this approach have reasoned that cognitive science has its own topic - cognition (Whitehouse & Laidlaw, 2007: 26), or that cultural data can't be *explained* at the level of cognition. Nevertheless, historians, like archaeologists, anthropologists, linguists, sociologists, philosophers, and scholars of religion have already started to use cognitive science as a strong methodology in their scientific repertoire.

---

2. See also (Johnson, 2004: 45-66). Here Johnson argues that evidence from archeological sites in southwestern Iran (fifth and fourth centuries BCE) provide evidence that Whitehouse's doctrinal mode appears before the earliest know writing.

3. This is a comment that I borrow from Archaeologist James Mallory (Queen's University, Belfast). In 1996, Professor Mallory argued that archaeology wasn't a selfish discipline; archaeologists would steal from any methodology they could, if it made the data and analysis a better science. I argue that historians should adopt a similar approach.

Two distinct examples might provide evidence for the cognitive sciences in the field of history. First I will look at the example of the *Tomb of the Shroud* in modern day Israel. Second, I will examine the case of the historical Buddha (Siddhartha Gautama), early Buddhism, and supernatural agency in Buddhism. Recently in Israel, archaeologists discovered a tomb which contained several ossuary (bone) boxes and the remains of a dead (male) body covered in a burial shroud (Tabor, 2006). The skeleton apparently had been left in the first phase of what anthropologists call 'twice burials' (Hertz, 1960). The second phase would have resulted in the skeletal remains being transferred to an ossuary box after full decay of the corpse had taken place. The burial shroud from the skeletal remains was carbon dated (carbon-14) to sometime in the first half of the first century of the Common Era (Tabor, 2006: 14). The placement of the names on the ossuary boxes and the probability of random occurrence of these particular names lend themselves to peculiar results. The skeletal remains also were found to have 'leprosy (Hanson's disease), and, microbiological tests, indicated, that he most likely died of tuberculosis' (12).

Cognitive anthropology and science may provide explanatory hypothesis as to why this skeleton was left in the first phase of the burial. According to Robert Hertz (1960), the first phase of a 'twice burial' involves the alerting of the community that something dangerous to the groups has happened. This may involve the ringing of bells, gongs, wailing and loud noises by the community, where the group is summoned together to handle the danger. Maurice Bloch (Bloch & Parry, 1982) argues that this is the where the group realizes that biological death is a danger to their survival, either at an explicit level or more importantly at an implicit level (3-5). The community must come together to recognize the danger of the corpse and, therefore, the corpse becomes polluted and must be separated in some way from the community. The second phase in Hertz's schema has very little to do with the body and its remains at all. Once the biological pollution of the first phase is completed the dead person's social roles and material goods must be redistributed back into the community. The skeleton in the Tomb of the Shroud never made it to the second phase. Recent experimental evidence suggests that dead bodies stimulate contagion avoidance but ritualized compulsions to dispose of the remains supersede the biological contamination of such endeavors (McCorkle Jr., 2007). If this body was infectious, the human drive to complete the 'twice burial' would have taken precedence over the contagion. In fact, contagion may drive humans to perform such ritualized behaviors (Boyer & Lienard, 2006a; McCorkle Jr., 2008: 290-291). Therefore, historians might deduce that something happened to the community associated with the skeleton that prevented them from finishing the 'twice burial'. The cognitive sciences may not provide proof of what happened, but it may likely provide a better hypothesis on what likely happened to this first century community.

The second example is the recurrent argument that the historical Buddha and early Buddhism were not defined by the intellectualist argument that religions are special because they contain representations of supernatural beings or agents. This argument is mainly based on Protestant (and Victorian) reconstructions of early Buddhism from the Pāli Vinaya (monastic code) by western scholars. Nevertheless, recent historical and archaeological research on early Indian Buddhism reveals that early Buddhism concerned itself with representations of supernatural agents *ad infinitum*, including supernatural agency of the historical Buddha (Decaroli, 2004; McCorkle Jr., 2007; Schopen, 1997, 2004). In fact this appears in the epigraphical record as some of the earliest historical evidence of Buddhist practice before the Common Era (Salomon, 1998).

James Laidlaw has recently argued that Buddhism presents a problem for a cognitive study of religion, since in his words 'Propitiation of deities is indeed common, but *no remotely reflective Buddhist*, including those who spend time and resources participating in such rites, would confuse them for a moment with following the *teachings of the Buddha*. And whatever else Buddhism is, it surely must include that. Buddhists will certainly maintain so' (Laidlaw, 2007: 220-221) (italics mine). The definition of *religion* notwithstanding, Laidlaw makes several crucial mistakes in his argument against a cognitive science of religion, by using several flawed historical arguments with regards to the history and practice of Buddhist traditions.

First, Laidlaw presumes that there are (and were) original 'teachings of the Buddha' (Skrt: *buddhavacana*). There is no direct historical or archaeological evidence of any *original*, *pristine*, or *early* Buddhist teaching that can be directly connected to the historical Buddha himself earlier than the second to third century BCE with material culture (Bharhut, Sanchi), and no historical text that can be dated prior to the common era, some five hundred or more years after the Buddha's death (Skrt: *Mahaparinirvana*) (Salomon *et al.*, 1999; Schopen, 1997). In fact, Carol Anderson argues that even the *core* of

Buddhist teaching (i.e. The Four Noble Truths, The Eight Fold Path, and even the concept of Nirvana) may come from a much later period (first and second century of the Common Era) than the historical Buddha lived (Anderson, 1999). Laidlaw errs because of a salient representation passed down by scholars that does not, at any cost, want Buddhism to be a religious tradition like others (Schopen, 1997: 1-22; Silk, 1994). And why not? Because Buddhism has provided a natural argument against an intellectualist definition of religion (Tylor, 1871) containing representations of supernatural agents for over a hundred years[4].

In fact what we appear to know about early Indian Buddhism (not the historical Buddha himself) provides a radically different portrait than the one argued by Laidlaw. It appears that Buddhists (monastic and laity) in early India were involved in a great deal of behavior associated with supernatural agents, including the historical Buddha. Much of the epigraphical data proffers evidence that monastics and the laity were giving money for votive offerings for good karma (Skrt: *karman*; to do or make fruit) for their *dead* ancestors made possible by the *dead* historical Buddha (Decaroli, 2004; Schopen, 1997: 62; Strong, 2004). The Pālī Vinaya, taken to be the authoritative text of early Buddhism by over a hundred years of scholarship appears to be edited heavily, translated from at least one language to Pālī (Schopen, 1997: 204-237), was a minority text until the second century of the Common Era in Ceylon (Sri Lanka) (Collins, 1990: 96), and was not closed as a text - as an extant tradition for followers of the Theravada in southeast Asia

---

4. In fact Laidlaw makes another crucial mistake in his argument with regards to Pascal Boyer's evolutionary theory of religious representations. Laidlaw presumes that Boyer is operating from an intellectualist position where supernatural agents are the core principle of religious belief and behavior. Actually, Boyer argues against an intellectualist definition for religion, because Boyer critiques this same position in chapter one of *Religion Explained* (2001) (many times religions bring up more questions than they report to answer). In fact, Boyer explicitly claims that religion is not a category in and of itself. Religion for Boyer is a 'hodge-podge' of the stimulation of an aggregate number of mental systems. These mental systems have evolved in humans over the last several hundred thousand years, where many times they produce *counterintuitive violations* of *intuitive ontological categories* (the bird swims, the bush hears, the dead man lives). Agents are important only because agents are more important for human survival. So, it makes sense that our cognitive architecture would produce counterintuitive representations concerning agents much of the time, rather than ones about 'teapots and kettles'.

- until the latter part of the fifth century CE (Akira, 1990: 125; Schopen, 1997: 23-24).

The cognitive sciences actually help historians to construct a realistic portrait of early Buddhism in India. The Buddha, his followers, and the early community actually had to compete with other religious traditions of the Gangetic plain. In fact, according the Pāli Vinaya itself the Buddha and his followers were engaged in competition for followers. Two of the Buddha's favorite disciples (Sāriputta and Moggallāna) left a rival sect to join the Buddha's community of believers (Sangha). Upon leaving the rival's group, all (250) the followers left with them to join the Buddha. The Pāli Vinaya states that the rival (Sañgaya) became angry and engaged the Buddha, resulting in Sañgaya's death. The text itself is not clear on exactly how Sañgaya's death occurred; however, the Pāli text states that 'Sañgaya began to vomit hot blood from his mouth' (Rhys-Davids & Oldenberg, 1996: 149). In other texts, we are told that the historical Buddha used 'skillful means' (Skrt: *upāya*) to confront rivals. Sometimes, this involved superhuman agency, where the historical Buddha did miraculous things or caused supernatural events (McCorkle Jr., 2007). It would seem that the supernatural character of the Buddha is simply dismissed by many western scholars in favor of the philosophical Buddhism that is found in idealized texts (Bachelor, 1997).

From the empirical evidence in the cognitive sciences, especially experimental research, we might deduce that it is natural for humans to reproduce counterintuitive representations naturally (Boyer,1994). This feature appears to be pan-human and across cultures, spanning historical and pre-historical time. By the time of the Buddha's death, according to tradition, two of his favorite disciples had preceded him in death (Sāriputta and Moggallāna). In Sañchi, dated to somewhere in the second and third century BCE, presumably Sāriputta's remains are buried beneath a large burial mound, or *stupa* (Harle, 1994: 34; Strong, 2004: 206). This would provide evidence that the Buddha may have sanctioned ritual burial of his close followers in his own lifetime. In the *Mahāparinibbāna sutta*, the Buddha details how his own body is to be ritually handled by his community of followers (Ling, 1981: 187). Elaborate mortuary behavior is requested by the historical Buddha for his closest attendant Ananda. This is a peculiarity in the text because Buddhist philosophy dictates that the body is a mere shell of existence, an illusion of the world of birth, death, and rebirth (*Samsara*).

Nevertheless, in many texts, especially the monastic codes of several different sectarian Vinaya (including the Pāli *Mahāparinibbāna sutta*) there are elaborate details and regulations on the ritual disposal of dead bodies (McCorkle Jr., 2007: 30-48; Schopen, 1997: 99-113).

Historians can utilize findings in the cognitive sciences to hypothesize that if humans are natural teleologists and producers of counterintuitive inferences, then with regards to Buddhism it would seem fairly implausible to construct early Buddhism as non-theistic or not connected to supernatural agents (at least at the minimum in regards to dead bodies, their ongoing representations, and hyperactivity of Theory of Mind for the living participants). Here Luther Martin adds,

> ...a more venturesome question for historians is whether this [cognitive] theory can offer historians an infrastructure for explaining their data with greater confidence. Historians have always been more comfortable with description than they have been with theory - even though the latter always lurks implicitly below the surface of the former. But, it can be asked, to what extent might formal theoretical predictions, such as those proposed...help clarify the complexities of the historical data? (Martin, 2004: 12).

Martin is correct in his assessment of the uses of the cognitive sciences and its role in the development of a science of history (especially the history of religions). By using modern empirical tools (experimental data), historians may be able to extract more tractable data from weaker ones, and develop more substantial hypotheses concerning past events based upon the knowledge of the human mind. The transmission of history it seems is really the communication of human minds to human minds, via material representations. History is then the science of historical cognition.

RELIGION BEFORE 'RELIGION'[1]?

RUSSELL T. McCUTCHEON

Preamble

I offer this essay - which moves across a number of areas of interest to him, though doing so in ways that are of interest to myself - in tribute to the influence that Luther Martin has not only had on the international field but on me personally. The North American field's links to Europe and beyond have been enriched beyond measure by Luther's entrepreneurial spirit and his never retiring, can-do attitude - an infectious attitude that sweeps others along, whether into collaborative projects (such as the role that he played in what I think to be among the more important books on Foucault's work), into new friendships and professional relationships (such as my ongoing work with Prof. Panayotis Pachis, whom I first met through Luther while he hosted Panayotis in Burlington, VT), or into really exceptional (sometimes costly but, so long as one is dining with Luther, always worthwhile) restaurants while attending conferences. I have benefited professionally, in countless ways, from the fact that I arrived on the academic scene long after such resilient and hard working people had begun trampling down what was once unwieldy phenomenological grass, helping to make a place for people like me, with interests such as those that animate this chapter.

****

Near the end of his first visit to the continent of Africa - or, taking the lead from the Vatican's designation of 'apostolic journey,' what the press simply called his 'pilgrimage' - Pope Benedict 16th relied on the well-established

---

1. Part of this chapter was delivered at the November 2008 meeting of the Society for Biblical Literature and a longer version was delivered at a workshop entitled 'Interrogating Religion' at the University of Ottawa in April 2009.

distinction between magic and religion when speaking to a group of Church bishops, priests, and nuns, during an invitation-only mass at St. Paul's Church in Luanda, Angola. As quoted in *The New York Times*, and as confirmed by the Vatican's website, the Pope closed his homily by focusing the faithfuls' attention on those at risk because they have not yet heard the Church's message, posing to his congregation the following rhetorical question: 'Who can go to them to proclaim that Christ has triumphed over death and all those occult powers?'[2].

The strategic pairing of licit/illicit, which Benedict used to distinguish those practices termed sorcery from what he termed the cause of Jesus Christ, has a long history of usage with which scholars of religion are more than familiar. Important to bear in mind when studying those who divide up their social world in this way is the principle that, after describing the participant's own use of such designators, scholarship requires us to redescribe all such first order classification systems, seeing them as instances of other, far wider, cross-cultural or historical processes - processes which do not necessarily share a common identity but which, instead, serve as analogies exemplary of something that the scholar finds curious. In the case of the Pope's distinction between false sorcery and true religion, I believe that following this principle would be relatively uncontroversial for the majority of scholars - those who would undoubtedly advise that we would be terribly mistaken to begin our study of the Pope's address by assuming that the categories 'sorcery' and 'religion' referred to something substantial, authentic, and thus distinguishable in the acts being signified; instead, such scholars would advise us to see in such classification systems evidence of a set of prior social interests that the speaker was putting into practice by means of such bounded pairs as pure/impure or magic/ religion - interests that, in this case, had something to do not only with the goal of reinvigorating the collective identity

---

2. Benedict cited the letter of Paul to the Ephesians (6:10-12) to support his point: '10 Finally, be strong in the Lord and in his mighty power. 11 Put on the full armor of God so that you can take your stand against the devil's schemes. 12 For our struggle is not against flesh and blood, but against the rulers, against the authorities, against the powers of this dark world and against the spiritual forces of evil in the heavenly realms'. See: 'Pope Tells Clergy in Angola to Work Against Belief in Witchcraft', http://www.nytimes.com/2009/03/22/world/africa/22pope.html?ref=world (accessed on March 28, 2009). For the text of the address, see the Vatican's site: http://www.vatican.va/holy_father/benedict_xvi/ homilies/2009/documents/hf_benxvi_hom_20090321_saopaolo_en.html (accessed on March 28, 2009).

of those Angolans in attendance but also inspiring them to work toward the elimination of this very social distinction through what is known locally as proselytization. For in this way, the superstitious 'them' would, God willing, someday be converted to a totalized 'us'.

As Durkheim might have concluded, through their participation in the ritual known locally as a mass, the so-called 'heroic and holy heralds of God,' as Benedict phrased it, became unified insomuch as they could understand those within eyesight and earshot as clearly distinguishable from those not in attendance, the ones living in the fearful grip of magic. When redescribed in this manner, the address that the Vatican and the press alike termed the homily ends up being but one species of the far wider genus popularly known as a pep talk, akin to the sort of talk that might be delivered by an inspiring coach prior to the big game, creating a shared sense of affinity among members of the home team by juxtaposing them to a caricature of their cross-town rivals. Given that the modern term 'homily' derives from the Greek *homilos*, meaning simply a gathering of people, even a crowd - or, more specifically, a gathering of soldiers (*homou* 'together' + *ile* 'troop') - then, once redescribed in this manner, we see that the Vatican and the press's shared use of this local, theological descriptor to name Benedict's address may have told us far more about the effects of the event than they had ever imagined.

But what made the report of the Pope's pep talk stand out for me was not that it illustrated so well that much is at stake in the designators that scholars use when describing people's behaviors - though we learn much by substituting 'visit', 'audience', and 'pep talk' for the participant's own choice of 'apostolic journey', 'congregation', and 'homily' - and not even because of how nicely its analysis demonstrates that seemingly unlike first order things can be profitably compared in light of a second order curiosity. Instead, what caught my attention was how difficult it would likely be to get those scholars of religion who so easily see through Benedict's transparent distinction between magic and religion also to historicize the conceptual distinctions that provide the enabling conditions of their own work, not least being the sacred/secular pairing that animates our use of the category 'religion'. For, much like the term 'apostolic journey', the category 'religion' is a local designator, peculiar to certain historical periods and specific groups of human beings - although, unlike 'apostolic journey', it is one that people worldwide have adopted (no doubt grudgingly in some cases) and elevated to the status of cross-cultural universal. That is, while many of us would judge it insufficient for a scholar to be content with merely describing the Pope's rhetoric, or worse yet, simply to adopt it and then identify the magical qualities of the illicit practices that so concerned him, I nonetheless find in the work of some who claim to be historicizing the category 'religion' no less troublesome assumptions concerning the stubborn significance that apparently remains even after we dispense with our Latin-derived signifier. Thus, despite the now commonly accepted practice of historicizing the word 'religion', I often find in such work that some prior, elusive concept ends up being naturalized, despite our supposedly rigorous attention to history; it is this strategically partial historicization that I wish to document and then problematize.

Although I could cite a number of examples, consider the conference in Bochum, Germany, that I participated in during the Fall of 2008. Like many in the field today, its organizers conceptualized religions not as static things but as ever-changing objects in motion - a theoretical shift signaled by the replacement of singular nouns by their plural form and an emphasis on studying people's observable practices rather than on such intellectual abstractions as their beliefs. This move from the one to the many, and this emphasis on local difference over abstract similarity, is by now quite common in our field - so much so that the turn toward studying what is now known as 'religions on the ground', 'public religions', and 'embodied religions' suggests that a generation's worth of critiques of essentialism have had their desired effect and that we will no more find scholars assuming that their object of study is unique, timeless, and self-evidently interesting. And so, for the critics in the field, we might now raise a mission accomplished banner and then get on with the business of studying religions - in the plural, on the ground, and on the move - knowing that a little theory, much like a pinch of salt when cooking, has offered a necessary corrective.

Or so one might think.... But after looking a little more closely at the work of some of those who are now investigating the category 'religion', it seems to me that troubles persist despite - or perhaps in the guise of - the so-called advances. This was made evident to me during the final session of the above-mentioned conference in Germany, for example, in a paper presented by José Casanova, the noted US sociologist and author of *Public Religions in the Modern World*[3], a book on the worldwide deprivatization, or better put,

---

3. José Casanova, *Public Religions in the Modern World* (University of Chicago Press, Chicago, 1994).

politicization, of religion over the past decades - an argument that has played a crucial role for many in how they now conceptualize religion and its relation to the state. Like a number of those who focus on the category 'religion', Casanova has become interested in the study of secularism, though, of course, this is hardly the old secularization thesis. Instead of predicting the eventual decline of religion in our modern world, Casanova described the manner in which such notions as church and state are binary pairs peculiar to that worldview that goes by the name of secularism - conceptual pairings that, once entrenched in laws and institutions, provide a framework in which modern social actors establish and negotiate their worlds. But it is a framework that, or so the argument now goes, the so-called 'return of religion' is now challenging - to such an extent that scholars along our field's cutting edge have begun to imagine what the role of religion will be in what they are confidently calling a post-secular world.

Now, I had no disagreements with Casanova's position on the binary nature of the sacred/secular pairing and how it provides the conditions in which we, as social actors, think and move[4]. Or, more correctly, I had no disagreement until, in my opinion, he waffled on thinking seriously about the historicity of binaries and how they function; for his analysis concluded by arguing that the notions of religion and the secular were early modern Christian theological concepts, as if those things that we today commonly identify as religious - such as the social group known by its participants and scholars alike as 'Christianity' and the literary genre known by its authors and scholars alike as 'theology' - somehow preceded, and thus caused, the subsequent ability to name those things that were *not* religious (i.e., the secular).

Now, I fully understand why some people persist in making the move of imagining religion - or perhaps we should say not religion but, instead, the sometimes preferred term, religiosity - to pre-exist the terminological distinction between religion and not-religion, somewhat as if, prior to the invention of cooking (and prior to Lévi-Strauss's rigorous examination of how binaries function) early humans simply had a natural sense that their food was raw; for imagining religion to precede the word 'religion' - what basically amounts to reviving the nineteenth-century notion of natural religion - enables

---

4. For example, see my own "'They Licked the Platter Clean': On the Co-Dependency of the Religious and the Secular", *Method & Theory in the Study of Religion* 19/3&4 (2007): 173-199.

scholars to retain the deeply felt baby (variously called experience, faith, authenticity, spirituality, meaning, the religious, and thus religiosity) while throwing out the merely contingent linguistic bathwater - bathwater that some of us happen to know by a Latin-based name that we (or so the critique goes) ethnocentrically export when naming babies in other people's hearts. The trouble, then, is that under the guise of progressive historicism a rather conservative argument concerning natural identity and transcendental value is nicely re-inscribed, for although we may no longer use a once dominant vocabulary to name certain discursive items, the items are left untouched and merely used for new, tactical purposes.

For example, look through the recent criticisms of the work of some prominent (though, as I have argued elsewhere, rather problematic[5]) U.S. Indologists - I have in mind here the critiques of the work of such writers as Wendy Doniger, James Laine, Paul Courtright, and Jeff Kripal. Many of their critics are rather upset with how so-called Western scholars - a category that, despite a generation of anti-Orientalist critiques, occupies a curiously central place in the work of these critics - use alien, imported theories to study what the critics know as their own, actual religion. For example, as phrased by Prof. S. N. Balangangadhara, writing in the foreword to a recent collection of essays entitled *Invading the Sacred* (2007), many Indian intellectuals 'realize that Western explanations of their religions and culture trivialize their lived experiences; by distorting, such explanations transform these, and this denies Indians access to their own experiences. It can thus be said to rob them of their inner lives'[6].

The corrective offered is simple enough: Indian scholars must get off the sidelines, as the book's Introduction phrases it, and take back their religion from the neo-colonialists - a move that will supposedly result in reclaiming their very selves. What I find interesting here is that the critics do not adopt the far more rigorous position by arguing that European-derived designators - such as the just quoted terms 'culture' and 'religion', as well as the now popular 'lived experience' - have no analytic utility when it comes to examining, in this case, life in the Indian sub-continent, either today or in the past. Instead, in a move reminiscent of that famous quotation from the

---

5. For example, see my own 'A Gift with Diminished Returns', *Journal of the American Academy of Religion* 76/3 (2008): 748-765.
6. Krishnan Ramasway, Antonio de Nicolas, and Aditi Banerjee (eds.), *Invading the Sacred: An Analysis of Hinduism Studies in America*. Delhi, India: Rupa & Co., 2007, vii.

opening to chapter three of Rudolf Otto's *The Idea of the Holy* - where those who have not had 'a deeply felt religious experience' are asked to put down the book - the charge is leveled that outsiders *misrepresent* what the self-described participants already know to be their true religion, their unique culture, and their lived experiences - all terms that, despite their obviously foreign pedigree, are curiously assumed by such critics to be in one-to-one step with Indians' most authentic and enduring selves.

What must not go unnoticed in this linkage of authenticity, self, and religion - and, for anyone familiar with the work of, say, Wendy Doniger, it will be more than apparent that this linkage is made by actors on both sides of this debate, but in the service of rather different conceptions of self and society - is that, despite what some politically liberal observers might read as a laudably critical attitude toward Western imperialism, such critics are surprisingly conservative in their responses, for they seem to have little choice but to play by an imported and internalized set of rules - rules that presume this inner thing that some of us happen to call religion to be a self-evident, eternal, and thus universal possession of all humankind. Rather than changing the rules of this scholarly game altogether - and, for instance, establishing in India and elsewhere the cross-cultural, scientific, study of, say, caste, conceived as the carefully nuanced, historical and theoretical study that adopts and then elevates this no more or less local concept to the status of a universal possession of all humankind - they instead try to beat those they are criticizing at their own hegemonic game (which is a quick but no less useful description of the postcolonial situation). The irony, then, is that even if so-called Western Indologists lose the battle over how best to define and study this thing we somehow all know to be 'Hinduism' they nonetheless win the imperialist war, for now their so-called Others are, for lack of a better term, Westerners too, for they can apparently only think themselves into identity and agency by naturalizing a technical discourse developed elsewhere and then defending not just the proper way to study 'their religion', but the proper way to *be* a 'Hindu' - whether in Bombay or Burlington.

My interest in examining the work of those who historicize the word 'religion' while presupposing some natural, pre-linguistic religious domain (what I am calling religion before religion) has to do with the fact that I think it is this two step process of social segmentation (i.e., seeing some human practices as naturally distinct from others) and linkage (i.e., seeing in the set apart zone something that, despite being buried deep in my heart, is nonetheless unified by its presence in all our hearts) that constitutes this thing that we know as modernity. For example, we see this identity formation process by which matters of difference and similarity are managed in the just read quotation from the foreword to *Invading the Sacred* no less than in the Pope's previously cited pep-talk - when, in response to some imagined relativist interlocutor, who, despite seeing the difference between witchcraft and religion, advises to 'leave them in peace ... [since] they have their truth, and we have ours', Benedict replied:

> If we ... have come to experience that without Christ life lacks something, that something real - indeed, the most real thing of all - is missing, we must also be convinced that we do no injustice to anyone if we present to them and thus grant them the opportunity of finding their truest and most authentic selves, the joy of finding life.

Despite the differences in the content and contexts of their texts - exemplified in references to some homogenously existing thing called 'Church tradition', on the part of Benedict, and in Balangangadhara's interest in identifying the 'properties [that] characterize India of today and yesterday' - both writers, as products of those social conditions that we know as modernity, employ the same techniques - the intertwined rhetorics of distinctly unique experience and enduring collective identity - to create an impression of ahistorical unity in contradistinction to those pursuing contrary interests and competing identities. And the point where this segmentation/linkage process is perhaps most evident in our field is in the discourse on religion before the category religion.

But I am getting ahead of myself; so, to return to documenting the problem of those who historicize the word yet naturalize the concept.... As already indicated, I find in such work a rather traditional, and to me problematic, theory of language, inasmuch as concepts are assumed to float free of language, and thus free of history. For a field that prides itself with taking history seriously - in fact, attention to the historical is among the most common ways that students of religion distinguish themselves from their object of study[7] - this presents us with a number of problems that deserve far

---

7. The opening thesis of Bruce Lincoln's 'Theses on Method' (*Method & Theory in the Study of Religion* 8 [1996]: 225-7) provides one of the strongest examples of this.

more attention than they have received in the work of those interested in faith's apparently recent escape from the interior confines where interfering secularism supposedly imprisoned it several centuries ago. For example, thinking back to Casanova's paper in Germany, we should more carefully consider the now commonplace claim that the category 'religion', along with the binary pairing of church and state, was invented by Christian theologians; as suggested earlier, this claim prompts me to ask to what the pre-category 'religion' signifiers 'theology' and 'Christian' refer? For without contemporary scholars projecting the modern distinction of religion and not-religion backward in time, and selectively naturalizing just one pole of this binary, I am not sure why we, as scholars, continue to see something called 'Christianity' as an item of discourse, much less conclude on the redescriptive level that it is a causal force in human history. For example, without using our modern concept religion, and all that comes with it, to distinguish among literary genres, I'm not sure how that group of rhetoricians we know as theologians stand out as any different from a host of other propagandists. Now, of course they have a first order vocabulary peculiar to themselves, talking about such things as 'sin' and 'salvation', and they work within their own institutional settings, etc., but so do all propagandists. If our job as scholars is not to sanction these discrete identities by taking for granted the means that create them, but, as proposed in my opening example, to study how these identities become possible and reproducible, then as soon as we move from first order description to second order redescription such terms as 'theology', let alone 'sin' and 'salvation', ought to be dropped from our analytic vocabularies, for we should see such first order domains as analogous to other culture-wide processes. In much the same way, I am not only unsure what to call 'Christianity' if we drop our category 'religion' - is it then a mass socio-political movement that uses philosophically idealist rhetorics and ritual references to ancient victimization and capital punishment to mobilize members? But more than this: without the discourse on religion I am not sure why we still assume some discursive 'it' will remain to correspond to whatever new name we may lend to it. This is why I am troubled by finding in the work of many who claim to historicize and theorize the category religion the intact boundaries of those things that we used to call religions, such as Buddhism, Hinduism, Judaism, etc. Would not a new, analytic discourse rearrange the human material - regardless how people themselves arrange and authorize their social identities? Sadly, regardless what we call it, many of us end up studying the same things because a focus merely on the *word* 'religion' leaves untouched what Michel Foucault might have termed 'the internal rules of the reasoning practice'[8], much as the no doubt well-meaning replacement of the word 'Negro' with 'Colored', then 'Black', and now, at least in the U.S., 'African-American' leaves utterly untouched the discourse on race that makes it seem sensible and thus natural to classify and organize people based on the presumably enduring relationship between identity and skin color.

Now, of course, I recognize that people long before us used the designations 'theology' and 'Christian', but - and here's the point that many too easily overlook - just because the same words were used does not mean that they signified in the way that they do for us today. For as my friend Bill Arnal put it in an email, when commenting on an earlier version of this chapter:

> Ignatius of Antioch, in the second century, uses the terms 'Christian' - and even weirder – 'Christianity'. [However], he is not opposing these labels to anything like secularity (which would be anachronistic); rather, they are identity labels for group affiliation. Thus to be a 'Christian' one need not have a concept of 'religion' but simply other [social] identities that are *not*-Christian.

Thus, without the category 'religion', and its relations to a series of other categories, behaviors, and institutions with which we moderns think and act our particular world into meaningfulness, I have no idea what we'll mean when talking about 'Christianity' and 'theology' nor do I know why we, as self-described scholars of religion, will feel compelled to be the ones doing this talking. Simply put, without the discourse on religion - that is, without scholars merely adopting and thereby sanctioning how some people understand and represent their own sense of social identity - would not the study of what we commonly know as early Christianity occupy a relatively minor place in the curriculum of a Department of Classics?

So, much like the role played by creative corporate accounting in our present global economic crisis, supposed gains in studying the category

---

8. Michel Foucault, 'Human Nature: Justice vs. Power' in *The Chomsky-Foucault Debate on Human Nature*, 5. John Rajchman (foreword). New York and London: The New Press.

'religion' hide undisclosed losses. For another example, consider the previously mentioned preference for plural nouns over singulars. The trouble here is that, in dropping such a singular noun as 'Judaism', and replacing it with the plural, the gain in arriving at a more nuanced approach to cross-cultural and historical difference is overshadowed by leaving unexamined just what it is about all those Judaisms that makes them members of some singular genus; for in the proliferation of the many a singular identity remains intact and untheorized. I think here of Talal Asad's influential critique of Clifford Geertz's definition of religion as a cultural system. If we grant to Asad that, as he phrases it in the Introduction to *Genealogies of Religion*: '[t]here is no single, privileged narrative of the modern world'[9], then we will not only critique the export of the concept 'religion', as he does so well, but we will also scrutinize how it is that, when such scholars do away with the universal, reified concept of, say, 'Islam', they somehow still know who the Muslims are and thus who they ought to be studying. That is, despite the proliferation of plural identities 'a single, privileged narrative of the modern world', to use Asad's words once again, but now ironically, is nonetheless retained - for even when we let our universalized categories go there somehow still remains a necessary and natural identity lurking just below the contingent surface.

And it is just this sort of naturalization that occurs when scholars privilege one part of what is supposedly a binary so as to conclude that theologians created secularism. We see in such arguments a form of faulty logic akin to someone assuming that people just knew that they were outdoors prior to the invention of sheltered dwellings - in both cases some identity is presumed to pre-date the discursive conditions necessary to think it into existence in the first place (much as if Adam just knew he was a man prior to the creation of Eve - to borrow a quick example from an ancient Semitic creation myth with which some readers may be familiar). When it comes to recent studies of religion and 'not religion', what we therefore often find is simply a repackaged version of the old, old story of how the so-called primitive world was once homogenously religious and, with the advent of modernity and its imposing opposition of church and state, how it was sadly tamed and disenchanted - we could go so far as calling this the *new* secularization thesis. It is a thesis in evidence, for example, when scholars critique the noun religion while yet relying on the adjective 'religious'. For instance, consider Muham-

---

9. Talal Asad, *Genealogies of Religion: Discipline and Reasons of Power in Christianity and Islam*. Baltimore: The Johns Hopkins Press, 1993, 9-10.

mad Zaman's interesting book, *The Ulama in Contemporary Islam*, in which the British colonial use of the category 'religion' is, in my opinion, rightly named as a socio-political management device. Yet having freed himself from the oppressive category 'religion', the author is still somehow able to speak of people's 'faith' as being different from their political orientation (going so far as to use the awkward term 'religiopolitical' as an analytic tool) and still somehow able to identify such items as religious history, religious scholars, religious authority, religious traditions. Somewhat reminiscent of a mythical phoenix, the adjective - which, unlike the reifying concept 'religion', has the supposed advantage of naming an authentic quality of people – miraculously arises from the ashes of the noun's critique. Yet if we agree with Zaman when he argues that 'British colonial officials routinely invoked what to them were familiar and often self-evident concepts and categories' to manage colonial populations[10], then why do we not also entertain that his and our own 'familiar and often self-evident concepts and categories' - such as the presumption that religiosity pre-exists our modern ability to name things as religion or not - is also up to something? For I would argue that, much like the Vatican's choice of the term 'homily,' so too the presence of the adjective 'religious' in the work of those who critique the word 'religion' tells us far more about the scholarship than its author may have ever imagined.

Although I think that the sleight of hand that produces the adjective's miraculous resurrection is what we ought to be studying - that is, we need to study the discourse on religion and not simply the word 'religion' - looking over the work that has appeared over the past decade or so, my view is not widely shared. In fact, reading such work sometimes puts me in the position of imagining myself to be participating in the academic study of sin - as suggested earlier, no less a local noun than 'religion', and thus one that it is easily possible to imagine being elevated to the status of cross-cultural universal. Moreover, I can imagine how some members of such a scholarly field might come to see, and then become critical of, this obviously imperialist move - critiquing the analytic utility of the category 'sin' as a scholarly tool and therefore dropping it entirely from their work. Yet the curious thing is why such scholars would then end up using the adjective 'sinful' to qualify the things that they study, as if dropping the noun 'sin' and then replacing it with, say, 'gluttony' or 'greed', would somehow suffice. If this imagined

---

10. Muhammad Zaman, *The Ulama in Contemporary Islam: Custodians of Change*, Princeton, N.J.: Princeton University Press, 2002, 62.

scenario is laughable then why do we tolerate the signifiers' faith, spirituality, authenticity, religious, and religiosity - let alone religion - in the work of those who claim to be doing something more than merely paraphrasing the claims already being made by the people whom they study?

What is perhaps most ironic of all is that not just humanists (the scholars whose work has occupied me throughout this paper) but so-called, social scientific reductionists equally presume the uniqueness of that which stands behind our word 'religion' - though they claim to find its origins not in the metaphor of the human heart but in our genes and neurons. For inasmuch as they are working to create specific theories to explain religion, theories distinct from those that account for other aspects of human behavior, they too naturalize first order classification systems. Case in point: in the opening pages to his influential book, *Modes of Religiosity*, the anthropologist, Harvey Whitehouse, devotes a section to the question, 'What is Religion?' After acknowledging on the opening page that '[t]he everyday meaning of the word 'religion' is not all that easy to pin down' he argues that, despite 'a range of exemplary features' often being called upon to name something as religious, '[n]one of these features is necessary for the attribution of the label, but almost any combination is sufficient'[11]. He concludes that this utterly vague use of this imprecise term is itself sufficient warrant for the development of a scientific theory of religion - that is, because the word is used imprecisely by those who employ it in their first order acts of signification, scholars ought to develop a more precise manner of using it as a transcultural universal.

Whitehouse seems to be suggesting that our scholarly task is to systematize the use of other people's folk taxons; although the people who use this term in their acts of self-designation at least had a pretty good intuition into the existence of some sort of cross-cultural universal, or so his argument seems to go, they were nonetheless sadly imprecise in their approach to the topic - an imprecision that scientists can apparently correct. The trouble is, of course, that any anthropologist can offer up a host of other imprecisely-used, first order, folk designators (developed and employed by groups of human beings), that we, as scholars, would hardly think approach the level of cross-cultural universal - no matter how useful members of the groups we study may find them in extending and thereby confirming their picture of the world. After all, as suggested earlier, many human beings not only think they are

---

11. Harvey Whitehouse, *Modes of Religiosity: A Cognitive Theory of Religious Transmission*. Lanham, MD: AltaMira Press, 2004, 1.

capable of sinning (although disagreeing significantly as to what constitutes a sin, of course - which would constitute the needed imprecision to warrant possible adoption as a scientific concept, perhaps) but also that those who are outside their group - and thereby outside the scope of their concept of 'sin' - are sinning without even knowing it; however, such a folk concept is hardly a candidate for developing a scientific study of sin! In fact, should the act of designating some social acts as 'sin' attract our scholarly interest, then our curiosities ought to be explored on a higher order level (after having done our ethnographic work and gathered our descriptive data, of course), whereby 'sin' disappears entirely from our analytic toolbox, to be encompassed by some wider instance of techniques used to (should our interest veer toward the social theoretical) regulate normative identity and behavior.

With this example in mind, why not, instead, theorize why some humans (though hardly all) use 'religion' to name aspects of their social world - and we might include in that group cognitivists themselves along with the humanists who equally see religion everywhere and in every time - thereby studying the various ways in which the discourse is used, and the practical effects of these uses? However, like other cognitivists, Whitehouse's work is premised on the old troublesome folk notion of religion: belief in superhuman agents and the actions grouped around these beliefs. The trouble is this: some of the modern people that we study say that superhuman agents exist, and that a collection of beliefs, behaviors, and institutions relevant to these agents are somehow necessarily inter-related and thus set apart from other aspects of culture, making them 'religious'. Apparently taking their research subjects' word for this set-apartness/grouped togetherness (what I earlier characterized as the modernist practice of simultaneous segmentation/linkage), scholars then busily set about looking for these connections in the realia of our own and other cultures and then try to account for the existence of this supposedly distinct domain - after all, they develop theories of religion and are not content to understand the thing that their research subjects call religion to be sufficiently explained by a higher order theory of something else commonly found in culture. But what if what attracted our scholarly attention was not the taken-for-granted distinctness of religion (thus requiring a specific theory to account for its existence) but, instead, the practitioner's compulsion to represent one part of their social world *as* distinct, *as* unique, *as* set-apart? Then we would work on developing a theory of the *discourse* on religion, which would be none other than a component of a wider theory of social classification - something toward which Maurice Bloch is moving, as evidenced in a provocative essay of his published last year, entitled, 'Why Religion is Nothing Special But is Central' (2007)[12].

Finding the folk discourse on religion naturalized in the work of humanists and social scientists alike[13], I feel compelled to conclude that, of the two best known Smiths in our modern field (purposefully excluding Houston from consideration), each of whom represents contending sides on some of the issues that I have too quickly surveyed in this chapter, Wilfred Cantwell Smith's work is, lamentably from my point of view, far more representative of the modern field than Jonathan Z. Smith's work. My hunch is that this is because of the strategic interests that can be accomplished when using the former's distinction between interior faith and observable practice - interests that are doggedly undermined by the carefully composed comparative studies of the latter Smith, for whom religion is but one instance of wider human practices. For by helping us to distinguish between the pristine, individual person, on the one hand, and his or her subsequent social situation, on the other (a distinction implicit in studies of religious beliefs), the scholarly tradition represented by Wilfred Cantwell Smith prevents us from ever examining the institutional practices that led to just this or that view of 'the individual'. Thus, a limitation is placed upon our scholarship, for while we can historicize the language in terms of which our meanings are exchanged, we can never historicize meaning itself or its presumed source.

What I am arguing is that, despite its seemingly progressive tone, a highly effective depoliticization of the social is taking place when we limit our critical work to just the word 'religion' and when we uncritically treat what is merely a widespread folk category as if it was a natural kind. Tied as it is to notions of interiority, autonomy, privacy, individual choice, authenticity, lived experience, and thus enduring personal identity and intentionality, the folk and scholarly category 'religion' is part of a wider discourse on the modern, universal subject - that which today finds its political home in the notion of citizenship, its economic home in the notion of consumerism, and its kinship

---

12. Maurice Bloch, 'Why Religion is Nothing Special But is Central', *Philosophical Transactions of the Royal Society* 363: 2055-2061.
13. I do not have the space to address how such writers as Daniel Dennett, Christopher Hitchens, and Sam Harris also accomplish this same surprisingly conservative naturalization of what is no more than a local folk category by means of their bestselling books criticizing religious beliefs as irrational and harmful.

home in the notion of race - to name but three sites where a typically modern individual is created. I therefore advise reconsidering our critiques of the category 'religion' and much of our current work on the secular, seeing the persistent habit of assuming people to have rich, active inner lives as but one more folk site where a specific form of what Foucault aptly termed gouvernmentalité has been globalized, a process that has legitimized certain types of subjects and specific types of social relations. For, as I see it, the mission of scholarship is not to reproduce and thereby normalize such local processes - whatever the political ends to which it is working. For it is a process that slips *sui generis* religion in behind the mask of its adjective and its plural, as if some enduring value mysteriously lurks in the background of our contingent vocabulary. Instead, regardless how the people we study may understand their own systems of classification, and regardless how close some of these systems are to each of our own non-scholarly personae, I presume that our work as academics requires us to take such processes seriously *as* history - and thus *as* contingent, *as* fleeting, and therefore *as* requiring labor if they are to appear to us as a permanent acquisition of this thing some call the human spirit. Recovering this labor is the task of scholarship.

THE DISCOURSE OF A MYTH: DIODORUS SICULUS AND THE EGYPTIAN *THEOLOGOUMENA* DURING THE HELLENISTIC AGE

PANAYOTIS PACHIS

It seems to me extremely difficult to write something in a *Festschrift* for the seventy year anniversary of an academic scholar like Luther H. Martin, especially as I consider him to be, due to his idiosyncrasy, an 'eternal adolescent' whose character and inquiring mind as well as long-lasting scientific experience and acute, critical approach to his scientific field, comprise determinative factors that formulate his overall scientific activity. Furthermore, Luther's character contributes to his unique approach to various aspects of the academic field he specializes in, to the control of those aspects, but also to the carving and demarcation of new paths in the complicated world of modern scientific research.

Moreover, we should not forget that Luther belongs to the generation of scholars who, at the end of the 20th century, deem that the transitional character of this era demands a different approach to matters pertaining to the object of the study of religion. This tendency played a central role in the creation of NAASR, the vehicle of expression of those opinions that provide a different perspective on the overall study of religion. This particular initiative is an indication of a tendency to go off the beaten track through new scientific proposals, but at the same time doubt the ideas that dominated scientific research on religion in the 20th century.

Luther's overall scientific approach is twofold: it is connected, on the one hand, to his proposals about the study of religions and the cults of the Hellenistic world and on the other, to his proposals regarding matters that deal with the scientific study of religion. His scientific enterprise is successful due to the tireless attempts that characterize his overall scientific inquiry, which can be compared with the investigations of a detective, who patiently and insightfully tries to find solutions to unresolved problems. We should not fail to mention here that Luther's detective tactics work in such a way that he always 'goes behind the familiar metaphors, typologies, or sets of concepts proposed on the modern historical assumptions' in order to achieve his goal (Martin 2005: 5).

Luther expresses a 'methodology' that lies at the core of the current scientific study of religion, in order to achieve 'a theoretical filling-in of the evidential gaps in our field that is based upon testable hypotheses' (Martin 2005: 5).

In addition, Luther's research can be compared with the construction of a sophisticated mosaic, in which we cannot completely represent the desired image unless every tessera is placed in the appropriate position. This is the only way to acquire the necessary basis for complete and objective research. The modern scholar, especially of the scientific study of religion, should always keep in mind that the different testimonies which shape the framework of his or her study are a part of a system and must always be studied as such. Viewing religion as a social system, which is justified by a reference to a superhuman power, constitutes a *sine qua non* factor of current research. The interdisciplinary method of research is a necessary feature in the study of human religious events. If one does not take into consideration all the elements that shape the period in which a particular religious phenomenon takes place, one will inevitably end up with generalizations and shallow conclusions.

This essay, a minimum token of gratitude to Luther (who is a constant source of inspiration), deals with the ideas of Euhemerus, a writer of the Hellenistic period, whose theoretical views challenge Diodorus Siculus' approach to the mythical world of Egypt and, through these ideas, the religio-political reality of that transitional period.

*

In the first book of his *Library of History* (ιβλιοθήκης Ἱστορικῆς) Diodorus presents a panorama of ideas and practices that dominated the Egyptian way of living during his own life time (1st century BC). Diodorus lives in an unstable and constantly changing world where everything is disputed. After Alexander's death, new social, political, economic and religious ideas and worldviews come to the surface —ideas that complete the image of the Classical world. The traditional principles and values of the city-state are gradually replaced by an ecumenical ideal. Cosmopolitanism is creating new conditions regarding the co-existence of different people within the borders of the vast ecumene. The expansion of the limits of the traditional world that existed within the city-states resulted in the questioning of the traditional cosmological way of thinking. This phenomenon is later amplified with the emergence, during the 2nd century AD, of a new cosmology created by Claudius Ptolemy (Martin1987. Pachis 2003a).

The information that Diodorus provides, regarding the religious practices of Egypt, is of particular importance since it adds to the knowledge about Egypt. At this point, 'Egyptomania' is the new trend that dominates the new reality and, as a result, everything that comes from this country is of great importance and constitutes a subject of examination and description (Ashton 2004). In Diodorus' work there are constant innuendos regarding the oppositions between the local and ecumenical character of the cults of the Egyptian deities. The first tendency is represented by the priests of Egypt who remain loyal to religious traditions. The second one is represented by the ecumenical and syncretistic character which is adopted in the worshipping of these deities already from the beginning of the Hellenistic era. The tendency towards ecumenism is directly related to the Ptolemaic attempts for a renewal of the traditional character of the Egyptian religion.

In order to understand more clearly the way Diodorus describes the land of the Nile, we should keep in mind the choices he makes. Every writer makes descriptions on the basis of the ideas and views of their time. This particularity constitutes an essential factor for understanding the writer's thoughts and intentions. Diodorus Siculus is influenced by the way of thinking and the intellectual tendencies that dominated the period he lived in. His description of Egypt is based on reports of earlier writers, mainly Hecataeus of Abdera, whose travel descriptions of Egypt are the main source most of the subsequent writers have drawn upon. Hecataeus, who lived in the last quarter of the 4th century BC, concentrated on matters relating to the origin of religion and kingship (*FGrH* 264 F25 = Diodorus Siculus I 13,1 in Burton1972: 8-9; 18-19; 15; 28; 32-33; 70-71. Fraser 1998 [1972]: IIa, 450-451, n. 815). He considers both phenomena to be divine benefactions towards humanity. His way of thinking is influenced by (early) Euhemerism, which is one of the dominant religio-philosophical ideas of the Hellenistic world (*FGrH* 264. Jacoby 1907:

969. Burton 1972: 70. Fraser 1998 [1972]: I, 293; 497. IIa, 453-454, nn. 827-828. IIb, 720, nn.15-16).

Euhemerism is being portrayed in the descriptions of Diodorus Siculus and constitutes the main feature of his point of view (Diodorus Siculus I 13,1 in Burton, 1972: 70-71). This specific tactic of Diodorus, which decisively shapes his mythical narration, deals with the presentation and description of the Egyptian divine world. It is indeed an audacious hermeneutical method, since it challenges the particularity of the divine world. This new theory is very important because while it is being shaped it also provides information about the religious ideas and worshipping practices that prevail in the capital city of the Ptolemies during the 3rd century BC. We should also take into account that Euhemerism is again in fashion during the 2nd century BC when it deals with the teachings about the gods. The uncertainty of the religious and political scene over the last decades of the 1st century BC is an additional factor for the dominance of Euhemerism, thus the influence on the work of Diodorus.

Euhemerus of Messene, the introducer of this theory, lives during the 3rd century BC (Jacoby 1907. Cooke 1927. Rostovtzeff 1941: II, 1132. Brown 1955: 59-60. Brown 1964. Spyridakis 1968. Nilsson ^{3}1974: II, 283-289. Green 1990: 55; 108-109; 132; 172-174; 189; 196; 207; 247; 264; 273; 339; 398-399; 402; 602; 622; 629; 632. Winiccrzyk 1991. Winiccrzyk, 2002. Fraser 1998 [1972]: I, 289-296; 298; 301. IIa, 453-457, nn. 800-843. Littlewood 1998. Pachis, 2003a: 61; 199; 339. Garstad 2004: 246-257. Angelis-Garstad 2006). He spends a long period of his life in the diplomatic service of King Cassander of Macedonia, but he spends the longest part of his life in Alexandria as a priest of Zeus Trifillios. During this period he writes a book called 'Sacred Record' (Ιερὰ Ἀναγραφή), in which he develops the ideas of this theory (Euhemerism), at least according to Diodorus. In this he shapes in a decisive way the 'spirit of his time'. According to his theory, the Olympian gods, or to be more specific the 'terrestrial ones', were charismatic people (heroes, kings etc) who particularly contributed to the dissemination of the 'gentle and useful fruits' to mankind and were generally responsible for many 'benefactions' (Diodorus Siculus, I 2,2; 12,10-13,1 in Burton 1972:70-71).

Many scholars believe that Euhemerus' theory coincides with the ideas of the sophist Prodicus (Henrichs 1984). Both writers hold a common base of thought. Euhemerus uses the same terms; like Prodicus, for example, his views are about the 'ferocious living' of mankind and the decisive influence of the spread of cultivated fruits on people's life. These ideas affect a group of writers and shape in a conclusive way their thought (Henrichs 1984: 141 and no 10; 142 and nos 14-15; 143 -147). Among them, Hecataeus of Abdera influences Euhemerus way of thinking (Henrichs 1984: 147-152). Besides, both writers live during the first years of the Ptolemaic dynasty and are affected by Prodicus' ideas as well as those of another writer, Leon of Pella, who lives during the same period. Leon is an Egyptian priest who is considered to be an authority on the divine world of Egypt (Hyginus, *de astronomica* II 20,4. Augustinus, *de civitate dei* V 12, 11. Minucius Felicius, *Octavius* XXI 3. Geffken 1925. Taylor 1975: 26-27. Spuerri 1979. Fraser 1998 (1972): II, 447-448 and no 847; 800-836. Nilsson ³1974: II, 283-289. Zumschlinge 1976. Henrichs 1984: 147-152. Ebach 1990. Walbank 1993: 307-308. Richter 1996: 638-640. Price 1998).

Such ideas, like Euhemerus', would be considered as insolence (ὕβρις) for people of previous periods, especially during the Classical era; in the Hellenistic period, on the contrary, such ideas determine the people's way of thinking. We should not disregard the fact that people who wander in the vast ecumene, and in particular within the newly established cities (Alexandria, Pergamos, Antiochia, Efessos etc), are feeling more and more abstracted from their homes; a tendency that leads to the emergence and dominance of similar innovative ideas such as Euhemerism (Martin 1987: 23-25. Martin 1994: 125-131. Pachis 2002. Pachis 2008). According to the 'spirit of this time' monarchs are deified and Euhemerism constitutes one of the most important means of propaganda about their absolute power, resulting in the establishment of ideas such as their divine descent (Martin 2003).

Euhemerus' approach offers the best possible explanation about the dominance of this particular political system. The limits that always separated the divine from the human world are now being minimized. It is possible that these ideas come from Protagoras (5th century BC), who claimed that 'man is the measure of all things' (Pl., *Cra.* 385 E6-a1. Sext. Emper., *P.* I 216, 1-2). On the contrary, the old Delphic motto 'know thyself' invited individuals to realize that they are mortal and have nothing in common with immortal gods, an idea that we have already seen in Pindar's work and in the great tragic writers of the 5th century BC (Plut., *de E apud Delphos* 392A6. Nilsson ³1976:I, 561;736. Wilkins, 1980 (1917). Schroeder 1989: 346-347. Martin 1994: 122-123). All these result in the novel ideas that are being shaped among the citizens of the cities of the Hellenistic world who are placing Tyche at the top ranks (Herzog-Hauser 1943. Kajanto 1972. Nilsson ³1974: II, 200-210. Martin 1987: 21-23. Martin 1995a. Martin 1995b. Sfameni Gaparro 1997. Pachis 2003a: 22; 56-57; 317-324; 336; 348) and deifying the monarchs.

This new tendency starts with Alexander and continues with the Ptolemies and the Seleucids. The emergence of kingship is one of the best known and characteristic political phenomena of the Hellenistic period (Habicht ²1970. Koenen 1993. Pachis 2003a: 247-289. Chaniotis 2004 [2003]). There is a transition from the local worshipping practices, that are connected with the family, to the broad kinship groups of the public cults of the city-state, and then to the kingship and legitimation of their power. While in the former two cases we find a collective distribution of power in the frame of social organization, in the latter power is restricted only to the king (Martin 2003).

The citizens of the ecumene suffer from the authoritarian expression of royal power in every aspect of their life and that is the reason why they are trying to find a way to justify its existence. They no longer participate in the public life of the city they live in. Their overall way of living can be compared with that of the citizens of the Classical period, who are not meddling in politics. The writers as well as the representatives of philosophical schools of this period act as supporters, through their works, of the established order of this period and as the par excellence *voice* of the idea regarding the deification of monarchs. It is the monarchs who cover the living expenses all those who reside in their courts, who in their turn declare that only the monarchs can resolve the problems that people face during this period (Green, 1990: 84-91; 173-182. Pachis 2003a: 48-49. Hazzard 2000. Stephens 2003). At the same time, the representatives of royal courts are considered to be responsible, with their actions and way of government, towards gods. Through the works and ideas of this period, the 'principle of the divine command' is established; a principle that constitutes the best way of justifying their absolute power.

This idea demonstrates the religio-political *status quo* of Euhemerus' age. It is a manifestation of the broader views of this time, according to which the monarchs are protectors and constant carriers of civilization in the whole ecumene. They are the representatives of harmony, order and stability –in other words the 'divine saviors' and 'benefactors' (Nock 1972a. Nock 1972b: 720-729. Vanderlip 1972: 15. Dunand 1983: 49. Köenen 1983: 152-170. Green 1990: 402. Pachis 2003a: 269-270). The monarchs are recipients of divine honors due to their political tactics as, according to Diodorus Siculus (I

13, 5), they did many things of service to the social life of man (πολλὰ πρᾶξαι πρὸς εὐεργεσίαν τοῦ κοινοῦ βίου). This particular tactic is directly related to the ideology of this period, according to which the king is deemed to be the constant representative of harmony and law of the whole country (*OGIS*, 56, 68. Fraser 1998 [1972]: I, 193-210. Green 1990: 402. Price 1998: 31. Pachis 2003a: 192-193). People know that the wealth and prosperity depend mainly on the abundant crops of the land. The fields of their influence are those in which laws rule. The monarchs' 'philanthropy' is directly related to their overall policies and constitutes the *par excellence* indication of their influence on their subjects, especially when the latter are in danger (Pachis 2003a: 186-206), as was the case during the disordered and uncertain years of the Hellenistic period. Testimonies for all these can be found in various governmental orders issued during that period, where the co-existence of religion and politics is even clearer (*OGIS* I 90. *SEG* 8, 463, 33; 1357. Pachis 2003a: 204-205).

This view is especially important if we take into consideration that the dissemination of a mythical narration in a specific historical period, according to Bruce Lincoln, 'will depend on a great many factors, many of which are contingent to the specific situation. In general three factors must be taken into account. First, there is the question of whether a disruptive discourse can gain a hearing, that is, how widely and effectively can it be propagated; this largely depends on the ability of its varied channels of communication – formal and informal, established and novel. Second, there is the question of whether the discourse is persuasive or not, which is partially a function of its logical and ideological coherence. Although such factors, which are by nature internal to the discourse, have their importance, it must be stressed that persuasion does not reside within any discourse, *per se* but is, rather, a measure of audiences' reaction to, and interaction with the discourse. Although certain discourses may thus be said to have (or lack) persuasive potential as a result of their specific content, persuasion itself also depends on such factors as rhetoric, performance, timing, and the positioning of a given discourse vis-à-vis those others with which it is in active or potential competition' (Lincoln 1989: 8-9). Furthermore, Lincoln maintains that 'there is the question of whether – and the extent to which – a discourse succeeds in calling forth a following; This ultimately depends on whether a discourse elicits those sentiments out of which new social formation, operating along rational (or pseudorational) and moral (or pseudomoral) lines, but it is also an instrument of sentiment evocation. Moreover, it is through these paired intrumentalities – ideological and sentiment evocation – that discourse holds the capacity to shape and reshape' (Lincoln 1989: 8-9).

Nevertheless, the innovation of the Hellenistic period regarding the beneficial influence of gods on people's lives lies in the transition from the mythical to the historical time; as B. Lincoln's mentions 'a dialectic interaction of past and present is evident in the myths' (Lincoln 1989: 28). The above position is further reinforced if we take into account the view of Russell McCutcheon who points out: '[W]e should not forget that despite the attempts to construct a past or future long removed from the present, mythmaking takes place in a specific socio-political moment and supports a specific judgement about the here and now' (McCutcheon 2000:204). Diodorus Siculus' narrations about the mythical world of Egyptian religion can be considered as applications of the above mentioned positions. Among the cultural goods that are offered to people by Osiris and Isis, the most important are: the creation of laws, the foundation of new cities, the learning of how to make new tools that will be useful in people's everyday jobs (like agriculture, hunting, etc.). In this cultural enterprise the divine couple is helped by one more god, Hermes-Thoth (Diodorus of Siculus, I 16, 1; 20, 6; 45, 6; 94,1; 96,6 in Burton 1972: 77-79; 272). Diodorus' tactic aims to ascribe to the monarchs of the country of the Nile, according to the *status quo* of this period, the benefactory attributes of the divine couple of Isis and Osiris, who spread agriculture to the world and thus created more prosperous conditions for the development of civilization (Diodorus of Siculus I 13,5 - 20,6 in Burton 1972: 73-89). This actually determines the monarchs' political behavior: as new gods, they travel around the ecumene and disseminate cultural goods to all people, who in their turn recognize them as providers of culture and benefactors of humanity.

The main feature of their civilizing project is the offering of laws, which constitute the cornerstone of the cities and everyday life of people. The meaning of laws, as prerequisites of culture, can be related to the whole ideology of the Hellenistic and Roman periods. In this way, the adjective 'law-giver' (*Thesmophoros*) (Diodorus Siculus I, 14, 4; 25,1. Burton 1972: 75. *Aretalogy of Kyme*, 16= Totti 1985: no 1, 2. *Aretalogy of Andros*, 20 = Totti 1986: no 2, 6. *Aretalogy of Thessaloniki*, 4 = Totti 1986: no 1, 2. *Aretalogy of Casandrea* [*Potidea*] 4) = *RICIS*, 113/1201 = Bricault 2008: 105-107) that is attributed to the goddess by Diodorus, shows the important role that laws play for the creation of appropriate conditions that contribute to the harmonious co-existence of people. This also applies to the divine adjective 'justice' that is also attributed to Isis (*ID 2079* . *CIG* II, 2295 = *ID* 2079 = P. Roussel 1916: 117= *RICIS*, 202/ 0282. *ID* 2103 = Roussel 1916: 122. *RICIS*, 202/288. Fraser 1998 [1972]: I, 221; 241. IIa, 335-336, and no 76); 392; 413. Dunand 1973b: II, 113. Skowronek – Tkaczow 1979: 135. Mikalson 1998: 276-277)[1].

The foundation of cities by Osiris, with altars for the worship of gods, indirectly implies the monarchs and their tactics (Diodorus I 15, 1-5 in Burton 1972: 75-76). Besides, the establishment of cities in the kingdoms of the Successors is another characteristic feature of this era. The civilizing activity of the monarchs is clearly established by specific actions: the monarchs raise monuments and temples with inscribed texts as commemoration of their benefactions to the citizens of the ecumene. Furthermore, we should not forget that Isis' temples are constructed in the whole Graeco-Roman world (Bricault 2001).

The spread of civilization by these specific deities can be connected to relevant theories about the origin of civilization, where emphasis is given to the cultural role of grains but also to the direct relationship between religion and politics. The latter holds, as we have already pointed out, an important position in Diodorus' work and is connected to the hermeneutical approaches made by Euhemerus. Osiris and his wife Isis are both responsible for the civilizing activity that benefits humanity as a whole. Isis is considered to be the one who discovers the grains while Osiris discovers the methods of agriculture (Diodorus I 14; 27 in Burton 1992: 74-75; 114-116. Frankfort 1978 [1948]:187-189). For this reason Diodorus characterizes Osiris, as the one who 'was interested in agriculture' (φιλογεωργόν) (I 15,6 in Burton 1972: 76), thus placing special emphasis, on the part of Osiris. In this way Osiris, always according to Diodorus, is the one who spreads (διαδούς) all the goods of civilization to people (Diodorus Siculus I, 14-20 in Burton 1972: 73-89). The correspondence between grain and civilization is also dominant in previous periods, especially in Athens during the 5th century BC. This phenomenon is connected to the 'cultural heores' (ἥρωες εὑρετές), who spread agriculture, the primal cause of every culture (Pachis 1998). What is really important among the cultural accomplishments of the 'cultural heroes' is the end of

---

1. All these adjectives are related to the diffusion of grains and they display the identification of Demeter with Isis (Diodorus Siculus I 14). Pachis 2004: 163-207.

cannibalism, which, according to Diodorus, was a usual phenomenon prior to the spread of agriculture to humanity (Diodorus of Siculus I, 14,1; I 27 in Burton 1972: 73-74; 114-116. *Aretalogie of Kyme*, 21= Totti 1985: no 1, 2). In this manner takes place the transition from the environment of the wild and out of limits nature to the civilized way of living (Diodorus of Siculus I, 14,1 in Burton 1972:73-74). Diodorus considers that each god plays a different role in this mission. Isis spreads wheat and barley, which are already known to other peoples, while Osiris teaches them how to cook them (Diodorus Siculus I, 14. Henrichs, 1984:142. Pachis 1998: 144). In this way, it is made clear that any savage habits of the previous periods (such as scavenging) are done away with; at this point we have a barrier between the 'thorny (i.e. uncivilized)» (ἀκανθώδη) and the 'cultivised (i.e civilized) way of living' (ἀληλεμένος βίος) of people (Suda, s.v. βίος ἀκανθώδης and βίος ἀληλεμένος. Henrichs, 1984: 142 and n. 15; 145-150. Detienne 1994: 117. Pachis 1998: 132-136; 157). The next step is the establishment of laws – another basic principle that enhances the way of living, making it more civilized. It is nevertheless necessary to point out that all the above-mentioned reports attribute to Osiris and Isis the discovering of agriculture, i.e. the exploitation of the vegetal world in favor of the human world, paying attention only to the use, not to the reason why plants exist. This omission is justified by the adoption of the Peripatetic teachings by Hecateus; this philosophical school teaches the eternity of the World, in which the elements of the cosmos (the stellar and planetary world) and the species of the animal and vegetal world participate. Thereby when Hecataeus of Abdera and Diodorus Siculus present Isis and Osiris as finders of agriculture and civilization, they continue an already formed tradition, which was not created for the first time by the spirit of the 4th century BC or by Hecataeus' personal inspiration. This adjusted concept is further emphasized in the thought of writers of this era (5th – 1st century BC) and clearly reflects the Greek ideas of the 5th century BC (Henrichs, 1984: 140-152. Pachis 1998: 125-140).

The combinatory activity of Osiris and Isis regarding the cultivation of grains and the spread of agriculture to mankind can be further explained by another interpretation which is closer to the reality of the writer's time. The eccentric activity of the two deities seems to correspond to the analogous task of Demeter and Triptolemus (Pachis 1998: 160-176). One should mention here that Demeter's cult acquires great importance during the reign of Ptolemus I the Soter (Schneider.1967:I, 552- 553. Fraser 1998 [1972], I: 198- 201.

IIa: 333-342 (nos 64-95). Dunand, 1973a: 85-92. Nilsson [3]1974: 94- 95. Thompson 1998. Herrmann, (1999) (2000): 74-75), mainly due to the influence of Eumolpides priest Timotheus, whose participation is decisive in the religious reformative project of the Ptolemies. (Laqueur 1938. Zielinski 1923. Schneider.1967-1969, I: 193; 490-493. II: 770; 834; 840; 841; 848; 872; 885; 929. Griffiths 1970: 76; 78; 84; [92]-93; 394-395; 403. Fraser 1998 [1972], I: 251; 254. IIa, 401, nos 476-478. Nock 1972: 38; 40-41. Dunand 1973a: 71. Nilsson [3] 1974: 94-95; 156; 641. Hani 1976: 75; 191; 196. Solmsen 1979: 23. Mora 1994: 116; 125-127. Gehrke 2000: 269. Chamoux 2003: 330; 345. Stephens 2003: 142. Parker 2002. Arnobius, *Adversus Nationes* V5). The special meaning that they attribute to the cult of the Greek goddess is also made obvious by the fact that the name 'Eleusis' is given, according to Polybius (Pol. 15, 27, 2; 29, 8; 33, 8. Calderini 1935, I: 110; 115. Adriani 1961-: I, 212-214; 218; 245), to a suburb of the new capital (Strabo, 17, 1, 16. Livius, *Ab urbe condita* xlv 12, 2 Schiff 1905. Calderini, 1935,: I: 1, 110. Adriani 1961-, I: 62-63, and nos 11; 159-160; 219. II: fig.5, 9; 11. Fraser 1998 [1972], I: 35; 200. IIa: 110, and nos 276-277; 338, and nos 81-82. Alföldi – Rosenbaum, 1976: 206-207; 215, fig. 21, nos 28-32. Skowronek – Tkaczow 1979: 132; 133, and fig. 1 map of Alexandria); 134, and no 13; 135. Dunand 2004 [2003]: 256. Fraser 1998 [1972], I, 35. IIa, 110, nos 276-277. Tondrieu 1948), as well as to a village in the area of Fayum (Dunand 1973, I: 73. *OGIS* 83. *SB*, 2674. *PPetr*. III. 41,1-7: 66B 1,5-8:. 5 *P.Cair. Zen*. III, 59350, 1-8 [Thesmophoria, Nesteia]. Callimachus, *Cer*. Polybius, XV, 29, 8, 33[Thesmophoria]), where was possibly established the Thesmophorion that the celebrations in honor of Demeter were taking place. The worship of the goddess is also portrayed in the feasts of 'Kalathos'[2]) and 'Demetria'

---

2. Fraser 1998 [1972], IIa: 339, and no 87, says: «In nay case it seems clear that the procession described is part of the Thesmophoria, and not of the Eleusinian mysteries as the scholiast (on line 1 of Callimchus, *Hymn* VI) suggests: ὁ Φιλάδελφος Πτολεμαίος κατὰ μίμησιν τῶν Ἀθηνῶν ἔθη ἵδρυσε ἐν Ἀλεξάνδριαι, ἐν οἷς καὶ τὴν καλάθου πρooδoν. ἔθος ἦν ἐν Ἀθήναις ἐν ὡρισμένηι ἡμέραι ἐπὶ ὀχήματος φέρεσθαι κάλαθον εἰς τιμὴν Δήμητρος. He (1998 [1972], IIa: 334-335, and no 71) also says: «I so not believe that the background of Callimachus' *Hymn to Demeter*, VI, which contains a description of the «Procession of the Basket» in the goddess's honour, if it is of any particular city, is Alexandria; A very strong case can be made out for Cos...». . See also Fraser 1998 (1997). IIb 915-916, no 290.

(*P.CairZen* I 59028 (Demetreia) (258 B.C.]; III 59350,1- 8 [Thesmophoria, Nesteia] [244 B.C.] *P.Tebt.* III, 2 1079, 1, 1-4 (Demetreia] (3rd -2nd c. B.C.]; III,2 826, 1, 3-5 [«Verenikes Thesmophorion]. Bilabel 1929: 2-3. Visser 193: 81; 82 and nos 7; 9. Dunand 1973a: 209 and nos 2. Dunand 2000 [1992]: 158. Dunand, 2004 [2003]: 255 Casarico 1981: 127-128. Sfameni Gasparro, 1986: 249. Perpillou-Thomas 1993: 78-81. Skowronek-Tkaczow, 1979: 135; 137), the two important festivals that were taking place during this period, as mentioned in Zeno's Papyrus (Edgar 1931: 79; 82; 90. Pestman 1980. Clarysse 1980. Reekmans 1983. Orrieux 1983. Franko 1988. Clarysse- Vandorpe 1995. Orrieux 2000 [1992]:175-186. Manning 2003: 16; 102; 110-118; 140-141. *P. Col. Zen* 1,8. 9. *P. Cair Zen* 1, 59056; 2, 59177; 59202; 59203; 59204; 59209). At the same time, during the reign of the Ptolemies, Triptolemus is receiving great honors in Egypt. His cultural activity sets a perpetual example for the activity of the Egyptian rulers (Fraser 1998 [1972], Iia: 340-341, no 95). Pachis 1998: 171-174).

Diodorus Siculus gives us, in this context, some information concerning Triptolemus (I, 14,1; 18,2; 20,3; 29,1, in Burton 1972:74;124). This particular report is directly related to the fact that, during the reign of the Ptolemies in Egypt, the cult of Triptolemus also spreads and gains great popularity during the Ptolemaic and Roman periods (Fraser 1998 [1972], Iia: 340-341, and no. 95. Pachis 1998: 171-174). The idea of the civilizing activity of the hero during this period is becoming particularly popular in the environment of Alexandria. The rulers of Egypt are identified with Triptolemus, a fact which results to the connection of the latter with Osiris (Diodorus Siculus I 18-19; 20, 3-4. Schwarz: 5-6), both of whom, according to Diodorus, are connected with the discovery of agriculture. In this case, Diodorus goes one step further by identifying Triptolemus with Osiris, without separating the Greek from the Egyptian tradition. The formation of such an idea, which defines Egypt as the birthplace of civilization, corresponds to the tendency of many cities to impropriate this primacy. Besides, Diodorus himself, at another point, maintains that the cultivation of grains began in Sicily and then spread to the whole world, propagandizing in this way in favor of his birthplace.

In the 5th Book of his *Historical Library*, Diodorus Siculus mentions the mythical traditions of his birthplace regarding the cults of Demeter and Kore. He focuses on the divine donation of the goddess to humanity, thus confirming the special bonds of Sicily with her cult (Diodorus Siculus V 1, 3,3. Pace 1946: 463-480; 487-507. Gentili, 1954. Gentili, 1959-1960. Miro 1963. Manganaro 1965. Viellefond 1972: I, 141-148. Vozo 1973. Vozo 1980-1981. Kilner 1977. Wegner 1982). Demeter's cult is particularly connected with the religio-political life of this island, since its emergence (7th - 6th century BC) until the Roman domination (211 BC), which during this period is becoming the symbol of state power (Grose 1923: I 347 and no 105,10-11. White 1964. Provitera 1980. Martin 1990. Pachis, 1998: 123-124). Diodorus, in his description about the divine donation, combines the two dominant mythical traditions (the Attic and the Sicilian). The Alexandrian writers use the same tactic (Diodorus Siculus V, 4,4. Nic., *Ther*. 483-487. Schol. in Nic., *Ther*. [ed. H. Bianchi] 484c. Montanari 1974: 109-137. Sfameni Gasparro, 1986: 151; 157. Pachis 1998: 122). In the Classical period, the accepted birthplace of civilization was the city of Athens, due to its cultural effulgence and its political power. But when, during the Hellenistic era, Alexandria is gradually getting acknowledged as a political and cultural centre, it is natural for the geographical position of the birthplace to adjust to the new data and Alexandria to start being considered the beginning of civilization (Diodorus Siculus V 69, 1-2. Pachis 1998: 176-196. Fraser 1998 [1972]:, I, 7; 20-21. Wycherley ^{2}1962: 35. Ferguson 1973: 29, fig. 16. Heinen1981: 3-12. Green 1990: 80-91; 153; 157-158; 160; 313-315. Pachis 2003a: 46 Strabo, XVII 1,7; 13). This opinion was amplified by the prevalence of the syncretistic spirit, which was fully expressed in Alexandria of the Graeco-Roman period, due to the favorable geographical and historical conditions for the conflation of heterogeneous cultural elements. The Egyptian descent of Triptolemus and the assignment of his project by Isis reflected these two tendencies: the appropriation of the beginning of civilization and syncretism. These beliefs resulted in the hero's identification with the monarchs. The relevant tradition affects the house of the Ptolemies; we should keep in mind that this is a period that the deification of humans, and mainly monarchs, is a practice that receives great attention. The new reality that is formulated will be maintained during the Hellenistic era and will reach its climax in the imperial age –mainly in the period of the late antiquity (mid 2nd century onwards).

At this point it is worth mentioning an innovation that we find in Diodorus' text regarding the common beneficial activity of Osiris and Isis. This practice seems to fade in Diodorus' narration as we can see in the so-called Aretalogy of Isis, which is among his descriptions about the Egyptian mythical world (Diodorus Siculus I.27 in Burton 1972: 114-116). This text is the most ancient testimony regarding the hymns in honor of the goddess, which accompany the spread of her cult in the Graeco-Roman world, where the goddess herself declares her benefactions towards humanity. The *I am* (ἐγώ εἰμί), that characterizes the personal style of these hymns, is for the first time presented by Diodorus Siculus and is related to the gods, the kings and the priests of the East. The same notion is a common *locus* in all subsequent Aretalogies, according to which the goddess, and not Osiris, takes over the obligation to perform all the civilizing actions in the Graeco-Roman ecumene (Pachis 2003b: 97-125). Diodorus, in the beginning of this particular hymn, characteristically mentions: 'the tombs of these gods (i.e., Osiris and Isis) lie in Nysa in Arabia, and for this reason Dionysus is also called Nysaeus. And in that place there stands also a stele of each of the gods bearing an inscription in hieroglyphs' (I, 27), in which are written, according to what Isis mentions, all her divine benefactions towards mankind. This report constitutes one more piece of evidence regarding the explicit influence of Euhemerus on Diodorus. This is made clear if we take into account the above mentioned passage, which comprises a clear analogy to Euhemerus' relevant text entitled 'Sacred Record'.

This idea is directly connected to the religious reality of Diodorus' time, according to which Isis starts to attract, during the second half of the 2nd century BC, bigger interest than her husband Osiris (Bianchi 1980: 36. Frankfurt, 1978 [1948]: 181-212. Griffiths 1980. Pachis 1988: 7; 79-80). This wider acceptance of the goddess is attributed, according to the historical facts of this period, to the increased role that the queens of Egypt played in the religio-political matters. This affects the overall way in which the goddess is being presented in subsequent periods (Diodorus Siculus I 22; 27. Bouche-Leclercq1903-1907: III, 53. Dunand 1973a: 34-35. 41. Dunand 1979: 112-113. Dunand 1983b: 87-88. Hazzard 2000: 111; 129; 135-136; 137; 158. Hölbl 2001: 85; 195; 206; 207-210; 219, and no 133)[3].

At the same time, we should keep in mind another factor of this period, i.e. the co-existence of tradition and renewal which constitutes a determinative factor that shapes the 'spirit of the era' of the Hellenistic world (Nilsson ³1974: II, 1-10. Stewart 1977: 517-519; 603-615, Mikalson 1998: 307. Mikalson 2006. Pachis, 2003a: 35 and n. 26). The spirit of renewal contrib-

---

3. We find the first relevant evidence, according to Hazzard (2000: 136, no 173), in an inscription that is dates back to 136 B.C. Cf. *BGU* VI 1249.

utes, within the ecumenical environment of the Greaco-Roman world, to the gradual retreat of the local spirit of traditional society. Diodorus knows, as we have already mentioned above, this reality and therefore it is natural to insert innuendos in his relevant narrations. By adjusting the Egyptian mythical world to this particular historical reality, we can argue that Osiris is, in this case, the model *par excellence* of the centripetal society of Egypt, while Isis on the contrary is completely identified with the ideals of the centrifugal ecumenical world (Smith 1993: 10; 131-132). The henotheistic character of Isis is an example of this reality – it expresses the religio-political ideals of the imperial age (Versnel 1990: 39-95).

Diodorus differentiates himself from the ongoing terminology of his time and this is another innovation regarding the adjustment of his narration to the demands of his era. The term 'benefactress' (εύεργέτρια), which is used in the 1st century BC, replaces in the Greek area the terms πρώτη εύρίσκειν and εύρέτρια, because the ideology of the monarchs' benefactions of this time influences the wider ideology of the period (*UPZ*, 81, 9-10. *I Hymn of Isidorus* [*Madinet Madi*, 3 = Totti 1985: no 21, 76 = Vanderlip 1972:22. *II Hymn of Isidorus* [*Madinet Madi*], 3; 8 = Totti 1985: no 22, 78 = Vanderlip 1972: 37; 38. Vanderlip 1972: 22-23. *Aretolgy of Kyme*, 7 = Totti 1985: no 1, 2. *Aretalogy of Thessaloniki*, 7 = Totti 1885, no 1, 2. *Aretalogy of Casandrea (Potidea)*, 7 = *RICIS*, 113/1201= Bricault 2008:105-107). But Diodorus uses the traditional term. Isis' benefactory feature is displayed in the context of these hymns and in the usage of terms such as κατέδειξε - έδειξε (Diodorus Siculus V 68, 1 [Demeter]. *Orphic Hymn* 76 [Muses]. 78, 3 [Themis]). 'Seeking' (ζήτησις) and 'finding (εύρεσις) constitute the basic features of the Eleusinian ideology, which influences greatly Isis' cult (Plut., *Is. Os.* 371B-372C in Griffiths 1979: 469-499. Vanderlip, 1972:, 22-23. Pachis 2008:389). The same stands for the presentation of the myth by Plutarch during the 2nd century AD while, in the same sense, Apuleius uses the terms *quaestionis* (*petionis*) and *repertus* (*inventio*) (Apul., *Met.* XI, 2 in Griffiths 1975: 116. Horace, *Stat.* I, III, 104).

The climax of the goddess' beneficial donations to humanity is the spread and establishment of the mystery rituals of her own cult. Isis is considered to establish the mystery rituals in honor of her husband, as it is proved by the usage of the terms 'mystery' (μυστήριο) (Diodorus Siculus I 29,2-3 in Burton 1972: 135-126) and 'mystery rituals' (Diodorus Siculus I 96,4) found in Diodorus' text. The Sicilian writer, in contrast to Herodotus' analogous usage of the term, relates it to the new form that it receives during his lifetime in the environment of the goddess' cult (Hdt. II, 170 in Lloyd 1988: 206-209. Sfameni Gasparro, 1985/ 1986: 130-150). The development of mutual relationships between the gods and humans and the support that the divine world offers to mortals constitutes one of the basic features of her cult during this period, a communication that results in their salvation. But at the same time, the notion of salvation is connected – both in this and in the subsequent periods– to ideas about the believers' posthumous bliss (Diodorus Siculus I 25, 6 in Burton 1972: 109); this expectation becomes real through their initiation rites that result in their complete transformation. The external and public character of this cult is also supplemented by the actualization of the mystery cults in general. This is accomplished through the influence of the eleusinian mysteries, which continue to be the *par excellence* alternative form of religiosity, even in this period.

A prevalent position in the descriptions of the Egyptian divine world by Diodorus Siculus holds the connection of Isis with Demeter and of Osiris with Dionysus (Diodorus Siculus I 11; 56,6; 17,4-5; 25). It should not escape our notice that the identification of those gods and goddesses by Diodorus Siculus is different compared to the one made by Herodotus (Hdt. II, 41; 42; 47; 48; 56–59; 121-122; 143; 155-156; 158. Kolta 1968: 31-41; 42-51; 58-70. Dunand 1973a, I, 9-71; 85-86. Dunand 1991: 238-240. Griffiths 1980: 123; 172. Sfameni Gasparro 1985/1986: 130-150. Mora 1986: 84-86; 96; 100; 107; 212-219. Herrman Jr., [1999] [2000]: 71 and nos 7; 73. Lloyd 1976: 218; 220-221; 269. Lloyd 1988: 57; 59; 110-112; 146), despite the fact that they both depend on the *interpretatio Graeca* (Martin 2000: 51. Ando 2008: 43-58). The point of view from which each writer makes the comparison is adjusted to the respective 'spirit of the time' which they live in. Herodotus deals with the Egyptian tradition as a Greek of the Classical period, while Diodorus is the carrier of a cosmopolitan spirit that is dominant during the Hellenistic era. In Herodotus, the contrast between Greeks and barbarians was the one that defined the ideological orientation, shaped the criteria of the cultural evaluation and eventually designated the points of agreement or disagreement with other people. The citizens' position within the Greek city-states and their democratic institutions ensured an *a priori* acceptance of the conditions and principles according to which the city-states were operating. One sees though, that Diodorus criticizes the way in which Herodotus provides his information about Egypt, as the former does not try to simply describe the Egyptian world but to understand it (Diodorus Siculus I, 37, 11; 39; 69 in Burton 1972: 26-28; 30; 100-104; 137; 160; 208); he also emphasizes the ethical and political principles that are related to law. This feature leads to a more intense worshipping of Isis – more than that of all the other deities, because she is the one who gave the goods of civilization to humanity. On the contrary, Herodotus primarily bases his inquiry on what he has heard or on popular beliefs (Hdt. II 42, 2; 47; 91; 56, 6. Lloyd, 1976: 77-140. Lloyd 1988: 145-146. Thompson 1988a: 28-29. Thompson 1988b: 705).

At this point one should not overlook Diodorus' narration about the connection of Osiris with Dionysus and those gods' actions regarding the finding and dissemination of culture (Diodorus Siculus I, 25; 27. II 38. Schwarz 1987: 5-6. Vernière 1990: 279-285. Pachis 1998: 171-172. Hölbl 2001: 283). This identification of the Egyptian with the Greek god constitutes, as in the case of Isis and Demeter, a continuation of the idea first found in Herodotus. But in this case, Diodorus' goals are a lot different from the ones of the Greek author of the Classical period. Diodorus aims to demonstrate or, better, praise the policies of the Egyptian monarchs, who he presents as gods, because he deems them to be the legal successors of the Pharaohs and Alexander. It should not escape our notice that Dionysus constitutes the god-protector of this dynasty. In Diodorus' descriptions, Osiris is identified with Dionysus and both gods undertake the task to civilize humanity. The local character of the Egyptian myth acquires in this way a clearly ecumenical character that is connected to Diodorus' overall way of thinking. The dissemination of this kind of ideas is expected during the Ptolemaic period because the humanitarian ecumenical activity of Osiris, which is a Greek influence, reflects the policies of the Egyptian monarchs. The culture-giver Osiris, according to Diodorus, is the alter *ego* of Dionysus (Diodorus Siculus, I 13). Osiris is presented not only as the master of grains but also of grapevines. These are the two goods that the god is spreading in the whole world with his cultural project. For this reason he is identified by the Greeks with Dionysus. At the point where Diodorus mentions that the god is spreading the cultivation of grapevine, he does not imply Osiris but rather Dionysus (Diodorus Siculus I 15, 8 in Burton 1972: 37). In this way, he refers to the spread of the cultivation of grapevine and the use of wine in Egypt by the Ptolemies. This tactic is part of the general enterprise of the new monarchs to renew the economic situation of the country. Their tactic is part of the framework of the co-existence of cultural ideas of the two peoples, leading to catalytic results in the way of thinking of this time.

There is no doubt that the Egyptians did not know, during the reign of the Pharaohs, wine but only beer, which they produced from barley, something that is confirmed by Herodotus as well (Hdt. II 77,4. Diodorus Siculus I 34, 10-11 in Burton 1972: 33).

Osiris continues his civilizing task in many countries, Arabia and India included (Diodorus Siculus I 19, 5-9 in Burton 1972: 87); then he assigns the task to Triptolemus (Diodorus Siculus I 19). These narrations are part of the 'spirit of the age' of the Ptolemies and imply the civilizing activity of Alexander (Verniere 1990). This theme receives great attention in the narratives of the 4th and 3rd century BC (see, for example Nonnos, *Dionysiaka*) but at the same time echoes the similar activity of the Successors. These points lead us to set forth some thoughts regarding the meaning of Diodorus' report. We can simultaneously see in the description of Diodorus an innuendo about the expansionary pretensions of the Ptolemies. India was always considered to be a *par excellence* source of wealth and that is why everyone had aimed, already from the beginning of the Hellenistic era, to control this region. The development of commercial transactions between Alexandria and India, through the Arabic peninsula, is mentioned in many sources that come from the Ptolemaic period (Fraser 1998 [1972]: I, 180-184. II, 310-317, nn. 380-415. Green 1990: 138; 329-330; 365; 370).

Undoubtedly, Dionysus is an amalgam of the Hellenistic era. The 'hellenization' of Osiris' cult takes place in the Ptolemaic Alexandria according to the teachings of the cult of Dionysus. Proof for the hellenization of the Egyptian myth is found in the narration of Diodorus regarding the dismemberment of the god's body, a choice that shows the direct influence from Hecataeus' narration, who was mainly interested in this kind of syncretism. According to the mythical description, Isis looks for Osiris' dismembered body. She manages to collect all of the pieces with the exception of his 'genitals', which the Greeks and the Egyptians (Diodorus Siculus I 21, 5; 88, 1 in Burton 1972: 92-94) used to worship during the rituals that comprise the main part of the god's cult. These rituals can be traced, according to Diodorus, in the 'feasts' that Isis establishes in memory of her husband (Diodorus Siculus I 27,6).

It is possible that our writer implies at this point the actualization of the *phallophoria*, which hold a central position in Dionysus' cult. Diodorus seems to follow two traditions regarding the identification of the two deities, the Greek and the Egyptian one. In the former, Dionysus is the leading figure while in the latter Osiris is the most important Egyptian god. The common denominator of the two cases is the direct and indirect influences that come from the environment of Orphism. Osiris' culture-giving and humanitarian activity is connected to the relevant feature of Dionysus (Orpheus), which constitutes a hint as to the special position of Orphism within the Egyptian environment –especially during the reign of the Ptolemies when Orphism was one of the main instruments of the Ptolemaic propaganda (Diodorus Siculus I, 11; 23. Gernet- Boulanger 1932: 408-420. Nilsson 1957. Nilsson ^{3}1974: 160- 163. Zuntz 1963 [1972]. Fraser 1998 [1972], I: 204; 205. IIa: 345-346, no 114; 350, no 128). Burkert 1987: 70-71. Burkert 2003. Diez de Velasco – Molinero Polo 1994. Bottéro 1991. West 1994. West 1997. Hordern 2000. Bernabé 1997. Bernabé 2003. Merkelbach 1999. Graf –Johnston 2007: 50-51; 57; 76; 142; 146; 150-155; 159; 177-178; 188-190. *P.Gurob* 1 = Kern 1922: 578. *P. Berlin 11774* (250-200 B.C.) = Kern 1922: 44. Kern1922: 377; 378. *BGU* 5, 1211= *SB* 7266 = *Paryri Selecti*. 208. Call., *Aet*. 3. *fragm.* 75, 4-8 [ed. R. Pfeiffer]. Theocr., *Eidyl*. 26). Diodorus seems to follow the same tactic when it comes to the identification of Isis and Demeter. Thus, Osiris' rituals are identified with Dionysus' (Orpheus'), while Isis' with Demeter's. There is only one differentiation and this is the deities' names, while the essence is common (Diodorus Siculus I 96, 4-9 in Burton 1972: 275-277).

* *

In conclusion, we should mention that this approach of the Egyptian *theologoumena* by Diodorus Siculus may comprise a further confirmation regarding the encouragement, especially in the case of examining mythical narrations, that we should always 'pay attention to the man behind the curtain' (McCutcheon 2000: 205-206). This is even more apparent, if we take into account the fact that the particular mythical narration could be considered 'as an ideologic activity' (McCutcheon 2000: 203), which is an 'ongoing process of constructing, authorizing and reconstructing social identities or social formations' (McCutcheon 2000: 202). This becomes clear if we take into consideration what Bruce Lincoln (1986: 164) so aptly remarks: 'ideology … is not just an ideal against which social reality is measured or an end toward the fulfillment of which groups and individuals aspire'[4] It is also, and more importantly, a screen that strategically veils, mystifies, or distorts important aspects of real social processes' (McCutcheon 2000: 204). All the above acquire a special significance if we consider that 'all human doing is contextualized within historical (social, political, economic, gendered etc.) pressures and influences, we must therefore understand all such doings, partial and linked to specific temporally and culturally located worlds' (McCutcheon 2000: 205).This is further emphasized if we approach them under the spectrum of a panorama of beliefs and practices, that are characteristic of the 'system' of the Hellenistic age. In this way, in the preface of the Greek translation of Peter Brown's *The Making of Late Antiquity*, Theodosis Nikolaides writes: 'all functions cohere; one cannot possibly isolate an event and understand it without viewing it, first of all, as part of the whole, of a system. The use of this common – at first sight – suggestion is explicit in the study of religious forms... The people around the Mediterranean Sea had a *Koine* (common language) that they used to describe their social experiences, their relations to one another, as well as their relations to the holders of power. Aspects of the *Koine*, however, were also used to describe their relations to the numinous power – as P. Veyne would say: tout se tient'. (Nikolaides 2001:18-19).

ABBREVIATIONS

ANRW	*Aufstieg und Niedergang der Römischen Welt*, ed. H. Temporini –W. Haase. Berlin –New York, 1972-
ASS	Archivio Storico per la Sicilia Orientale.
ASNP	Annali della Scuola Normale Superiore di Pisa.
BASP	Bulletin of American Society of Papyrologists.

---

4. According to L. H. Martin, 'the primary goal is to maintain itself (while its stated goal is secondary to that requirement). Consequently, anything produced by any group will be related to that primary goal, i.e., will grow out of its own self-interest to propagate itself. Thus, myth (or ideology, etc.) is always, at least to some extent, propaganda' (personal communication, 30/08/2002).

BSRA	Bulletin de la Société Royale d' Archéologie, Alexandrie.
CASA	Cronache di Archeologia e di Storia dell'Arte.
CIG	*Corpus Inscriptionum Graecarum*. Berlin, 1827-1877.
CP	Classical Philology
EPRO	Études Préliminaires aux Religions Orietales dans l' Empire Romain, ed. M. J. Vermaseren. Leiden:Brill, 1961-1990.
FGrH	*Die Fragmente der Griechischen Historiker*, ed. F. Jacoby.
HThR	Harvard Theological Review.
HSCPh	Harvard Studies in Classical Philology
ID	*Inscriptiones Deliacae.*
JDAI	Jahrbuch des Deutchen Archäologischen Instituts.
NSc	Notizie degli Scavi di Antichità. Roma Accad. Nazionale di Lincei.
OGIS	*Orientis Graeci Inscriptiones Selectae.* Supplementum Sylloges Inscriptionum Graecarum , ed. W. Dittenberger, vols I-II. Leipzig, 1903; 1905.
P. Petr.	The Flinders Petrie Papyri, Pt. 1, ed. J. P. Mahaffly (Royal Irish Academy, Cunninghan Memoirs, No 8]; Pt. 2, ed. J.-P. Mahaffly(Royal Irish Academy, Cunninghan Memoirs, No 9); Pt. 3 , ed. J. P. Mahaffly – J. G. Smyly (Royal Irish Academy, Cunninghan Memoirs, No 11). Dublin, 1891-1905.
P. Cair. Zen.	Zenon Papyri, ed. C. Edgar, vols I-IV. 1925-1931.
P.Tebt.	*Tebtunis Papyri*, ed. B. P. Grenfell-A.S. Hunt-J. G. Smyly-E.J. Goodspeed. London-New York vol. I, 1907. Pt. 1933, pt. 2 (ed. A.S. Hunt-J. G. Smyly-C.C. Edgar . London& University of California Press), 1938.
P. Col. Zen	*Zenon Papyri: Business Papers of the 3rd century B.C.*, ed. W. L. Westermann- E.S. Hasenoehr I (Columbia Papyri), Greek Series,vol.III). New York, 1934.
RAC	*Reallexikon für Antike und Christentum.* Stuttgart, 1950-

RE	A. Pauly – G. Wissowa, *Realencyclopädie der klassischen Altertumswissenschaft.*
RICIS	L. Bricault (ed.), 2005. *Reciueil des Inscriptions concernant les cultes Isiaques (RICIS)*, préface par J. Leclant, tom. I-III. (Mémoires, XXXI). Paris.
RHR	Revue d' Histoire des Religiones.
SB	*Sammelbuch griechischer Urkunden aus Ägypten*, ed. F. Preisigke, then F. Billabel-E. Kiesslingh.-A. Rupprecht, vol.I-: 1915-
SEG	*Suplementum epigraphicum Graecum.* Leiden, 1923-
ZPE	Zeitschrift für Papyrolgie und Epigraphik.

THE HISTORY OF RELIGIONS AND EVOLUTIONARY MODELS: SOME REFLECTIONS ON FRAMING A MEDIATING VOCABULARY[1]

WILLIAM E. PADEN

The way we think and teach about patterns in comparative religion will sooner or later be deeply influenced by evolutionary worldviews that interpret religious behaviors in light of millions of years of ecological processes, a deep history shared by all humans and that has built our brains and our worlds out of the stuff of that process. At the same time that the comparative, historical study of religion—the 'history of religions' field - begins to evolutionize its thematizations it will also realize that it possesses a level of analysis that is different from the empirically based natural sciences. This is not only because of the rich heritage of its own but also because its natural interest and level of knowledge is human behavior in social environments - not genes, not molecules, not neurons. Speaking from the history of religions tradition, and its body of evidences of panhuman thematics, this brief exercise in stocktaking - it is hard to resist the opportunity - tries to sort out some ways of modeling consilience with evolutionary ideas. Because of its short scope and general tone, I offer it mainly in essay rather than research form. I will emphasize that comparative religion needs to build a vocabulary that accommodates evolutionary science though it is not beholden to become a natural science.

---

1. I take this opportunity to thank Luther Martin for forty-five years of lively collegiality and friendship, beginning at Claremont University as graduate students, and then as co-conspirators in developing a department of religion at the University of Vermont. I am grateful not only for Luther's inimitably challenging role as a conversation partner and mover of issues, but also for his encouragement and key support at many stages of my own academic career.

If the academic study of religion is not theology and not experimental science, still it does follow the criteria of disconfirmable historical evidence and analytic, controlled comparison - hence 'the science of religion' as conceived by the founders - and it continues to draw ideas of theoretic application from whatever domains seem pertinent to the intelligibility of the subject matter. Connecting the categories of comparative religion with the categories of the human sciences is not new. We have always tried to redescribe the data of religion in terms of the pertinent philosophies, sociologies, and psychologies of the time and have done so since the late nineteenth century, showing ways that religious life reflects patterned ways that the mind, language, and sociality work. To describe religious behaviors as playing out or improvising upon evolutionary socio-mental infrastructures is then in itself not revolutionary but part of a perennial pursuit of frames that help organize and decipher an otherwise chaotic mass of data. So in principle, translating religious patterns into evolutionary patterns is not out of methodological character, given the way we have routinely raided the toolboxes of other explanatory and interpretive theories - think Durkheim and Freud, existential phenomenology and neo-Marxism, Lévi-Strauss and Foucault. While none of those were based on Darwinian evolution, they were all about inherent ways that humans deal with the world - and by extension, about ways that religious life deals with the world.

At the same time, integration with the evolutionary sciences does pose some new problems. Insofar as those fields address human life at all they are both technical - adaptivity issues go back to genetic research - and far from scientific agreement. Within the sciences issues about the existence, number or nature of cognitive modules is contested, as is the question of the biological basis of cultural evolution. As well, within evolutionary science there are numerous fields, including behavioral ecology, evolutionary psychology, sociobiology, not to mention neuroscience, and these each have their own agendas and levels of analysis (Laland and Brown 2002). The historian of religion by definition is not a professional in any of these fields - all of which depend upon extensive professional training in empirical methodologies - and is then left with the role of picking models on the basis of convenient or theoretic applicability. Normally this will mean keeping to a few thematic applications to religion, selected from an endless repertoire of potentially relevant models.

The professional gulf between the history of religions and the natural sciences has its bridges, which I will focus on, but the most telling point of difference is that there is no direct explanatory route from genes to specific cultural behaviors - all behavior is a response of genes/brains to specific environments. It is behavior-in-environments that interests the historian.

I organize my discussion around four conceptual matrices that I maintain are mediatorial between comparative religion and evolutionary science broadly understood: the concepts of behavior, environment, in-group marking systems, and the role of thematization. I explain these as ready-made conceptual hooks, from the history of religions side, for consilience.

Behavior and environment as levels of analysis
Behavior is the primary level of analysis and interest for historians, where behavior includes practices and productions of physical, mental, linguistic and social kinds. Thus, ritualizing is behavior and so is representing and building; emoting is a behavior and so is writing; collective formations of institutions and artifacts are behavior and so is an individual dreaming. Behavior is all the things humans do; it is all the things genes 'do' when they find themselves in environments. In classic terms, it is the subject matter of phenomenology; in biological terms it is the realm of the phenotype and especially in the human case, phenotypic plasticity.

Behavior is thus a common ground between biological and human sciences. It is a place where theories and knowledges from either side can be communicated, sorted out, tested, applied. It is a matrix that links our biological inheritance as populations that build and inhabit environments with our cultural inheritance as peoples who form lived historical worlds of language and indulge in culture-specific meaning making. Such world-forming amounts to niche-construction where humans collectively build habitats, analogous to the extended phenotypes we call beaver dams, snail shells and hives. Religious systems are then only one more version of this ancient, three-billion year old activity that characterizes the life of organisms. Niche construction, viewed by its biology proponents is 'not just an end product of evolution, but an evolutionary process in its own right' (Laland, 36). I take this to be a major mediating concept between behavioral ecology and the concept of worldmaking in comparative religion.

The concept of environment is a critical component in Darwinian natural selection, where it is the environment, in effect, that is doing the selecting regarding the survival value of the organism. The human brain itself, such is its own adaptive nature, is built to be responsive to different environments, and this is how diverse cognitive infrastructures have formed. Cosmides and Tooby note that the developmental mechanisms of several organisms 'were *designed* by natural selection to produce different phenotypes in different environments' (1997, 13). For example, they point out that certain fish can change sex:

> Blue-headed wrasse live in social groups consisting of one male and many females. If the male dies, the largest female turns into a male. The wrasse are *designed* to change sex in response to a social cue - the presence or absence of a male (13).

At the more complex level of human ecologies, 'cultural variability', as Joseph Bulbulia puts it, 'selects for developmental plasticity - capacities to usefully interact with variable cultural resources and to build functional, behavioral dispositions from these local, varying resources' (2008, 78).

Humans build, rebuild, and manipulate environments. Environments are not just savannahs and rain forests and not just cultural artifacts in general, but any specific object, frame, setting, situation or medium that forms a horizon or stimulus activating behavior. Where any mental or physical object can function as an environment for a subject, the object could be words, ideas, sounds or images; dreams, stories, habits, delusions or lies; ideologies and mythologies; other people's opinions and invisible persons such as gods, demons and ancestors and their variegated masks. The phenomenology tradition here emphasized how objects and subjects mutually constitute each other in distinctive kinds of experiences and consciousness. Every living brain has a piece of environment in play. In religious worlds, those 'pieces', for example, include representations of gods and notions of how to behave in relation to them.

Religious environments are specialized spaces involving interaction with gods, hence they are both a language environment and a social environment. They put an interactive face on the universe. At the cultural level, any of these god-systems represents an ecology just as natural as the shifting seasonal flora on the Galapagos. They can have the force of a physical object and generate commitments that have deep social consequences; as mental objects they can motivate any emotions and actions. I might jump into a volcano if the god tells me to do it.

Cultural evolution is marked by emergent, ratchet-effect platforms of social and intellectual institutions, values and innovations, and while these play out and upon cognitive architectures, the behaviors they induce are not predictable from the default infrastructures. It is then the specifically religious versions or aspects of those environments and the specifically religious versions of behaviors activated by them, that are the domains of interest of the historian of religion.

Systemic in-group environments

Because sacred values and markers are group-specific, there can be no general science of religion without attention to group-level systems. In-group affiliations are particularly strong social environments. Evolutionary psychology is observant that 'culture' or 'society' in general underdetermine behavior and do not automatically download their norms into individual minds, but culture and society are not the same kind of social category as an in-group, a crucial distinction. Groups have membership constraints and a dynamic of the benefits of cooperation as social programs, drawing for their acceptance on deep cognitive intuitions for loyalty, mutualism, imitation, kin recognition, coalitionism, conformity, relevance, and status recognition in individual minds. Culturally constructed 'kin' groups, such as teams, gangs, nationalisms, fraternities and religions then may trigger, play upon and channel various of those evolved dispositions. This kind of group is not the same thing as an anonymous, low-stimulus backdrop of 'culture' in general.

Where 'religion' may be well used as an umbrella term for any variety of interaction with gods, there is another term that is more specific: it is 'religions', for which '*a religion*' would be the singular. A religion here is not religion in general, which is only a semantic, classificatory phenomenon, but a historical group-level system marked by system-specific sacred beliefs and practices. It is, in Durkheimian terms, a set of persons that define themselves in terms of a common origin, symbols of that ancestry, and practices derived from it. These sets may be small clan-like associations or something like the billion-member Muslim Umma, or a sect or denominations within a larger cultural grouping. Popular or folk religious traditions sometimes lack this normative element, to be sure. The differentiating factor for religious groups is that the norms are understood as ordained by superhuman revelations.

Historians and anthropologists of religion will be on their home turf in examining mythic niche-construction. The human niche is one that configures life and habitation according to certain versions of time, space, authority, objects and practices but religions endow these parameters with a superhuman basis. Membership in the religion is marked by certain ritual performances and practices - such as adult baptism, obligatory missionary work, or a bar mitzvah - markings that are signals of affiliation to other insiders as well as outsiders.

Seeing the connections of religions with forms of biological group-centeredness helps historians reorganize large swaths of data. The obvious example would be a shift from thinking of religions in terms of a world religions model, where beliefs and practices are lined up as representing ideas about reality in general, to thinking of religions as on-the-ground affiliations of 'kin' bound populations for whom their religion is a lineage not to be violated, a sacred history or mythic past that is specific to that group, a set of practices that reflect that chain of memory, and an ability to recognize kin by explicit markers. While the social nature of myth is a standard model in religious studies and in socio-anthropological work on religion, it remains to make the bridgework to evolutionary thinking. 'Beliefs' then become kin behaviors in the form of language acts, recitation markers, speech performances, indicating affiliation. Religions thus make norms in the midst of an otherwise complex environment; among other things they reduce social complexity 'by strengthening and disambiguating signals of co-operation' (Bulbulia 2008, 87).

The kinship paradigm points to the theme of conformity to 'kind' by way of signaling and other communicative markers, including phenotypic matching, which so thoroughly configure religious systems. Stereotypic signals, practices of discriminating kin and non-kin, or in-group marking, run throughout the natural world - there, some are genetic, and some learned or environmental. An example of the latter would be nest location (some animals show high degrees of 'site fidelity') or brood odors (E.O. Wilson, 1987: 15).

The markers are ensconced with a certain inviolability both because they are mechanisms of survival - the individual's survival being linked in these normative cases with the survival of the group - and because of the authoritativeness of their superhuman origins, the honor of the group and of the god here being coterminous. It is an honor that must be maintained against the forces of subversion and transgression and maintained by systems of punishment, exclusion, policing, and shaming. Where the norm is under threat, the notion of sacred order applies (Paden, 2000).

Inviolability here is synonymous with one of the root meanings of sacrality, that is, things dedicated and protected from violation. As such, the concept of the sacred - in other contexts often taken as a term for the transcendental object of religion - is brought into alignment with evolutionary concepts of defense, territory, homeostasis, and survival, explaining why upholding the honor of one's sacred things can be a life or death matter. In the Durkheimian tradition, in which the social and the sacred are not different categories, *le sacré* has always been a secular, behavioral and analytical term without any resonance as an epithet for the divine; Rappaport (1999) and then D.S. Wilson (2002, 225-27), have moved that tradition into conjunction with group-level evolutionary models.

If it is not possible to differentiate the honor of the god and of the group, the focal protective points of survival are marked objects and practices themselves. If a group of Episcopalians breaks off from its parent church on the issue of gay marriage, but the authority of the god - presumably the same god - is the same for both groups, it would appear that a precipitating reason would be the schismatic group's belief that the original church had broken the markers of tradition, broken the faith, violated the scripture - to the point that group identity had to be reconstructed. But this has nothing to do with the superhuman being - it has to do with breaking the magical circle of the community's identity, of destroying or tampering with the 'signs' of identity. We see it at all levels of groupings in society (families, nations) where gods may not be salient but *not* breaking the tradition and its flags is.

Prestigious objects are in-group specific in their attractive power. Their prestige is a construction of the group. Gods and their vehicles command attention and respect, eliciting deference, loyalty, sacrifice and devotion. Along with marked borders, then, there is also the magnetic power of sacred entities and commitment to them. Religious prestige here would be a form of social capital within the group and systemic in that sense. When the status is divine or superhuman, and thus hypertrophied, 'supernormal sign stimuli' (Tinbergen 1951, 44-46) might cue the most egregious forms of commitment and expenditure along the lines of the ethological phenomenon that herring gulls 'will ignore their own eggs when presented with appropriately painted wooden models so large that they cannot even climb on top of them' (E.O. Wilson 1999, 252). Gods may function as such painted wooden models, inputs that can drive their human servants to uncommon behaviors. As with the notion of inviolable social markers, the notion of the prestige of the gods when linked to evolutionary perspectives will add new thematic interest to the concept of the sacred.

Thematic bridges

Another link or space between the study of religion and evolutionary science is the activity and function of thematization. Thematizing, normally related to theorizing in some way, has been the primary activity of comparison and comparative religion. With the now abundant litanies of evolutionary concepts as an emerging lingua franca, the comparativist's repertoire of themes is instantly expanded and hence sets of cross-cultural data otherwise separated or kept insulated by the 'world religions' map can take on affinities. The science of religion has rarely found religious phenomena of interest simply as things to collect or look at. Rather, the data are of interest insofar as they illustrate or instantiate some trait, idea, topic, concept, theme that is itself understood to be important or real, thus giving significance to the particular facts. Consilience here - adaptationist theories aside - means testing through comparative reference points and historical analysis the degree to which religious data can legitimately be organized by a particular evolutionary category.

The study of religion has always gone hand in hand with thematizations, typologies, comparative categories. Classical themes were non-evolutionary: concepts of deity, rites of passage, origin myths, sacred space and time, sacrifice, the high god, sacred kingship, totemism, animism and so forth. The history of religions could then be read in terms of these topoi and historical examples of them. G. van der Leeuw's influential *Religion in Essence and Manifestation* found 106 types of religious action and representation, Mircea Eliade's *Patterns in Comparative Religion* similarly provided a thematic handbook by which to read groups of material, though with a different interpretive orientation, and William James' *Varieties of Religious Experience* differentiated several psychological dispositions for religious susceptibility. Anthony Wallace, to take but a single example from anthropology, held that religion should be understood in terms of verbs, rather than nouns, behaviors instead of institutions, and proposed a list of 13 nuclear actions that he thought 'alone or in combination' accounted for religious life. One could take these as pre-evolutionary attempts to parse 'religion' into its natural joints. To some extent these theme-based phenomenologies seemed to parallel chemistry's chart of elements, a list that identified the basic, irreducible building blocks of anything in the material world. Typologies have always been the coin of the realm and will continue to be, albeit in new theoretic languages.

At one end of the spectrum of its functionality, a theme is organizational, a place holder for a collection of examples that have something in common. At the other end, a theme takes on theoretic, explanatory functions, for example 'kin selection', 'reciprocal altruism', or 'the totemic principle'. In between, functions themselves may be the theme, as in 'the paradigmatic function of myth'. Themes have different currencies: they do the work of linking data at specific levels of analysis and purpose.

Like its typological predecessors to some degree, a great benefit of what I am calling evolutionary thematics is the way it can parse into small units the otherwise clumsily essentialized notion of religion. I say it 'can', even though it does not always do this, so ingrained is the essentialism of 'religion'. But religion is just a word, commonly now an umbrella term for the part of culture that involves interaction with gods, and it has no general explanation or origin other than the history of word itself. If so, religion's 'origin' is found in the *Oxford English Dictionary*, not in societies of chimps or in this or that adaptive human mechanism. It may be countered that it is the 'gods' part that we seek the origin of, but then that concept (gods) too breaks down into the thousands of behaviors and types of representations that attach to superhuman agents (which are themselves of endless genres - are gurus gods?), so that any explanation would have to be an account of one of those behaviors or types, or at best one combination of them, and not some imaginary whole. Categories subdivide (to put it positively) - or crumble (to put it cynically) - when held to the complexity of behavior.

Because of the complexity of religious formations, good comparativism in religious studies means having at one's command the widest repertoire of thematic frames and their theoretic pertinence, and the best judgment and versatility in their application. This latter also means knowing the limits of any theme, as the theme is never the whole: themes are always reductions from a whole. As evolutionary models produce greater and greater differentiation of their themes, those sub-themes and differentials (costly signaling breaks down into many sub-aspects, for example, as does reciprocity, or kin preference) become co-opted conceptual capital for the comparativist's pursuit of more refined depictions of religious behavior, one of its essential tasks.

Evolutionary study as a whole provides a thematic alphabet that represents the inherited capacities of homo sapiens. It is a very long alphabet indeed, ultimately something that would include every evolved human system that carried a genetically functional load or program. On the thematic surface religious data could be selected to roughly align with these ABCs. For example, the study of submission to gods might line up with the study of submission/dominance behaviors including the phenomenon of alphas, the study of dealings with gods would line up with the study of reciprocal altruism and exchange relationships, and the study of group cooperation might line up with kin selection or inclusive fitness. But the default or hardware column would include parental investment, conformity and prestige biases, imitative learning, status signaling, predator detection, intuitions about other minds, bonding and attachment, cheater detection, swarming intelligence, mate attraction and selection, metarepresentation, moral judgment, fear of contagion, kin recognition, coalitionalism, for starters, but also human capacities for every type of emotion (awe, shame, wonder, fear), memory, language, intelligence, altered states of consciousness and sociality. Each of these is an ongoing research area for science, and as I am proposing, a thematic resource for expanding religious studies.

It remains that the actual historical productions that at first appear to correlate with these hardware systems seem to get out of hand. They do not seem to just fill a priori slots but appear to also be theaters in their own right with rules of their own games. The history of religious systems takes on its own ratcheted-up possibilities in terms of social organization, innovative cultural scaffolds (print, for example), and imaginative improvisation. Single-letter explanations do not explain the novel. As well, as Luther Martin notes, the relation between biological/cognitive archives on the one hand and culturally idiosyncratic meanings on the other is complex, for 'the development of self-reflexivity among human beings means that biology is not determinative but that its default programming may be overridden' (2001, 303).

Reduced to themes - and what is it that is not reduced to a theme? - the rules of comparison apply regardless of conceptual domain: close control of analogies, awareness of the role of prototype effects, aspectual selection, and generally critical checks on converting a notion about a piece of behavior into a generalization about the whole. Every phenomenon is an instance of more than one conceptual reference point - a fact that makes one to one causal explanation difficult. The new evolutionary forms add to but do not replace the thematic analysis of language forms, aesthetic modes, and institutional types. Comparative work on sacred space, for example, retains its value as a general category but stands to be enhanced and reworked with evolutionary perspectives. Comparative religion thus accrues multiple sets of grammars, criss-crossing the subject matter of religious history with conceptual material from every region of the human experience. In mentioning group-level symbol systems as an example, I have selected only one of many aspects of sociality - one theme, as it were.

I end with a nod to some thoughts on history by the modern creator of the history of religions field. Mircea Eliade contrasts mythic and humanistic history. Both, he says, are a form of self knowledge, of recapitulation, of memory or *anamnesis*. In the mythic version, the past is laden with the work of superhuman agencies and their creation-defining deeds that set the supposed human condition. In contrast, in the 'historiographic' version of history or time we have a 'vertiginous widening of the historical horizon' since the goal 'is no less than to revive *the entire past of humanity*' which is all that took place in historical and prehistoric societies. Through this new historical *anamnesis*, this new version of memory, 'man enters deep into himself' (1968, 136):

> A true historiographic *anamnesis* finds expression in the discovery of our solidarity with these vanished or peripheral peoples. We have a genuine recovery of the past, even of the 'primordial' past revealed by uncovering prehistoric sites or by ethnological investigations. In these last two cases, we are confronted by 'forms of life', behavior patterns, types of culture - in short by the structures - of archaic existence (1968, 136-37).

In a post-religious world, Eliade characteristically avers, the cosmic dimension is denied to us, but the modern man though living in the dominion of time and 'obsessed by his own historicity', opens himself to reconstruction by increasingly broader understandings of history and thus himself.

These remarks and the new 'deep history' shown by evolutionary perspectives on culture (Smail 2008) show a certain unexpected continuity. I find evolutionary science to be like such a prodigious effort of anamnesis and finding ancient patterns within ourselves - and that, ironically, if one plays out this line of thought, our new history can even be said to have its own cosmic dimension.

RELIGION AND MODERN CULTURE

PETROU S. IOANNIS

This article is dedicated to Professor Luther H. Martin, friend and colleague. Luther H. Martin is worldwide renowned for his multi-faceted contribution to the study of Greco-Roman world on the one hand, and on the other to the creation of a theory for the scientific study of religion in combination with cognitive theory. His work is characterized by a strict scientific methodology, which he applies to the research and study of religion as a system that is formed within a society through the cumulative effect of various social and cultural factors. Martin especially challenges the views of those who claim that religion is a particular phenomenon that calls for a special way of study. For this reason, as he characteristically notes, religion should be studied just like any other cultural reality, that is, scientifically[1].

---

1. L. H. Martin, 1987. *Hellenistic Religions: An Introduction.* New York: Oxford University Press. Idem,1996. 'The Post-Eliadean Study of Religion and the New Comparativism', in: *idem* (ed.). Introduction to a symposium on 'The new comparativism in the study of religion' *Method & Theory in the Study of Religion* 8.1. Idem,1997. 'Rationality and Relativism in History of Religions Research,' in: Jeppe S. Jensen and L. Martin (eds.). *Rationality and the Study of Religion.* Aarhus: Aarhus University Press. Idem 2000. 'Secular Theory and the Academic Study of Religion,' in: Tim Jensen and Mikael Rothstein (eds.). *Secular Theories on Religion. A Selection of Recent Academic Perspectives.* Copenhagen: The Museum Tusculanum Press. Idem, 2001. 'The Academic Study of Religion during the Cold War: The Western Perspective', in: *idem* - I. Dolezalová - D. Papoušek (eds.). *The Study of Religion during the Cold War, East and West.* New York: Peter Lang Press. Idem, 2003. 'Kingship and the Consolidation of Religio-Political Power during the Hellenistic Period', in: L. H. Martin and P. Pachis, (eds.). *Theoretical Frameworks for the Study of Graeco-Roman Religions*, Thessaloniki: University Studio Press. Idem - H. Whitehouse (eds.), 2004. *Theorizing Religions Past: Archaeology, History, and Cognition.* Walnut Creek, CA: AltaMira Press. Idem, 2004a. 'The Very Idea of Globalization: The Case of Hellenistic Empire', in: *idem* - Panayotis Pachis. *Hellenisation, Empire and Globalization: Lessons from Antiquity.* Thessaloniki: Vanias

Over the last fifteen years there has been a revival of the scientific debate about religion and the role it plays in the social and cultural realm. Following this debate one may distinguish three different tendencies in approaching religion. Firstly, there are those who approach religion on a confessional basis, many times in a clearly apologetic mood, usually restricting their interests to the study of the particular religion they belong to. Religious officials form part of this category, especially those who are interested in the study and promotion of the role of their own religion with a view to improving and upgrading their own position and role in society. On the opposite side stand those scientists who examine religion from a philosophical and anthropological point of view, as well as from the standpoint of evolutionary biology. Those scientists challenge, either indirectly or in several cases directly, the role of religion and especially its supernatural dimension. As examples for this approach one could mention the studies of D. Dennett (1992, 2007); P. Boyer (1990, 1994, 2002) and R. Dawkins (2006). A third category comprises scientists interested in the interdisciplinary study of religion as a social system, who, among others, put emphasis on the special role the implementation of cognitive theory may play to this effect (H. Whitehouse 2004, 2005, 2007; P. Antes 2004; D. Wiebe 1991, 1999; R. McCutcheon 1997, 2001, 2003, 2005; P. Pachis 2009).

Press. Idem, 2004b. 'History, Historiography, and Christian Origins: The Case of Jerusalem', in: R. Cameron and M. Miller (eds.). *Redescribing Christian Origins*, Leiden: E. J. Brill. Idem, 2004c. 'Redescribing Christian Origins: Historiography or Exegesis?' in: R. Cameron and M. Miller (eds.). *Redescribing Christian Origins*,. Leiden: E. J. Brill. Idem, 2005. 'Cognitive Science of Religion', in John Hinnells (ed.). *The Routledge Companion to the Study of Religion*. London: Routledge. Idem, 2008. 'The Academic Study of Religion: A Theological or Theoretical Undertaking?', in: P. Pachis, P. Vasiliadis, and D. Kaimakis (eds.). *PHILIA KAI KOINONIA. Cultures in Contact: Essays in Honor of Professor Gregorios D. Ziakas*, Thessaloniki: Vanias Publications. Idem, 2008a. 'Can Religion Really Evolve? (And What Is It Anyway?)', in J. Bulbulia, R. Sosis, E. Harris, R. Genet, C. Genet and K. Wyman (eds.). *The Evolution of Religions: Studies, Theories, and Critique*, Santa Margarita, CA: The Collin Foundation Press. Idem, 2008b. 'What Do Religious Rituals Do? (And How Do They Do It?): Cognition and the Study of Religion', in: R. McCutcheon - W. Braun (eds.). *Introducing Religion: Festschrift for Jonathan Z. Smith*, London: Equinox. Idem, 2008c. 'Daniel Dennett's *Breaking the Spell*: A Response to Its Critics', *Method & Theory in the Study of Religion* 20,1. This is a selection of his literature. For an analytical description see in this volume in Martin's CV.

Indisputably, all these analyses and attitudes are connected with the developments and formations of religion in modern reality, which have been greatly affected by modernity. As a result, we will start our analysis by making reference to basic data of modern culture and their implications for religion. This reference is essential for understanding the role of religion in the context of the modern world as well as for the developments and the questions that are raised in the field of the scientific study of religion.

It is a fact that modern culture created a new situation in the world. It is not possible, however, to present analytically in this single article all the developments of modern culture that impinge on our field. We will limit ourselves to a concise presentation of some basic elements that pertain to the role of religion in society. First, modern developments have led to the gradual overcoming of the traditional world, a given fact mainly in Europe as well as in North America, and have created the foundation for human emancipation. By the term emancipation we mean the liberation of man from authorities that dominated the pre-modern world, and generally speaking the liberation of society and people from incontrollable "authorities" (A. Renaut 2004) that usually lead to various forms of authoritarianism. Second, among the changes brought up by modernity one can see the separation of politics from religion and the emergence of new forms of politics that are expressed in the context of modern pluralistic democracy, which is widely accepted as the modern mode of governing[2]. Third, National Constitutions and International Conventions gradually established Human Rights which in modern reality acquired global power and validity. The right to freedom of religion is one of the basic rights; it means that the human beings are free to be religious or not, to change their religious stance and to maintain at will their relations to a religious community or not. Any stance towards religion is reliant upon their will and ought not to be imposed on them by the state. Fourth, there was a rapid development of the sciences that are currently the providers of knowledge, having replaced the traditional knowledge that was closely connected to religion and afforded a religious character. In this way, religious authority lost its power and the role it used to play in pre-modern society in the field of knowledge. Fifth, a lot of traditional social conceptions have been altered giving their place to other ones that refer to human freedom, human relationships, the relation between humans and society or community, the

---

2. For modern political theory see Manfred G. Schmidt (2000). *Demokratietheorien. Eine Einfuerung.* Opladen: Verlag Leske & Budrich.

position of woman in society etc. All this made it possible for humans to understand that it is they who create social relations and structures, and consequently that they are free to choose their own relationships, to develop their personality and to alter social relations and structures. The above social and cultural developments are particularly important as, on the one hand, they created the appropriate presuppositions for humans to realize that they themselves are responsible for the constitution and condition of society and, on the other, they resulted in the demystification of the society itself thus rendering it free from religious legitimation. This last dimension especially, paved the way to social change and released the society from the long-lasting inertia, which was a characteristic of traditional society (I. S. Petrou 2005; T. B. Bottomore 1987; A. Giddens 2006).

These modern choices led to intense ruptures and more particularly they challenged the religious basis of traditional culture. These changes were caused by the emergence of innovative ideas with an intensely anti-religious character that, in several cases, even rejected the existence of God. As it appears, this was a necessary choice for society to be freed from religious authority. These ideas did not, of course, remain consistent but changed a lot of times depending on the problems faced by society. At this point, we should make some observations that will underline the difference between the reasons that led to the atheism of the 18th century and those which result in the non-theistic positions of modern people. Then, the problem was directly connected with politics, which afforded a theocratic nature, and the power of the Catholic Church. The effect of these reactions was immediate and led, on the one hand, to the change of political perceptions and structures and, on the other, to the gradual restriction of religion to the so-called religious field.

In modern reality religion is free and its activity is constitutionally protected. Despite this fact though, several believers, with highly conservative attitudes, and religious bureaucrats find it difficult to accept the basic principle of modern pluralistic society that various groups are free to develop their perceptions, but cannot impose those as a general rule on society (Rawls 2001). By ignoring this principle they pursue by all means to impose their views on others. These demands of the believers provoke the reaction of all those who do not share their religious convictions. This reaction takes the form of challenging the basic facts of religion. Another issue that comes up is the content of basic education, part of which is religious education; religion should be taught at school as a social phenomenon in a non-confessional way, while at the same time the knowledge provided should contribute to the removal of prejudices and cultivation of tolerance among students[3]. Moreover, as Martin maintains, it is essential to safeguard the study of religion at university level and ensure that it is carried out in a scientific mode. Other differentiations will be made further down that might provide explanations as to the different approach of issues as such in Europe and N. America.

It is well-known from the social history of our civilization that big cultural changes are accompanied by the emergence of radical ideas, which in their turn bring about reactions. With the passage of time though, views that initially appeared diametrically opposed become moderated and gradually lead to new formations and configurations. Hegel's view of "thesis, antithesis, synthesis" is a schematic articulation of this situation. This change though, that triggered the change of the social role of religion, provoked, and continues to provoke, reactions on the part of religious officials that try to maintain the position and role they used to have in traditional society. It should be noted here that neither those who adopted a totally negative attitude towards religion nor its apologists saw their pursuits become accomplished in the context of modernity, due to the very important changes, briefly mentioned above, that were brought about in society and culture. The role of religion, despite certain exceptions, has been to a great extent limited, at least in the developed world, to the so-called religious field[4]. Religious organizations, of course, try to promote the importance of religion claiming that it is an event beyond cultural change, thus it should be studied in special ways as it constitutes a *sui generis* phenomenon. Contrary to all these, the

---

3. See for example the Recommendation 1720 (2005) of the Council of Europe and the Guidelines of OSCE (2007) for religious education.
4. The term religious field denotes that the role of religion, which was broad in traditional society and covered a wide range of social activities, is limited in modern society and refers mainly to issues that are characterized as religious. The same applies to other fields, such as the economic, social, scientific etc. Within those fields, special perceptions were developed as well as ways of justifying the activities that take place within their frame. This separation is not, of course, that clear so as not to permit any interaction among them. The important thing, though, is that religion stops providing an ideological back-up to those fields, as was the case in traditional society. The fact that religion acts within the limits of the so-called religious field bears no relation to the view held by some that religion is a sui generis phenomenon and cannot be examined by common scientific means of analysis. But it is known that this position is not valid.

condition of religion in the developed world denotes that in the context of modernity it has lost the biggest part of the role it used to play in traditional society. Another characteristic fact is that those who act on the basis of a confessional, apologetic and a-historical dimension first try to clear religion of any negative historic facts, by making reference to the founding texts of their religion and, secondly, support the positive functions religion may have in the social field[5]. But they avoid examining the true facts about religion within history, which in many cases invalidate de facto the views they support. This is due to the fact that religion does not remain unaffected by the cultural events of each era, but is in constant interaction with various social factors. In the same way religion had adapted to the needs of traditional society, it similarly appears to react to the impact of other cultural factors in modern reality. Taking this into consideration, one comes to understand that religion has not functioned through history in the way that is presented in the basic texts of the religion in question. So, when examining religion one is obliged to study the real context in which it exists and not its idealized form that appears either in the texts of the founders and their disciples or especially in those produced by their successors. At this point one can discern an essential difference between those who study religion or religions as a cultural event or as an event that interacts with culture and those who approach confessionally the specific religion they belong to. Considering, therefore, the interaction between religion and culture, one avoids ending up with a study of religion that is confessional in character, or which examines religion in an idealistic and, consequently, unreal past and present.

In modern reality, as it has already been mentioned, the discussion about religion has been renewed. The circumstances in the developing, as well as in the developed world, but especially the interferences of religious elements in the field of politics[6], have led to the renewal of the relevant discussion. The aforementioned texts of religious scholars constitute examples of this discussion. The most astringent criticism on religion comes from R. Dawkins, who uses various arguments with a view to proving that religion does not produce ethics, whereas humanism does, and that the evolutionary biology invalidates basic facts of religion. Of course, regarding ethics, he could prove

---

5. See M. Eliade's views on the "terror" of history and his a-historical theoretical thinking. M. Eliade (1989); B. Rennie (2006).
6. As an example one could mention the views expressed by President George W. Bush and the Christian Right in the U.S.A., as well as the perceptions of Islamic countries.

what he supports by using data that comes from the history of civilization of the last two centuries. Religious pluralism, an aftermath of the Protestant Reformation, made it impossible for a specific religion to be used as the basis of ethics, which was necessary for facing various problems of social life. This difficulty necessitated the formation of ethics independent of religion. Religious authority was replaced by human reason and the widely accepted moral values (I. Petrou 2005; G. Patzig 1983)[7]. To the same direction, a bit milder, is the analysis of D. Dennett (2007), who combines his basic theoretical arguments with the cognitive theory as well as the theory of biological evolution. Even milder is the analysis of P. Boyer (2002) who maintains that human beings create their gods, bringing to the forefront the basic view that had been expressed by Xenophanes in antiquity[8]. In praxis of course, religion is not only expressed through concepts, but it mainly comprises practices, habits and customs that are culturally disseminated and has formed institutional structures that have gradually acquired comparative autonomy and assert their role in different ways and various arguments. Both practices and institutional structures derive from cultural adaptations; they are not self-evident results of religious concepts. In several cases the adaptation of those concepts to the pursuits of religious institutions is more than obvious. In addition, the establishment of religious freedom made it possible to accept that each person is free to adopt a stance of their choice towards religion. This resulted in the formation of a variety of attitudes towards religion that characterize modern reality.

Sociologically one could distinguish four different types of people's attitudes towards religion. These attitudes appear mainly in developed societies. The first type comprises those who declare their faith and maintain a relationship with a specific religious community. To the second belong those who do not categorically express a non-theistic stance and may have a loose relationship with a religious community. Besides, it is common for people, given the fact of free choice, to form their personal, of syncretistic nature, composition of religious elements that draws on the religious pluralism of

---

7. The work of Kant constitutes a characteristic example. See I. Kant, *Die Religion innerhalb der Grenzen der blossen Vernunft*. Hrsg. Karl Vorlaender. Hamburg: Meiner, 1978 (1793).
8. As Xenophanes characteristically notes: ' If oxen or horses or lions could paint, they would depict their gods as oxen or as horses or as lions respectively'. Quotation 14 B 21 Diels-Kranz. 'If Aethiopians or Thracians depicted their gods, they would definitively depict them black or white respectively'. Quotation 16 B 21 Diels- Kranz.

modern society or the so-called religious market. The third type consists of people who are indifferent towards religion, simply participating in various traditional celebrations that have been obviously secularized but remain religious in name only. A large number of citizens in developed countries belong to this type. Finally, the fourth type comprises all those who are categorically negative towards religion.

Let us now return to the issue of the study of religion. In relation to the three approaches of religion that were mentioned above, Martin appears to follow the middle course, interpreting religion by employing a variety of data as well as using conclusions of various scientific disciplines. At this point we should make reference to a basic difference we think exists between the European and North American study of religion, which might, to a certain extent, account for the theory formed by Martin. This differentiation is due to the different conditions within which the relevant study was realized. Europe, facing religion as a social event and taking into account its predominant position in traditional society, examines religion on a social basis and approaches it from the angle of religious freedom and religion's separation from politics. Therefore, religion is studied as a social event that has lost a big part of the role it used to play in traditional society. On the contrary, in North America, without disregarding the fact that religion is a social event, the study is dominated by an epistemological question, i.e., in what way one should examine religion in the framework of the modern university and what will be the object of scientific study and research. As it was claimed by J. Smith, 'religion' does not really exist in the way it is presented, but this term is a modern scientific product which acquires its content on the basis of the data each researcher chooses to present (J. Z. Smith 1982, 2004)[9]. What played an important role in these developments was, on the one hand, the opposition to the methodology of M. Eliade, who supported the idealistic approach to religion (M. Eliade, J. M. Kitagawa 1959; M. Eliade 1989; B. Rennie 2006), and on the other, the challenges posed in contemporary times to the

---

9. J. Z. Smith's point of view may sound provocative, and exaggerated as well, if one considers that the term religion comes from the first Christian texts and was used later by the representatives of the Enlightenment in contrast to the faith professed by the catholic church, or in any way the predominant church of that time. Moreover, Smith's is affected by the method of research adopted mainly by anthropologists who study various traditional societies and then, when they depart from the place of their study, record the results of their research.

humanistic studies asking them to prove that their research bears scientific status. All this plays an important role in the theory of Luther Martin, especially the increasing tendency in N. America to overcome the positions and perceptions held by M. Eliade, thus creating the presuppositions for a scientific study of religion (L. H. Martin 1996). More specifically, religion cannot simply be considered a set of perceptions for the sacred and the profane or just an esoteric event, as maintained Eliade[10], who, based on those views, tried to draw general exegetical rules for religion starting from personal experience. Religion is not only a set of beliefs; it is formed through institutional structures and practices. Eliade's standpoint is unfortunate, despite the fact that it dominated the post-war period, as it ignores the history and the reality of culture. Luther Martin, in his theory, goes beyond all this and places emphasis on the value of the scientific study of religion. This dimension has led to his taking into consideration the cultural developments and adaptations of religion as well as to his effort to use the cognitive sciences to explain the ways in which religion is disseminated (L. H. Martin 2003).

The essential conclusion of all this is that religion should be studied scientifically, and to be more specific, in an interdisciplinary way. This study is not subjected to emotional or confessional limitations or influences; neither does it obey institutional intentions or pursuits. Moreover, one cannot describe the real functions of religion, as they were formulated through its historical course, by making assumptions for its genesis. The search for its origins, despite the fact that it is a rational philosophical question, does not suffice to explain the further historical course and formation of religion[11]. On the contrary, what is important is to record its true facts and formations through history, its stance towards social and cultural changes, and to interpret the reasons why religious bureaucracies avoid accepting the formations and conditions of modern culture.

---

10. Similar views can be found in the work of J. Wach, although he combined the sociological study of religion as well, who says that what can be examined is the institutional aspects of religion and not its personal aspects that belong to the depth of human existence. See especially his work *The Sociology of Religion*, Chicago: University of Chicago Press, 1944.
11. Further formation means the ritual, concepts, superstitions, adaptation to cultural data of later times, political use of religion, absolutization of past knowledge through religion, myths for the creation of the world, authority and institutional pursuits of religious officials that are combined with the cultivation of fear etc.

Exhausting one's research on a topic such as the search for the origins of religion, which is doubtful as to the final outcome, does not help to understand the formation and role of religion within society and culture. In addition, it is a fact that no matter if one advocates the existence or non-existence of God, there is no point in talking about proof; this will not help to explain the existence, formation, or role of religion in society. Religion may exist regardless of the existence or non- of God. One way or the other, the representations of the sacred through history are so varied, that it is impossible for everyone to claim that all of them are all true despite the fact that in modern times each religion claims to possess the 'absolute' truth. In many cases the way of formation and representation of the sacred obviously afforded no truth or was anthropomorphic or served human needs. No matter what the representation was, it had been combined with practices, rituals, customs and perceptions which in reality constitute the content of religion. Of course, it is important to examine the cultural implications of various perceptions of the sacred in relation to the human activities and attitudes. In short, what kind of actions, activities and relationships did the perception of the sacred produce? This question is open to scientific examination because it has to do with the study of real data.

More specifically, one may study the institutional formation of religion; the role of religious authority in the legitimation of a 'supernaturally' established power; its adaptations to social and cultural events or their presentation under the cover of religious legitimation; the role of religion in the interpretation of the creation of the world and therefore its function as a worldview; the coverage of social knowledge and its function as the keystone of traditional 'scientific knowledge'; the justification through religion of the social practice and structure as well as perceptions of the human, of human life and its purpose.

It goes without saying that it is impossible to follow up all religions in a single article. Using modern culture as a point of reference, however, one can briefly make mention of the Christian traditions. In this way, things become clearer and more specific.

It is a given that in the traditional world, especially in its medieval and pre-modern version, religion adapts itself to the cultural circumstances and plays the most basic role in the legitimation of authority, both political and religious; it covers with religious status the social structures and perceptions of society; it presents narratives that provide explanations for the creation of the world; it contributes to keeping society stable; it interprets disease with reference to the supernatural and cultivates a sense of guilt in humans; it defines evil and the prohibitions in life; it reacts to material nature of life identifying it with the evil etc. These functions and perceptions do not necessarily correspond to what one finds in sacred texts; they usually acquire a dimension reality of their own through their adaptation to cultural data and their independence from their origins. All these constitute a framework that regulates the life of human beings, creating psychological complexes and a sense of commitment.

Modern culture opposes all this. It is a reaction of human beings to the suffocating obligations, imposed upon them by tradition. The result was intense reaction, assertion of freedom, demystification of politics and society but, in many cases, overcoming religion itself. The change in perceptions and the developments in science brought on the collapse of the traditional worldview. The religiously legitimized politics is demystified and replaced by secularized politics. Humans consider themselves able to set the rules of the game; to define what is ethical or not, what is allowed and what forbidden. They do not need religious authority for this, as it leads them to wrong choices, oppresses them and deprives them of their freedom.

The attempt to attribute these developments to the biological factor makes them rather difficult to understand. If all the aforementioned were biologically defined, how could they change? On the other hand, the cultural approach may provide a more convincing interpretation, as it may explain the changes brought on by human beings. It is they who are capable of producing culture, disseminating it but changing it as well. Modern culture, and culture in general, is a human creation; it is not something that remains firm and unchanging. On the contrary, it has undergone a lot of changes in the course of history. Humans judge it, alter it, find its faults and seek to correct them; sometimes their efforts are successful. But there is change, as they have broken free from religious authority and legitimation, which attributes their creation to another, non-human, level. Of course, it has always been the case that culture and society are human creations. And it has always been the powerful members of the community who imposed the specific structures, perceptions, practices, rituals and principles that characterized society. The members of traditional society, however, believed that all this was a 'divine' creation; something that was demystified by modernity.

One modern accomplishment, among others, is the establishment of religious freedom as a fundamental human right. This means the liberation of human beings from the imposition of religion by the state; the right to change one's religious convictions according to wish; the disengagement from the obligatory accession to a religious community, that makes accession a matter of personal choice; the option to change one's religious community. It also entails, not only the protection of the right of people to choose their way of expressing or not their religious beliefs, but the right to practice their religion either individually or collectively as well. For the religious communities, no matter which ones, this means freedom of action but at the same time restriction as to the potential they used to have in traditional society to impose as obligatory the accession of people to them. Moreover, it gave people the opportunity to form their identity freely, considering religion not a compulsory but an optional element of one's identity.

Religious communities, despite the fact that they are obliged to accept these developments, usually face religious freedom positively when they are minorities but not so when they are majorities, simply because this makes it difficult for those communities to attract new members. In Europe, especially, centrifugal powers regarding religions become all the more obvious. This development which is essential for human beings, as it gave them the opportunity of free choice and liberated them from the obligatoriness of religion, is treated cautiously by religious institutions because it can undermine their authority.

If one considers religion to be a biological aspect of being human[12], then this attitude towards religion remains inexplicable. On the contrary, the connection of religion and its reproduction with culture and social developments provides an opportunity to interpret this phenomenon. It is a fact of course, that one will explain biologically the ability of man to perceive the socially formulated religion through the higher brain functions and disseminate it (Martin 2003), as is the case with various other cultural data.

Now, what is the attitude of religious institutions towards modernity? Some characteristic elements may be mentioned here as an answer to this question. Religious institutions reacted defensively against these

---

12. In this way, talking about immanent religiosity, apologetics of the 19th century supported religiosity. See for example the works of two apologists: Ioannis Skaltsounis and Ernest Luthardt. This view is used in many cases even today for the apologetics of religious institutions.

developments. They projected their role as keepers of tradition, maintaining that tradition is essential to human life (Giddens 1996). However, when one approaches all these issues from the perspective of human freedom and the humans potential to create culture, one can easily realize that argumentation of this kind does not convince a large number of citizens in developed countries. Nevertheless, the institutionalised religions try to describe reality in a way that justifies their own position and function as an indispensable element to the smooth function of society[13]. But they fail to see that in many cases the views they support reflect social perceptions and structures that have been long overcome. One could perhaps support, with a touch of irony, that the way in which they describe reality creates the feeling that it is reality's 'problem' the fact that it does not fit its religious descriptions. Surely this is easily understood not to be true, and it is further proved by the attitude of people towards society. People, even those who maintain their relationship to religion, accept as self-evident the social structure and description of reality as they live in it; they shape practices, perceptions and actions within its context, and participate, either actively or passively, in the formation of modern reality. On the contrary, religious institutions attempt to impose, directly or indirectly, a reversion to a religious description of society, considering that this will produce a justification for their role. In this framework they avoid taking into account human attitudes and the way in which modern people face society, its problems, and even religion itself.

The whole attitude of religious institutions tells of their negative attitude to modern culture.

A characteristic example of this kind of attitude towards modern culture is the way religious institutions face science and democracy. They usually try to challenge the rationalism of science, its developments and accomplishments as well as the position it has taken in people's life. Religions see that what is questioned is not only the religiously-cloaked knowledge of traditional society, but also a lot of religious perceptions, as they are expressed in

---

13. See for example the text for the climate change of the World Council of Churches, an ecclesiastical body, which could of course be considered the most modernized religious body compared to all others and to its churches-members. In this text they try to justify the claim that climate change should be approached and dealt with on the basis of religious categories. Reality, though, proves that such an approach cannot have any practical results in facing this problem. On the contrary, what is needed is a rational approach that will point to the necessary action to be taken.

modern times. Of course nowadays it is a fact that religion blames science, for authority is something that does not stand to reason. This happens because science is constantly developing and nobody can tell what its future will be, as scientific research invalidates or goes beyond perceptions taken for granted at any specific time. The fluidity of this situation does not negate the value of science; it simply points to the fact that acquired knowledge cannot at any point be considered definitive and final. New knowledge that is constantly produced by means of scientific research may supplement, amend or negate previous knowledge. This fluid situation in the field of scientific knowledge does not fit the views of traditional past for the world, man, life, disease, society etc, that are held by religious institutions. This attitude stems from their static perception of reality. But human freedom and continually renewed knowledge lead to constant modifications of social and cognitive data. It is through this ever-changing data that people face life and reality. Furthermore, even those who hold perceptions of religious nature, when in need, in case of disease for example, rush as it is normal to the doctor and do not limit themselves to any obsolete views expressed by religious officials or especially religious texts.

The example of the attitude towards science and social perceptions is a characteristic one. It shows that there is a great divergence between religion and its believers. Believers, who are simultaneously members of society, by inference accept modern developments contrary to what religious officials and those who are confessionally involved with religion profess. It is not possible, of course, to know if those are only verbal claims or if they actually believe what they say they believe. But, no matter what they say, when they face health problems they also seek medical care.

The issue, though, that optimally depicts the attitude of religions towards modern culture is that of democracy. They surely cannot, with the exception of fanatics, openly express themselves against democracy, as pluralistic democracy appears, in our modern era, to be the basic way of political structure and governing. But it is a fact that religions still hold to their traditional, non-democratic, hierarchical structures. Yet it is essential at the present time for them to operate within a democratic structure, so as to avoid the appearance of non-transparency and authoritarianism, to limit economic scandals, and to persuade people that their huge property does not belong to closed groups of religious officials.

Despite all the above it is true that by promoting some perceptions found in their scriptures, religions have the potential to project positive attitudes for life and cultivate hope. Their practices, though, as recorded in everyday life, do not always depict this attitude. It should be mentioned here that when one would like to study religions, one should search for the practices and views that they project on modern life and examine the social implications of those views. That is, one should seek to find the true role of the religious authorities and not project the ideals of their Scriptures so as to justify their role in modern society.

Finally, if one reverts to the methodological issue of study of religion, one is faced with the ideological research tactics that were employed over the previous decades. It is now obvious however, as I have shown in the preceding analysis, that such a methodology is wrong and can lead to misunderstandings. The scientific study of religion cannot but include the dimensions mentioned above, and examine religion interdisciplinarily, taking into consideration not only its ideas but its practices as well. This is an essential perspective for the study of religion. In this way one avoids idealisation and searches for the real contribution of religions, instead of the picture religious bureaucrats try, in an apologetic way, to project. This fact clearly shows the inability of religions to adapt to modernity or to persuade us of their value to society. Such an apologetic stance raises the question for their role and constitutes evidence that religion is not undeniably recognized by all people.

In this case it is vital to distinguish between religious communities in Europe and those in societies where elements of the pre-modern world are well-preserved. Examples of such societies are those dominated by Islam. In spite of the interest that such an analysis presents, this cannot be realized at this point, as it calls for writing another article. This is the reason why we limit ourselves to comparative observation only. Religious bureaucracies in Europe in many cases try to see life and society through the perspective of the traditional era, which took religion and society to be closely connected. Such an approach does not, of course, correspond to modern reality. The majority of European citizens, no matter if they preserve relationships with religion or religious communities, have a view of life and society that derives from reality, has demystified society and politics and has restricted religion in a special and demarcated sphere. On the contrary, in societies that, generally speaking, still experience a pre-modern situation, in which religion is closely connected with society, members face life through a religious perspective. So in this case, it is not rational to try to separate their society from religion and deal only with society in a judgemental and negative way, as such a society preserves relations and conditions that are regarded as negative when judging by the standards of modern life and the values that have acquired ecumenical validity, such as human rights; on the contrary one should evaluate their religious traditions positively dissociating them from social conditions. This approach, that fictitiously separates religion from society in traditional societies, is wrong for the simple reason that it is the perceptions of their traditional religion that legitimate the social status quo and deprive people of their right to emancipation. It is therefore significant to examine society and religion in societies that are governed by standards of the pre-modern era. This approach is called for by a fair scientific study of both these societies and the religion that is directly connected to them.

We consider that this analysis brought to the forefront some elements that characterize the attitude of religious traditions towards modernity and democracy, as the latter is understood in the developed world. At the same time attention was drawn to the differentiation of societies that are governed by standards of traditional religions. All that attests to how much religion is affected by social conditions and culture does not allow for examining it as a special phenomenon that is solely defined by a supernatural dimension and functions separately from society and culture. On the contrary, it necessitates, as it is obvious in Martin's analysis of religion, that what is needed is an interdisciplinary study of religion as a social system that functions within the context of culture.

DO RELICS DO? MAKING A PLACE FOR 'PRESENCE' IN THE COMPARATIVE STUDY OF RELICS

DOUGLAS ROBINSON

> 'There's really no such plant as a weed.
> A rose bush, growing in my cornfield, is a weed.'
> - Jonathan Z. Smith (1978, 290-2)

I would like to begin by restating a second-hand account of a remark made by Jonathan Z. Smith during a panel discussion dedicated to his book 'To Take Place' on the occasion of its twentieth anniversary[1] in which Smith evokes the Pearl Harbor Memorial as illustrative of 'a fixed sacred space, built upon the very bones of the victims of the attack' (Thomas, 2008, 776 n. 2). The 'locative specificity' or 'thick associative content' (Smith, 1987, 86) of this place, signaled, as it were, by the memorial's historical significance and inclusion of actual bodily relics of those killed, stands in contrast to Smith's example of the Vietnam War Memorial in Washington, D.C. which is principally 'constructed at an arbitrary place, wherever happened to be room in the park system' (2004, 108). For Smith it is this selfsame arbitrariness of place that affords the latter memorial with autonomy from the distractions and exigencies of 'locative space' and highlights, instead, the Memorial's *systemic* role as a unique place for social 'clarification'. In effect, Smith writes, '[t]here was nothing to interfere with the Vietnam War Memorial's pure social representation' (Ibid.; cf. 1987, 83). The inclusion of bodily remains in the first memorial, on the other hand, the fact that they are *materially* 'there' is

---

[1]. The panel was organized for the 2007 Annual Meeting of the Society of Biblical Literature under the group, 'Space, Place, and Lived Experience in Antiquity;' the proceedings of which have subsequently been published as a Book Review Symposium in the *Journal of the American Academy of Religion*, 2008 (Sept.) Vol. 76.

what signals the fixity of the sacred space. Rather than pausing here to critically engage with the issues raised by Smith's theoretical proposal in *To Take Place*, I would like instead to use the above comparison as a context for thinking specifically about the concept of 'relic' as a cross-cultural category. What *constitutes* a relic?

While the study of relics has amassed a considerable body of scholarship within the history of Christianity and Buddhist studies in particular, considerably less work takes the form of analyses where 'relic' is treated as a theoretically formulated category informing the study of religion more broadly (see however: Schopen, 1997; Sharf 1999; Strong, 2004). One might attribute this general dearth in comparative research to the reigning area studies epistemology of the academy which emphasizes context and local knowledge. Yet one need not look that far! As the Pearl Harbor Memorial makes plain, relics, by virtue of their intrinsic locative specificity as the bodily remains of significant people with specific biographies, resist any and all attempts at delocalization. Yet, while this may be a nice characterization of relics at an emic level (albeit for different reasons than those suggested above), the localizing tendency of relics does not mitigate the raison d'état for establishing what constitutes a 'relic' as a comparative category. Rather, such tendencies to localize the 'nature' of relics in both religious and scholarly discourse alike make the present theoretical proposal all the more necessary. Writing on the processes of theory and category formation in the study of religion Martin has observed that:

Theory is but a kind of generalization; both attempt to explain the greatest about of data in terms of the fewest number of principles... Unlike the invariable laws of nature sought by the natural sciences, the human sciences...can attempt, however, to differential between *valid* and *invalid* generalizations (2000, 137-8).

Following Louis Gottschalk, Martin affirms that a valid generalization should 'at least conform to all the known facts so that if it does not present definitive truth it should at any rate constitute the least inconvenient form of tentative error' (Ibid.). Consequently, a generalization is only as powerful as the evidence supporting it, and is thus a thoroughly heuristic exercise (Paden, 2000, 186).

When comparative generalizations are proposed in the study of relics, they most often affirm the significance and centrality of the 'presence' of relics. What differentiates the remains of the 'very special dead' (Brown,

1981, Ch.4, *et passim*) from ordinary bodily remains, it is argued, is that relics are understood by religious insiders to retain a powerful 'presence', a type of intentionality, even after death. However, as both a point of comparability (Smith, 1990, 51) affording scholars with a 'reason to compare' (Poole, 1986) as well as a taxonomic discriminator distinguishing relics from other material objects, the category of 'presence' has only recently begun to precipitate serious theoretical reflection. The present inquiry, then, attempts to advance this body of work through a consideration of the salient *generative mechanisms* involved in the representation of relics as conferring a power of presence.

I

On meeting a Buddha slay the Buddha
- Lin-chi, 9th cent[2]

The popularity of the notion of 'presence' - in Latin, *praesentia* - as a distinctive characteristic of relics is often attributed to Peter Brown's seminal work *The Cult of the Saints: Its Rise and Function in Late Antiquity* (1981) in which Brown, reflecting on the shifting location of grave sites which crept inside the city limits by the 6th cent. CE, explains, 'this was because the saint in Heaven was believed to be 'present' at his tomb on earth' (3). While the present inquiry is restricted the field of Buddhist relics in particular, much of the same language of presence is identifiable. Before the study of Buddhist relics emerged as a thriving discipline, Paul Mus, for instance, is quoted as having said, 'The *stūpa* is the Buddha, the Buddha is the *stūpa*' (Kinnard, 2004, 135 n. 2). Traditionally, the debate over the Buddha's presence in his relics has taken place on a continuum in which the Buddha's pure ontological presence and absolute absence act as two polarities around which a constellation of different theoretical proposals have emerged (see Strong, 2004, 4-5).

The most popular position - following Brown's critique of the two-tiered model between elite and popular forms of religiosity (1981, 1-12) - has been to emphasize that the Buddha's material relics, the objects associated with his life, as well as the images found at Buddhist ritual sites (constituting the tri-

---

2. For a discussion of this quote in its original context see Dumoulin (1988, 196).

fold classification of Buddhist relics which has been operative since the 5th century Pali commentarial tradition) all act to denote the Buddha's presence in light his post-*parinibbana* absence. Consequently, these scholars understand the establishment of the Buddha's relics as a form of worship to be what Smith has called a 'response to a situation' (1979), a technology for overcoming the problems associated with the possibility of generating merit in light of the Buddha's absence. While these scholars are right to emphasize the centrality of the Buddha's presence, and thus alleviate the study of Buddhist relics from traditional rationalist paradigms which marginalize the importance of Buddhist piety (Sharf, 166), others have noted that the language of 'presence' is often 'inherently vague and carries with it significant and sometimes troubling philosophical and theological overtones' (Kinnard, 2004, 117).

Circumventing the cryptic language of 'absent presences' and 'present absences' with its language of 'zero signifiers,' 'chronotopes', and 'heterotopias' (Strong, 2004, 4), others have emphasized the *functional* equivalence of the Buddha and his bodily relics in ritual practice. Scholars have demonstrated, for instance, that the Buddha's relics continued to act as the owner of monastic property (Schopen, 1997), that 'a monk's failure to worship at a *caitya* [relic monument] [was] equivalent to neglecting to attend upon the Buddha' (Trainor, 1997, 93), and, most notably, that the destruction of a *stūpa* constituted an atrocious act of violence classified alongside with murder (Schopen, 1996). As Trainor corroborates following a commentary by Buddhagosa, a fifth century Pali commentator:

[t]hose who destroy a *caitya*, cut down a Bodhi-tree, or attack a relic are guilty of a grave offense equivalent to an *anantariya* act, i.e., an action so heinous that it inevitably results in the perpetrator's being reborn in one of the Buddhist hells in his or her next life. There are five such offenses, including patricide, matricide, killing an *arhant*, shedding the blood of a Buddha, and causing division in the sangha' (1997, 93).

Functional accounts of the 'presence' of relics provide a context for critically engaging with the political and economic appropriation of relics (the basis of Geary's *Futra Sacra*, 1978) while simultaneously affirming the emic reality of ritual actors; however, such accounts often bracket questions asking *how* or *by what mechanisms* relics are attributed such religious and social functions. As forms of explanation, then, functionalist accounts tend to presuppose the 'presence' of relics rather than treating the concept as a problem in need of theoretical clarification (yet see Lawson, 1984). Finally, when scholars have set forward explanations of the Buddha's 'presence' in his relics, such proposals often come very close to 'attributing to Buddhists a kind of Lévy-Bruhlian 'pre-logical mentality' that senses a 'mystical participation' between the Buddha and his relics' (Strong, 2004, 4; Sharf, 167). It is my contention that such a theoretical framework is inadequate as an explanatory foundation for the category of 'relic,' and, having briefly recapitulated and critiqued this proposal, I will venture to offer an alternative.

In *How Natives Think* (1979) Lucien Lévy-Bruhl offers a reading of the Arunta and their use of sacred objects known as *churinga* which closely echoes certain interpretations offered for the 'presence' of Buddhist relics. We are told, for instance, that the *churingas* are decorated with signs and are kept in a sacred place where 'women and children dare not approach' (1979, 92). Most tellingly, however is when Lévy-Bruhl observes:

A man who possesses such a *churinga*...gradually comes to feel that there is some special association between him and the sacred object - that a virtue of some kind passes from it to him and also from him to it. Can we be surprised, therefore, that the *churinga* is represented, or rather, felt, to be a living being? It is something very different from a piece of wood or stone [of which it is created]. It is intimately connected with the ancestor (93)[3].

Thus, just as the ancestors of the Arunta are intimately connected with the *churinga*, the functional accounts cited above likewise preserve an interpretation of the Buddha's presence as continuing to 'participate' in his relics despite his final passing away in Nirvana. While most scholar emphasize Lévy-Bruhl's notion of 'primitive mentality' which, as a mode of thought distinctive among 'primitive' non-literate cultures, is characterized by: a 'veritable indifference to "contradictions" ... an absorption in the "mystical" ... and a certain marriage between what we call the cognitive and the affective in which the latter is much the dominate spouse' (Saler, 1997, 45). Saler has also convincingly argued that to understand this 'pre-logical' mode of thought more emphasis must be focused on Lévy-Bruhl's notion of 'participation'. Participation 'accounts for the togetherness of elements of

---

3. Compare with his discussion of Chinese attitudes toward the dead: 'The primitive finds no difficulty imagining the dead as sometimes constituting a community in the other world quite distinct from living communities, and again, as intervening on all occasions in the life of those here in every respect their intercourse bears an active character...the dead are believed to be alive in their graves' (1979, 304-5).

diverse ontological type in the essential unity of a single instance' (Ibid. 49). As Tambiah explains, 'participation...signified the association between persons and things in primitive thought to the point of *identity* and *consubstantiality*. What Western thought would think to be logically distinct aspects of reality, the primitive may fuse into one mystic entity' (1990, 86). Thus, for Lévy-Bruhl in primitive thought 'what is given *in the first place* is participation' (Saler, 51); such 'participation is not represented but felt' (51), it is thoroughly affective, and, consequently, through participation primitive man is guaranteed 'that one and the same time he may be both a human being and something else' (48). This is certainly a provocative argument; however, as a theoretical proposal for 'presence' the outcome of the notion of 'mystical participation' remains decidedly circular let alone in opposition to contemporary theories of mind.

Durkheim's reading of the *chruinga* in *The Elementary Forms* makes this point explicit (Smith, 2004, 108, 115 n. 32). In particular, he criticizes Lévy-Bruhl's explanation of the *churinga* which functions 'as the residence of an ancestor's soul and that it is the presence of this soul which confers these [sacred] properties' (Durkheim, 1954, 122). Since 'in the constitution of these pieces of wood and bits of stone...there is nothing which predestines [the *churinga*] to be considered the seat of the soul,' (1954, 123), Durkheim admonishes that Lévy-Bruhl's explanation '[has] obviously been made up afterwards, to account for the sacred character of the churinga' (Ibid.). In a similar vain, Swearer has noted that proposals seeking to explain the 'presence' of the Buddha's relics are most often formulated as 'doctrinal solutions to a seeming paradox' (2004, 113). While we might rightly question the validity of our analogy here between relics and the churinga given that Buddhist relics are understood by believers to have the forty-two major and eighty minor physical marks of a Buddha's body, it is Durkheim's conclusion that is most pertinent:

This explanation...resolves the question only by repeating it in slightly different terms; for saying that the churinga is sacred and saying that it has such and such a relation with a sacred being, is merely to proclaim the same fact in two different ways; it is not to account for them (123).

Thus, as an explanation of the 'presence' of Buddhist relics, emphasizing the functional equivalence between the Buddha and his relics proves, in the final analysis, to be inadequate in our search for what *constitutes* the 'presence' of relics. In all fairness, scholars working within this paradigm are often not engaging with such explanatory questions, and, consequently, emphasizing the functional equivalence of relics does not detract from the significance of the questions and types of data that have emerged from such analyses. With that said, I would like to propose that it is principally through ritual performance that relics come to embody the presence of the Buddha. We may take as our privileged example the account of King Duṭugāmuṅu enshrinement of the Buddha's relics in the Mahāthūpa (Great Relic monument) at Anuradhapura recorded in the Sri Lankan chronicle the Sinhala Thūpavaṃsa (the History of the Buddha's Relics).

II

> Sacrality is, above all, a category of emplacement.
> - Jonathan Z. Smith (1987, 104)

The Sinhala Thūpavaṃsa was composed in the late thirteenth century and is generally attributed to Parakrama Pandita (Berkwitz, 2007, 3-8). Textually, the Thūpavaṃsa is part of a larger genre of Sri Lankan historical writings known as *vamsa* which is often translated as 'chronicle' or 'history' (for extended discussion see Collins, 2003, 654-7). The historicity of the *vamsa* literature, however, has come under serious scrutiny. We know, for instance, that the Mahāthūpa, which is the subject of the text, was constructed over a millennium prior to the writing of the Sinhala Thūpavaṃsa (Berkwitz, 3). Thus, whether or not such rituals were actually performed at Anuradhapura remains contested; however, what seems clearer is that the text preserves something of the Sri Lankan representation of the significance of ritual and its relationship to the enduring presence of relics. Such issues come to the fore in the rituals involved in the construction of the relic shrine as well as the enshrinement of the Buddha's relics. To briefly recapitulate the narrative:

King Duṭugāmuṅu upon reading the stone inscription which tells of his eventual establishment of the Relic Shrine of Golden Garlands is soon visited by a goddess who spreads the news of his plans to construct the shrine among the Six Divine Worlds. Soon after, villagers from all areas of the tributary lands of Anuradhapura send word of their miraculous discovery of various building materials which they then donate to the king. Having collected the bricks, golden shoots, copper, gems, and every other imaginable type of precious object, King Duṭugāmuńu begins to meticulously lay the foundation of the Mahāthūpa on the full-moon day of Vesak. The entire city is decorated and the Sangha assembles for the laying of the eight golden ceremonial bricks. The king, who is fully adorned in royal regalia and is accompanied by a retinue of thousands of warriors, officials, and dancers, lead the ceremonial procession to the building site. Thousands of arhants residing in diverse parts then assemble with the most venerable arhants orienting themselves in the cardinal directions.

Having established the parameter of the Great Relic Shrine, King Duṭugāmuńu then has brick masons begin the construction of the relic chamber in which a Bodhi Tree is placed in the center along with a silver bed resembling the one used by the Buddha in his final Nirvana as well as the 550 narrative depictions of the Buddha's previous lives and depictions of his biography. After completing the relic chamber the king has the Sangha assemble and declares that he will enshrine the relics the following day which falls on the full moon. Thereupon, the novice monk Sonuttura travels down to the abode of the Nagas and proceeds to steal the Buddha's relics which were in possession of the Naga king but were destined to be enshrined in the Shrine of Golden Garlands. The myriad of gods then rejoice of their possession by chanting, venerating and making offerings to the Buddha's relics. At the enshrinement ceremony King Duṭugāmuńu places the relics upon his forehead and upon entering the relic chamber followed by a retinue of arhants, he attempts to lay the relics upon their bed in a position similar to that of the Buddha's final passing away; however, instead the relics rise up into the air and perform the miraculous site of the Twin Miracle in front of the entire assembly. The relics then descend back onto the kings' forehead, and he proceeds to enshrine them permanently within the relic chamber so that the relics do 'not succumb to any danger from anyone for five thousand years' (Berkwitz, 2007, 194-244).

At an important level, this narrative is concerned with the inevitability of the Buddha's relics and their enshrinement at this particular place. King Duṭugāmuńu, we are told, establishes the site of the Golden Garland Relic Shrine on a location foreordained to enshrine one drona of the Buddha's relics that were distributed upon his death. The site is thus enmeshed in the 'locative specificity' and 'thick associative content' of the history of the Buddhist institution in Sri Lanka. More specifically, King Duṭugāmuńu's decision to enshrine the relics at this particular location is understood to be the fulfillment of one of the Buddha's final resolutions (Trainor, 1997). This does well to mitigate the ethical impropriety of the monk Sonuttura who abruptly and remorsefully robs the Naga king of his relics and thus legitimates his possession of relics by attributing his actions to the intentions of the Buddha (Trainor, 1992). So, there is an emphasis within the text itself on the 'localization of the Buddha's presence' (1997, 96-117), in which the Buddha's resolutions, a type of karmic intention or agency, are played out through the entire ritual sequence from the establishment of *stūpa* and relic chamber to the acquiring of the relics and their enshrinement. Yet, as I first framed the question, much like Smith's example of the Pearl Harbor memorial, this rhetoric of localization obscures the ways in which the Buddha's relics only become 'present' within the ritual context.

Returning to Smith's arguments in *To Take Place*, then, there are ways in which the Buddha's relics are *present*, by virtue of their 'emplacement' within the *stūpa*. Through 'emplacement' sacred space becomes a focusing lens that directs a ritual agent's attention, but attention to what? For Smith, places marked off as sacred are sites of social clarification where idealized representations of a social system are magnified and ritually enacted in direct contrast to the way things are in reality. Thus, in contrast to Kantian theories which conceive of place as 'a particular location with an idiosyncratic physiognomy or as a uniquely individualistic node of sentiment' [i.e. home place] Smith emphasizes instead that space, and emplacement in particular, is best understood as 'a social position within a hierarchical system' (1987, 45). In this way sacred space is best conceived as 'contested space' caught up in creating and adjudicating, to use Dumont's typology, 'systems of purity' and 'systems of power' (See Knott, 2005, 99). With the Thūpavaṃsa, for instance, we see the Buddha's relics orchestrate a complex system of ritual behaviors in which King Duṭugāmuṅu affirms his royal power by placing the relics upon his forehead in front of the assembly of monks who also perform various acts of veneration. The presence of the relic and the attendant system of 'order' that it highlights, then, is guaranteed only when such sacred objects 'occupy the places allocated to them' (Levi-Strauss, 1962, 10; Smith, 1987 122 n. 2). But it seems to be equally important to consider, then, the character of the physical place allocated to the relics.

Smith notes well that architecture too serves as a focusing lens. Yet unlike the structural character of a temple, the *stūpa occupies* rather than *encloses* space and is thus more akin 'to a monumental sculpture rather than an architectural form' (Bandaranayake, 1974, 136), thus making the term 'enshrinement' all the more appropriate. The relic chamber itself is described in the Thūpavaṃsa as containing all of the objects, symbols, and narrative depictions that are of significance to the Buddha's biography. John Strong has convincingly argued that the relic chamber constitutes something of a 'biorama' in which the enshrinement of the relics is understood to be a continuation of the sequence of the Buddha's life story (2006, Ch.6). It is also significant to note, as Strong does, that the relic chamber and much of the *stūpa* is constructed even before King Duṭugāmuṅu or the Sangha has possession of the relics (166). Thus, the text seems to support our argument that it is place which confers sacrality, which is here taken to be synonymous with 'presence', onto relics rather than vice versa.

This is further supported by the fact that the Buddha's relics perform the 'Twin Miracle' only after the king holding them enters the relic chamber. Here is what I have been leading up to: if we are to speak of 'presence' as an analytical category, we cannot do so by deferring our explanations to the materiality of the relic. By themselves, relics are void of all signification; they possess no external marks of identity to distinguish them from other bones. This is to contrast bodily relics with those of images which can suggest presence through operations of verisimilitude. Further, although the Buddha's relics are understood by the tradition to possess the thirty two major and eighty minor marks of the Buddha's body, upon enshrinement the relics are sealed off in the *stūpa* not to be seen again. Shorn of their ritual context, then, 'relics resemble so much dirt' (Sharf, 2004, 170). As Schopen qualifies his arguments for the functional equivalence of relics based upon prohibitions against their destruction, 'that which is in a church or *stūpa* is alive; that which is dumped in a cemetery or thrown away is dead. Relics, then, are defined as much by *where* they are located and what people *do* with them as they are by what they physically are' (1997, 260; emphasis added). Those attune to post-Eliadian scholarship on the category of the 'sacred' will have noted that I am much more partial to the perspective that the category 'relic' is best conceived as 'relational' and, better yet, 'situational' rather than ontological and 'substantive'. One might raise the pointed criticism that I have yet to propose an answer to the question I began with; namely what are the *generative mechanisms* that constitute the 'presence' of relics, and I have, much like functionalist arguments, simply deferred this discussion by replacing one problematic term for another, i.e. presence for 'place.' It is with this in mind that I turn to the final section, which both marks a departure from Smith's work and a greater engagement with discussions of ritual in light of cognitive and evolutionary theory.

III

> 'Place is a practiced space'
> - Andrew Merrifield (1993, 522)

For there to be an academic study of religion, says Martin, we need to be able to distinguish between valid and invalid generalizations. The problem, however, remains that scholars have yet to agree upon what exactly constitutes evidence. More recently, scholars have begun to focus attention on a growing body of cognitive psychological research to provide an empirical foundation upon which theoretical generalizations may be formulated and tested. Unfortunately, as William Paden observes, this new trajectory has tended to overplay the dichotomy of group-agency versus the information-processing brain (2008, 410) to the extent that the cognitive has been emphasized at the expense of the social. Nevertheless, I do agree that the cognitive sciences provide a new context in which to ground theoretical explanations in ways that were unavailable to Kant, William James, and Levy-Strauss, and thus serves as an effective response to Smith's own skepticism about the priority of individualistic theories in the study of religion (1987, 35; 2004, 162). As Martin pointedly notes, 'Smith's emphasis upon the comparative value of place neglects…*how* ritual and place do what he claims for their ritual subjects' (2008, 312). Martin proceeds to reverse Levi-Strauss' dictum that an object is sacred by virtue of 'occupying the places allocated to them' by citing the Pawnee informant Tahiroossawichi who states, "before man can build a dwelling [in which to place his sacred articles] he must select a spot and *make* it sacred" (311). Grimes makes a similar critique when he observes that for Smith, 'Ritual does not make sacred space; rather, placement renders actions sacred,' and consequently 'The 'where' of ritual becomes theoretically more important than the 'how', sometimes occluding the where entirely' (1999, 264). For the study of relics, I welcome the shift from *taking* place to *making* place for the latter understands ritual, as mode of paying attention, to be the means by which 'particular places become identified and represented' (Martin, 315) by social groups even if such groups are the "perpetually reconstructed output fictions of individual minds" (Paden, 2008, 410).

At a theoretical level, we must ground our generalizations concerning the presence of 'relics' in what people 'do'. As a level of analysis, behavior provides empirical data that has connective tissue with evolutionary psychology - in particular, human ethology - with its language of 'status, submission, communicative signaling and displays, kin selection/recognition,' (Paden, 2008, 407) etc. in ways that 'remove the ostensibly unbridgeable gulf that one intuitively wants to place between culture and nature' (Paden, 2004, 408). From the view of adaptive behavior, then, the power of a relic is inextricably linked with the acts of submission and deference which, as forms of communicative display, are costly behaviors that signal and thus imbue the relic with a 'prestige bias' (Paden, 2008, 413). Whatever theoretical camp we choose to frame our analysis on relics, in the spatial, the cognitive, or the ethnological, it remains clear that each of these approaches underscore that relics are the by-product of religious ritual which focuses attention by 'setting apart' certain places and behaviors thus creating a 'system interrupt' from the flow of sensory perception that we label 'experience' (Martin, 2008, 316). This also entails an epistemological claim. For the power of a relic, under this theoretical framework, is only meaningful as a relationship expressed in terms of human behavior. As such, I have intended to situate myself within that tradition that emphasizes representation over presence, expression over experience, and in this way shifts the talk over the presence of relics out of the quasi-theological and into the anthropological realm (Smith, 1987). Thus, one answer the title's question 'do relics do' is to say that relics don't 'do' anything. Shift the valence entirely I have suggested that it is humans who make meaningful places in which to 'inhabit'. Yet, while I have valorized *homo faber* over that of *homo fabricatus* we must also recognize that both of these work together to create a relic's presence in both the relationship between ritual and architecture as well as in the relationship between the outsider who understands the ritual actor to be creating and recreating the 'presence' of a relic, and the insider who perceives him or herself as responding to a given environment/situation. Our 'redescription' of the 'presence' of relics has revealed some salient linkages with recent analytical reformulations of 'the sacred' in the context of cognitive and evolutionary theory (see also, Anttonen, 1996, 1999, 2000). As such, relics, by virtue of their materiality as contextless 'signifiers,' have become a privileged example for thinking about the constitution and formation of religious objects generally.

MITHRAS IN THE MAGICAL PAPYRI.
RELIGIO-HISTORICAL REFLECTIONS ON VARIOUS
MAGICAL TEXTS[*]

SANZI ENNIO

During the "Second Hellenism" theologians and magicians accumulated the names of deities from different panthea in order to indicate the multiple powers of divinity, the same divinity that is analyzed by the great philosophers of this epoch (see: Gordon – Marco Simón 2010; Piranomonte – Marco Simón 2012; Tardieu – Van den Kerchove – Zago 2013).

The PGM V 1-53 sets out how to obtaining an oracle by invoking Sarapis, under the names Zeus, Helios, Mithras, Sarapis, unconquered, Meliouchos, Melikertes, Meligenetor: "... I call upon you, Zeus, Helios, Mithras, Sarapis, unconquered one, Meliouchos, Melikertes, Meligenetor, *voces magicae*, the great, great Sarapis *voces magicae* appear and give respect to him who appeared before fire and snow, *voces magicae* for you are the one who introduced light and snow, hurler of shudderful thunder and lightning *voces magicae*... : (Pronounce) the A with an open mouth, undulating like a wave; the "O" succinctly, as a breathed threat; the IAO to earth, to air, and to heaven; the E like a baboon; the "O" in the same way as above; the E with enjoyment, aspirating it; the Y like a shepherd, drawing out the pronunciation. If he says: «I prophesy», say: «Let the throne of god enter, *voces magicae*, let the throne be brought in». If it then is carried by 4 men, ask: «With what are they crowned, and what goes before the throne?». If he says: «They are crowned with olive branches, and a censer precedes», (the) boy speaks the truth. Dismissal: «Go, lord, to your own world and to your own thrones, to your own vaults and keep me and this boy form harm, in the name of the

---

[*]During an "anomalous" like special Vermounter Fall spent like visiting scholar at UVM, Department of Religion, in 2005, I had the opportunity of discussing so friendly very often with my great friend Luther Martin on difference between "method and theory" in historical-religious studies. This little text (an increased chapter of E. Sanzi, C. Sfameni "*Magia e culti orientali. Per la storia religiosa della Tarda Antichità*", Cosenza 2009) is a little gift to thank him again for his generous friendship.

highest god *voces magicae*». Do this when the moon is in a settled sign, in conjunction with beneficial planets or is in good houses, not when it is full; for it is better, and in this way the well-ordered oracle is completed..." (Tr. W.C. Grese, in Betz 1992, 101-102).

It may be useful here to quote an inscription from Rome dated 146 AD, whose opening lines read as follows: "The holy Roman brotherhood of paean-singers (*Paianistai*) of Zeus, Helios, great Sarapis and of the Augustan gods (*Sevastoi*) have honored Embes, the prophet, the father of the brotherhood aforementioned..." ..." (RICIS, n. 501/0118). It is important to point out that: a) this inscription honors a *prophetes* and *pater* of a *taxis* dedicated to Sarapis; b) the god is called *megas*, Helios and Zeus as in the magical papyrus quoted above. This inscription clearly has the purpose of emphasizing Sarapis' oracular significance, since a prophet of this god is celebrated. It may be useful to remember here a passage of Macrobius' *Saturnalia*, in which we find Sarapis answering king Nicocreon: "I am a god to be known in the way I want to describe it. My head is the heavenly cosmos, my stomach the sea, the earth my feet, my ears rest on the ether, my eyes make the light of the sun shine from afar" (Macr. *Sat.* I, 20, 17 [ed. Willis 1970, 115]). Given such a claim to a cosmic dimension, Macrobius can easily conclude: "So it is possible to demonstrate that the nature of Sarapis and that of sun is only one". Before him, Aelius Aristides stressed the henotheistic and soteriological significance of Sarapis. In his *Oratio* dedicated to the Alexandrine God, the famous rhetorician founds the soteriological dimension of Sarapis on the omnipotence of god: "The citizens of the great city in Egypt even invoke him as «*heis Zeus*», because he has no deficiency through his abundant power, but he has passed through everything and has filled the Universe. The powers and the honors of the other gods are divided, and men invoked different ones for different reasons. But he, as it were the chorus leader of all things, holds their beginning and their ends. He alone is ready to help when one has need of anything... since he possesses the powers of all of them, some men worship him in place of all the gods, and other also believe in him as being a special universal god for the whole world" (Ael. Ar. *In Sar.* 21-23 [ed. Keil 1958, 358-359; tr. Behr 1981, 265-266]). It is possible to find an echo of this devotion in epigraphic and papyrological documents. For example, there are two inscriptions more or less contemporary with Aelius Aristides. The first: " ... I, Statios Kadratos, the most excellent, neocoros, often saved by big dangers, have dedicated (this offer) expressing my gratitude to Zeus Helios great Sarapis and the gods who share his temple (*synnaoi*)... (RICIS, n. 501/0145); the second one: "Lytôn has dedicated for his daughter Phlaouia (this altar) to Zeus Helios great all-embracing divinity Sarapis" (RICIS, n. 703/0110; about the *formula* Zeus Helios *megas* Serapis, see: RICIS, *Index*, 1. *Dieux et déesses*, 1.1. *Divinités isiaques*, 1.1.1. *Les dieux et leurs épiclès*, s. vv. *Sarapis, Zeus, Helios megas; Sarapis Iuppiter Sol*). Moreover, in a letter dated about 200 AD, we read: "May the lord Serapis be thanked, since, while I was in ranger in the middle of the sea, he saved me immediately. When I reached Misenus, I received a viaticum from Caesar of three gold coins. I was really very fortunate!" (Lietzmann 1934, no. 1). The Oxyrhynchus papyri report two requests dated about 200 AD addressed to *megas* Sarapis evoked as oracular god and invoked as Zeus and Helios. The first: "To Zeus Helios great Sarapis, and the consociate gods. Nike asks whether it is expedient for her to buy from Tasarapion her slave also called Gaion. Grant me this (sheet)" (P.Oxy. 8, 1149); the second one: "To Zeus Helios great Sarapis and the associated gods. Menandrus asks, is it granted me to marry? Answer me this (sheet) "(P.Oxy. 9, 1213). The Christian Egypt will continue this *habitus*. For example, Saint Colluthus will answer to his devotees by *sortes* addressed to him (see: Sanzi 2008); it is matter of " oracular notes" (see: Papini 1992) where it is easily to read sentences like these: "God of my lord, holy Colluthus the real doctor, if you order to me to wash my feet, then bring out for me the sheet"(Donadoni 1964); "God of holy Colluthus, if you want that I place my daughter in your *topos* (i.e. sacred place, church, monastery etc.), your desire will be fulfilled, according to your will" (Papini 1985, part. 249-250).

It is interesting to note that the quoted invocation Heis Zeus Sarapis concludes a magical prayer that consecrates a phylactery giving glory, honor, favor, fortune and power to those who bear it. The prayer, addressed to Helios, "greatest god, eternal lord, world ruler", finishes with these words: "I conjure earth and heaven and light and darkness and the great god who created all, Sarousin, you Agathon Daimonion the helper, to accomplish for me everything (done) by the use of this ring or (stone). When you complete (the consecration), say: «Heis Zeus Sarapis»" (PGM IV, 1708-1715: [tr. M. Smith, in Betz 1992, 69]).

Michel Ange de La Chausse quotes a gem depicting Sarapis with the same henotheistic invocation around the god: HEIS ZEUS SARAPIS (*Idem* 1700, I, 47-48, tab. 63). Even if this fact does not prove that this gem is a magical amulet, the fact that this henotheistic invocation is attested also on a gem is proof of the wide diffusion of this henotheistic sensibility during the "Second Hellenism" (see: Peterson 1926; Versnel 1998; Mitchell – Van Nuffelen 2010a; Mitchell – Van Nuffelen 2010b; Sfameni Gasparro 2010).

Coming back to our magical papyrus and the association of Sarapis and Mithras attested by epigraphy,

it is interesting to quote a pillar inscribed on two sides from Mithraeum in the Baths of Caracalla. In front: "Heis Zeus Sarapis (after changed in Mitras [sic]) Helios *kosmokratôr* unconquered"; in back: "thank-offering to Zeus Helios great Sarapis saviour giver of riches listening to prayer benefactor unconquered Mithras" (RICIS, n. 501/0126 = CIMRM, I, n. 463, 1-2). The fact that in the first inscription the name of god Sarapis has been replaced by that of Mithras and in the second the same name has been kept, stresses once again the henotheistic religious sensitivity typical of late Hellenism. Mithras is substituted for Sarapis because both are *heis theos, summi dei, kosmokratores*. With reference to this inscription, Ugo Bianchi observes that "chiamare Serapide Zeus significa che, come Zeus è sommo dio, così lo è Serapide, il quale è universale e domina su tutto. Egli inoltre è chiamato Helios, Sole... nel senso in cui (sc. questo dio) appariva in età ellenistico-romana, quando il Sole era divenuto o si appprestava a divenire una di quelle figure in cui si poteva concentrare l'essenza della divinità... Nella nostra iscrizione ... Serapide è piuttosto il sostantivo, mentre Zeus ed Helios hanno funzioni di apposizioni, che qualificano la grandezza del dio. Il quale è appunto *kosmokratôr*, «dominatore del mondo» e tale lo qualifica ulteriormente l'attributo di *aneiketos*, «invitto», che non si riferisce ad un dio bellicoso, ma a un potente signore del mondo. Infine il dio è detto *heis*, «uno solo», per affermare la sua eccellenza, unica: una proclamazione di fede enoteistica"(Bianchi 1975, 247).

In PGM XIII, 343-646 the epithet *kosmokratôr* as an attribute of Sarapis recurs twice (PGM XIII, 622, 640). Even if Sarapis is not important in this text and his name is probably interpolated, this does not prove that the whole section in which the name occurs is interpolated. In any case, this complex spell evoking a god able to protect the practitioner from astral destiny and to prolong life finishes by saying: "Protect me from all my own astrological destiny; destroy my foul fate; apportion good things for me in my horoscope; increase my life even in the midst of many goods, for I am your slave and petitioner and have hymned your valid and holy name, lord, glorious one, ruler of the cosmos of ten thousand name (?), greatest, nourisher, apportioner, [[Sarapis]]" (PGM XIII, 633-639 [tr. M. Smith, in Betz 1992, 187-188]; see: Fauth 1995, part. 74-78).

Mithras appears in a Coptic magical amulet with the Jewish god Iao, Sabaoth, Adonai, angels and archangels, divine "furnishings", and various words of Christ, such as the dramatic *eloi eloi lama sabachatani*: ".... holy .... Iao, Sabaoth, Adonai........................... arath. Mithras XC The praise of the song of David. Who dwelleth in the help of the Most High shall abide in the shadow of the God of heaven. He shall say to the Lord: «Thou art my protector and my refuge. My God, I shall trust in Thee». The book of the generation of Jesus Christ, the son of David, the son of Abraham. In the beginning was the Word and the Word was with God. Forasmuch as many have taken in hand. The beginning of the Gospel of Jesus Christ. I adjure you by your powers and your names and your blessed virtues (?), I adjure you by Orphamiel, the great finger of the Father, I adjure you by the throne of the Father. I adjure you by Orpha, the whole body of God, I adjure you by the chariots of the sun, I adjure you by the whole host of angels on high, I adjure you by the seven curtains that are drawn over the face of God, I adjure you by the seven Cherubin who fan the face of God, I adjure you by great Cherubin of fire whose name no one knows, I adjure you by the great name of God whose name (*sic*) no one knows except the camel, I adjure you by the seven archangels, I adjure you by the three words which Jesus spoke on the cross: *Eloi Eloi elema sabakthani*, which is: «O God, my God, why hast Thou forsaken me?» that, whosoever wears this amulet, you keep him from all harm and all evil and all witchcraft and all star-blight and all evil spirits and all the works of the ruthless Adversary that you guard the body of Philoxenus (?) the son of Euphemia, from all these things. Holy Holy Holy Amen Amen Amen" (Drescher 1950; see: Harrauer 1987, 12-52, part. 28-29).

The name of Mithras also occurs in other magical papyri. Leaving aside the so-called "Mithras Liturgy" (see: Betz 2003), the god is invoked in the "cat ritual for many purposes" (PGM III, 1-164, the name of Mithras appears in line 80), and in a "charm that gives foreknowledge and memory" (PGM III, 424-466, the name of Mithras appears in line 462). In both spells, the name of Mithras is found alongside the name of Helios. On another occasion we have the same situation as PGM V, 1-53: Mithras is not invoked *simpliciter*, but as part of an ensemble of various gods, more or less related in order to obtain ends that only magic can accomplish. In a specific section of the so called "cat ritual", where the name of Mithras appears, there seems to be a "heliacal" link among various deities invoked: "Proceed toward the sunset and, taking the right-hand and the left-hand whiskers of the cat as a phylactery, complete the rite by saying this formula to Helios. Formula: «Halt, halt the sacred boat, steersman of the sacred boat! Even you, Meliochous, I will bind to your moorings, until I hold converse with sacred Helios. Yea, greatest Mithra, *voces magicae*, holy king, the sailor, he who controls the tiller of the lord god *voces magiacae*, before (you attain to) the southwest of the heaven, before (you reach nightfall?) in flight from the outrages committed against you. Hearken to me as I pray to you, that you may perform the NN (deed), because I invoke you by your names *voces magicae*. Perform the NN deed – add the usual, whatever you wish – for it is those same people who have mistreated your holy image, they who have mistreated (the holy) boat, wherefore for me…, that you may return upon them the NN deed – add the usual –. Because I call upon you *voces magicae*. Perform the NN deed – add the usual –, I conjure you in the Hebrew tongue and by virtue of the Necessity of the Necessitators *voces magicae*. Accomplish this for me and destroy and ravage in the coming dawn, and let the NN deed befall them – add the usual, whatever you wish –, immediately, immediately; quickly, quickly. Pleasant be your setting»" (PGM III, 94-124 [tr. J.M. Dillon, in Betz 1992, 21]). By the way, we should stress that in this same magical papyrus the name of Mithras occurs among various *voces magicae*: "For I conjure you by Iao, Sabaoth, Adonai, Abrasax, and by the great god Iaeo *voces magicae* holy king, the sailor, [who steers] the tiller of the lord god" (PGM III, 75-82 [tr. J.M Dillon, in Betz 1992, 20]). The same translator notes that: "This formula… may well be garbled Greek for *Damameneu, Zeu chthonie*, identifying Helios-Mithras with Hades" (*ibidem*). The section of the second magical charm quoted where the name of Mithras appears is full of lacunae, but in any case we can see that the name of Mithras once again appears alongside the name of Helios. The fact that this association is very common in official Mithraic epigraphy also demonstrates that magic utilizes *proprio more et pro domo sua* various religious elements.

Without re-examining the various magical Mithraic gems presented in another study (See: Sanzi 2002), we will now consider some other examples of magical amulets where the name of Mithras appears as a *vox magica* or a divine name (re-)used.

In his catalogue, Campbell Bonner describes "a black stone... bearing no figure design, has an inscription in which the name of Mithra is inserted into a formula of *voces magicae* which is found many times... SALAMAZAXA MEITHRAS BAM[AIAZA]; the common form is SALAMAXA BAMAIAZA. Here, MEITHRAS is simply inserted as *par inter paria* in a group of magical words" (Bonner 1950, 33). In fact, these magical words can be found alone or in association with others in magical gems and the magical papyri without quoting the name of Mithras.

For example on the obverse of a magical gem is carved Harpocrates wearing a solar crown seated on a lotus, between Sol and Luna; on the reverse side: SALAMAXA DOS CHARIN PORON EPITYCHIAN IOE (Bonner 1950, n. 192). In this case, SALAMAXA seems to be an appellative of the god invoked. In another magical gem, also described by Bonner the invocation SALAMAZA BALAIZA AMATROPAITHER IAO IAO SOZE TON PHOROUNTA TOUTO TO PHYLAKTERION appears inside an ouroboros among *charakteres* and Greek letters (Bonner 1950, n. 271); this time SALAMAZA seems to be an appellative of IAO. In a protective charm studied by R.W. Daniel and F. Maltomini the *vox magica* SALAMAXA appears among various common magical words; these must be recited to provide protection from all shivering and fever, tertian, quartan, quotidian, daily or every-other-day. In this papyrus, in a pot-pourri of *voces magicae* such as vocal series and palindromes, the name of the Jewish god, angelical names and so on, the *vox magica* SALAMAXA returns once more (Daniel – Maltomini 1990-1992, I, n. 10). It is clear that, in this case, this *vox magica* has no association with Mithraism.

A lead tablet from Hermoupolis inscribed on both sides and dated 3rd-4th century AD contains a lesbian erotic charm: Gorgonia is to be made to love Sophia (Daniel – Maltomini 1990-1992, I, n. 42). The editors note that "a special feature of the present charm is that a daemon is commanded to bring Gorgonia to a bath-house and then become a bath-woman in order to inflame Gorgonia with love. As the bath-woman heats Gorgonia with hot water, so Gorgonia should become heated with love for Sophia" (*ibidem*). It may seem somewhat strange, but in a lesbian erotic charm there is also a place for the Mithras. In fact, on side A, lines 56-62, we read: "*Voces magicae* and *nomina magica* drive Gorgonia, whom Nilogenia bore, to love Sophia, whom Isara bore; burn, set on fire the soul, the heart, the liver, the spirit of burned, inflamed, tortured Gorgonia, whom Nilogenia bore, until she casts herself into the bath-house for the sake of Sophia, whom Isara bore; and you, become a bath-woman" (tr. *Iidem*).

In this erotic charm we find the same evocation *meliouche meliketor meligenetor* already encountered in PGM V, 1-53, where it is used to obtain an oracle of Sarapis. In this same PGM V, 1-53, the name of Mithra also occurs flanked by the names of Zeus, Helios and Sarapis, that is, flanked by the names of the highest god such as Mithras himself. If we compare these two charms, we can repeat that a feature of the magic of the "Second Hellenism" is the ability to "line up" various deities coming from different panthea to indicate the enormous *potestas* of the godhead.

The name of Mithras is again used as *vox magica* in a phylactery to protect patients from podagra recorded by Alexander of Tralles: "After having taken a gold lamina, when the moon is setting, engrave on this as follows, and after having tied it to the nerves of a crane, make a small tube as big as the lamina, close it [inside] and take it around the astragali. *Voces magicae*. As the sun becomes strong with these words and becomes new every day, so do you give strength to this body/these limbs as it was before, now, now, quickly, quickly. See that I am saying the great name in which the things which are weakened become strong *voces magicae* give strength to this body/limb as it was before, now, now, quickly, quickly" (Alex. Trall. XII [ed. Puschmann 1879, 583]).

The *voces magicae* MITHREU and MITHRAO are present among other *voces magicae* and vocalic series also in an *ostrakon* from Oxyrhynchus dated to the 2nd century AD. It contains an erotic charm to cause the separation of a certain Allous from Apollonius: "*Voces magicae* let burning heat consume the sexual parts of Allous, [her] vulva, [her] members, until she leaves the household of Apolloniois. Lay Allous low with fever, with sickness unceasing, starvation — Allous — (and) madness! Allous"; inside: "Remove Allous from Apollonios her husband; give Allous insolence, hatred, obnoxiousness, until she departs the household of Apollonios. Now, quickly" (Amundsen 1928; [tr. Gager 1992]). On the *voces magicae* MITHREU and MITHRAO John G. Gager suggests that it is "a rendering of the god Mithras, apparently to exploit the sound of his name rather than evoke his power" (Gager 1992, 111, note 109).

A spell for favor and love is carved on a gold amulet from Thessaloniki, dated to the 2nd century AD and originally rolled up: "*Voces magicae* Aphrodite's name *voces magicae* make favor, success with all men and women, but particularly with him whom (she) herself wishes" (Kotansky 1994, n. 40). Again the name of Mithra is involved in an erotic charm, and looks like an amplification of Aphrodite's name invoked at the beginning of the spell. In Roy Kotansky's opinion, it closes the divine names following that of the goddess of love. He adds that "the name of Aphrodite is not otherwise apparent among the divine names given and must represent a title added on"(*ibidem*). In fact, this charm is intended to assure *epicharis* and *euodia* in the field of love, and in the magical papyri the name of Aphrodite very often occurs in spells that are used to obtain the same kind of success requested from the gods of this *lamella*. In this way, the names that follow Aphrodite's look like the gears of a complex machine. The product of this machine is the woman's success in love. In any case, in this magical amulet the name of Mithras appears more as a divine name than a simple *vox magica*. In fact, the name of Mithras is the last in a series of divine names. As the same Kotansky observes, "The naming of «Mithra» is clearly intentional in this spell: the Old Persian naming Mithras means friends... hence the name is particularly significant on a love-charm or friendship-spell" (*ibidem*). Obviously, we must always remember that the *dea princeps* in this series is Aphrodite.

So, it seems correct to suppose that the name of Mithras and/or the god himself are current in the magic of "Second Hellenism", a) as one of the names of the highest and enotheistic god invoked by magicians; b) simply, as a *vox magica*. In any case, it does not seem possible to assert that Mithraism tout-court was engaged by magic; it seems more likely that some specific element of this Oriental cult was utilized by magic. In fact, the very particular form of religion represented by magic during "Second Hellenism" is characterized by its use of various elements from different religious and philosophical currents to produce something new in the contemporary religious panorama. So, the bull-slayer god met at the beginning of this historical-religious analysis has his legitimate place also in the field of magic. Obviously, here it is not a matter of some magician who believes in Mithras or some believer in Mithras who is also a magician; rather, it is a matter of some magician who does not ignore the classical, Oriental and Jewish deities when he is forming his magical and particular gods. In this way, not only he can invoke the god Mithras *magico more*, but he also can use the name of the god as a simple *vox magica* that in isopsefic meaning recalls the entire annual cycle (see: Merkelbach 1984, 223).

349

THE USE OF EGYPTIAN TRADITION IN ALEXANDRIA OF THE HELLENISTIC AND ROMAN PERIODS: SELF-DISPLAY AND IDENTITY OF RULERS, IDEOLOGY AND FURTHER POLITICAL PROPAGANDA

KYRIAKOS SAVVOPOULOS

Egyptian elements of religious character can be found in several types of material evidence, such as coinage, monumental art and architecture. All these types of material are related, in multiple ways, to the public life in Alexandria's multicultural society, throughout the Hellenistic and Roman periods. Although the archaeological evidence during the first half of the 20th century was relatively poor, fortunately, the more recent archaeological discoveries of the 1990s and 2000s -mainly coming from the submerged water area of the Pharos island and the Eastern (royal) harbour- managed to reinstall the interest in the ancient history of Alexandria, since they offer a much more comprehensive and wider picture into the city's material culture.

For the sake of a thorough insight into the material evidence, the discussion will be divided in the following three chronological sub-periods:

1. The early Ptolemaic period (Ptolemy I- Ptolemy V)
2. The Late Ptolemaic period (Ptolemy V- Cleopatra VII)
3. The Roman period (From Augustus to the end of the 3rd century AD/ beginning of the 4rth century AD)

1. The early Ptolemaic period (Ptolemy I- Ptolemy V)

Beginning from the coinage, the repertoire of the early Ptolemaic period consists of the Alexander the Great "cycle", as the Ptolemaic way of thinking used to conceive it. This cycle includes Alexander himself and his divine father Ammon-Zeus. The choice is relevant to Ptolemies' internal and international cultural and political aspirations. For a better understanding on the political and cultural aspects, we need to attentively look into the attributes of these figures. As a matter of fact, letting out Ammon's horns, none of the attribute in the two coinage figures is Egyptian. More spesifically:

- Alexander the Great: The Macedonian king is presented wearing the skulp of an elephant, bearing also the horns of Ammon on his temples and *tainia* on his forhead (Fig. 1).
- Zeus-Ammon: The syncretic God is presented in profile as bearded with curly hair and curly horns in his temples. This image was produced thoughout Ptolemaic period (Fig. 2).

Regarding the latter, Ammon-Zeus was known in the Greek world much before Ptolemies' chronicles. He was conceived as the syncretic version of different Gods and became timeless mostly through his famous oracle in Siwa, in the Lybian dessert. Already in the 5th century BC, the human figure of Ammon with horns of Ram was included in the Kyrenian coins, representing Kyrene as the main port of access to the oracle[1]. Alexander himself visited it in order to ask for a prophecy and right after, he came to claim himself as the son of Ammon-Zeus. Hence, in the first place, Ammon Zeus has been included in Ptolemaic coinage as the father of Alexander, who was the symbolic, mythical founder of the Ptolemaic dynasty[2].

Ammon is the first of a series of certain multi-dimensional figures, because of their identity and the audience they refer to. He represents the local, Egyptian version of Zeus and he figures, at the same time, the universal, Hellenistic, version of Ammon. On the one hand, this figure represents the major god recognized by the whole Hellenistic world, Zeus, with the attributes from the major Egyptian –once state- god, Ammon. On the other hand, he represents a human image of Ammon in Hellenised style, recognizable by the Greek emigrants of Egypt and the rest of the Hellenistic world. Therefore, apart from the religious and stylistic syncretism between Ammon and Zeus, there is also a combination of the local and international

---

1. Steward, 1993, p. 234.
2. The same idea was supported though the Legend of Alexander Romance. See the following discussion on "Pharaonica".

perspective found in the socio-cultural and political context of Ptolemaic Egypt.

Alexander the Great was also a popular image of the early Ptolemaic coinage. According to a local point of view, Alexander's conquest of Egypt legitimized Ptolemy's own governorship in Egypt. Ptolemy Soter spread out several stories on his blood relationship with Alexander the Great, boosting, in that way, his credentials and prestige. According to these "rumors", Ptolemy Soter was not the son of Lagos, but of Phillip. That placed him as the brother and legitimized successor of the great conqueror, who was the son of an Egyptian god and a recognized king of Egypt[3].

Referring to Hellenistic politics, after the final fragmentation of the empire and the declaration of Ptolemy as the King of Egypt in 305 BC, the image of Alexander had further messages to carry on to the rest of the successors. This might have been the main reason for changes on Alexander's repertoire of insignia. Until this period coins minted in Alexander's years depicted the king "identified" with Hercules, wearing on his head a lion skin. In addition to this type, around 305 BC, Ptolemy circulated coins, depicting Alexander with the horns Ammon wearing an elephant scalp on his head. In contrast, the One-eye Antigonos and Cassandros maintained the Hercules version. The latter occurred possibly because Antigonides claimed Hercules as their ancestor[4].

Both heroes, Alexander and Hercules, were semi-gods, sons of Zeus with mortal women. Both of them were great conquerors but Alexander's *Athlos* was regarded as greater, since he conquered an even bigger territory than Hercules did[5]. Both types of insignia, Elephant scalp and Lion skin, relate to power and universal hegemony. The former is the symbol of Hercules, the powerful semi-god, who according to the mythology of the 4th century BC had conquered most parts of the known world. Similarly, the elephant scalp was a symbol possibly related to glorious memories of Alexander the Great's expedition to Asia where his army had to fight against the elephants of the Indian army. Therefore, it became the equivalent of the lion skin of Hercules, the symbol of the world conqueror. Compared to Antigonides, the advantage of Ptolemies was attributed to the fact that their proclaimed ancestor was

---

3. For a complete study on the Ptolemy Soter's policies of legitimising himself as heir of Alexander the Great see: Steward, 1993, p. 229-262.
4. Steward, 1993 p. 261.
5. Ibid, p. 236.

resting in their own territory. Furthermore, "Ptolemaic" Alexander was "physically" further oversized, since only a head of a superhuman size could wear the scalp of an Elephant. Hence, the new image of Alexander reflects how Ptolemy visualized his own place in the ongoing struggle of power among the successors.

During the Reign of Ptolemy IV, two major products of Greco-Egyptian cultural interaction are firstly introduced, as developed in Alexandria during the Ptolemaic period. This is the case of the divine couple of Alexandria, Sarapis and Isis (fig.3). While both figures are presented in their Hellenised forms, they preserve the original Egyptian insignia. Sarapis is crowned with the Atef crown of Osiris, and Isis with ear corns, revealing her identity as fertility goddess. Both of them, apart from the Egyptian origin, promoted their royal identity. Sarapis was the king Osiris, the mythical Pharaoh of Egypt, who became the king of the underworld. Additionally, as patron god of the Alexandrian royal house, he was the divine manifestation of the composite Ptolemaic ideology[6] Therefore, in terms of Ptolemaic political propaganda, the presentation of Sarapis and Isis as a divine royal couple in Alexandrian coinage is related to the promotion of the royal ideology and its representatives.

Both figures represent the double nature of the Ptolemaic ideology, as this was promoted in Alexandria, concerning the past, the present and the feature of Egypt. The Ptolemies seem to acknowledge the long history and culture of Egypt, as this was developed through millennia, until the Late period, when the land of the Nile was "inherited" to Alexander the Great. At the same time, they intend to promote an updated, Hellenised, universal image for their kingdom, both locally and internationally, taking into account the newly formed Political, social and cultural conditions of the Hellenistic world. Therefore, aspects of the Egyptian culture would be now known and received not only by the Egyptians, but also from Greeks and others in Egypt, not to mention those anywhere in the Hellenistic world.

It is worthy to mention that the two gods continued to be displayed by Antiochos even during the intermediate period of the Syrian occupation. The continuation of the divine protectors of the Ptolemies in the repertoire of Hellenistic coinage corresponds to Antiochos' plans and acts in Egypt. In 168 BC, after defeating Egyptian army, he conquered a big part of the Lower

---

6. See the following discussion on Sarapeion.

Egypt, arriving finally to Memphis in order to establish his rule in Egypt. It is even possible that he was crowned according to the Pharaonic tradition. Still, it seems more likely that the Syrian king tried to establish a Seleucid protectorate over Egypt, in which his nephew Ptolemy VI was to be the official ruler.[7] Therefore, nothing will change not even Ptolemaic cultural products and instruments of the royal propaganda, such as Sarapis. The Alexandrian god will continue to represent the Ptolemaic authority in Egypt, even if the latter has been actually applied by Seleucids.

It is interesting to go on by pointing out certain aspects of the Sarapeion of Alexandria: the major field of public interaction between the Ptolemaic authority and Alexandrian society. In the first place, it is worthy considering the positioning of Sarapeion in the site of the city. It was located in the heart of the Egyptian district (Rakhotis), and not in the city centre or the royal quarters, together with all the rest of the important Ptolemaic institutions. Furthermore, it can't be an excuse anymore the idea of constructing Sarapeion on the Rhakotis low hill in order to be visible from other areas of the city, since Ptolemies were quite capable in the erection of extremely high buildings in Alexandria, for instance the Pharos lighthouse, as well as to create artificial interventions, like the Heptastadion[8]. In contrast, it seems that this sanctuary aimed to act as a point of contact between the Greek and Egyptian parts of the city, so as to avoid the picture of two isolated ghettos. Therefore, the participants in the several rites of Sarapeion, such as priests, kings and Greek elites, will have to cross the whole Egyptian neighborhood in order to reach the sanctuary, while the Egyptian population will also be present in the whole process, as it will be occurred in the heart of the Egyptian district. At the same time, the Sarapeion might have served as a means to inspect the Egyptian population of the city.

The Sarapeion of the Ptolemaic period was mostly Hellenic[9] in appearance, still, with several elements inspired from the Egyptian tradition; there is a narrow, rectangular colonnaded court, unusual for religious structures in Greeks so far, but common in Egyptian ones, underground

---

7. Hölb, 2001, p. 143-148.
8. Heptastadion was the two kilometres granite dyke, which connected the mainland of Alexandria with the Pharos island.
9. With the term Hellenic instead of Greek, it is attempted to describe a general Greek decorative architectural style, still without following a specific prototype which existed in Greece.

galleries for burials of sacred animals and a Nilometer. Additionally, foundation plaques from Sarapis and Harpocrates temples have been discovered, having both Greek and hieroglyphic inscriptions[10] (fig.4), a fact which is reminiscent of the importance of God's Egyptian identity, evident, also, in Alexandria. Finally, further decorative elements, such as lotus-form column capitals and sphinxes (fig.5 and 6), add a further Egyptian "touch" in the sanctuary[11].

Within the –mostly- Hellenised architectural environment of Sarapeion, Egyptian style media were used in order to present members of the Ptolemaic dynasty to the visitors of the sanctuary. This was the case of several statues, some of them donated by elite Alexandrians, as indicated by their inscriptions[12].

All of them were executed in granite, a common Egyptian material for the execution of monumental art and architecture. Unfortunately, all of them have been discovered in fragments. Still, it could be assumed that their style could be either typical Egyptian, like in the case of the Anfushi triad[13] (fig.7), or Egyptianising, an alternative, which is attested in the case of the faience oinochoai of the Ptolemaic period (fig.8)[14].

The collection of royal statuary in Sarapeion could lead us in two suggestions. Firstly, Ptolemies promoted an Egyptian image (or merely Egyptian) for themselves in terms of proper kings of Egypt, who acknowledge the Egyptian tradition. This message was to be delivered both by Greek and Egyptian Alexandrians. Secondly, Egyptian style statues were considered as appropriate medium for Alexandrians to express their loyalty to the kings of Egypt who reside in Alexandria[15].

---

10. Greco-Roman museum, P. 10052.
11. This is the case of the Lotus-form capital, and several sphinxes discovered in the site, which are dated in the Ptolemaic period.
12. Three fragments statues of Arsinoe II: Greco-Roman museum nos. 14941 and 14942, the statue base dedicated by Thestor, son of Satyros (in situ) and 1 statue base dedicated Ptolemy IV Arsinoe III and Ptolemy V dedicated by the chiefs of the palace guard.
13. The triad of Anfushi presents Ammon between Ptolemy II and Arsinoe II, in Egyptian style. Once more, the message is clear: Ptolemies are the legitimised kings of Egypt, and more over they respect and follow the Egyptian tradition as proper pharaohs. Greco-Roman museum no. 11261.
14. For a detailed figured faience oinochoai see: Thompson, 1972.
15. It is worthy to note Alexandrians seem to acknowledge Ptolemies as proper mediators not only between themselves and Egypt, but also between humanity and the realm of divine.

2. The Late Ptolemaic period (Ptolemy V- Cleopatra VII)

The above picture was meant to change, during the late Ptolemaic period, due to major political, and socio cultural developments. Among others, we could distinguish three points:

- Civil unrest in Egypt, and international decline of the Ptolemaic power and influence[16].
- At the same time, Greco-Egyptian interaction in lower levels of society, who have reached an advanced stage.[17]
- Since the middle of the 2nd century BC and onwards, Egyptians were able to participate in higher administrative and military positions after a process of Hellenisation in terms of name, education and possibly lifestyle[18].

It is a logical assumption that these socio-cultural and political developments affected respectfully the public life of Alexandria. In several points of the late Hellenistic period, there is an attempt from the side of the Ptolemies to revise their ideology, a fact, which is reflected in the appointment of new points of public interest within the city, and in major developments concerning the use of the Greek and the Egyptian traditions. This new picture seems to have aimed to further emphasize the Pharaonic identity of the Greek kings, as a reaction to the cultural and political developments both in local and international aspects[19].

Such a role is indicated by the Alexandrian Faience Oinochoai of the Ptolemaic period, on which surfaces Egyptianising figures of Ptolemaic queens are, sometimes in an Isis priestess dress. Those vases were used both for cosmic and funerary purposes, implying the wide range of meanings that the royal image with Egyptian references could incorporate.

16. Hölb, 2001, p. 125-159.
17. Fraser, 1970, p. 70-73 and 75-77.
18. (La'da, 2003, p. 166-167).
19. The revised picture of the Ptolemaic ideology corresponds to the developments in the expressions of identity in Alexandrian society. People of Anfushi and other elite necropolises of the late Ptolemaic city, represent a new cast of elite people, mixed or Hellenised Egyptian, which definitely added a new color in Alexandrian public life and activities. Such would be the case of priest Hor (Cairo, Egyptian museum no. 697), which represents one of the 4 cases of statues, which belong to elite people of Egyptian origin.

First of all, there is a clear disinterest for Sarapeion, both from the side of the Ptolemies and the Alexandrian society[20]. Instead, several Pharaonic colossal statues, belonging to the Late Ptolemaic dynasty have been discovered in the area of the Pharos Island (fig. 9), in the eastern port, but also in the Hadra district, in the eastern part of the city[21].

Such monuments could have been installed in various occasions, when the Ptolemies had to promote their Pharaonic identity and "Egyptian" self-image. For instance, this could have been the case of the restoration of Ptolemaic authority, after the rebel in Thebes in the late 3rd century/early 2nd century BC[22], and from this period onwards, it could be related to the abnormal replacement of various kings in the Ptolemaic throne down to the reign of Cleopatra VII. Concerning the latter, Cleopatra used extensively Egyptian style media in order to promote the revised, more Egyptian ideology of her potential Ptolemaic empire. Such could be the case of the coronation and declaration of Caesarion as heir of the Egyptian crown, in a ceremony organised by Marc Anthony in the Gymnasium of Alexandria. Furthermore, this event was to promote, especially in the eyes of the Romans, the "rebirth" of the Egyptian empire. In this ceremony, Cleopatra performed in an Egyptian style, Isis-dress[23].

Isis was in favor of Ptolemaic queens, a series of women with strong personality and extensive activity, inside and outside the palace. This cycle starts with Arsinoe II, wife of Ptolemy II Philadelphus and comes to an end with the most famous of all, Cleopatra VII. Most of these queens are presented with attributes of Isis not only in monumental sculpture and

---

20. There are no royal statues of the late Ptolemies in the site, while there is also an absence of dedications by the Alexandrians.
21. In the Pharos Island: Colossal statue of a young Ptolemaic ruler (Alexandria, Bibliotheca Alexandrina), two Heads of Ptolemaic rulers (Alexandria, Kom el Dikka 121 and 1321), Colossus of Isis or Isis-style late Ptolemaic rule (Alexandria, court of the Maritime museum), upper part from the statue of an Isis or Isis style queen(Alexandria, Kom el Dikka 1005); In the eastern port area: head of young Ptolemaic ruler, possibly Caesarion (Goddio and Clauss, 2006, no.463); in the eastern district of Hadra: the Hadra couple (The male: Alexandria, garden of the Greco-Roman museum 11275; The female: Belgium, Mariemont museum, B.505 (=E.49). For an overview of the Ptolemaic statuary in Alexandria in Egyptian style, see: Ashton, 2004, 15-40.
22. The two rebel leaders were Herwennefer (206-200 BC) and Anchwennefer (200-186). See: Hölb, 2001, chapter 4, p. 127-152.
23. For a short description of this ceremony see: Pollard and Reid, 2006, p. 171.

coinage, but also in other types of material of more private nature, such as Gems and faience Oinochoai[24]. This might be related to the fact that they share common characteristics and responsibilities with the goddess. Both Queens and Isis are wives, and many times sisters of the present Pharaoh, and mothers of the future one. This correlation is visually displayed in the style of Cleopatra's I head in the coinage of Ptolemies V and VI (fig. 10). The queen is presented with the hairstyle of Isis, following the style of Cleopatra's portraits in monumental sculpture. Her Isis-style appearance must be related to her role as mother and official regent of the king, after the murder of Ptolemy V. Her supremacy is further noted in the dating formula of this period, where she is named before her son and takes the title "goddess". This association constitutes a direct link to the earlier Egyptian queens, who ruled with their sons, and to the promotion of "kings mother"[25].

Still, from Arsinoe II to Cleopatra II, queens promote a strong relation to Isis rather than sole identification. Cleopatra III was the first queen, who was declared the living personification of the goddess, mother of the Pharaoh, the living embodiment of Horus. Cleopatra VII also took the title "Nea Isis", and probably built a temple dedicated to the Goddess. The colossal statues of Isis-style queens from Pharos water area and Hadra represents the best examples of the late Ptolemaic period queens, whose upgraded status is further implied by the incorporation of the Hathoric crown of Isis in their images[26].

Finally, in the coinage of the 2nd and 1st centuries BC, the head of Isis becomes the most popular theme of Egyptian origin, reflecting the popularity of the goddess, directly or though associations, with the Ptolemaic queens, as discussed above. The Isis figure, whether represents the goddess her self or indicates an Isis-style queen, becomes a trademark, among others, for the late Ptolemaic Alexandria and Egypt. The same occurred with the symbols of Isis such as the Hathoric crown, which represents the Egyptian aspect of the Ptolemaic state. An example from the coinage of Ptolemy XII would support such an assumption, where an interesting allegorical scene is depicted in the reverse side. An eagle is presented, holding in his feet, a thunderbolt and an

---

24. For faience oinohoai see footnote 9; for gems and cameos of the Hellenistic period see Plantzos, 1996; 1999.
25. Ashton, 2003, p. 62.
26. Ibid, p. 115-122.

Isis crown (fig.11). The eagle[27] constitutes a typical symbol of the Ptolemaic dynasty, while thunderbolt represents Zeus and the Hathoric crown, Isis. Thus, a Ptolemaic ideology in the end of the Ptolemaic period can be clearly noticed, where in fact Egypt itself, has a double, Greco-Egyptian identity, which is symbolically represented by a certain Greek and Egyptian symbol. Both of them were chosen as the most representative of their relevant traditions. It is in this chronological point, that both components are promoted side by side, sharing the importance of the scene. Therefore, Isis did not represent only the Egyptian aspect of Ptolemaic ideology, but also the rise of the Egyptian cultural element as an almost equal counterpart during the late Hellenistic period.

This was the assessment of the role and perception of the Egyptian tradition in Alexandria, during the Ptolemaic period. It is interesting now to see how the process of incorporation and adaptation of Egyptian elements was developed in Alexandria, after the fall of the Ptolemaic state, within the socio-political context of the Roman Empire.

3. The Roman period

Concerning the Roman period, Alexandrian coinage represents the best type of material for initiating a discussion on the use of Egyptian tradition in issues of ideology and further political propaganda. Egyptian gods, symbols and buildings are presented in various styles and forms[28]. In total, 60 emperors published 33 or more types of coinage themes, involving Egyptian elements, either in "Greco-Roman" or Greco-Egyptian or even more traditional Egyptian forms[29]. Among others, Sarapis, Isis, Harpocrates, but also Falcon of Horus, Nilus, Osiris Canopus, but also Osiris in a more traditional version, in a Naiskos on a sacred bark, Hermanubis, Ammon, Ptah Sokaris but also Ptah-Hephaistos, Sobek, Sphinxes, ureaoi, Egyptian crowns

---

27. According to the Greek tradition the eagle is related to Zeus. Eagle is a mostly popular theme in the Ptolemaic period coinage.
28. The following picture derives from a catalogue of 483 Alexandrian coins of the Roman period. This catalogue was based on the following collections, projects, and older catalogues: Svoronos, 1904; Dattari, 1901; Milne, 1933; Poole, 1896; SNG (online project); RPC(online project); Geissen, 1983.See forthcoming Savvopoulos, *Alexandria in Aegypto*.
29. With the terms Greco-Roman we try to describe interactive forms resulting from the Greco-Egyptian religious interplay as this was developed thoughout the so-called Greco-Roman period.

and Egyptian temples compose the multifarious repertoire of reverse-side themes in the Alexandrian coinage of the Roman period.

In terms of density, the highest point starts from the reign of Domitian until the days of Commodus, while the next era is mainly marked by the figures of Sarapis and Isis.

In terms of repertoire, the vast majority of themes present popular religious figures of Alexandria and the Egyptian chora, all related to the concept of fertility, death and rebirth both for the nature and humans. Sarapis is the most popular figure here, which has been further used by Romans as a means of political propaganda having though a different direction to the one of the Ptolemaic period. Several emperors such as Vespasian, Caracalla and Hadrian are presented next to the image of the god (fig.12). As a matter of fact, Sarapis incorporates identities of several Egyptian and Greek deities, resulting in the Sarapis-Pantheos version (fig. 13). This figure came to represent, in a sense, the extensive religious syncretism that occurred between Greek and Egyptian religious systems, and not only, resulting in such conclusive forms.

Similar was the case of Isis, who was the second most popular figure. Most of the cases describe some of the most important capacities of Isis among others, being the provider of fertility, both for nature and humans. These capacities have been known since the indigenous Pharaonic and Ptolemaic periods, mentioning for instance the forms of Isis Suckling Horus (Harpocrates) – known from the Ptolemaic period as Isis *Galaktotrophousa*, Isis-Tyche, Isis-Therenouthis, the typical form of Isis with Hathoric crown and also some distinctive Alexandrian versions such as Isis Euploia and Isis Pharia[30].

Isis Euploia and Isis Pharia are related to the capacity of the goddess as the protector of Sailors. The sea capacity seems to have been a development derived from the Greco-Roman period, and especially from Alexandria, being the major port of Egypt. Yet, Isis has been known since the indigenous dynastic period as a great sailor herself, as she had travelled to Byblos, in the cost of Levant in order to bring back the sacred Ark of Osiris[31].

In the Alexandrian coinage, Isis Pharia (fig.14) is presented next to the lighthouse of Pharos. Her image represents the so-called Isis Pelagia type,

---

30. For a complete overview of the identity and names of Isis as mistress of the sea see: Bricault, 2006, p.101-112.
31. Witt, 1971, p. 166.

another name related to her capacity as the protector of Sailors, well known all over the Mediterranean and especially in ports. A respected stylistic parallel is the statue of Isis Pelagia from Messene, which shares identical aspects and posture with this of Isis Pharia (fig.15).

In the Euploia version of the Roman period coinage, Isis is depicted holding or pulling a ship from its bow, between two shorter figures of Nilus and Euthenia, who incarnate the fertility of the Egyptian land and its rich production (fig.16). The allegory seems clear enough. Isis is presented as a coordinator of Egypt's richness[32] while at the same time, she is responsible for the safe transfer of products and sailors. Therefore, Isis incarnates the role of Alexandria itself during the Roman period: the products of Egypt should be distributed from the Alexandrian port into the rest of the Roman world, in great quantities, in order to feed several parts of the empire or even Rome itself.

In some other cases, Alexandrian coins present Isis Pharia next to another figure of Isis, holding a scepter (fig.17). This figure is also presented on the top of a Pylon style temple (figs.18 and 19). Therefore, a comparison of the older evidence with the most recent one will lead to the assumption that after it could be the well-known case of Isis of Akra Lochiados. Recently, HIAMAS[33] discovered a monolithic Pylon, in the water area of Akra Lochiados, about 4 meters high, a miniature of the huge pylons of the Egyptian complexes (fig. 20). According to excavators, the pylon was not transferred from any site, but was abandoned in situ, just like the case of the Pharos Colossi. It is clear that this pylon belongs to the Greco-Roman period, since Egyptian temples would never have such monolithic piece in such small dimensions. Additionally, it shows an immense similarity with the coin structure.

For all these reasons, it seems that both the temple structure and pylon of Akra Lochiados piece could belong to the temple of Isis Lochias. While it is known that this temple existed since the late Ptolemaic period, the Pylon style structure of the coins as well as that of the Cape of Lochias could represent a

---

32. Isis is mostly responsible for the fertility of Egyptian land, and even for the existence of the Nile water. According to pyramid texts of the Old kingdom, Isis would fill Nile River with her tears. Moreover, she would held the forward cable of the sacred bark (Witt, 1971, p. 166).
33. Hellenic Institute of Ancient and Medieval Alexandrian studies, *Ta Alexandrina* (forthcoming).

Romanian influence on the temple. Besides, Romans were keen on promoting their own structures exclusively, rather than those of Ptolemies, whose legend was possibly their most dangerous enemy in Alexandria.

The promotion of these topics in the coinage of the Roman period can be explained in terms of the Roman policies concerning Alexandria and Egypt. It seems that Roman authority further utilised the process of adaptation of Egyptian elements in order to conciliate the Alexandrian society and, furthermore, to make Alexandria and Egypt adapt to the standards of the Roman Empire. Alexandrian society should continue its life according to its own multi-cultural rules. Only now, the Ptolemaic-political part has to be forgotten for Alexandrians. Egypt has to become a common province of the Roman Empire, and Alexandria its capital. In this direction, the Egyptian tradition was used in many ways.

In the Sarapeion of the Roman period, there are important works in Egyptian style such as the "Roman portico"[34] and the statue of Apis bull[35] (fig. 21), dedicated by Hadrian. Hence, during the Roman period, Sarapis finally meets and cohabits with his Egyptian root, Apis. Of course the Hellenised image of Sarapis remained popular, since four or more Greek style statues of the god, dated in the Roman period, have been discovered in Sarapeion[36].

Yet, this was not the only "monumental" use of Egypt in the Roman - period Alexandria. According to the most of the present-day scholars[37], the new authority managed to transfer and reuse monumental structures from the Egyptian chora, such as Heliopolis, dated in the indigenous Pharaonic period. This type of material is often described as "Pharaonica". Such material was found in great quantities in the area of Sarapeion, in the city centre, in the eastern port and the Pharos Island (fig.22). While their exact use is unclear, it is profound that they represent the most glorious dynasties of Egypt, such as the 12th, 18th, 19th and 26th Dynasties.

---

34. Some fragments of the portico are still in Sarapeion. See: Tkaczow, 1993, p. 276, no. 242, Ashton 2003, p. 31; 2004, p. 9. For a recent overview of the site of Sarapeion see: McKenzie et al, 2004.
35. Alexandria, Greco-Roman museum 3512.
36. Alexandria, Greco-Roman museum no. 3914, 3912, 22158, 3816 (From Attarin).
37. Pr Paolo Gallo is the major reader of the Pharaonica of Alexandria, whose study will result in a complete forthcoming study. Also see: Ashton, 2001; Stanwick, 2002.

The transfer and reuse of monumental structures seem to have been a common Roman policy not only in Egypt, but also in Greece and Italy[38]. In contrast, Ptolemies were famous for their extensive and expensive new sacred building activities all over the country, which would not excuse a systematic reuse of Pharaonica in Alexandria, being their own hometown and capital of their state.

The only case, where there is an evidence of a Ptolemaic date of transfer and re-installation, is the small group of the 30th Dynasty Pharaonica, belonging to the last indigenous rulers. Those pieces have been found in limited numbers in the royal quarter and the Soma area exclusively[39]. There is also a reference in the literary sources that a huge obelisk of the 30th dynasty was re-erected in the Arsinoeion of Alexandria[40]. Their re-use in Alexandria could have been related to issues of legitimation of the Ptolemies in the Egyptian throne. According to the famous legend of *Alexander Romance*, the last native Pharaoh of Egypt, Nectanebo II, was hosted in the court of the Phillip II, right after the second conquest of Egypt by the Persians. This king was known also as a great magician. Therefore, he managed to be united with Olympias, the wife of the Macedonian king, in the form of Ammon. The product of this "meeting" was Alexander the Great. It seems clear, once again, that the Ptolemaic dynasty was keen to legitimise itself, whenever possible, as (followers) continuers of the last native dynasty[41].

Still, the majority of the Pharaonica must have been brought in Alexandria during the Roman period. In this way, through the re-installation of "Pharaonica" in several public areas of Alexandria, Romans could reframe the monumental image of the city, in a wider chronological and historical context. Alexandria would represent not only the recent and in a sense glorious Ptolemaic past but also the long indigenous history. "Pharaonica" seems to compose, among others, a "museum" of the indigenous Pharaonic history in Alexandria, which pieces were "exhibited" in various public areas of the Roman period city. Hence, the addition of Pharaonica could blur the

---

38. The same occurred with the temple of Ares in Athenian agora, which was transferred during the Roman period from a suburban are of Attica, while several classical and Egyptian antiquities were transferred in Rome.
39. London, The British museum, EA 22 and EA 10.
40. Stanwick, 2002, p.16, McKenzie, 2007, p. 51.
41. For an overview of Alexander Romance see Fraser, 1970, p. 675-681.

monumental public appeal of the Ptolemies, in several public areas, without totally extracting it.

As a final assessment of this work, it would be interesting to suggest the use of a term, which seems appropriate to express all the aspects of this single, continuous, but also multidimensional process: that of the perception and adaptation of Egyptian elements in the public life of Alexandria. Therefore, we could suggest the term *Alexandrianisation*, as a process and result of perception and adaptation of Egyptian cultural elements in the cultural social and political standards of Alexandria, as were formed thoughout the Greco-Roman period[42]. As it is reasonable, the quality and the density of this phenomenon vary, thus we could re-describe our case study from such a point of view.

Alexandrianisation could be described as the policy of Ptolemies, concerning Sarapis and Isis: the adaptation and systematic promotion in a Greek visual vocabulary of the old Egyptian values, while also preserving their Egyptian identity.

Moreover, Alexandrianisation was the perception and adaptation of Egyptian references in funerary structures of Greek elites, which now plays the role of the frame, which would define the quality of the Greekness, as Alexandrian.

Even form that point, we could consider Alexandrianisation as a more specific term compared to Hellenisation, since these adaptations were formed according to the Alexandrian needs, addressed to the capital of Egypt, and to those who have been born lived or either died in Egypt; and not for any Greek city-state citizen.

From the 2nd century BC onwards, the Alexandrianisation of Egyptian elements and values, aimed to support the revised Ptolemaic ideology, and also to express major socio-cultural developments. Therefore, Egyptian statues like those of the Pharos Island and the eastern port area have been installed within the already existing public monumental environment of the Ptolemaic city.

In the Roman period, Alexandrianisation was a popular process in public and private religion of Alexandria. This process was further supported by Romans, resulting in the richest repertoire of Egyptian elements both in terms

---

42. This term concerns not only issues of ideology and political propaganda, but also the process of Greco-Egyptian interaction within Alexandrian society. See: (forthcoming) Savvopoulos, K. *Alexandria in Aegypto*.

of content and style. Finally, Alexandrianisation could be described as the reuse of the "Pharaonica" in Alexandria, which during the Roman period had to serve as representatives of long indigenous history in the provincial capital of Egypt. In other words, they constitute a monumental manifestation of Egyptian cultural memory, contributing to the relocation of Alexandria in the cultural-political map of the Roman Empire.

In this paper an attempt has been made to describe and further interpret all the possible aspects of the *in Aegypto* perspective. It has been assumed that Egyptian culture constitutes an integral component of the Ptolemaic and Roman Alexandria in all periods of its cultural history. *Alexandria in Aegypto* existed, as much as its *ad Aegyptum* counterpart, using Durrell's expression both "real and imagined", contributing from its side to the formation of this major Mediterranean Cosmopolis.

PICTURES

1. Silver Tetradrachm of Ptolemy I,
(Svoronos, 1904, 104)

2. Bronze coin of Ptolemy II
(Svoronos, 1904, 462)

3. Silver Tetradrachm of Ptolemy IV

4. One of the foundation plaques (bronze) of Sarapeion
(La Gloire, 1998, p. 95, no. 51)

5. Lotus-form capital of Sarapeion
(taken by the author)

6. The two Ptolemaic sphinxes of Sarapeion *in situ* (taken by the author)

7. The Anfushi Triad (Stanwick, 2002, no. A10)

8. Figured faience oinochoe with figure of Arsinoe II (Thompson, 1972, no.1)

9. One of the Pharos Collossi (Stanwick, 2002, no. C 22)

10. Isis-style queen in a bronze coin of Ptolemy V (Svoronos, 1906, 1233)

11. Silver Tetradrachm of Ptolemy XIII

12. Bronze coin of Commodus (Geissen, 1983, 2212)

13. Figure of Sarapis Pantheos in a Bronze coin of Antoninus Pius (http://rpc.ashmus.ox.ac.uk/coins/15860/)

BUDDHIST HYMNS AND MEDIEVAL PLAINSONG: SOME REFLECTIONS ON THE LINKS BETWEEN NEUROSCIENCE, MUSIC AND RELIGION

KEVIN TRAINOR AND ANNE CLARK

Music and neuroscience have been much in the public eye of late. A number of popular books that examine musical cognition from a neuro-scientific perspective have appeared in the last several years, including William Benzon's *Beethoven's Anvil*, Daniel Levitin's *This Is Your Brain on Music*, and Oliver Sack's *Musicophilia* (Benzon, 2001; Levitin, 2006; Sacks, 2007). What has received less attention, however, is the connection between neuroscience, music and religion. R. Stephen Warner's 2007 presidential address to the Society for the Scientific Study of Religion suggests that this may be changing.

After some brief introductory remarks, including comments about his recent recovery from cancer surgery, Warner began his formal address with a song. Joined by his wife and six friends, the ensemble performed a shape note hymn called 'Poland' from *The Sacred Harp*, a venerable collection of music widely used by shape note singers. Warner's decision to include personal reflections on his struggle with cancer and to begin with a performance in which he was a participant is noteworthy. The point he was making to his audience was 'not that Warner sang a song because it was meaningful to him' but that 'the song was meaningful to him because he sings it'. In other words, rituals that incorporate hymns such as that one he had just performed 'do not so much *express* meaning as *create* it' (Warner, 2008, 178)[1].

Highlighting Randall Collin's work on 'interaction ritual,' Warren identifies several key factors in such performances, including 'bodily co-

---

1. See Catherine Bell's useful summary of theories that highlight ritual's performative character (Bell, 1997, 61-89).

presence' and a shared focus of attention. Taken together, these musical rituals generate a sensual, preverbal, nondiscursive form of interaction ('rhythmic entrainment') that is simultaneously highly sociable and deeply embodied within individual participants. Warner links this peculiar power of ritualized musical performance with human evolution and suggests that it helps explain the persistence of religion in modern American society. He concludes his address by exhorting his fellow sociologists to move beyond their prejudices toward sociobiological research (Warner, 2008, 187).

Judith Becker's *Deep Listeners: Music, Emotion and Trancing*, published in 2004, offers one innovative model for exploring the interaction of biological and cultural factors in the emergence of trance states. Trained as an ethnomusicologist, Becker makes effective use of current neurobiological research and ethnographic studies to explore the phenomenon of trance in a wide diversity of cultural contexts. Becker's work is exemplary in its promising synthesis of research methods that are typically kept apart. Like Warner, she highlights the emotional power of musical experience in the context of religious ritual, and seeks to understand how particular qualities of such musical experiences are rooted both in the biology of the human brain and in diverse social milieux that are culturally inflected. As she puts it: 'Emotion and music and trancing viewed as evolving together in the interaction of each individual with performances dissolves intractable dichotomies concerning nature versus culture, and scientific universalism versus cultural particularism' (Becker, 2004, 129).

By focusing on the biological foundation of musical experience, in particular its links with parts of the brain associated with emotional arousal, Becker illuminates a continuum of experiences, ranging from what she calls 'deep listening' to trance. All of these experiences are characterized by high levels of emotional arousal, and in some cases this arousal is linked with feelings often characterized as religious, including the loss of a clear separation between self and other, feelings of wholeness and unity, and a sense that one has approached a presence felt to be sacred or transcendent (Becker, 2004, 54). At the same, Becker stresses important differences between some forms of deep listening marked by the absence of kinetic activity, and trance states that typically occur in conjunction with dance and other forms of vigorous bodily movement. All of these states depend upon the confluence of biological and cultural factors. She writes:

Religious trancers differ from secular deep listeners in the degree of arousal and in differing social milieu. Trancers have been socialized within a community for whom trancing is valued as a means of interaction with the holy. They have learned to trance within that community, based on trance models of the community. Both trancing and deep listening are physical, bodily processes, involving neural stimulation of specific brain areas that result in outward, visible physical reactions such as crying, or rhythmical swaying or horripilation. Deep listeners may stop there and remain physically still. Trancers seem to experience an even more intense neural stimulation that may be expressed in some form of gross physical behavior such as dancing. Both, I suspect, are initially aroused at a level of precognition that quickly expands in the brain to involve memory, feeling, and imagination. Deep listening and trancing, as processes, are simultaneously physical *and* psychological, somatic *and* cognitive (Becker, 2004, 29).

Thus Becker's research points to the value of an integrated approach that takes seriously the dynamic interaction of universal biological factors with diverse social environments shaped by local cultures. These culturally inflected social environments make it more likely that those who are raised within them will tend to listen in particular ways, creating what Becker, drawing on Pierre Bourdieu's work, calls a *habitus* of listening, 'an inclination, a disposition to listen with a particular kind of focus, to expect to experience particular kinds of emotion, to move with certain stylized gestures, and to interpret the meaning of the sounds and one's emotional responses to the musical event in somewhat (never totally) predictable ways' (Becker, 2004, 71). Thus the *habitus* that shapes the behavior of a Western classical music audience is quite distinct from the one informing the participants in a Balinese Rangda and Barong performance. In the remainder of this article, we will follow Becker's lead and explore some connections between cognition, music and religion through two case studies which, taken together, testify to the interpretive force of an integrated analytical approach that moves beyond the presumed chasm between scientific universalism and cultural particularism.

Paul Carus' *Buddhist Hymns*:

Paul Carus (1852-1919) has become widely recognized as one of the most important figures in the popularization of Buddhism in the U.S. during the late nineteenth and early twentieth centuries. While he never formally converted to Buddhism, preferring instead to advocate for the 'religion of science'[2], he published numerous articles and books on Buddhism, a religion that he believed was uniquely compatible with a scientific perspective (Tweed, 2000, 65)[3]. Carus was born and educated in Germany, where he earned a Ph.D. in classical philology at Tübingen. Carus' father was a distinguished Lutheran minister who ascended through the administrative ranks of the church to become superintendent general for Eastern Prussia (Henderson, 1993, 4). Carus himself had intended to pursue a clerical vocation, possibly as a missionary, until he suffered a crisis of belief. He entered the Prussian civil service as a teacher, and taught at military high schools in Dresden for several years. His publication of a pamphlet that denied the literal truth of the Bible brought him into conflict with his superiors and he was given the choice to either resign from the service or apologize for his religious views. He resigned and, after a period of residence in England to improve his English, emigrated to the U.S. in late 1884 or early 1885 (Henderson, 1993, 9). In 1887 he became editor of *The Open Court*, a new journal 'Devoted to the Work of Establishing Ethics and Religion Upon a Scientific Basis' (Henderson, 1993, 35), a publication founded by a fellow German immigrant named Edward Hegeler who had made his fortune smelting zinc in LaSalle, Illinois. Hegeler

---

2. The advancement of the Religion of Science was the fundamental purpose of Open Court Publishing. What this meant to Carus is suggested by the following quote from the preface of his *The Religion of Science*: 'The Religion of Science is the invisible church, and its members are all those who, like ourselves, believe in the religion of truth, who acknowledge that truth has not been revealed once and once only, but that we are constantly facing the revelation of truth, and that the scientific method of searching for truth is the same in religious matters as in other fields' (Carus, 1896c, v).
3. See, for example, Carus' essay, 'Buddhism and the Religion of Science', where he asserts: 'We do not mean to sink the Religion of Science into Buddhism, but on the contrary, understanding that Buddhism in its noblest conceptions is in strong agreement with the principles of the Religion of Science, we set forth Buddhistic doctrines because they anticipated some of those important truths which we are in need of emphasising to-day in the face of the dogmatic assertions of traditional religion' (Carus, 1896a, 4845).

and Carus shared a commitment to a monistic philosophy, and were likewise zealous advocates for a new kind of religion purified by scientific truth. As editor of *The Open Court* and, four years later, of *The Monist*, a more scholarly philosophical publication, Carus published a vast collection of articles in service to the cause of monism and the religion of science. By the time of his death in 1919, he had written 74 books and nearly 1,500 articles, including a number of works focused on Buddhism (Henderson, 1993, 1). No doubt his most influential work in the sphere of Buddhist studies was his *Gospel of Buddha, According to Old Records*. First published in 1894, this work sold in excess of three million copies and was translated into more than ten languages (Verhoeven, 2004, 1). Less well known was a slim volume that he published in 1911 entitled, *Buddhist Hymns: Versified Translations from the Dhammapada and Various Other Sources, Adapted to Modern Music*[4].

What motivated Carus to publish his collection of 'Buddhist Hymns'? As an outspoken rationalist who believed that religion must be purified of any beliefs or practices that were in conflict with a scientific perspective, he would appear to be an unlikely composer of hymns, whatever the religious sentiments to which they gave voice. Thomas Tweed, whose *The American Encounter with Buddhism, 1844-1912* details the history of Buddhism's emergence as a cultural force in this period, identifies Carus as a 'rationalist' in contrast to the 'esoterics' and the 'romantics' who constituted the two other major categories of Buddhist sympathizers in this period (Tweed, 2000). Carus was also a critic of what he saw as superstitious and debasing religious rituals, including Buddhist relic veneration[5]. Moreover, in advocating for the importance of music in the practice of Buddhism, Carus had to contend with elements of the Buddhist textual tradition that were explicitly critical of the sensual power of music. In his introduction to *Buddhist Hymns*, for example, he responded to the criticism of Ananda Maitreyya, an English convert to

---

4. Carus' *Buddhist Hymns* has been little remarked upon. For an attempt to locate the work in the context of American religious history, see Stowe, 2004.
5. I discuss Carus' dialogue about relics with the Venerable Alutgama Seelakkhandha, a Sri Lankan Buddhist monk and Sanskrit scholar, in my introduction to *Embodying the Dharma* (Germano & Trainor, 2004, 6-8). In his letter to Ven. Seelakkhandha, he expressed his preference for the 'spiritual' words of the Buddha in contrast to his material body.

Buddhism, who argued that music was inappropriate to the practice of Buddhism[6].

Carus' commitment to the incorporation of music in his religion of science becomes more understandable when placed in the context of Becker's integrative approach to musical experience; such an approach foregrounds some key features of musical cognition that contribute to its importance for the human species, as well as several distinctive cultural factors that shaped his response to music. In fact, Carus' positive valuation of music went beyond his composition of hymns. He addressed the subject of the arts, including music, in his second English language publication, an essay entitled, 'The Principles of Art, from the Standpoint of Monism and Meliorism'. Published in Boston in 1886 by the Industrial Art Teachers Association, this work lays out a theoretical foundation for the importance of the arts, including music, according to Carus' monistic philosophy. He divides the arts up into two categories, those characterized by a spatial orientation such as architecture, painting and sculpture, and those that have a primarily temporal dimension, including music, dancing, song and poetry (some arts, including ballet, opera and drama bridge the two categories). Invoking Lessing's distinction between the arts of painting and poetry in his *Laocoon*, Carus observes that '…the arts of space deal with coexisting materials which produce permanent and visible beauty, while the arts of time refer to successive and transitory impressions of sound, sentiment or motion' (Carus, 1886, 14).

These two categories parallel another binary classification, his distinction between monism and meliorism. He associates the former category with a static ideal of unity and beauty, and the latter with progressive movement or change, including the human aspiration for improving the world. Both of these are essential for Carus, just as for him the pure and detached rationality of science finds its necessary complement in religion, which shapes human action by engaging the emotions. While science operates through inductive and deductive analysis, 'art, intuitively grasping the idea of the universe and representing it in single examples, gives a clew to the enigma of the world'. Thus the poet (and by extension the composer) should be seen as a 'priest of humanity' (Carus, 1886, 19).

---

6. See discussion below. In a 1904 letter to Carus, Ananda Maitreya observed: 'Here I think you forget that to the Buddhist music is one of the gross abstractions of the senses, which his Religion teaches him to abstain from as one of the fertile causes of the rise of emotional feeling' (Rangoon, 5 Sept. 1904), cited in Fields, 1992, 142.

Carus reflected further on the importance of music in a later essay, 'The Significance of Music', published in 1895. Referring to music as 'the most abstract art', he emphasizes that it is not a reflection or imitation of nature, but rather constitutes a world of its own, one that embodies 'purely abstract law'. Music, in his view, also exerts a powerful influence over emotion, and while rhythm provides music's backbone, pitch brings forth its beauty. He goes on to make an analogy between the rhythmic patterns that characterize music and the fabric of human life:

> ...if we could analyse all the throbs of our life, we would find nothing but motion. Our pulse is rhythm, our breathing is rhythmic, our walk and all our doings, our loves and hates, our hopes and fears, our pains and pleasures, in a word, all our emotions are rhythms that are scanned in the vibrating functions of the organs of our bodies. Our physical life, in all its details, is a sonata which we perform without being able to hear its music. We know nothing of the metre, we only feel it, or, better, our life-actions are the changeful metre itself, and we live on in its perpetuation and constant repetition (Carus, 1895b, 407).

This passage could be seen as evincing a kind of romantic mysticism, a perspective quite at odds with recent neuroscientific research on music. When viewed in the context of Carus' monistic epistemology, however, it can also be regarded as an affirmation of a scientific analysis of music that would include close attention to the biology of musical experience. While he rejected exclusively materialistic views of causality as one-sided and inadequate to the task of accounting for important aspects of human subjectivity, he was also confident that human experience was best understood through scientific analysis and that human culture should be placed in an evolutionary perspective.

Carus wrote a great deal on monism, since it provided the foundation for his religion of science. At one point he noted that the animal psychologist C. Lloyd Morgan[7] best grasped the full significance of monism for understanding

---

7. C. Lloyd Morgan (1852-1936) is known in the field of animal psychology for Lloyd Morgan's Canon, an interpretive principle important for the development of behavioral psychology. On the proper interpretation of this principle, see Thomas, 2001.

the natural world (Sheridan, 1957, 13). Morgan, writing in *The Monist* in 1895, defined the essential character of monism in these terms:

> Its cardinal tenets are: that nature is one and indivisible and is explicable on one method, the method of reason; that experience is one and indivisible, though we may distinguish its subjective and objective aspects; that man is one and indivisible, though our analysis may disclose two strongly contrasted aspects, body and mind. It contends that man in both aspects, biological and psychological, is the product of an evolution that is one and continuous; and, combining the results of its theory of knowledge with those of its analysis of man, it identifies the mind, as a product of evolution, with the subject, as given in experience (Morgan, 1894, 332).

Viewed in the light of this monistic epistemology, Carus' above description of the music of life appears less a mystical vision than a poetic rendering of his conviction that subjective and objective reality are fundamentally congruent. For Carus, the scientist's rational, empirical inquiry into the ordered working of natural phenomena and the composer's intuitive orchestration of rhythm and pitch into an emotionally powerful musical composition were two equally valid and complementary approaches to the search for truth.

The importance of musical experience for Carus' own reflections on Buddhism appears clearly in an article he published in 1905 on the 'three characteristics', an important set of Buddhist concepts that define the nature of phenomenal reality (in Pāli, *anicca, anatta,* and *dukkha*). Carus begins the article by informing the reader that he had been reciting Buddhist texts to a friend when he was struck by a textual statement of the importance of these three truths, which all Buddhas realize when they gain enlightenment. After some philological reflections and an analysis of the meaning of the terms, Carus concludes the article with a hymn about the three truths set to the main motive of the second movement of Beethoven's Seventh Symphony (including an arrangement of the musical score)[8], and the following account of how he came to compose it:

---

8. This hymn appears in his *Buddhist Hymns* collection published in 1911.

It was under these impressions that I listened in the evening to the powerful strains of the Andante from Beethoven's Seventh Symphony. The master exhibited here the full power of his genius and was preaching a religion. He emphasized his precepts with a serious conviction and vigorous earnestness, repeating the motive three times just as old Buddhist monks repeated their formulas three times in order to give emphasis to a truth and to inculcate its moral applications. The melody was almost a monotone, repeating the same measure again and again, without any attempt at embellishment; and the harmony consisted of a few changes in the accompaniment, apparently serving no other purpose than to lay stress on that one motive which was the main theme and the sole burden of the composer's thought. Without shaping my thoughts into definite words, I felt that Beethoven was a prophet who revealed the selfsame truths that had been explained by the Buddha. There was the same stern attitude, the same simplicity in propounding the doctrine and the same accentuating repetition, so that almost unconsciously the melody of the master's melodramatic theme spoke to me in words expressive of the Buddhist Dharma.

As in a dream I saw a Buddhist congregation, and a choir sang *sotto voce* the following formula three times successively:

'All conformations
Always are transient,
Harassed by sorrow,
Lacking a self'.

A solo rendered in firm notes expressive of conviction sounded the answer in threefold repetition as follows:

'This is the doctrine
Taught by all Buddhas;
This is a fact and
Always proves true'.

Finally the chorus of the whole congregation repeated the melody with the following words:

'Words of the Buddha
Never can perish;
They will remain for
Ever and aye.

'Words of the Sangha
Set up a standard,
Point out salvation,
Teach us the way.

'Words of the Dharma—
Truths are immortal,
Errors and passions
Will they allay'.   (Carus, 1905, 566)

This passage is noteworthy in several respects. Carus' identification of both Beethoven and the Buddha as religious prophets is striking, as is his assertion that they teach the 'selfsame truths'. One is also struck by his vision of the monks of the early Buddhist saṅgha participating in chanting ceremonies that parallel in their liturgical form the contours of Beethoven's symphonic composition. The manner in which Carus says he experienced this insight is also telling: he felt the essential congruence of the messages of Beethoven and the Buddha '[w]ithout shaping my thoughts into definite words'. The passage as a whole speaks strongly of Carus' sense of the power of music: it helps define a community of shared sentiment, one that through shared musical performance comes to embody a coherent moral perspective. Music, as the old saying goes, is a universal language, one that communicates fluidly across the entire human species, transcending the cultural differences that separate particular communities. But, contrary to the saying, it does this in part by not being a language, i.e., by inducing experiences that transcend linguistic expression. This perspective clearly parallels Becker's stress on the silencing of internal linguistic activity, what she calls 'inner languaging', in both deep listening and trance, an experience for which she finds support in the results of neurobiological research (Becker, 2004, 145)[9].

---

9. Becker draws upon the research of Antonio Damasio, in particular his distinction between core consciousness and extended consciousness; the latter is associated with inner languaging, and tied to what Becker calls the autobiographical self. In trance, the

If Carus greatly valued music for its universal capacity to elicit and integrate human emotions that conduce to moral behavior, he also clearly valued some forms of music over others. His evolutionary views are apparent in his disdain for popular music ('rag-time melodies'), and in the contrast he draws between the origins of music, which he associates with savages clapping and beating drums, and more advanced, intellectual forms of music. Regarding the tastes of the great majority of people in the United States, he observes that they are vulgar, though no more so than those of the masses in Europe. In keeping with his confidence in the 'law of evolution', however, he expresses optimism that American intellectual life will gradually progress from a focus on the practical to a more elevated intellectualism. Toward that end, he calls for the creation of independent art centers that can 'organise the better elements constituting an intellectual aristocracy' (Carus, 1900). He also expresses the hope that the technology of the player piano will make good music available to the masses and prepare them to enjoy orchestral concerts by familiarizing them with the classics (Carus, 1906). His interest in the education of children led him to devise a new notation system to facilitate their progress in learning to play the violin, an instrument he felt was superior to the piano for early music education because it required students to create manually the proper notes, thus training the ear (Carus, 1907).

These views serve well to highlight some of the cultural factors that shaped Carus' attitudes toward musical experience. Becker's notion of a distinctive *habitus* of listening becomes relevant in this context, where culturally specific ideas, values and practices combine to create a diversity of embodied dispositions that predispose individuals to experience and respond to music in culturally distinct ways. As she notes, high levels of emotional arousal under the influence of music do not necessarily result in trance behavior. Carus' ideal of highly intellectual music, best experienced by a properly educated audience that has learned to sit with very limited body motion and in perfect silence, contrasts sharply with the dramatic kineticism associated with trance behavior. If classical concertgoers 'lose themselves' in musical experience, this differs in important ways from what trancers experience amidst the frenetic physical activity that Becker associates with trance.

---

autobiographical self is replaced by the trance persona, and is often linked with a kind of temporary amnesia (Becker, 2004, 145-146).

In fact Carus was highly critical of occult forms of religion, including spiritualism, which was strongly associated with another community of Buddhist sympathizers, a group that Tweed identifies as the 'esoterics'. This latter group included H. S. Olcott and H. Blavatsky, co-founders of the Theosophical Society in 1875. Becker provides a helpful historical overview of the idea of 'trance', and the generally negative associations it has carried in Europe and North America (Becker, 2004, 13-24). She notes both the tendency to feminize trance (with the implication that women lacked rationality and the capacity to distinguish adequately between self and other) and to pathologize it (associating it with hysteria and multiple personality disorder). A dominant theme that emerges from her survey is a deep cultural preference for guarding the integrity of the unified self and for gaining and maintaining self-control, at least in European and North American academic circles. Carus' disdain for what he termed 'spiritism' (Carus, 1888), the superstitious longing for contact with departed spirits through medium possession, seems consistent with this broad cultural ethos.

It is also striking that we find very little discussion of Buddhist meditation in his accounts of Buddhism, despite the centrality of these special states of consciousness in Buddhist canonical texts. Carus' Buddha was 'the first positivist, the first humanitarian, the first radical freethinker, the first iconoclast, and the first prophet of the Religion of Science', not a master of esoteric, trance-like, altered states of consciousness (Carus, 1896a, 4845). Carus' apparent ambivalence toward Buddhist textual accounts of the Buddha's extraordinary meditational attainments is clear in the glossary of Buddhist terms that he included in his *Gospel of Buddha*. For example, he translates *samādhi* as 'trance, abstraction, self-control', and then quotes a statement by the Buddhist scholar T. W. Rhys Davids that highlights ancient Buddhism's rejection of dreams and visions, and its diminishment of the importance of *samādhi* when compared to the noble eightfold path, even if it was unable 'to escape from the natural results of the wonder with which abnormal nervous states have always been regarded during the infancy of science' (Carus, 1895a, 255). Carus translates the term *jhāna* as 'intuition, beatic [sic] vision, ecstasy, rapture, the result of samādhi,' but then observes that 'Buddha did not recommend trances as means of religious devotion' and explains that the Buddha understood the state as 'not losing consciousness but a self-possessed and purposive eradication of egotism' (Carus, 1895a, 247-248). A similar distrust of trance-like states appears in a brief article in *The*

Open Court by the Sri Lankan Buddhist Anagārika Dharmapāla, with whom Carus maintained a regular interaction from the time of their meeting at the World's Parliament of Religions in 1893. In an article that takes up the question, 'Is there more than one Buddhism?' Dharmapāla points to the centrality of the four noble truths, and characterizes the fourth truth as the 'essence of all Buddhism', noting: 'It proclaims that not by asceticism, not by methods of the Brahmanical yoga (hypnotical trances), not by looking out for our happiness, but solely by walking on the noble eight-fold path of righteousness can Nirvāna be obtained' (Dharmapala, 1897, 83).

This relative lack of interest in Buddhist meditational practices helps set the context for understanding the conflict between Carus and Ananda Maitreya (born Allan Bennett) on the place of music in the practice of Buddhism. As noted above, Carus and Ananda Maitreya corresponded on this question, and Carus addressed their disagreement in his introduction to *Buddhist Hymns*. Interestingly, Ananda Maitreya, found his way to Buddhism after serious engagement in occult practice. Bennett was a member of Order of the Golden Dawn, and was apparently well known for his psychic powers. This initial interest in esoteric religion led him to the practice of yogic meditation. In 1900, Bennett traveled to Sri Lanka in order to bring his severe asthma under control, and it was there that he undertook his first serious study and practice of Buddhism, and continued his practice of yoga, including the mastery of deep trance-like states (Harris, 1998, 4-8). In 1901 Bennett was ordained as a Theravāda Buddhist monk in Burma, taking the name Ananda Maitreya, which he later changed to Ananda Metteyya, the Pāli form of the name. What is relevant for our purposes here is his serious and sustained engagement in Buddhist meditational practice and his embrace of monastic renunciation, two aspects of the Buddhist tradition for which Carus showed little enthusiasm.

As a Theravāda Buddhist monk, Ananda Maitreya was critical of visual and aural sensory stimulation; this reflects aspects of both the monastic discipline (*Vinaya*) that restricted monks from watching performances of dance, singing, and instrumental music, as well the significant motif in Buddhist canonical literature warning of the dangers of pleasurable sensory stimulation (Carter, 1993)[10]. This was consistent with his commitment to a life

---

10. Carter's essay identifies a countervailing appreciation for the religious value of music among Sri Lankan Buddhists, particularly in lay Buddhist settings.

of monastic renunciation, including celibacy, fewness of possessions, and a degree of disengagement from society. Music, from Ananda Maitreya's perspective, was a distraction that interfered with proper monastic conduct and the practice of meditation. In a 1905 letter to Ananda Maitreya, Carus articulated his defense of music in the context of Buddhist practice:

> ...Music is a language in which you can express high and noble as well as low and vulgar thoughts, and it is like poetry which can be applied to most opposite uses. I claim that no one who knows anything about music will not regard Beethoven as the composer of most sacred and deeply philosophical music, the suppression of which would mean the suppression of noble and elevating thoughts. If a reformer denounces poetry, meaning thereby some frivolous lines, he cannot at the same time reject all poetry wholesale, and with it also verses like the Dhammapada. It is not poetry that is objectionable, but the wrong poetry, and so it is not music that is objectionable, but frivolous music such as I grant people are mostly exposed to. I wish you would show more consideration for the publication of art, and also of the significance of music, in this light, and if you are possessed of the spirit of consideration, you will easily understand that in this world many things are imperfect and can not at once attain to the perfection which you are aspiring to... (Sheridan, 1957, 90).

Given Carus' commitment to depicting Buddhist practice in terms that emphasized its affinities with his Religion of Science, a religion characterized by a commitment to an active and positive engagement with the world[11], it is

---

11. In his *Open Court* essay, 'Goethe the Buddhist', Carus notes the widely held view that Buddhism is a religion well adapted to the 'passive nations of Asia' and was therefore not suited to the 'energetic races of the West'; he responds that this is true only if Buddhism is identified with quietism and indolence and concludes that nothing could be further from the Buddha's teachings (Carus, 1896b, 4832f.). Carus' interest in Buddhism goes back well before his encounter with Buddhists at the World's Parliament of Religions in 1893. One of his first publications in Germany was his 1882 collection of poems entitled, *Lieder eines Buddhisten*. In this collection, Carus speaks from the perspective of a Buddhist monk who has rejected the world because of the pain that arises from transience and death; at the conclusion of the set of poems, however, the monk returns to the world and embraces a life

not surprising that Carus downplayed the centrality of monastic renunciation and retreat from the world in his version of Buddhism, turning instead to music as an important means for defining his ideal community of Buddhists. As we have seen, Carus was well aware that music (along with poetry) could serve effectively to stimulate the emotions and give the abstract philosophical ideals of his Religion of Science a more fully embodied force supportive of ethical behavior. In the absence of some of the fundamental institutional forms and practices that have historically characterized Theravāda Buddhist communities, there was a pressing need to find a means to integrate and galvanize the followers of his nascent Religion of Science, to create, in other words, a church. This awareness inspired him to publish several hymns in *The Open Court* as early as 1898; a year later he published a freestanding hymn collection, *Sacred Tunes for the Consecration of Life* (Carus, 1899) with verses celebrating his Religion of Science. His first hymns in a Buddhist idiom appeared in *The Open Court* in 1904, based on his translation of three verses from the *Dhammapada*, a very influential canonical collection of the Buddha's sayings (Carus, 1904)[12]. Tellingly, his translation of *Dhammapada* 387 drops out a reference in the original text to meditative absorption ('a brāhmaña [in this context, referring to an enlightened *arahant*], meditating[13], shines brightly':), rendering the line, 'And the sage in his thought shineth bright'. Carus captions this hymn, 'Buddhist Doxology'.

Becker's research on trance in the light of current neurobiological research on musical cognition thus directs our attention to the ways that music, particularly in ritualized settings, can serve as a powerful force for shaping human experience along a continuum of mental-somatic states that range from 'deep listening' to trance. Her careful attention to the diversity of cultural contexts also directs our attention to the distinctive '*habitus* of listening' that informed Carus' musical experience and the experiences of those he hoped would sing his hymns.

---

of engagement (Carus, 1882). His introduction frames the collection in terms of his alarm at the influence of Schopenhauer's pessimistic philosophy, inspired in part by Buddhism, and expresses his hope that the youth of his day will recover from this sickness.

12. One of these hymns is based on *Dhammapada* 183; I discuss this importance of this *Dhammapada* verse in Trainor, forthcoming.
13. The Pāli verb *jhāyī*, 'meditating', is cognate with *jhāna*, often translated 'trance'.

The Divine Office in Medieval Christianity:

Becker's discussion of the learned dimensions of trance and Paul Carus' attempt to create a church of the Science of Religion highlight the relationship between music and social formation. Although cognitive science may at first glance look like the study of how the individual mind processes information, the mind, of course, only exists as part of a much larger system: a community of embodied human beings living in a physical environment. Citing the work of William H. McNeill, William Benzon emphasizes that the creation of human society requires coordinated rhythmic activity: 'Music and dance are not mere luxuries consuming resources; they are every bit as fundamental as hunting or child rearing, for example, but fundamental in a different way' (Benzon, 2001, 6). The social function of music in coordinating bodies in the construction of community is an important point, one that is inescapable when religious ritual is taken seriously. The cognitive processes by which this works are not fully understood as yet. Given music's fundamental role in social formation, it is perhaps not surprising that cognitive processes necessary to musical engagement - e.g., interactional synchrony in which one person unconsciously synchronizes his or her body movements with the rhythm of another person's verbal expression - are generally associated with the phylogenetically old core brain structures. What is surprising then, is that this ability seems to be unique to human beings. Although neonatal humans synchronize (being exposed to sound in utero), chimpanzees and other primates do not, nor do they ever learn to keep time to music or a drum beat (Benzon, 2001, 27-28).

The power of music to create community seems to have been understood by the great monastic legislator, Benedict of Nursia. Benedict may have grasped - as we easily do - the social piece of this picture. But I suspect he also grasped - at least intuitively - the cognitive dimension. In what follows, I will try to support this suggestion by considering Benedict's *Rule* as well as some evidence about the experience of medieval monks and nuns.

Not much is known of the life of Benedict, who died in the mid-sixth century. It is his one text, the *Rule for Monks*, which made him so immensely influential in medieval Europe. The *Rule* was adopted and adapted by countless male and female monastic communities throughout Western Europe (and beyond, with the later expansion of Latin Christianity into the New World, Asia, and Africa). Although other orders were founded and other monastic rules written, Benedict's *Rule* often still served as the foundation for organizing new patterns of monastic life[14].

The purpose of the *Rule* was to provide a fairly flexible structure to organize the life of communities of adults and children to best enable them to succeed in their ongoing struggle against the devil and thereby gain eternal salvation. The overarching metaphors for the monastery and its life are battle (against demonic forces) and school (for the Lord's service). Both battle and school are communal: the monastic is a warrior *in an army* not a knight errant, and the school is never envisioned as the life of a solitary scholar. The individual chapters of the *Rule* provided guidelines that over the course of centuries were used to design the built environments of monastic complexes, to structure the passage of time into liturgical seasons and ritual observances throughout the day, to organize monks and nuns in hierarchical ranks, to determine what could be eaten, what could be drunk, what could be worn, what could be said, what could be read, what could be owned, and ultimately, what could be thought. Most of the specific rules within the text can be easily seen as very effective techniques for building a human community out of a group of individuals. These techniques, from the prohibition of private ownership, to absolute obedience to the abbot or abbess, to the requirement of kitchen service by each member, orchestrated the enactment of what Benedict referred to as the giving up of one's own will[15]. In the place of an identity grounded in a sense of one's own will, the monk or nun cultivated an identity as a member of community, a body who with one voice coming from its mouth, sang to God.

Sang to God? Indeed, the single aspect of monastic life to which Benedict directs the most attention in the *Rule* is the Divine Office, the 'work of God (*opus dei*), over which nothing was to be preferred'[16]. The Divine Office was comprised of eight services of communal worship in the monastery's chapel.

---

14. For example, in the Gilbertine Order, founded in 1131, the nuns followed a modified version of Benedict's *Rule*. This tendency for new orders to use the *Rule* was enhanced by the Fourth Lateran Council in 1215, which prohibited the founding of new orders and mandated the use of existing approved rules for any new monastery.

15. See, for example, *RB* ch. 5.6, where immediate obedience to the abbot's order means abandoning what one is doing and giving up one's own will: *relinquentes statim quae sua sunt et voluntatem propriam deserentes*; *RB* ch. 4.60, where it is an instrument of charity to hate one's own will: *voluntatem propriam odire* (Fry, 1981).

16. *RB*, ch. 43.3: *Ergo nihil operi Dei praeponatur* (Fry, 1981).

Nightly sleep itself was interrupted for one of these services. And this communal worship was fundamentally musical: it was structured around the communal chanting of the Psalter, supplemented by other hymns, sequences, and antiphons. Much of the education of children in monasteries was directed toward training them for ritual performance: memorizing the psalms and hymns and learning how to chant. For medieval oblates (i.e., children offered to monasteries), the monastic experience of corporeal discipline (both in terms of training and punishment) and liturgical training produced a ritualized body and mind[17].

Benedict's chapters on the Divine Office offer rather bare enumeration of what psalms, hymns, antiphons, and prayers are to be chanted at each service. He spends little time trying to express the quality of the ritual experience. He offers a few quick biblical passages to remind the monks to sing wisely and with fear of God, mindful that they are singing in the presence of God and the angels. For this reason, he concludes, 'let us stand for singing the psalms so that our minds are in harmony with our voices'[18]. As is so evident in the rest of the *Rule*, Benedict well understood the need to orchestrate inner and outer experience. But this still rather laconic description belies the nature of the ritual involved. The Psalter is a collection of one hundred and fifty songs of often wrenching emotional expression, set to melodies which were also emotionally evocative. By Benedict's day, the use of the Psalter as the foundation for ritual was already embedded in monastic practice. The *Conferences* of John Cassian, one of the two texts Benedict specifically mandated as required reading for monks, gives a detailed description of the type of experience that was to be cultivated in using the Psalter. For Cassian, the monk who is continually ingesting ('grazing'; 'pasturing') the words of Scripture:

> will begin to repeat them and to treat them in his profound compunction of heart not as if they were composed by the prophet but as if they were his own utterances and his own prayer... When we have the same disposition in our heart with which each psalm was sung or written down, then we shall become like its author, grasping its significance beforehand

---

17. See Boynton, 2000, 2007, 899; Boynton & Cochelin, 2006.
18. *RB* ch. 19.7: *Stemus ad psallendum ut mens nostra concordet voci nostrae* (Fry, 1981).

rather than afterward. That is, we first take in the power of what is said rather than the knowledge of it (Cassian, 1997, 384).

Thus the monk sings the terror or grief or joy of the psalms to embody those emotions. Owning those emotions (in our pop psychological jargon) leads, according to Cassian, to a knowledge that is not perceived as heard but as seen. And, it leads to ecstasy: 'Once the mind's attentiveness has been set ablaze, it is called forth in an unspeakable ecstasy of heart and with an insatiable gladness of spirit, and the mind, having transcended all feelings and visible matter, pours it out to God in unutterable groans and sighs' (Cassian, 1997, 384-385)[19]. This is a theory of mental experience - that mindful repetition of the psalms leads to ecstasy - that Benedict does not enunciate so much as take for granted.

It's easy to suggest that Cassian and Benedict, as idealists and/or propagandists, would have a lofty view of the possibilities of mental experience enabled by a potentially wearisome life structured by a potentially wearisome ritual. But is there any evidence for the effectiveness of such techniques in producing 'ecstasy'?

Ecstasy in the sense envisioned by Cassian fits neatly on the continuum of deep listening/trance experience that Judith Becker describes. The chanting of the psalms and other hymns - simultaneously performed and heard by the monastic - serves the function that Becker ascribes to music in trance experience: 'by enveloping the trancer [or in this case, the monk or nun] in a soundscape that suggests, invokes, or represents other times and distant spaces, the transition out of quotidian time and space comes easier' (Becker, 2004, 27). And the deep emotional response to music created by the monastic *habitus* of participating in the liturgy should, not surprisingly, lead to what Becker identifies as effects of deep listening: 'feelings of nearness to the sacred, loss of boundaries between self and other, experience of wholeness and unity', or sometimes feelings of anguish or pain, as well as gnosis (Becker, 2004, 54-55). The monastic life in general, and the divine office in particular, promoted both the social magic, what Warner refers to as the solidarity created by making music together (Warner, 2008, 180), and the interior magic, what Cassian refers to as the ecstasy of heart, spirit, and mind.

---

19. For a brilliant example of the ecstatic use of the psalms, see Otter, 2008, although Otter distinguishes perhaps too sharply between vision and meditation.

Warner, Benzon, and Becker would suggest that these social and cognitive aspects of experience are deeply entwined: the rhythmic entrainment of the members of the monastic choir to each other produces emotional energy that is perceived by the participants as a transcendent experience[20]. And although Benedict of Nursia had no language of coupled nervous systems or supra-individual biological processes, his *Rule* provided the framework for generations of monks and nuns to move in sync, to spend their lives performing the chanted cries of the Divine Office, ideally to feel the diminishment of their own wills, and yet to feel heightened shared emotional arousal.

The repetitiousness of this ritual enacted over a lifetime could of course be monotonous. The monotony of this wordy ritual, along with its ostensible lack of high arousal pageantry (in comparison with other more sensorily dramatic rites) has led to some speculation that monks and nuns were, on the whole, mentally passive. In this view, monastic ritual tended to suppress personal intellectual engagement, instead inculcating an orthodox script through its incessant verbal repetition[21]. Medieval monastics and those who tried to control their lives were aware of the potential for monotony. Sloppiness in performance due to boredom or haste was a frequent enough complaint in monastic visitation records. A typical remedy for monotony-induced performance is instructive in this regard. In addition to exhorting monastic choirs to greater devotion, monastic overseers sometimes also instructed stricter adherence to the musical structure of the psalms. Proper chanting required a ruminative pause in the middle of each psalm verse, and monastic choirs were urged to observe the pause[22]. The structure of the verse and the style of intoning were thus learned tools for creating mental space for personal meaning-making. Such space was surely not always used by every monk and nun, but the structure was there, as was the recognition of its potential.

---

20. Warner (Warner, 2008) draws on Randall Collins' Durkheimian analysis in Collins, 2004. Benzon theorizes 'coupled nervous systems' as a way to suggest the biological basis for patterns of shared neural activity that could explain heightened emotional connection among people musiking together (Benzon, 2001). Becker uses Maturana and Varela's idea of 'structural coupling' and Nuñez' language of 'supra-individual biological processes' (Becker, 2004).
21. See the modes theory in Whitehouse, 2000, 2004. For more extensive consideration of its application to medieval Christianity, see Clark, 2004, 2007.
22. Anne Bagnall Yardley cites several examples of criticism of nuns' failure to observe the proper pauses in chanting the psalms (Yardley, 2006, 82-84).

Another aspect of this highly repetitive ritual is its susceptibility to 'autopilot' performance. After mastery of the chant repertoire, isn't it likely that monks and nuns went through the motions without much attention to the words they chanted, the moods evoked by the intervals of the simple melodies of the chant, the bodily comportment Benedict mandated, the festivals of saints and salvation-history events that marked the passage of the year? Of course this is possible, yet again, our contemporary impatience with repetition shouldn't blind us to its possible emotional satisfaction, and the cultivation of awareness of nuanced variation. Also, autopilot mastery of basic procedures - bodily or mental procedures - is exactly what enables improvisation[23]. It is the ability to continue with the performance without explicit attention to it that allows the monk or nun to take advantage of that ruminative pause, or to follow a chain of associations triggered perhaps by one word in the psalm verse, to invent - to discover meaning, to feel something, to reach out, or to go within, to 'make thought'[24].

Although the language of Cassian suggests the accessibility of ecstasy, and the provisions of the *Rule* of Benedict provided the ritual structure to enable a community of monastics to dissolve the boundaries of self and share transcendent emotional experience through musical expression, there was nonetheless a strong sense in the Middle Ages that such experience was not in fact the lot of every monk and nun. It was the extraordinary monk or nun who testified - at least in ways preserved for posterity - about his or her ecstasy. 'At least in ways preserved for posterity' is an important caveat, considering that many aspects of monastic culture remained in the realm of oral discourse. Unless the ecstatic monk or nun felt compelled to share his or her experience beyond the walls of the monastery, there would be no evidence of this experience produced. But even with this important caveat that speaks to the unknowable lives of most medieval monastics, there is evidence of monks and

---

23. See, e.g., Benzon on the creative ability to pay attention to new ideas while executing well known complicated musical and bodily maneuvers (Benzon, 2001, 14) and on how extensive practice within an ensemble allows the possibility of improvisation (Benzon, 2001, 5-6).
24. The monastic idiom of *making* ideas, not *having* ideas, recalls the insights of cognitive theorists who emphasize the mental activity that takes place in cultural acquisition: people activate cognitive resources that make other people's utterances or gestures or artifacts relevant in the sense of producing inferences. See, e.g., Boyer, 2001, 2002; Sperber, 1996. For medieval monastic life as designed for pursuing the making of prayerful thought, see Carruthers, 1998.

nuns who claimed to have gone beyond the boundaries of normal cognitive experience, who were 'lifted up' or who 'saw the heavens open' or who felt invaded by light, or spoke words they did not feel were theirs, or who felt embraced by warmth, or claimed any number of other 'paramystical' experiences. Often disdained in earlier scholarship as not genuinely mystical experiences because of being so obviously embedded in the ritualized life of the monastic, these claims of extraordinary experience are now better examined in light of cognitive theory of music and trance. While not fully conforming to Becker's profile of trance - there is, for example, no strong kinetic dimension to monastic ecstasy - the monastic claims of extraordinary experience are more readily understood when approached through a theoretical perspective that takes into account the social dimension of cognitive experience.

These two case studies, drawn as they are from disparate historical and cultural contexts, merit a more detailed examination than is possible within the confines of this article. Despite the brevity of their treatment here, however, they illustrate the promise of cognitive science for understanding the functions and dynamics of music in a diversity of religious communities. Theoretical approaches such as Becker's, grounded in both biological research and cultural analysis, provide a model for future comparative analyses of music in the context of religious practice. Such approaches help us understand why Paul Carus was motivated to compose Buddhist hymns and why Benedictine monks and nuns have spent so much time chanting the Psalter, even as they illuminate the different cultural contexts to which these distinctive forms of musical practice were attuned.

CITATIONS OF BIBLICAL TEXTS IN GREEK, JEWISH AND CHRISTIAN INSCRIPTIONS OF LATE ANTIQUITY: A CASE OF RELIGIOUS DEMARCATION*

EKATERINI TSALAMPOUNI

In Deut 11:18-22 Yahweh orders his people to write his words in their hearts and souls, to bind them as a sign on their hand and fix them on their forehead, to teach them at home and elsewhere, to bear them in mind every moment of their life, and to write them on the doorposts and the gates of their houses (cf. also Deut 6:6-8, Prov 6:20-22). God's commandment has been interpreted by his faithful among Israel either literally or metaphorically, but in all cases according to the spirit of the text: the words of Yahweh – in the narrow sense, as the words in Deuteronomy itself, and in a broader one, as those written in other books of the Hebrew Bible, too – occupied an important place in Israel's liturgical and every day life. These words became the object of study in synagogues, rabbinic schools and private homes, were recited in the liturgical assemblies of Israel not only in Palestine but also in the Diaspora, were worn in phylacteries on arms and forehead, and were written on doorposts of houses. According to Josephus (*Against Apion* II 175) a main characteristic of all Jews was the regular reading of the Torah (νόμος), which was the only eternal value that regulates the life of every Jew (*Against Apion* II 277). It is not the purpose of this paper to present a detailed discussion of the Bible's role in the life of ancient Israel since this has already been the subject of many studies. As an instance only of this historical reality one should mention some Jewish inscriptions from the Diaspora of the Roman

---

* This article is an expanded version of a paper read in the joint session of "Early Christianity between Christianity and Judaism Group" and "Early Judaism and Rabbinics Group" at the Annual Meeting of the European Association of Biblical Studies (Lisbon 2008).

times that bear witness of persons who were praised after death because they had been either students or teachers of the Torah (νόμος)[1]. Their devoted occupation with the study of the Law seems to have been highly evaluated by their environment and worth mentioning on their gravestones as a component of their identity and as an example for the rest of the community[2].

On the other hand, from the very beginning the early Christian Church understood itself as the real heir of ancient Israel and recognized in the person of Jesus the end and fulfilment of the Law and of the prophecies of the Bible. The New Testament writings contain a wide range of messianic reinterpretations of the biblical texts to the person of Jesus (e.g. the Epistle of Hebrews or the Gospel of Matthew) and therefore reveal the prominent place given to the Old Testament writings within the theological discourse and self-understanding of early Christians (cf. Justin, *Dialogue with Trypho* 24.1)[3]. The same texts also provide us with some information regarding the liturgical use of the biblical texts in the various early Christian communities (Acts 4:24 - 26, Eph 5:19; Justin, *First Apology* 67, 3-5). Beside the Old Testament writings early Christians started using another group of texts as God's authoritative word, those of the New Testament[4]. Old and New Testament were related to each other through the person of Jesus Christ and on the basis of the pattern 'promise – fulfillment.'[5] This Christian Bible was read in liturgical assemblies, served as a fundamental element of the theological argumentation of the ancient Church, functioned as the prior text of many liturgical texts, was the subject of the exegetical preaching of the Church

---

1. *JIWE* II, nos. 68 (νομομαθής, Rome); 270 (νομομαθής, Rome); 374 (νομομαθής, Rome); 390 (νομομαθητής); 544 (μαθητὴς σοφῶν); *JIWE* I, nos. 186 (didascalus, Terragona, Spain); 48 (διδάσκαλος, Venosa, Italy); 12 (ἁγίων τε νόμων σοφίης τε συνίστωρ = knowledgeable in the holy laws and in wisdom, La Botaccia Italy). The abbreviations of epigraphic corpora used in this study are those proposed in Horsley / Lee 1993.
2. Cf. Rutgers 1995, 199-200 for a similar function of community-related offices occurring in Jewish funerary inscriptions from Rome
3. A more radical position is held by the author of *Letter of Barnabas* who denies any fulfillment of OT prophecies within the Old Testament itself and declares that the Old Testament is the inheritance exclusively of the Christians, 14,4 (*PG* 2, 768A-B).
4. See for example the 59th canon of the Synod of Laodicea (4th century), which ordered that in the church only the canonical books of the Old and New Testament should be used for the liturgical purposes.
5. Cyril of Alexandria, *De adoratione et cultu in spiritu et veritate* 3 (*PG* 68, 268A); Justin, *Dialogus cum Tryphone* 34,2 (*PG* 6, 548A-B).

Fathers and of private study at home (Harnack 1912), became the 'touchstone' (Ellis 1988, 692) for the whole life of the Christian communities (Markschies 2006, 94ff.) and therefore part of their religious identity. At the same time the early Christian appropriation and interpretation of the Hebrew Bible seems to have been an area of Jewish-Christian interaction and theological debate in Late Antiquity (cf. Hengel 1992, 51-55). Christian texts like the *Epistle of Barnabas*, Justin's *Dialogue with Trypho*, Melito's and Irenaeus' works in the 2nd century and those of Origen in the 3rd from the Christian side and passages from the rabbinic writings (Alexander 1992, 6-15) and probably the *Collatio Legum Mosaicarum et Romanarum* from the Jewish side (Rutgers 1995, 213-268) contain direct and indirect evidence of the Jewish-Christian dispute over the interpretation of the Hebrew Bible as a whole or of particular passages in it.

As it has already been observed all these writings are certainly literary constructs and therefore the need to relate them and the historical situation they reflect to concrete pieces of real-life of their time has already been stressed (Rutgers 1995, 287-288). A welcome example of such data seems to be a group of inscriptions, both Jewish and Christian, dating from the last pre-Christian centuries and up to Late Antiquity, which quote verbally, paraphrase or allude to passages from the Scriptures thus placing them within new contexts (e.g. liturgical, funerary, civil etc). Despite their unique character these inscriptions have not attracted the desirable attention of biblical scholars yet. These documents offer us insight into the way individuals and whole communities, both Jewish and Christian, understood, interpreted and applied particular biblical passages in their everyday life, perceiving them as an inseparable part of their religious identity. The purpose of this short study is therefore twofold: first, to provide a systematic presentation of the geographical and chronological diffusion of these epigraphic texts, of their contexts and of the Biblical books and passages they mostly quote; secondly, by comparing the Jewish and Christian epigraphic evidence and placing it within its broader socio-historical and religious context to discuss a particular aspect of it, namely that in some cases the use of these Biblical quotations served as religious demarcation of individuals or communities.

A review of the previous research

The existence of such inscriptions has been known since the end of the 19th century when the first studies appeared (Böhl 1881; Nestle 1883;

Deissmann 1905). In the years that followed the inscriptional citations of scriptural passages were mostly treated sporadically by editors of various corpora of inscriptions or were briefly discussed within broader contexts (Felle 2006, 10-12). The scholarly attention was attracted primarily by the Christian inscriptions (e.g. Gensichen 1910; Jalabert 1914; Leclercq 1914), which were much more numerous. In the 1980's and 1990's the interest in these inscriptions was revived (Malunowicz 1982; Feissel 1984; Pietri 1985; Mazzoleni 1989, 346-352; Boffo 1996). The editions of the Jewish inscriptions from the Western part of Europe (Noy 1993), Rome (Noy 1995), Egypt (Horbury & Noy 1992), Beth She'arim (Schwabe & Lifshitz 1974), Cyrenaica (Lüderitz 1983), the reprint of Frey's *Corpus Inscriptionum Judaicarum* by Baruch Lifshitz (Frey 1975), and the systematic edition of the inscriptions from Asia Minor in the series *Inschriften griechischer Städte aus Kleinasien* gave access to a number of Jewish inscriptions and rekindled the interest in the Jewish Diaspora of the Hellenistic and Roman times. Some of these studies also devoted brief sections to the biblical citations in Jewish inscriptions (e.g. van der Horst 1991, 37-39). What is however actually missing is a systematic presentation of all Jewish inscriptions that make use of biblical citations and a comparison of them with the evidence gathered from the Christian epigraphic texts with biblical citations, a need that has already been pointed out by some scholars (Fine and Rutgers 1996, 22-23).

On the one hand the recent publication of the Jewish inscriptions from the Eastern part of the Roman Empire (Noy, Panayotov & Bloedhorn 2004; Ameling 2004; Noy & Bloedhorn 2004) along with the previous critical editions of Jewish inscriptions from the Western part of the Roman world and from Egypt and on the other the excellent Italian edition by A. E. Felle of the corpus of Christian inscriptions with biblical citations in 2006 (Felle 2006) make this task much easier. These inscriptional corpora supplemented by the most recent inscriptions published in *Supplementum Epigraphicum Graecum* and *Bulletin Épigraphique* each year offered the material that will be discussed in this present paper.

Some methodological considerations

At the outset the methodological background of this survey and the criteria for selecting the inscriptions discussed should be shortly explained. These considerations regard a) the geographical and chronological diffusion of the material, b) the types of artefacts included in the databank, c) the criteria for deciding whether an inscription is Jewish or Christian, and d) the types of use of biblical texts in inscriptions that should be taken into consideration.

a) For practical reasons only the Greek inscriptions – Jewish and Christian – of the Eastern part of the Roman Empire will be presented in this paper; they are much more numerous than those of the Western part and therefore offer a quite satisfying sample for our discussion. A further reason for not examining the evidence of both parts of the Roman Empire together is the fact that the socio-historical circumstances, in which the Jewish and Christian communities in the Western and Eastern part of the Empire lived and developed, were quite different, a fact that makes the separate treatment of the epigraphic evidence methodologically necessary. The earliest chronological limit for this study is fixed in the 2nd century B.C.E., when the first biblical citations appear in Jewish epigraphic texts, and the latest in the 6th c. C.E., when the end of the rather arbitrarily coined period of 'Late Antiquity' is usually placed. The 6th century has been chosen for another reason, too; the number of Jewish inscriptions declined considerably after this century and Jewish epigraphic texts reappeared later in the 9th and 10th centuries written now in Hebrew (Ameling 2007, 279-280).

b) In the database were included all types of monuments, i.e. architectural parts from various kinds of edifices, mosaics, personal and liturgical objects, jewelry and funerary monuments. However, the magical texts were excluded because they are a category on their own and should be discussed separately.

c) A further important methodological issue is that of the criteria that should be applied when determining the possible Jewish or Christian origin of an inscription. The definition of such criteria has been an issue of discussion over the past years. Larry H. Kant offered a list of such criteria regarding Jewish epigraphic texts acknowledging at the same time that "this way of determining Jewishness, however, by no means points to a perfectly clear boundary between Jews and non-Jews in the ancient Mediterranean" (Kant 1987, 683; cf. also Gibson 1999, 6). Many scholars have pointed out the difficulty of identifying a Jewish or Christian inscription as such when only one criterion can be established (Kraemer 1991; van der Horst 1991, 16-18; Bij de Vaate & van Henten 1996; Ameling 2004, 10-21), although in some cases the presence of some of them, even when they are not accompanied by other indications, may be of decisive importance (cf. Bij de Vaate & van Henten 1996, 17, n. 8 who mention as such the menorah symbol or the presence of an inscription in a Jewish catacomb; for some doubts, however, see Kraemer 1991, 151-154). In the present study the following criteria for deciding the Jewish identity of an inscription were applied: a) the use of Hebrew, b) the presence of Jewish symbols (e.g. menorah, shofar, ethrog, loulab, oil vase, Torah ark), c) typical Jewish formulae, d) typical Jewish names, e) reference to Jewish religious customs, f) an evident Jewish context (e.g. Jewish catacomb of synagogue), g) mention of a Jewish synagogue or of Jewish designations and titles, h) chronological indications (especially when deciding whether an inscription can be Jewish or Christian), i) association with other monuments or evidence of Jewish presence in the place of provenance, j) reference to a Jewish writing or citation of a biblical text, and, finally, k) reference to Samaritans. Regarding the Christian inscriptions of the present database similar criteria were also determined: a) typical Christian formulae, b) the presence of Christian symbols (e.g. cross etc), c) reference to Christian ritual acts and customs, d) an evident Christian context, e) typical Christian names (in contexts where their use seems more likely to be Christian), f) reference to a Christian edifice or church office, g) self-identification, h) chronological indications, i) association with other monuments or evidence of Christian presence in the place of provenance, j) reference to or citation of a biblical text (cf. also Felle 2007, 355). The comparison of the two criteria lists shows that some features are common to both kinds of epigraphic monuments, which is due to the fact that the line between Jews and Christians in Graeco-Roman and Late Antiquity was not as clear as it is assumed to be today. This intersection of the criteria as well as the correspondence of many monuments to those of the non-Jewish and non-Christian environment make it clear that an inscription should be identified as Jewish or Christian if it met more than one of the aforementioned criteria, a suggestion made by P.W. van der Horst (van der Horst 1991, 18) that is adopted in the present study as well. However, including or excluding an inscription in such a corpus can be a very complicated procedure since one has to face two dangers; on the one hand, a rigorous application of the above principle, which could lead to a minimalistic evaluation of the evidence and to a possible exclusion of valuable material, and on the other hand, a methodological slackness that could lead to a rather maximalistic inclusion of material that actually does not belong to this corpus (van der Horst 1991, 18). Since the golden mean is rather an ideal goal than an objective achievement, one should also leave some room for doubt and speculation and regard the identification of some of the texts that are discussed in this study as probable and not certain.

d) A last point to be dealt with is that of the different types of use of biblical texts in the inscriptions that were included in the present epigraphic corpus. A still unsolved problem among scholars is that of the precise definition of the terms that refer to various types of scriptural citations or references (Porter 1997, 79-96, esp. 81; Porter 2006, 98-112). This ambiguity is due to the quite different conception of citation in ancient cultures, which were more orally-oriented than our own and applied different standards when borrowing from other sources (Stanley 1997, 20-23). It is not possible to enter into a full discussion of the issue here, though. The typology that has been developed for the present study took into consideration the previous discussion of the issue (Porter 2006, 106-109 and Felle 2006, 17-20) as well as the particular nature of the epigraphic texts (e.g. the informal character of many of them, the question of access to written sources, the existence of clichéd text exemplars, the literacy level of both client and mason etc). In order to exemplify these various types of biblical citations an example for each case– either Jewish or Christian - is provided:

- verbatim quotation: e.g. 'ἐπὶ ἀσπίδα καὶ βα|σιλίσκον ἐπιβήσῃ'[6] (Alexandria, painted Christian inscription in a tomb, end of 3rd – beginning of the 4th c., verbatim quotation of Ps 90:13)

- paraphrase, i.e. a citation 'typified by the use of words from the same semantic domain, or similar words in differing syntax, as a recognizable passage' (Porter 2006, 108): e.g. 'Σοφία Γορτυνί|α πρεσβυτέρα | κὲ ἀρχισυναγώ|γισσα Κισάμου ἐν|θα. μνήμη δικέας | ἰς ἐῶνα. ἀμήν'[7] (Crete, Jewish funerary inscription, 4th or 5th c. AD, paraphrase probably of the LXX text of Prov 10:7)

- allusion, i.e. a citation that 'involves the invoking of a person, place, or literary work' (Porter 2006, 109): e.g. '(...) ἔσται αὐτῷ αἱ ἀραὶ | ἡ γεγραμμέναι ἐν τῷ Δευτερο|νομίῳ'[8] (Yenice of Phrygia, Jewish funerary inscription, 248/9 AD, reference to the book of Deuteronomy)

---

6. 'On asp and cobra you will tread'; all English translations of the Septuagint texts come from the *NETS* of Septuagint, http://ccat.sas.upenn.edu/nets/edition/ (10/2/ 2009).

7. English translation in Noy, Panayotov & Bloedhorn 2004, 252: 'Sophia of Gortyn, presbytera and archisynagogissa of Kissamus (lies) here. The memory of the righteous woman (be) forever. Amen.'

8. "the curses that are written in Deuteronomy should be applied to him."

– echo, i.e. involves 'the invocation by means of thematically related language of some more general notion or concept' (Porter 2006, 109): e.g. the expression 'δοῦλος Θεοῦ'[9] in Christian inscriptions.

In the present study echoes and texts that contain simple words or phrases whose biblical provenance is not secure or whose proverbial use makes their direct citing from the original biblical text rather improbable were not taken into account. Texts that seem to be citations of the liturgical versions of some biblical passages were excluded, too; although they give evidence of the influence of many biblical passages on the formulation of the liturgical texts, their direct dependence on the Scripture is doubtful.

The database

The epigraphic database compiled contains 506 inscriptions with Bible citations: 16 Jewish and 490 Christian. This apparent disproportion between the two groups is not so great when these inscriptions are compared with the total of Jewish or Christian inscriptions from the eastern part of the Graeco-Roman world; both groups of inscriptions cover about 1% to 2% of the total of the Jewish or Christian documentation. Two major problems should be taken in consideration: a) we do not have the complete corpus of the inscriptions, since many have been lost or are to be found in the future and b) the sample is not representative for the entire community, since the poorest strata of the population, which was also the most numerous, did not probably leave any epigraphic traces. Despite these two restrictions it should be observed that the low percentage does not seem to be accidental but indicates that the phenomenon of the epigraphic biblical citations was not a widely accepted trend. The majority of Jews and Christians were rather hesitant to write their holy texts on perishable materials.

The geographical diffusion of these texts also betrays that the practice of using citations in epigraphic texts was intense in particular areas, a phenomenon that was probably due to local traditions. The greatest concentration of Christian citations of the Bible is found in monuments from areas with a long Christian tradition like Syria, Palestine and Egypt, followed by Asia (including the Aegean islands), the Balkan and Pontus, whereas in Scythia the smaller sample is attested. The Jewish inscriptions paint a quite

---

9. 'God's servant'.

similar picture: the majority of the evidence comes from Asia and the Aegean islands, an area where a great number of Jewish Diaspora communities had flourished from the Hellenistic times until Late Antiquity, while from the Balkans come two inscriptions, one from Phoenicia and one each from Crete and Palestine.

Regarding the chronological diffusion of the monuments both groups of inscriptions – Jewish and Christian – display the same pattern. The majority of the datable monuments can be placed in the 3rd and 4th century. The inscriptions before the Severan era show a tendency to assimilate to those of the common praxis (Felle 2007, 360), whereas from the second half of the 3rd century AD onwards there is a general tendency of self-identification of individuals and groups (Felle 2007, 359) that conforms to the general epigraphic trend, too.

Another interesting aspect, and a point of dissimilarity between the two groups, is the question of the monuments that bore such citations. While most of the Christian texts are found on architectural parts of various edifices (cult, private or even public), the majority of the Jewish inscriptions belong to funerary contexts. Since the inscriptions of some synagogues from the Diaspora are now at our disposal (e.g. Sardis, Delos etc.), and in only two cases do we have (from Caesarea in Palestine and probably from Nicaea in Bithynia) a text quoting a scriptural passage which suggests that the Jews in the Diaspora avoided inscribing such texts on their buildings. It should be noted, though, that the funerary use of various biblical passages is a practice common among the Christians, too; about one sixth of the inscriptions – among them the oldest ones – belong to funerary contexts. Unlike Jews Christians inscribed their favorite scriptural passages not only on funerary monuments but also on the lintels of their churches and houses, on the walls of castles or cisterns and on their personal objects and jewels. The function of these texts in both Jewish and Christian monuments was not ornamental; they rather offered protection, instruction or authority (Young 1997, 103).

Jews and Christians did not prefer the same texts in their inscriptions. This is also a significant point of differentiation between them. Although the Christian inscriptions show a clear preference for passages from the Old Testament (about 359 out of 490 contain Old Testament citations), a phenomenon – as it has already been mentioned in the beginning of this paper – that is consistent with the general Christian attitude towards the Old Testament: the variety of the passages used by them is significant. More than two thirds of these citations come from the Psalms, a favorite book among the Christians that occupied a prominent place within the early Christian liturgies. The book of Isaiah and especially the passage 6:3, the so-called 'Trisagion' hymn, that was integrated in the Christian liturgy and played an important role in the theological debates of the Ancient Church, is the second Old Testament book that is often quoted in Christian inscriptions. The book of Isaiah does not seem to have been popular among Jews when quoting in their inscriptions; it appears only once. The Jewish quotations from the Psalms are more frequent (4 times); in two cases – Ps 90:1 and 45:8 – the same passage is used by both Jews and Christians, although in most cases in a quite different context. The book that is most often used is Deuteronomy (5 times) – especially the curses that are contained in chapters 27-28 – and always in a funerary context. These monuments will be discussed further later. It should be noted that Deuteronomy as well as the other books of the Pentateuch are very thinly attested in the Christian inscriptions. Proverbs 10:7 also seems to have been a text that Jews often used on their funerary monuments. Most of these inscriptions come from the western part of the Graeco-Roman world (*CIJ* 86, 201, 370, 625, 629, 661); however, the fact that there is also one case attested in the eastern part (cf. also the Hebrew inscription from Jaffa quoting the same text, *CIJ* 892), which is dated in the same period as those from the western, led to the conclusion that this text was commonly used by Jews in funerary contexts perhaps because this text was recited in the funerary liturgy of the Jewish Diaspora (van der Horst, 1991, 38). A probable liturgical use of Ps 135:25 is also assumed in the case of the inscription found in Nikaia (Ameling 2004, 322-324; Fine and Rutgers 1996, 10).

This uniform use of some passages in particular contexts is not an exclusive trend of Jews in Diaspora, though. The Christian monuments confirm that the same practice was well known among the Christians, too. For example Ps 120:8 ('God bless your coming in and going out') is usually found on lintels and doorposts of churches, private houses and castles (e.g. Felle 2006, 89, nr. 108, Arabia; 169, nr. 328, Syria), while Ps 117:20, is found mainly on cult monuments from Syria and Palestine (e.g. Felle 2006, 115, nr. 172). Ps 28:3 ('The voice of the Lord is upon the waters') is usually attested in buildings and objects that are related to water (e.g. cisterns, vases, jugs etc.; see for example Felle 2006, 116, nr. 174) and Ps 45:8 on various public and private buildings (Felle 2006, 184-185, nr. 380).

This application of particular passages in specific contexts brings into the discussion the issue of interpretation and of the Jewish and Christian reception history of biblical texts, a question that is however beyond the scope of this paper. It should only be observed that most of the inscriptions containing biblical citations are, in spite of their brevity and the singularity of their nature, excellent examples of Jewish and Christian reworking of the scripture and application of them in new situations. In this respect the citation of Biblical texts in epigraphic texts could be regarded as an example of ancient intertextuality in the way that this term has been defined by J. Kristeva (Kristeva 1966; Weise 1997, 41). This practice of reworking material already found in previous documents is well known in ancient composition. The three criteria that are usually employed in judging whether a text depends on another, i.e. a) external plausibility, b) significant similarities beyond the range of coincidence, and c) intelligible differences (Brodie 2001, 105), could also be applied to Jewish and Christian inscriptions quoting the Bible as the two examples in Table 1, one Jewish and one Christian, clearly demonstrate. Their dependence on biblical passages is plausible since the Scriptures played a prominent role in Jewish and Christian communities. The use of motifs, vocabulary and ideas from the biblical text is also evident. The new text differs from the reference text or texts regarding genre, content, function and theological message.

Besides their presenting paradigms of Jewish and Christian intertextuality these inscriptions are also instances of ancient textuality. It is generally accepted that there is a relationship between identity and textuality. Texts in the Mediterranean always have constructed a sense of 'who we are' (Lieu 2004, 30) and this seems to have been the case with many of the epigraphic documents under discussion. In his 1996 article Greg Woolf demonstrated the monumental character of the inscriptions and their function as medium for denoting the identity of the individual who had them carved on stone (Woolf 1996, 28-29). Jewish and Christian inscriptions are not an exception to this rule. Their chronological diffusion and their typology attest that Jews and Christians also followed the 'epigraphic habit'[10] and adopted the epigraphic culture of their pagan environment. This tendency of assimilating with the prevalent culture of their environment is also made evident by the fact that it is often very difficult to decide about the Jewish or Christian provenance of an

---

10. The term was coined by R. MacMullen in an article in 1982; cf. also Meyer 1990.

epigraphic monument or differentiate it from those of the pagan environment. This is not coincidental but it should rather be related to the dialectic nature of 'identity' in the sociological sense of the term and the way a minority group usually behaves within a broader social context. 'Identity' is usually understood as a dialectic relation between the subjective and the objective, between self-understanding and the opinion of the environment (Jenkins 1996, 20.86-87). In this process dynamics of social integration and aggregation develop and very often social groups select certain artifacts from their material cultural repertoire to act as their signs of boundaries. Socio-historical studies have shown that members of minority groups – like the Jewish Diaspora or early Christian groups – develop a dynamic and ambivalent double consciousness: they exhibit a tendency of assimilation to their environment while they often adopt symbols that denote their identity (Cohen 1989, 14-16). It seems that some of the inscriptions containing biblical passages could be explained using these models thus providing paradigms of religious demarcation. The Jewish inscriptions of Acmonia that refer to the curses of Deuteronomy or paraphrase a Zechariah text and use it as a threat for the potential grave desecrators (Table 2, text 1) or a Christian epitaph of a deacon from the same area that cites together with Phil 1, 21 a common pagan expression regarding death, could be named here as an example (Table 2, text 2). The persons that had these monuments erected seem to have adopted the common type of funerary monuments and expression used in the gentile environment while at the same time they used a text of their Holy Scriptures to declare their religious identity.

What happens however when two groups use the same significants – in this case the Bible – to notify their identity? The preceding discussion of the biblical passages used by Jews and Christians has shown that in this case different texts were preferred. A Jewish inscription from Thessalonica seems to offer an alternative answer to this question (Noy, Panayotov & Bloedhorn 2004, 93-94, Mac 13). It is a citation of Ps 45:8 ('The Lord of hosts is with us') carved together with the depiction of a menorah on the walls of a presumably Jewish tomb in the eastern necropolis of the city. Together with the tomb next to it they are the only Jewish monuments in a cluster of tombs that seem to have been used by Christians. The biblical phrase is not unknown to the Christians but it is often used by them in funerary contexts. Nevertheless they prefer the New Testament version of it (Mt 1:23) where the verse of the Psalm is used as an explanation of the name Emmanuel and is related to the person of Jesus (e.g. *SEG* 35:650).

Boundaries play an important role in the construction of identities or as an old English saying observes: 'Fences make good neighbors.' Boundaries and demarcation should not however be understood as expressions of seclusion and negative attitudes towards the environment but rather as means of self-affirmation and a mechanism of self-understanding. Homi Bhabha remarks that boundaries or cutting-edges between different groups and cultures are those that carry the burden of the meaning of culture (Babha 1994, 38-39). The last example of a biblical citation seems to point to this direction. In this case, the Bible is used by someone who is actually acting in this in-between space, by a pagan sympathizer. It is a funerary monument from Chalkis dated in the 2nd c. AD (*IG* XII, 9, nr. 1179). A certain Amphicles erected a monument for his son and in order to protect it from the tomb violators he recited among a number of other curses some that can be found in the Deuteronomy (28:22, 29). In this case the biblical threats are evoked in a way similar to that on the aforementioned Jewish monuments from Phrygia functioning at the same time as a means of declaring Amphicles' sympathy or attraction to the religion of his Jewish neighbors.

Although the biblical citations in Greek, Jewish, and Christian inscriptions are a rather marginal practice within the Jewish and Christian epigraphy they offer valuable insights into the '*Sitz im Leben*' of Jewish and Christian communities and their interaction in Graeco-Roman and Late Antiquity. They remain expressions of personal piety and community self-affirmation, monuments of the way ancient Jewish and Christian communities understood and adapted their scriptures in their personal and communal lives.

Appendix
   Table 1

Jewish inscription from Rheneia, 2nd-1st c. B.C.
(Noy, Panayotov & Bloedhorn 2004, Ach70)

ἐπικαλοῦμαι καὶ ἀξιῶ τὸν θεὸν τὸν	Sir 46:5,
ὕψιστον τὸν κύριον τῶν πνευμάτων	Jer 7:16
καὶ πάσης σαρκός, ἐπὶ τοὺς δόλωι φονεύ-	Gen 14:22
σαντας ἢ φαρμακεύσαντας τὴν τα-	Ex 21:14
λαίπωρον ἄωρον Ἡράκλεαν, ἐχχέαν-	
τας αὐτῆς τὸ ἀναίτιον αἷμα ἀδί-	Dtn 19:10
κως, ἵνα οὕτως γένηται τοῖς φονεύ-	Gen 9:6
σασιν αὐτὴν ἢ φαρμακεύσασιν καὶ	
τοῖς τέκνοις αὐτῶν, Κύριε ὁ πάντα ἐ-	
φορῶν καὶ οἱ ἄγγελοι θεοῦ, ᾧ πᾶσα ψυ-	Job 34:24
χὴ ἐν τῇ σήμερον ἡμέραι ταπεινοῦτα[ι]	
μεθ' ἱκετείας ἵνα ἐγδικήσῃς τὸ αἷμα τὸ ἀ-	Lev 23:29
ναίτιον ζητήσεις καὶ τὴν ταχίστην.	

Christian inscription from Seleucia Syria
BullÉp 1996, 470

--- ΒΑΝΩ	
[---] δοκιμάζε	Sir 2:5
[ται ἐν πυ]ρὶ χρυσὸς καὶ ἄν	
[θρωπο]ι δεκτοὶ ἐν καμί	
ῳ πτωχίας. οὐαὶ τοῖς ἀπ	Sir 2:14
ολελόκωσιν τὴν ὑπομο-	
νὴν καί τ' εἰ πυήσωσιν {²⁶ποιήσωσιν}²⁶ ὅτα-	
ν ἐπισκέπτεται {²⁶ἐπισκέπτηται}²⁶ ὁ Κ(ύριο)ς·	Mt 11:12
ὅτι βιασταὶ ἁρπάζωσιν {²⁶ἁρπάζουσιν}²⁶ τὴν βα-	
σιλίαν τῶν οὐρανῶν καὶ διὰ θλείψεων	Acts 14:22
{²⁶θλίψεων}²⁶ δι' ἡμᾶς εἰσε-	
λθῖν {²⁶εἰσελθεῖν}²⁶ εἰς τὴν βασιλίαν τοῦ	
<θεοῦ>	

Table 2

1. Jewish inscription from Acmonia 2nd / 3rd c. Ameling 2004, 368-370, nr. 174	2. Christian inscription from Synnada 4th c. *MAMA* 4,33
[..................]ιν [ἐξ]έσ- ται ἑτέρῳ ἀνῦξαι τὸ κάθετον ἢ μόνον ἐὰν συνβῇ τοῖς παιδίοις αὐ- τοῦ Δόμνῃ κὲ Ἀλεξανδρίᾳ· ἐὰν δὲ γαμηθήσονται, ἐξὸν οὐκ ἔσται ἀνῦξαι· ὃς δὲ ἂν τολμήσει ἕτε- ρον ἐπισενένκαι, θήσει τῷ ἱερωτάτῳ ταμίῳ Ἀττικὰς ͵α κὲ οὐδὲν ἔλαττον ἔσται τῷ τῆς τυμβωρυχίας ἐνκλήμα- τι ὑπεύθυνος· ἔσται δὲ ἐπικατάρατος ὁ τυοῦ- τος κὲ <u>ὅσαι ἀραὶ ἐν τῷ</u> <u>Δευτερονομίῳ εἰσὶν γε-</u> <u>γραμμέναι αὐτῷ τε κὲ</u> <u>τέκνοις κὲ ἐγγόνοις κὲ</u> <u>παντὶ τῷ γένει αὐτοῦ</u> <u>γένοιντο.</u>	<u>+ ἐμὺ τὸ ζῆν Χριστὸς κὲ τὸ</u> <u>ἀποθανῖν κέρδος.</u> Ἀμάραντο[ς] διάκων ἐλεεινὸς δοῦλος Κυρίου ἐνβλέψας τὴν τοῦ προσκέρου βίου ζοὴν πέντε μὲν δεκάδας τελέσας βουλιθὶς σὺν τῷ ἀδερφῷ Κυρι- ακῷ εὐξαμένης τῆς μητρὸς αὐτῶν Σωφρονίης ἐπύησαν τὸ μνῆμα τοῦτο ἑαυτῦς κὲ τὲς συνβίυς αὐτῶν Παπιανῆς κὲ Πανχαρίης κὲ τῶν γλυκυτάτων πεδίων Δόμνης κὲ πάντων τῶν πεδίων τῶν κληρονμούντων τὸν πε- νιχρὸν βίον. + βλέπε δὲ ὁ ἀναγινώσκων ὅτι <u>ὁ θάνατος</u> <u>πᾶσιν ἠτύμαστε.</u> εὔχεσθε [τὸ]ν θεὸν ὅπως ἰαθῆτε ἀπὸ τῶν ἁμαρτιῶν. + δίκευ χέροιτε. +

MEMORABLE RELIGIONS:
TRANSMISSION, CODIFICATION, AND CHANGE
IN DIVERGENT MELANESIAN CONTEXTS

HARVEY WHITEHOUSE

Nearly twenty years ago I wrote a paper that, although this could not be foreseen at the time, would lead to extended collaboration with historians, archaeologists, cognitive scientists, biologists, philosophers, and fellow anthropologists. One of the most significant and enduring of those collaborators has been Luther H. Martin, whom this volume honours, an historian fascinated by the religions of European antiquity. Yet the paper I had written was about religious traditions as remote from the Graeco-Roman world in space and time as it is possible to travel, focused as it was on rites of initiation and cargo cult activities in twentieth-century New Guinea. The reason why Professor Martin was interested in the paper and in the more elaborated hypotheses which it spawned[1], was that it pointed to some fundamental recurrent features of the way organized religion the world over, and stretching back into the distant past, is codified, transmitted, and remembered, as well as the structure, cohesion, and scale of resulting coalitions. What Professor Martin realized was that the small initiation cults of New Guinea had much in common with the mystery cults of the Hellenistic world and that some of the cargo cults of modern Melanesia were organized in ways forcibly reminiscent of the ancient state religions of Roman antiquity. These resemblances stemmed from the fact that certain features of the workings of the human mind have not changed appreciably over expanses of time and territory that separate the many complex civilizations of recent millennia. It was an appreciation of this, rather than the details of local cultural practices and representations, that brought us together, eventually

---

1. See especially Whitehouse 1995, 2000, 2004.

leading us to publish a series of edited volumes aimed at testing our hypotheses against the findings of historians and archaeologists studying religiosity from periods as far back as the late Pleistocene and as recently as the early twentieth century (Whitehouse and Martin 2004a, 2004b, 2005)[2]. The paper that triggered this journey of discovery is reproduced, with only light editing, below[3].

* * * *

According to anthropologist Jack Goody, the fact of storing cultural materials in memory, rather than in text, lies "at the heart of the nature of the process of 'cultural genetics'" (Goody 1987: x). In drawing attention to the need for deeper consideration of the role of memory in cultural transformation and divergence, Goody's point is well taken. But it seems to imply that the 'cultural genetics' of non-literate traditions are, at least in the respect of their characteristic dependence on memory, all of a kind. Goody's remark introduced a thought-provoking monograph by Fredrik Barth, entitled *Cosmologies in the Making* (1987), in which it was also seemingly suggested that the process of cultural transmission and transformation in non-literate cultures may be generally contrasted with cognate processes in literate traditions, where texts are used as repositories of cultural materials (e.g. Barth

---

2. These generalizing ambitions led to similarly ambitious publications in comparative ethnography (Whitehouse and Laidlaw 2004) and cognitive science (McCauley and Whitehouse 2005; Whitehouse and McCauley 2005; Whitehouse and Laidlaw 2007).
3. The following text was first published in Man (N.S.) 1992, Volume 27, Number 3, pages 777-797 and is reproduced here by kind permission of the Royal Anthropological Institute. An earlier version of the paper was originally read at the Department of Anthropology in the London School of Economics (Intercollegiate Research Seminar, October 1991). I should like to thank members of that department, along with Fenella Cannell, Mike O'Hanlon, Buck Shieffelin, and Christina Toren for many constructive criticisms. I owe a special debt of thanks to Gilbert Lewis and Alfred Gell for their comments on earlier drafts and for their long-term support and guidance. I should also like to thank the two anonymous reviewers for *Man* for their insightful suggestions. I am grateful to the ESRC for funding my Ph.D. research at Cambridge, to the Smuts Memorial Fund for the purchase of audio-visual equipment, to the PNG authorities for permission to conduct fieldwork and, above all, to the people of Dadul, Maranagi, Sunam, Riet, and Malmal for their generosity and tolerance.

1987: 75-6). It is argued below, however, that the really crucial distinction is not between dependence on texts and dependence on memory *per se*, but between two fundamentally different types of dependence on memory, one of which, though prevalent in many oral traditions, corresponds closely to dependence on textual materials.

Variations in the frequency of cultural transmission from one oral tradition to the next correspond to variable demands on memory. In the non-literate society of the Baktaman of inner New Guinea, Barth has described a situation in which large portions of the religious tradition are transmitted as infrequently as once every ten years. Thus, for long periods of time, the details of initiation rituals in particular are stored in memory and, since these materials are considered dangerous to contemplate and certainly to communicate, they are rarely discussed. If one also takes into account the elaborate nature of religious communication, it becomes clear that the system places considerable burdens on memory. Indeed, Barth infers that each reproduction of the initiation rituals is bound to differ in various ways from the last as a direct result of failing and distorting memories. Thus, according to Barth, most culture change occurs unconsciously, that is to say unrecognized by the Baktaman themselves (e.g. 1975:240). Moreover, in his more recent analyses, Barth has come to envisage the process of memory distortion in a particular way, likening the period of storage to a "melting pot of consciousness" (Obeyesekere's phrase quoted in Barth 1987:29), which is said to produce a process of cultural transformation by "incremental steps" (Barth 1987:31).

The pattern of gradual or 'incremental' culture change documented by Barth among the Baktaman is perhaps more constructively understood, not in terms of a 'melting pot' theory of memory, nor in terms of the "blocking" of certain kinds of innovation as Barth conceived of this process (1975:244; 1987:36), but rather as resulting from the emergence of a distinctive structure of ritual communication which is peculiarly adapted to infrequent transmission, and which effectively prevents innovations from contaminating the religious tradition as a whole. An instructive point of comparison is provided by the Kivung religion of East New Britain, at the other 'end' of Papua New Guinea, where I conducted fieldwork between 1987 and 1989 (see Whitehouse 1995). The institutions of the Kivung appear to be subject to far greater logical integration than those of the Baktaman and are further distinguished by conditions of frequent transmission and wide dissemination.

These features are related to the demands placed on memory in Kivung religion. The reproduction of this tradition closely resembles that of a literate tradition where successful cultural transformation necessarily takes the form of major systemic change, and cannot proceed piecemeal as among the Baktaman.

Baktaman religion is a fertility cult performed by the men, who are gradually introduced to the secrets of the religion by means of seven successive degrees of initiation. Barth stresses the communicative aspects of initiation and describes the construction of largely non-verbal messages through positive ritual acts and taboos. A very large portion of Baktaman religious knowledge is transmitted exclusively through initiations, each of which is ideally performed roughly once every decade. In the intervening period, very little discussion of initiation takes place, even in secrecy between initiates, and the cultural materials are silently stored in memory. The main ritual contexts for applying religious knowledge outside of initiations are the rites of worship performed by fully initiated men in three categories of temple, to which I shall presently return.

The Kivung of East New Britain has often been described as a 'cargo cult', originally established among the Mengan-speaking peoples in the South of the province, and subsequently adopted also by Baining groups to the North. My knowledge of the Kivung is based mainly on research among the Baining, although I have visited the Mengan region and would expect most of my observations on the Kivung here to apply generally. Kivung ritual is directed towards the production of a miracle, in which the ancestors will be reincarnated in the bodies of white people and confer upon their descendants all the wonders of western technology and the supernatural means of endlessly renewing them. The main categories of ritual action are the presentation of offerings to the ancestors in three kinds of temple, the cultivation of moral strength through various meetings and monetary donations, and the endurance of God's punishment for original sin through abstentions and gardening rituals. In this oral tradition, all ideas and practices are frequently repeated and their intricacies are widely disseminated within the community.

The essence of worship in both Baktaman and Kivung religions is to be seen in the cultivation of bonds with the ancestors for the purpose of securing benefits for the living, whether these benefits are conceived in terms of fertility as among the Baktaman or 'cargo' in the Kivung. Moreover, in both societies it is claimed that religion takes the specific form that it does because the ancestors so desire it. Thus, ideally, each expression of religion should be an exact reproduction of the last, in an unchanging tradition. Any interference in these conditions of continuity must be demonstrated to carry the blessing of the ancestors.

More specific similarities in the themes of Baktaman and Kivung religions are to be seen in the temple rituals of both societies and the idioms within which relations with the ancestors are pursued. In Baktaman society, worship is exemplified by the rites conducted in the temples known as *Katiam* and *Yolam*, (the third kind of temple, known as *Amowkam*, being somewhat inessential (Barth 1975 247) to the needs of worship among fully initiated men). It so happens that a similar division prevails in Kivung communities where the essentials of worship are provided by the rituals of two categories of temple, known as the 'cemetery temple' (*haus matmat*) and 'family temple' (*haus famili*) respectively. (The third category of temple associated with the prophet Bernard (*haus bilong Bernard*) need not be considered here.)

Worship in the Baktaman *Katiam* closely parallels Kivung worship in a family temple. As Barth explains: "*Katiam* ritual is... largely decentralized and individual. Any person initiated to seventh degree is an authorized priest with authority to approach his clan ancestors; so each clan shrine is the object of independent cult" (1975:249). In the Kivung, the religious equivalent of the seventh degree initiate would be the married man or woman, who similarly offers food to his or her own particular ancestors in the 'independent cult' of the family temple. And, like seventh degree initiates, married persons in the Kivung seek personal benefits through this kind of ritual which, in both societies, would commonly be success in subsistence activities. In such rituals, bonds with the ancestors are cultivated in the idiom of kinship or marriage stressing common material concerns and the moral importance of cooperation. This is particularly reflected in the intimacy of invocations made by Kivung worshippers on such occasions. But Barth's description of the nature of rites in the *Katiam* would serve almost as well as a summary of family temple ritual in the Kivung:

> Man relates to Ancestor through the offering of game in a relatively continuous, everyday flow of prestations, not just on special ritual occasions. In return he obtains taro fertility and plenty. This is a close metaphor to the man-woman relationship

of exchange and mutual interdependence in daily subsistence (Barth 1975:250).

If *Katiam* and family temples are in some sense theologically 'equivalent', then this is no less the case with respect to communal temples in the two societies. At the heart of communal worship in the Kivung is the cemetery temple, where the most exalted ancestors of the transcendental world are said to congregate in a body known as the 'village government'. The same ancestors who come to receive offerings in family temples can only be approached in the cemetery temple with extreme reverence and on behalf of the living community as a whole. This closely parallels the role of the *Yolam* temple in Baktaman society. As Barth puts it: "The *Yolam*... is always the scene of communal cult - even when the leader alone makes small offerings through the sacred fire, he does so on behalf of all" (1975:249). The sort of exchange that takes place in *Yolam* and cemetery temples is far too solemn and sacred to be cast in the idiom of kinship and everyday transactions, as in *Katiam* and family temples. Offerings presented to the ancestors in *Yolam* and cemetery temples are aimed at bringing the living and the dead together, by making the living more sacred, that is to say 'more like an ancestor'. In the cemetery temple, the relationship between living and dead is not analogous to the relationship between living persons. Rather, living persons express the sacredness immanent within them, to achieve solidarity with the ancestors based on common spiritual bonds, and not the shared interests of kinsmen, which are construed as pre-eminently material and which emphasize cleavages rather than a unified community. Barth explains that in *Yolam* ritual "the commensal meal becomes a communion in which man partakes of the deity" (1975:251) whereas, in Kivung ritual, the living strive to share their most sacred or divine aspects with the ancestors. But the effect in both societies is to make the living more like the dead, and thereby to create and express the unity of the living community as a whole.

Memory and the Codification of Religious Materials

Despite the presence of some common themes in Baktaman and Kivung religions, there are fundamental differences in the ways in which these two traditions are cognized, codified, and transmitted. Religious understandings are cultivated by quite starkly contrasting techniques in the two regions. These differences need to be unravelled with care, but a useful point of entry is provided by a basic intellectual disparity: the fact that, for the Baktaman, the persuasiveness of religious insights derives from a moving but confusing experience of partial recognition and mystery whereas, in the Kivung, it is more deeply rooted in a sense of comprehension and intellectual revelation.

Barth has shown that Baktaman religious understanding is not produced through verbal explanation or exegesis. On the contrary, religious insights seem to be constructed out of the withholding of explanation and the cultivation of mystery. The Baktaman system of initiation into successive grades is such as to impress upon the novice that what he does not know is more powerful and dangerous than what he does know. Intense secrecy and taboo surround all sacred knowledge, and the novice at every stage is given to understand that behind the veils of deceit and partial truths, lie ever deeper mysteries. Moreover, the fully initiated men, and even the masters of initiation who know more than any other men alive, are apparently humbled by the unknowable mysteries of existence and fearful of unleashing powers which they only vaguely comprehend. The mystery, secrecy and danger surrounding Baktaman religious life are clearly associated with the absence of casual speculation and exegetical discussion.

This state of affairs contrasts quite starkly with Kivung religion. It is true that there are elements of mystery in this tradition as well but they emerge out of rather different conditions. Kivung religious knowledge is distributed more or less evenly throughout the community and exhibits a high degree of logical integration. Not only is exegesis available for every detail of ritual action, but these explanations and their logical implications are discussed at great length almost every day, whether in the formal conditions of a public meeting or the more casual context of amicable conversation. The experience of religious understanding tends to be focused on what one explicitly knows and can articulate rather than on what one dimly conceptualizes, profoundly fears, and cannot verbally express. A mystery in the Kivung is something which is logically implied, but not authoritatively confirmed, whereas a Baktaman mystery is something which is authoritatively confirmed, but inaccessible and indifferent to logical constructions. This statement requires, and will receive, more detailed clarification, but for now I merely wish to draw attention in very general terms to some differences in the way religious understandings are cultivated in the two societies[4].

Besides the difference of intellectual orientation - that is the emphasis on logical explanation versus the contrivance of mystery - there is also a difference in the sensual character of religious experience. Baktaman rituals bombard the senses from all directions, as Barth shows in his discussion of sacrifice (1975: 223). In the Kivung, by contrast, ritual action tends to be at turns cerebral and routine, and rarely does it construct meaning out of physical sensation or seek to excite or encourage a diversity of such experiences. In this regard, Kivung rituals have more in common with Christian worship than with Baktaman initiation. There is an ideological emphasis, in the Kivung, on the necessity of cultivating particular emotional states in the course of rituals ranging from contrition, guilt and sorrow, to generosity and altruistic love. However, the predominant experience as far as one can make out is either one of intellectual concentration, when a technically complex sequence is in motion, or else boredom and discomfort when actions are routine and automatic. Certainly mainstream Kivung rituals do not assault the senses to the degree found in Baktaman initiation, where anything from sudden pain and extended torture to peculiar aromas and loud noises, are unleashed upon the unprepared novices.

Barth himself laid particular emphasis on a distinction between what he called 'analogic' and 'digital' communication (cf. Bateson 1972: 372). Digital processing refers to the computational principle of binary opposition or polarity, central to the widely exploited idea (in formal linguistics, and structuralism generally) that values derive from arbitrary contrasts. In the 1970s, Barth was pulling against the tide of structuralism (e.g., 1975: 212-14) and wanted to stress the limitations of an approach which, for example in relation to animal symbolism, would want to envisage 'sets' of natural species as "reciprocally arranged in structures which are isomorphic with social arrangements or other features of reality" (1975: 189). Rather, Barth wanted

---

4. This general characterization of the divergent paths of Baktaman and kivung religions recalls the debate over 'misconstrued order in Melanesian religion', instigated by Brunton (1980). Kivung religion is more 'ordered' than Baktaman religion in the specific respect that it is subject to far greater logical integration. However, Brunton's criteria for order (1980: 122) amount to a constellation of variables in terms of which both Baktaman and kivung religions would be located towards the 'ordered' end of the continuum. Indeed, Brunton explicitly places the Baktaman in this position.

to demonstrate that the symbolism of animals used in rituals was to be understood in terms of an analogic code, such that natural species "enter individually into larger ritual contexts, each of them as a separate, more or less dense symbol carrying an aura of connotations" (1975: 189). In an analogic code, Barth argued, the meanings of symbols are not arbitrary but derive from a resemblance between the inherent characteristics of the symbols and their referents. In other words, such symbolism is essentially iconic. A familiar example of the use of an analogic code is the conventional 'stick figure' by which the human being is commonly depicted in the West. Although this affords the possibility of binary contrast (as between the stick man and the stick woman in the labelling of public conveniences!), there nonetheless remains an independent relationship between the stick man and the 'real man' based upon shared, inherent characteristics of symbol and object. Barth's argument is that Baktaman religion is predominantly cast in an analogic code, and indeed he does seem able to demonstrate, for example, that there is a link between many of the religious understandings of growth, increase, removal and loss and the natural characteristics of symbolic materials such as dew, fur, running water and so on.

It should be emphasized, of course, that Barth has never subscribed to a facile view of Baktaman religion, which would see it as something really constructed out of local equivalents of stick figures. What Barth seems to be saying, at least in part, is that ideas of growth and increase need to be communicated effectively in a fertility cult which is concerned with the growth of humans and taro. A useful way of conceptualizing this process happens to be the 'miracle' of dew, which appears on leaves apparently out of nowhere. The link between maturing taro, and water which 'grows' on leaves, is cast in an analogic code. The relationship between the two types of 'growth' is in some sense self-contained and can occur independently of other codifications. However, the Baktaman have many complementary ways of conceptualizing growth, for example through the symbolism of domestic pig fat, fur and hair. Thus Barth (1975: 200) writes:

> The Baktaman seem to be groping for something only diffusely understood, and the metaphors used are such as can provide a minimal cognitive grasp of it: Dew accumulates on leaves. Fat grows inside the pig and makes its skin hard and tight. Hair grows out where it is cut off. Fur covers the body, like

vegetation covers the ground. All these are images that can evoke
the idea of increase.

These sacred images, however, are not really comparable to stick figures. Each metaphor seems to trigger multiple connotations which often have a strong emotional or sensual character. The terrifying experience of being forced into the nocturnal forest at the beginning of first-degree initiation will perhaps always be associated with the symbolism of dew, which is first applied on this occasion. Each metaphor of growth and increase will carry its own peculiar associations, and each will 'harmonize' in varying degrees with related metaphors and connotations. Thus, the image of dew is far from being merely another way of communicating the same idea as that conveyed by pig fur, namely the idea of growth. To treat each image as simply standing for an idea would be to reduce Baktaman religious experience to the comparatively sterile medium of language. For taken on its own, there is nothing particularly persuasive about the proposition that taro growth and dew are manifestations of a single process. When this insight is cultivated in ritual, however, it is through the contrivance of ambiguity and multivocality based around emotionally charged connotations. Many anthropologists have drawn attention to the persuasive character of this process, and to the fact that verbal statements, such as in an exegetical commentary, could never present a substitute for the communicative functions of ritual symbols. Indeed in the Baktaman case, Barth's analysis suggests that exegesis would undermine the persuasive power of condensation and multivocality by simplifying, trivializing and de-sacralizing the act of revelation. And it could surely add little to the experience of partial comprehension, awe, fear and, above all, mystery which Baktaman religious symbolism sets out to accomplish.

In contrast with this state of affairs, the kivung religion transmits its knowledge through predominantly linguistic and exegetical communication. Every ritual act has a generally recognized meaning which is integrated into a complex system of logical implications, ultimately derived from a finite set of absolute presuppositions. Consider, for example, the ritual preparation of food for presentation to the ancestors in the cemetery temple. These offerings have to be prepared every day on a rota system by a team of three women. If, on one occasion, happen to notice that only two women are preparing food, then I am likely to set off a string of questions and answers along the following lines: Why are there only two cooks? Because the third is menstruating. Why should menstruation prevent her from cooking? Because her menstrual blood is a punishment from God. Why should God punish her? Because Eve had sexual intercourse with Adam ... and so on. It will be noted that each meaning associated with ritual behaviour is logically implied by other meanings which can readily be unravelled in the medium of language.

Drawing on Collingwood's terminology, Hanson has called this principle of institutional integration 'the logic of question and answer' (Hanson 1975: 11-13). If one were exhaustively to pursue the above line of questioning along all the various paths that are logically implied by the answers received, then one should discover that it is the moral sentiment, and not the food itself, which is 'consumed' by the ancestors or the fact that only some moral sentiments are acceptable to the ancestors while others are unacceptable and even cause offence. Eventually, one should discover that the miracle of returning ancestors is desirable and possible, but that its achievement depends in part on the establishment of solidarity and reciprocal relations between the living and the dead. If one were to ask why this is so, then the response would be simply that it *is* so. This idea is, in Hanson's (and Collingwood's) terminology, an 'absolute presupposition'; "Absolute presuppositions are not themselves answers to any questions; they are the ultimate assumptions which give rise to all questions" (Hanson 1975; 12). Hanson has gone so far as to assert that the logic of question and answer, emanating from absolute presuppositions, provides 'a paradigm for the institutional approach to social science' (1975: 12). At any rate, it provides a useful framework for analyzing kivung-type religious materials, but *not* those of Baktaman-type traditions.

The different principles of codifying and structuring meaning, in Baktaman and kivung religions respectively, can be related to the differential demands that are placed on memory in the two traditions. The analogic coding of Baktaman religious materials is consistent with conditions of irregular transmission. The tradition is insulated from the dangers of memory failure in two major respects. On the *one* hand, the separateness of metaphors implies that modification or elimination of any particular symbolic process, whether due to memory failure or some other specific cause, does not substantially undermine continuity in the religious tradition as a whole. This would not be true of a logically integrated structure, where interference in anyone sector of the religion can have profound consequences for the entire system. On the other hand, the bombardment of various senses and the cultivation of pictorial and emotionally charged associations, which occur in Baktaman ritual, may be regarded as powerful mnemonic devices, greatly reducing the risk of forgetting and thus of the unconscious or unintended introduction of innovations in subsequent performances. In these respects, the principles of Baktaman religion may be considered to constitute an adaptation to the considerable burdens which are placed on memory by its conditions of reproduction. The specific mechanisms of this adaptation are considered presently.

The continuity of kivung religion, however, is not threatened by memory failure. The very frequent repetition of all the sacred rituals of the kivung is sufficient to ensure a high degree of standardization. Moreover, it is not merely the technical side of kivung ritual that is rigorously standardized through continual repetition. The exegeses of ritual, indeed all the cosmological intricacies of kivung religion, are likewise fixed by convention as a result of frequent repetition. Two afternoons of every week are set aside in all kivung communities for the purpose of public meetings which all members of the cult (except for sick or menstruating persons) are expected to attend. At these meetings, three orators along with their assistants are charged with reiterating the details of kivung religious ideology. Admittedly, they cannot cover all the ideas at every meeting, nor indeed in the course of one week, but they do cover considerable ground and probably set out almost all the religious ideas in any given five-week period. It happens to take exactly five weeks to repeat all the doctrines surrounding the ten central tenets of the movement before the cycle begins again and the tenets or 'laws' are repeated nom the beginning. Since attendance at these meetings is compulsory for everyone, regardless of age. The effect is to 'drum home' every detail of the religion to the community at large. The explicit goal is to create a single, unified system of ideas within each individual mind.

Regular public meetings stand out among kivung rituals as a principle forum for transmission but temple rituals also provide a daily forum for public speech-making. On these occasions, religious teachings are rehearsed on the model of the sermon and transmission takes place through a combination of collective rote-learning and schema-based memorization. But whereas a church sermon is not supposed to repeat substantially another recent sermon but rather to reinforce or illuminate in 'new' ways principles enshrined in a written text (and is not, in consequence, particularly memorable), kivung speeches *are* the 'text', in the sense that they are constitutive of authoritative religious ideology and are required to sustain it accurately through regular repetition. The effectiveness of this mnemonic technique was demonstrated in the course of my fieldwork by the possession of Lagawop (Whitehouse *1990a:* 184-7), a young girl whose exposure to kivung doctrine, in common with other young girls, was based upon predominantly passive attendance at meetings and other kivung rituals. In a state of possession, Lagawop was able to repeat extensive logical strings that comprise the kivung orthodoxy for literally hours at a time in the manner of an experienced orator. This would suggest that all kivung members effectively assimilate an elaborate ideology as a result of its frequent transmission.

The degree of standardization of such an elaborate and logically integrated system as that of the kivung is only possible, in the absence of written guidelines, because of the frequency of repetition. It would be quite impossible to maintain an oral tradition of this kind given Baktaman conditions of reproduction. If a significant proportion of kivung rituals and ideas were communicated only once every decade or so, then each performance would require elaborate preparation with reference to recorded materials. Even an undergraduate student, who may be under considerable pressure to memorize the contents of lectures and who revises over an extended period, employing written records and a battery of mnemonic devices to maximize recall, is nonetheless liable to forget a very considerable proportion of this material over a subsequent decade of non-academic employment. To ask somebody to reconstruct even the most basic elements of kivung religion after a single experience of exposure a decade ago, would be equivalent to asking an ex-student to repeat a lecture from memory ten years after graduating.

Such exegetical complexity and richness as may be found in the kivung is only possible under conditions of frequent reproduction. The fact of reliance on memory in an oral tradition means that kivung-type institutions are ruled out, unless transmission occurs routinely. But my argument goes further than that. I would anticipate strong correlations between predominantly analogic coding and irregular reproduction, and between elaborate, logically integrated exegesis and frequent repetition. A plausible mechanism for the production of these correlations suggests itself, and at the heart of this mechanism are the differential demands placed on memory in kivung and Baktaman traditions.

At first glance, it may seem that the issue of memory is only relevant if one presupposes a uniform ideological emphasis on *continuity* in religious transmission. In the absence of this apparently ideological requirement, memory would seem to have little bearing on the matter. I do of course recognize the ideological importance that both kivung and Baktaman traditions attach to the accurate reproduction of religious institutions at each performance. I stressed this point at the outset, and it does indeed seem to introduce an element of inflexibility, opposing innovation and making legitimate religious expression contingent upon accurate recollection. This, however, is really only part of the story. One cannot say that ideological complexity, for example, is ruled out in Baktaman religion, merely because of the emphasis on continuity based on faithful recall (the argument being that too much ideology is too hard to remember accurately over a substantial period). In many ways, a much more important factor relates to the subjective experience of the validity of religious understandings.

Although complex ideology could not be accurately reproduced in an oral tradition once every decade, it could conceivably be invented or re-created in a profoundly modified form at each, rare performance. But in an imaginary system of this type, which would not emphasize continuity, it is hard to imagine how the authority of religious ideas could be generated or upheld. This is not a question about ideology itself-it is obvious that a spontaneously invented system of religious thought is no less capable than any other of invoking transcendental legitimation. Nor is it a question of what drives the innovator to make certain pronouncements, although it may indeed be unrealistic to hope that every decade some charismatic prophet will step forth to achieve feats of intellectual creativity comparable to the ideological intricacies of the kivung religion. Even that, however, is not beyond the bounds of possibility. What is really hard to envisage is how such experiences of ideological transmission might sustain any sort of religious life through the long years which separate performances. In a pattern of very infrequent reproduction, systems of logical relations will fade into virtual meaninglessness before some replacement is constructed. This would imply an oscillation between intense intellectual religiosity and a sort of expanding secular void, or at least a predominant experience of religion as a set of vague and rather dull ideas - much as an aging alumnus remembers his university courses. Such a system could scarcely be called a religion at all, and certainly it is hard to think of an ethnographic case which comes anywhere close to it.

The reason why analogic communication is so well adapted to Baktaman conditions of infrequent reproduction is that the messages which are transmitted are not concerned with something so dull and forgettable as the

logical implications of ideas, but with the intense experience of mortal
danger, mystery, pain and other extreme or abnormal sensory stimuli. All
these elements are agonizingly cultivated in Baktaman initiation, and they
haunt the initiate throughout life, by forming a complex harmony of
associations with the objects and processes of everyday experience. Barth
vividly describes, for example, how the otherwise 'dull business' of taro
cultivation is transformed into 'something of meaning and value' through the
remembrance of initiations and the powerful images, emotions, sensations and
other associations which they evoke (1975: 236).

If it is accepted, therefore, that analogic communication is an adaptation
to the excessive demands placed on memory in a non-literate tradition which
is very infrequently reproduced, then arguably the converse is also true: that
the systematic, logically integrated character of kivung institutions is an
adaptation to conditions of frequent reproduction. A religion which transmits
its messages through continual repetition is unlikely to rely on analogic
communication. In order to understand why this should be so, it is first
necessary to appreciate that the emotive quality of Baktaman-type symbols,
and what makes each symbol differrent from another, is the uniqueness of the
original set of associations between symbols and their referents. This requires
a momentary digression.

In Baktaman religion, hair is not different from fur as a symbol of growth
simply because hair is hair and fur is fur. Nor, to refer to an old joke, is the
difference between hair and fur the fact that they are spelt differently. Barth is
emphatic that hair and fur are not alternative symbols for the same thing,
namely growth. In fact, the first set of ritual associations with hair, as these
are cultivated among novices during fourth-degree initiation, relate to male
sexuality and potency. In the course of fourth-degree rituals, shredded
pandanus leaves are tightly bound to the hair of novices to create pigtails
arranged to suggest strongly the act of coitus. Besides the phallic symbolism
of pandanus and hornbill, which are both used in the head decorations, other
symbols of male potency are employed. The faces of the novices are
extensively stung with nettles and they are forced to dance for several days
and nights before the women, during which time they are deprived of sleep,
food and water. The extreme sensations evoked by these rituals provide
unique and powerful associations with hair and sexuality. The heat and
swelling of the face, the exertion and intense craving for physical satisfactions
– all these unique experiences come together in the original appreciation of

hair as a sacred image. In subsequent initiations, a deeper understanding of hair is provided and the significance of fur is introduced as a complementary but different way of conceptualizing the process of increase and growth. But hair could never be the same thing as fur, after the experience of fourth-degree initiation, and fur could never adequately replace it as a sacred symbol.

The transmission of Baktaman-type messages depends upon the unique and intense quality of ritual experience. It is not conducive to the cultivation of such messages to repeat them very often. Repetition deprives the experience of its uniqueness. Meanwhile, the intensity is largely generated by suffering which nobody would be anxious to repeat and which, if it were repeated, would not yield up its original fruits of revelation. In conditions of regular reproduction, the essence of religious understanding - the experience of revelation - has to be cultivated by some other means. The obvious and most accessible means is logical persuasion, the construction of an intellectual system of absolute presuppositions that are interwoven by the logic of question and answer. Such is the essential character of kivung religion.

In an oral tradition, persuasion by the logic and coherence of cosmology and ritual at the same time necessitates frequent repetition. That is to say, the two features are mutually reinforcing. Logical structures become dull and disconnected if they are not regularly contemplated and their persuasive capacities are in no small degree a function of the extent to which they can be preserved as an entirety through frequent transmission. This is less of a problem in literate traditions, where neglect of some portion of religious knowledge can be repaired through revision of appropriate texts. But in an oral tradition such as that of the kivung, all areas of the institutional system must be continually reproduced in order to preserve the logical integrity and comprehensiveness of the religion, through which its potential to reveal and persuade is realized[5].

These issues are raised by the different mixtures of statement and action in

---

5. Another effect of frequent repetition apparent in kivung practices, but which I do not focus upon in this article, concerns what Connerton (1989: 22-3) has called 'habit-memory' (in contrast to 'cognitive memory' by means of which logical constructions are recalled). The habitual character of kivung ritual undoubtedly contributes in important ways to the religious experience of cultists (Connerton 1989: 65) and provides yet another basis for contrast with Baktaman initiation. I have elected to focus on divergent uses of cognitive memory, however, because (as indicated below) these constitute particularly salient features as far as the analysis of transformations and political strategies is concerned.

Baktaman and kivung rituals. Clearly, although kivung religious knowledge is for the most part capable of being transmitted verbally, in the form of logical persuasion, this does not imply that exegesis could be substituted for ritual performance. The kivung is about not just knowing, but also *doing*. The standard exegetical commentary on the presentation of offerings in kivung temples has to do with bringing about the miracle of returning ancestors, but simply 'knowing' this meaning is no substitute for going through the actions. In this respect, kivung rituals are like productive, technical procedures, and are explicitly likened to them by participants. It goes without saying that the statement 'I am felling a tree' is no substitute for the act itself. But whereas the felling of a tree does not require any verbal commentary in order to communicate the meaning of the act, kivung ritual undoubtedly does. That an attempt is being made to expedite a miracle could not possibly be inferred from observations of non-verbal behaviour. Thus, so far as the indigenous appreciation of religious insights is concerned, the most important communicative capacities of kivung ritual action largely result from, and correspond to, the standard verbal commentaries available.

To varying degrees, kivung members undoubtedly engage in some intellectual speculation, and experience certain emotions and sensations in relation to rituals, so that there is bound to be some variation in the subjective appreciation of the meanings of these rituals. But my impression is that these variable responses are of much less importance in the construction of kivung religious knowledge than they are in the context of Baktaman initiations. I have attempted elsewhere to outline the probable range of associations evoked by the symbols of Western culture in kivung ritual *(1990a:* 67-9). These sorts of connotations, however, are under-exploited due to the continual repetition of rituals. For example, placing flowers on tables in the temples may originally have been associated with colonial domestic habits, but when these flowers are arranged daily in the temples, year in and year out, they come to be taken for granted. Thus, potentially diverse metaphorical connotations are not in fact exploited and the individual's appreciation of religious knowledge tends to converge on the standardized interpretations which are endlessly articulated at kivung meetings.

This suggests another way of expressing the contrast between kivung and Baktaman religions. Just as kivung rituals ought to produce the millennium, so Baktaman rituals are supposed to 'do' something of material and practical value: above all, to promote the growth of humans and taro. But in one respect, Baktaman rituals are slightly more like the productive, technical act of felling a tree, because novices receive messages about growth and increase, for example, without verbal clarification. Thus, ritual symbolism conveys religious understandings in the way that felling trees, under certain conditions, conveys the intention of clearing a garden. Let me put the contrast in the following condensed form: whereas, in the kivung, the most important things which rituals 'say' cannot successfully be communicated without words. In Baktaman rituals the most persuasive aspects of religious statements can only be conveyed by the actions themselves. The result, for the Baktaman, is considerable heterogeneity in individual interpretations of symbols. Under such conditions, verbal commentary becomes more or less redundant and, even if it were elaborated, it would fail to represent the variety of religious understandings which the symbolism is able to evoke in different individuals. The infrequency of ritual performances, coupled with the cultivation of emotions and sensations at the moment of first exposure to sacred symbols, produces an ongoing sense of revelation as symbolic connotations are evoked in the everyday experiences of life. 'Doing' and 'saying' in such rituals are so deeply interconnected that one can never definitively say what one has done, or share the resulting knowledge through the medium of language.

In summary, despite similarities in the themes of Baktaman and kivung religions, messages are cultivated, structured and transmitted by two contrasting techniques. These techniques constitute particular adaptations to differences in the frequency of reproduction and hence in the demands made on memory in the two societies. The implications of this divergence for cultural transformation are profound.

Codification and cultural transformation

Barth has long taken the view that culture change among the Baktaman is 'largely unacknowledged' (1975: 240). According to this view, initiation rituals are unconsciously changed at each performance, due to memory failure and the fact that only a very small number of experts are responsible for remembering the details of rituals. The minds of Baktaman cosmologists are thus likened to 'melting pots' (e.g., Barth 1987: 29), in which cultural materials are unwittingly remoulded. The title of Barth's recent book, *Cosmologies in the making*, echoes the point.

Apparently Barth has never published any direct evidence of unconscious memory failure among the Baktaman and there is perhaps a risk of exaggerating the extent of unacknowledged innovation. Studies by psychologists of long-term memory retention suggest that memories tend to evaporate rather than melt (see Cohen 1989: 156-9), and that such innovation as occurs may be primarily a function of the procedures of recall. Neisser (1988) has described the optimal conditions for accurate remembering, and these happen to correspond quite closely to the procedures used in the planning of Baktaman initiations (Barth 1987: 26), where ritual experts adopt the technique of 'free recall' at leisure, and are explicitly guided by a concern with 'verity', apparently in the absence of motives that might produce distortions (Neisser 1988). It should also be reiterated that the materials concerned are, by their very nature, memorable. Ritual experts are not required to recollect a great volume of verbal exegesis or mythology, but rather a limited number of graphic actions rendered all the more vivid by their association with powerful emotions and sensations.

Thus, unacknowledged innovation may well be the exception rather than the rule and, in the absence of empirical evidence, certainly cannot be invoked to explain the gradual and piecemeal character of culture change among the Baktaman. But Barth also advances a theory of incremental change in relation to documented instances of deliberate or conscious innovation in Baktaman rituals. In this context, the argument turns on a concept of 'blocking' rather than 'melting'.

Of the nine cases of *conscious* innovation in Baktaman ritual that Barth collected (1975: 239-40), three concerned the symbolism of wild pig. Traditionally, it would seem that the Baktaman have been reluctant to make use of the image of wild pig in fertility ritual, because the boars in particular are notorious for ravaging crops and are thus, as Barth puts it, construed 'as primarily an anti-taro and anti-gardening force, an enemy and rival to Baktaman male success' (1975: 241). For this reason, it had long been expressly forbidden to bring the meat or bones of wild boars into the *Yolam* temple. However, Barth cites an occasion in the 1950s when a group of warriors, initiated merely to third degree, undertook a successful raid while carrying with them the mandible of a wild boar which had recently been killed in the act of copulation. Following this incident, the mandible was incorporated into the *Yolam* sacra, where it was still to be found some fifteen years later, at the time of Barth's fieldwork.

It would no doubt be interesting to discover why this instance of innovation was successful, and why a comparatively large proportion of innovations generally seemed to focus on wild pig. Barth's explanation begins with the fact that the attitude towards wild boar in everyday life is somewhat ambivalent. Whilst this beast is undoubtedly a villain, its meat has a high value and most poignantly domestic sows depend upon the wild boar for impregnation (1975: 201). Thus, the wild boar could quite convincingly serve as a vehicle for ideas about male virility and aggressiveness (1975: 241). In this light, the introduction of the mandible to *Yolam* sacra becomes intelligible. However, Barth seems to regard the ambiguity of wild pig as something approaching an explanation for the tendency to focus on the symbolic potential of the animal in ritual innovation. Thus, he formulates the general rule that 'elaborations ... occur where problems and discrepancies are felt and *require* a resolution' (1975: 244). This argument raises problems, in view of Barth's foregoing demonstration that the Baktaman go out of their way to cultivate discrepancies and ambiguities in many areas of their ritual life. A good example would be the failure to use the most symbolically 'fitting' marsupial for the sixth-degree initiation, in place of the much less suitable species of marsupial which is actually used (1975: 185-6). As Barth subsequently puts it: 'messages are often made increasingly cryptic in the service of mystery by the veiling of insight behind layers of symbol substitution' (1975: 189). However, it would be facile to see this as a straightforward contradiction. Barth clearly envisages Baktaman religious experts as being pulled in opposing directions. On the one hand, they are concerned with clarifying cosmological themes, and on the other hand their task is to cultivate mystery. Thus, the temptation to innovate in pursuit of clarity may be counterbalanced by the need to confuse, a need which Barth associates generally with the 'blocking ... [of] interpretive elaboration' (1975: 244).

The tension between clarification and mystification, which expresses itself in 'blocking', cannot be used to explain the gradual or incremental character of culture change among the Baktaman, because 'blocking' and gradual change constitute two aspects of the same 'thing' for which a suitable explanation is required. I have shown that the problem is not resolved or clarified by the idea of unintended memory failure. Rather, the impact of memory on culture change is, I believe, largely indirect. I have attempted to demonstrate that it is due to a combination of the strengths and weaknesses of human memory that a religious tradition, so infrequently transmitted as is that of the Baktaman, is cast in an analogic code. Where this principle of codification is used, culture change is likely to be piecemeal as Barth has intuited. But this pattern of gradual or incremental change is not to be readily explained in terms of ideas about melting pots, clarification, mystification or blocking. What it comes down to is the fact that all the kinds of innovation which Barth has documented tend to have very limited symbolic ramifications. Since the relationship between symbol and object is somewhat self-contained in an analogic code, interference in any particular symbolic process has only a minimal knock-on effect for the religious tradition as a whole (1975: 210-12). In particular, the lack of logical integration serves to isolate innovations and protect the continuity of the religion overall. Consider, once again, the introduction of a wild boar mandible to the *Yolam* sacra. Barth categorically states that this instance of creative elaboration occurred "without changing the basic rule banning all other ... boar meat or bones from the temple. The various codifications of ... boar have still not been brought into harmony, nor its connotations as a concrete symbol clarified" (1975: 241).

By contrast, kivung religion is far more sensitive to interference. Innovations, in almost any area of the tradition, are likely to have immediate and profound ramifications for the system as a whole. I have elsewhere documented this sort of thing in detail (Whitehouse *1990a*), showing how the possession of a young kivung member led, by a series of logical steps, to a succession of radical transformations in the institutions of two Baining villages. It might have been otherwise, had the possession been attributed to diabolical forces rather than to the 'true ancestors', but in that case the original innovation would simply have been unsuccessful.

Consider the following example. The taboo on bringing wild boar products into the *Yolam* can be seen as roughly equivalent to the taboo on chewing betel nut in the kivung. The ban on betel nut is logically connected to an origin myth in the kivung, linking the red substance produced by betel chewing with the first flow of menstrual blood. Thus, chewing implies ignorance of God's punishment for original sin, and is equivalent to imbibing the very menstrual blood which God inflicted upon woman as a punishment for her sexual incontinence. However, as with wild boar among the Baktaman, Baining members of the kivung have an ambivalent attitude towards betel and menstruation. Unlike the Mengan-speaking originators of the cult, they have no traditional concept of menstrual pollution and few adults grew up with a fear of contamination. On the contrary, I was secretly informed that Baining ancestors used betel nut *and* menstrual blood in traditional acts of worship. It would however have been quite unthinkable for Baining communities to adhere to the Mengan model of kivung religion whilst leaving out the business about betel nut. Certainly, it is hard to imagine that such a discrepancy could have persisted for fifteen years without resulting in a re-ordering of other ideas, including the origin myth so as to make the chewing of betel logically consistent with the wider ideology. It so happens that, during a period of institutional transformation, the ban on chewing betel was lifted for certain newcomers into the kivung (although it was still upheld by long-term members). This dispensation received ample logical justification in connexion with the peculiar circumstances of restructuring which were taking place at that time within the entire religious tradition. Thus, in contrast to the Baktaman religion, the success of any given innovation in the kivung logically implies further and more sweeping changes. Cultural transformation thus becomes an all-or-nothing business, a process of ideological revolution. This is the price which a religious tradition must pay for logical integration, and it is not simply a concomitant of literacy.

Political pressures on patterns of transmission

The different types of religious experience available to the Baktaman and to kivung members ensue from divergent principles of codification, which in turn represent adaptations to differential demands on memory. But the causes and consequences of this divergence may be explored a good deal further. The fact that kivung institutions are frequently reproduced and based around extensive exegetical commentary is to be understood not merely in terms of the strengths and limitations of memory but also in terms of the political goals of religious leaders and activists. In a recent article, Barth (1990) has set up a number of contrasts between the transmission of religious materials through the teachings of Balinese Gurus on the one hand and Melanesian initiators on the other, corresponding rather closely to the kinds of contrasts which I have drawn out in the juxtaposition of kivung and Baktaman religions. It is of particular interest that Barth associates the Guru complex with verbalized, decontextualized and logically integrated materials, and the initiation complex with non-verbal analogical codification. One implication of this is that the religious materials of the Guru, being encoded and transmitted by individual operators, are well adapted to transportation over substantial distances. By contrast, the religious materials of initiators are transmitted through infrequent collective performances and thus 'the initiator is linked to his context, and his knowledge is untransportable except to immediately neighbouring groups or through the movements of whole populations' (Barth 1990: 647). This is an extremely important point, and helps to account for the various kinds of divergence which I have described in Baktaman and kivung religions.

The kivung, it will be recalled, is a popular movement, formulated with the goal of expansion in mind. The founder of the movement, Michael Koriam Urekit, travelled extensively in the region in an attempt to spread his ideas and, since the death of this leading Guru, new standard-bearers have taken on the task of maintaining ideological uniformity over a wide area by means of messengers, personal patrols and the regular invitation of local leaders to the central seat of power in the movement. The same could be said of Yali's movement in the Madang area of New Guinea, which was likewise transmitted in the eminently transportable medium of language, along channels of dissemination opened up by the Administration (Morauta 1972: 435). It was not conducive to the goals of either Koriam or Yali to confine the transmission of their ideas to the iconic symbolism of infrequent collective ritual. Their programmes of proselytizing demanded logically integrated and persuasive ideologies which could be transported by individual activists and transmitted by word of mouth.

Despite his many important insights, a problem with Barth's recent article is that it takes the initiation complex to be characteristic of Melanesia generally and even goes so far as to assert that the Guru complex is entirely absent. I would argue, to the contrary, that the Guru complex has deep roots in Melanesia, where it has long presented a viable alternative to initiation as a technique of transmitting revelations. Lacey has recently commented on the traditional Melanesian preoccupation with leaders who resemble pioneering myth-heroes, in so far as they travel over great distances bearing elaborate religious knowledge (Lacey 1990: 187). Not only is the theme of transportable ideology commonplace in Melanesia, but such linguistically encoded revelations are also the stock-in-trade of the many prophets, mediums and messianic leaders who form such an important part of the Melanesian heritage. What I have attempted to show is that the region presents not one but two principles of religious persuasion, with divergent implications not only for patterns of transformation but for patterns of diffusion as well.

Morauta comes close to recognizing the last point in her discussion of the contrasts between Yali's movement and the many smaller cults of the Madang area. She considers, for example, how the emphasis in smaller cults on traditional rituals inhibited their expansion (1972: 434-6). She mentions the restrictions imposed by reliance on highly localized sacred objects, and the limited opportunities for interaction with outsiders afforded by traditional ties beyond the village. The issue of language is also raised, yet only to show how smaller cults were confined by the reliance of their exponents on local languages, whereas Yali's ideas were conveyed in *tok pisin*. Above all, Morauta's emphasis is on different channels and opportunities for communication. What she does not consider is the possibility that the *types* of communication might have been radically different in Yali's movement and in the many smaller cults. The members of small-scale cults presumably spoke *tok pisin* as fluently as Yali's supporters, and their opportunities for using this language in interactions with outsiders were surely quite varied during the post-war period. If opportunities and channels for communication were really there for a competent leadership to exploit, then the fact that the small cults stayed small requires a fuller explanation. One hypothesis is that the smaller cults codified their religious materials according to the model described for Baktaman initiation, namely in the form of iconic symbolism cultivated in irregular ritual. The fact of their reliance on traditional ritual techniques is suggestive in this regard, but the matter requires further investigation.

My second point of disagreement with Barth concerns his argument that 'Gurus seek knowledge from far places out of an acquired religious and intellectual interest and so as to achieve fame, but not so as to homogenize a regional tradition - although that will be a conspicuous result of their activity' (1990: 651). In so far as Koriam exemplifies the Guru model, all accounts of his career which I collected from his close associates suggest that he very deliberately intended 'to homogenize a regional tradition', and there seems to me no reason why this motive should be denied, *a priori*, to other such leaders. Indeed, the possibility of such a motive is an important factor in the explanation of divergent religious practices in Melanesia. Kivung members are well aware of the opportunities afforded by initiation systems for the cultivation of religious revelations, and some of them personally endured the agonies of such rituals before the kivung prohibited them and replaced them with its programme of painless, if unremitting, daily repetitions. In part, these new institutions took the form that they did precisely because they offered opportunities for proselytizing and unification which the initiation rituals did not.

There is also of course the converse question of why some societies adopt the techniques of codification and transmission characteristic of initiation systems. But first the problem must be re-stated more clearly. Baktaman initiation is based around iconic symbolism expressed in the performance of rituals by groups of novices under expert direction. In part, the persuasiveness and enduring impact of this type of communication derives from the infrequency of transmission, implying the avoidance of exegetical commentary and the rigorous observance of secrecy and exclusion. In part, the power of the initiation complex lies in its remorseless assaults on the physical senses, contrasting and confusing pleasure and pain but above all bombarding the novices with surprising stimuli from multiple directions. I have argued that these features are bound together as creative adaptations to the demands which infrequent transmission inevitably make on memory in a non-literate tradition. What needs to be explained is the *pressure* for such adaptations.

One way of approaching this topic is to focus on the capacity of the initiation complex to generate political solidarity. An undoubted concomitant of the sharing of secrets, agonies, revelations and status among novices is an intense and enduring bond based around common identity and solidarity. In Papua New Guinea, this bond is highly 'gendered'. A number of theories have been advanced which suggest that the historical pressure behind the elaboration of initiation systems is to be found in strategies of male domination (for example Feil 1978; Lindenbaum 1984). Such arguments have recently taken a new turn in reacting to Godelier's (1986) theory of the 'alternative logics' presented by 'great-men' and 'big-men' systems (Godelier & Strathern 1991)[6]. For Godelier, male initiation is a political necessity in societies practising sister exchange (1991: 277) and, along with inherited or ascribed leadership, warfare and revenge killing, these features are said to form part of an organic complex of institutions encompassed by the logic of 'equivalent' exchange. Godelier argues that 'the direct exchange of women between *men* ... requires the construction of a collective male force to stand behind the individual (his sisters or daughters). This collective force is what is created by male initiation' (1991: 294).

I do not propose to examine here the details or merits of this argument. What is important is that the male solidarity which ensues from a common experience of initiation represents a potent political force, and consequently

---

6. For a sustained critique of Godelier's (1986; 1991) arguments, see Whitehouse (1992).

this feature of the institutional complex has become a focus of attention in the anthropological literature. On the one hand, the initiation complex is a result of experiments in religious persuasion, and my primary goal in this article has been to account for their success. On the other hand, it is necessary to address the question of why experimentation along these lines occurred in the first place, and it is here that theories of male strategies of domination may be fruitfully explored. The notion of, strategies' raises the question of motivation, and this is inherently problematic when one is dealing with prehistoric developments. In the case of the kivung, I could present evidence that Koriam deliberately pursued a 'Guru complex' as part of his expansionary strategy; however, it would be another matter to demonstrate that Baktaman male ancestors borrowed and developed the 'initiation complex' with the domination of women in mind. Rather, what is suggested is a process of social selection in which the exploration of new kinds of religious experience coincided with, and was reinforced by, other social institutions founded around the control and domination of women.

Conclusion

The broad distinction between two processes of 'cultural genetics' probably does not correspond to the distinction between oral and literate traditions in the way that Goody has suggested. This is so for a number of reasons beyond those which highlighted above. For example, Connerton (1989: 72-9) makes a distinction between 'incorporating' practices, which are capable of sustaining communication only through physical performance (such as speech or 'body language'), and 'inscribing' practices, in which communication takes place through recorded materials (such as texts, magnetic tapes, and so on). Connerton observes that inscribing practices usually (if not invariably) entail an element of incorporation. This leads naturally to a critique of theories of culture change in literate societies which "consider inscription to be the privileged form for the transmission of a society's memories" and therefore tend to overlook the "mnemonic importance and persistence of what is incorporated" (1989: 102).

My contribution to this topic approaches the claims concerning literacy from a different angle. I have argued that at least some aspects of the 'cultural genetics' of literate societies ensue from certain forms of codification and transmission also encountered in *non-literate* traditions or, to use Connerton's terminology, under conditions where 'incorporating practices' prevail. Thus the frequent repetition and extensive logical integration of kivung materials introduces a striking rigidity of structure, such that any departure from convention is readily apparent and, if endorsed, requires substantial modification of existing institutions for the purposes of logical reintegration. These are features that Goody long ago (1968: 2) associated with literate as opposed to oral traditions. And after roughly two decades his view remains essentially unchanged (1986: 9-10):

> In the literate churches, the dogma and services are rigid ... If change takes place, it often takes the form of a break-away movement ...; the process is deliberately reformist, even revolutionary, rather than the process of incorporation that tends to mark the oral situation.

Nonetheless, these two patterns of transformation are both discernible under non-literate conditions, in which the critical variable is frequency of transmission. I have also explored the implications of divergent techniques of transmission for the construction of religious life, suggesting that techniques of communication and persuasion adapt and diverge to meet the limitations of memory and to exploit its multi-dimensional capabilities. The contrasting experiences of the Baktaman and of kivung members are the outcome of different political pressures and strategies. At the same time, these two traditions have developed out of a highly creative process of experimentation in the cultivation of religious insights and in the search for revelation.

RECOVERING 'RELIGIOUS EXPERIENCE' IN THE EXPLANATION OF RELIGION

DONALD WIEBE

According to Robert H. Sharf's widely accepted views on the notion of "religious experience" (Sharf, 1998) that concept is of dubious value in the scientific study of religion. The idea of religious experience, he claims, is used "to valorize … the subjective, the personal, [and] the private" and, therefore, is clearly used rhetorically "to thwart the authority of the 'objective' or the 'empirical'" (94) in "the interests of personal and institutional authority" (107). Although it is true that religious devotees, and perhaps even some 'scholars of religion', invoke 'religious experience' in this fashion, I disagree with his implicitly negative general assessment about the value of the notion for the study of religion. Indeed, I believe that his general critique of 'religious experience' has itself been detrimental to the study of religion. The religious uses of the concept of religious experience, that is, do not exhaust the full import of the notion. As Sharf himself notes, he does not deny that people do have subjective experiences (113) nor does he deny that we have access to them. People, that is, talk about their experiences; they engage in conversation with others who lay claim to similar experiences in an effort to make sense of their experience. And I think one can justifiably think about such "conversations" as attempts to negotiate the meaning and significance of wholly private interior experiences – experiences without some external "originary event" that "grounds" them but which might nevertheless be "interpreted" (re-experienced) as encounters with an "alternate reality". In fact, it seems to me that it might well be the case that it is precisely such psychological, wholly interior experiences, and the social negotiation of their meaning and value, that can account for the birth of what today we call religion. Indeed, I believe that David Lewis-Williams's attempts over the last three decades to account for the meaning of the ancient rock art of the San of

South Africa in light of the experience that lies behind them ultimately provides us with the foundations for a general theory of the origin of religion.

Lewis-Williams is professor emeritus of cognitive archaeology at the University of Witwatersrand in Johannesburg South Africa and founder, former director, and now mentor of the Rock Art Institute at that University. In an earlier essay (Wiebe, 2009) I have provided a brief account of the neuropsychological theory of religion that he and his colleagues have elaborated in several books and numerous articles. My aim in this essay is twofold. I will first provide an account of the emergence and development of that theory from two methodological innovations Lewis-Williams brought to the field of rock art studies that not only made possible a more sophisticated interpretation of the artistic creations of the San (and our more distant forebears), but also revealed the essential character of their art as expressions of the religious beliefs they held. I will then show how those beliefs are intimately tied to what we commonly conceive of as "religious experience".

Rock Art Studies and Religion

Lewis-Williams's first methodological innovation in rock art studies involved applying a comparative analysis of artistic expression to bodies of information that had only received isolated attention and consideration by his predecessors. In his judgment, accounts of San rock art as art for art's sake, or as simply representational were, along with structuralist and other interpretations of this material, seriously flawed or, at best, unpersuasive. This modification in method in the interpretation of ancient rock art was adopted by Lewis-Williams in his Ph. D. work on San rock art and San spirituality which was published as *Believing and Seeing: Symbolic Meanings in Southern Rock Paintings* in 1981. Keeping in mind the ethnographic data – the nineteenth-century interviews with San informants and the writings of early travelers who occasionally noted aspects of southern San belief, as well as the contemporary ethnographic data on the Kung in Namibia and Botswana – Lewis-Williams drew on multiple sources of information that made possible the development of a two-way exegetical process between the rock art and the ethnographic data which revealed the paintings to be a system of signs, with many of them functioning as symbols. From this perspective, argued Lewis-Williams, one could see that the paintings and engravings did not primarily signify natural objects in the world but rather expressed meanings and values; that they were essentially expressions of the deep thoughts and spiritual beliefs of the San.

Lewis-Williams goes on to claim, in fact, that the paintings depict "realities" (eland, hartebeest, locusts, and girls at puberty) with supernatural potency, derived from altered states of consciousness, whose main purpose was to cure illness (1981:75). San rock art, therefore, was not meant simply to be admired or contemplated as Lewis-Williams and D. G. Pearce put it in their *San Spirituality* (2004), but rather was self-involving and participatory. As Lewis-Williams argues in *Believing and Seeing*: "Trance performance facilitates contact between the terrestrial world and the spirit world" (1981:83), and the relevance of that trance experience to understanding San rock art, he insists, has been greatly underestimated (1981:114). Trance he claims, enabled the San "to see what they believe", (1981:83) and the paintings enabled them "to see what they believed they could see" in the trance state: that is, "[the] rendering visible of a complex set of beliefs ..." (1981:131). The medicine men (sorcerers, magicians, shamans), that is, were manipulating metaphors rather than physical or ritual objects and the paintings rendered them in visual terms; they were, therefore, "reified visions of the spirit world" (2004:180,181). Thus, paintings of an eland buck are, in effect, paintings of a "trance buck", thereby transforming the buck into a symbol of something else of importance to the San (1981:103). San rock art, Lewis-Williams claims, is simply an element in a "shifting pattern of thought which moves from believing to seeing, [from] the conceptual to the visual" with the visual representation simply being "another part of the complex and subtle web of San thought and belief" (1981:131). In drawing attention to the role of trance experience in achieving a more persuasive understanding of San rock than those already on offer, Lewis-Williams provides a glimpse of a second methodological innovation he was to adopt which, in effect, provided the foundation for a theory of prehistoric religion (and prehistoric social life) that he developed in his later work.

Lewis-Williams's second methodological approach to understanding prehistoric art and thought was spurred on by his interest in the cave art of Upper Paleolithic Western Europe. Ever since reading André Leroi-Gourhan's book *The Art of Prehistoric Man in Western Europe* (1968), he was "persuaded", as he puts it, that there must be a way "of moving from southern African rock art to these images so remote in time and space", no matter how unlikely that appeared at the time (2002b:163). With the assistance of Leroi-

Gourhan, Lewis-Williams had opportunity in 1972 to visit a number of those ancient cave sites in the Dordogne and to think more seriously about bridging that gap of time and space. However, he also recognized the dangers in drawing upon ethnographic analogies between "twentieth-century ethnographically known art to such [prehistoric] images". Following that route ran the risk of "turning the past into a carbon copy of the present" (2002b:163) and he was compelled, therefore, to seek alternative approaches to understanding the Upper Paleolithic material. In the 1980s, Lewis-Williams, together with T. A. Dowson, thought it possible to draw on the resources of neuropsychological research to bridge the gap between the ethnographic and the prehistoric past. As they put it:

> The brains of anatomically modern Upper Paleolithic people were wired in the same way as those of all people living in the twentieth century. Once the explanatory potential of neuropsychological research on the brain and altered states of consciousness was evident, it became necessary to construct a model of those experiences that would facilitate reaching beyond San rock art and indeed beyond all those rock arts for which there is relevant ethnography (2002b:163-164).

The point, claims Lewis-Williams, is that both we and the San have much in common with our Upper Paleolithic forebears: our anatomical and neurological identity. Consequently, we have a bridge to the deep past, and it is in this deep past (that is, a past one finds in the evolution of the human brain), he is convinced, that we shall find the hypothetical original religion of humankind (Ur-religion). He is fully aware of the vast differences between historical moderns and the San, and between the San and us and, for example, the Cro-Magnons of the Upper Paleolithic period, but we can nevertheless, says he, "point to the underlying, physical functioning of the brain that is molded, but not eliminated, by the culture and community in which it develops from birth" (2004:29).

It is in *The Mind in the Cave: Consciousness and the Origins of Art* that Lewis-Williams attempts at length to account for "the efflorescence of Upper Paleolithic art" and the religious beliefs (he thinks) it expresses (2002a:95). Between the world of the Middle Paleolithic and the Upper Paleolithic there is, he maintains, "a cognitive leap" (2002a: 77). Modern human consciousness, he argues, is more advanced than that of the Middle Paleolithic mind and of Neanderthals, in that it "includes the ability to entertain mental images, to penetrate mental images in various states of consciousness, to recall those mental images, to discuss them with other people within an accepted framework (that is to socialize them), and to make pictures" (2002a:94). As with the San, he argues that these cave paintings were reified visions of the spirit worlds. Image-making clearly requires more mental abilities, and therefore indicates the existence of a different type of consciousness from that of earlier hominids. It appears, therefore, that the "neurological deficits" of the Middle Paleolithic mind and the Neanderthals accounts for the fact that there are few, if any, indications of a rich set of experiences and beliefs in those communities that one might consider religious. Lewis-Williams admits that the "neural advantages" of Upper Paleolithic peoples cannot of themselves explain the "creative explosion" of the Upper Paleolithic period. And he rightly points out that this creative explosion should not be confused with the origin of *Homo sapiens sapiens* (2002a:99) for, as he points out, the (anatomically) modern mind and modern behaviour had already evolved in Africa. The potential for image-making and other symbolic activity, therefore, was already in existence before *Homo sapiens* communities arrived in Europe. Therefore, he maintains that we must look elsewhere for the cause of that "explosive event". As he states it: "This pre-existing potential means that we should not seek a neuronal event as the triggering mechanism for the west European "Creative Explosion" (2002a:99). We should rather, he argues, look to social circumstances, and especially social divisions (social conflicts, stress, and discrimination), and the role that art may have played in those situations to account for this development. Nevertheless, without the (anatomically modern) neuronal potential of Upper Paleolithic *Homo sapiens* there would have been no Upper Paleolithic revolution in experience and belief; a revolution now seen to be expressed in its artwork. Therefore, understanding the electro-chemical activity of the (anatomically) modern human brain, and the consciousness (mind) it "produces", constitutes a bridge by which we can achieve an understanding of our prehistoric forebears. "[I]f we are to attempt to cross the neurological bridge that leads back to the Upper Paleolithic", he writes, "we need to look more clearly at the visual imagery of the intensified spectrum and see what kinds of percepts are experienced as one passes along it" (2002a:126). This is possible because we have the same brain as they and can experience the full spectrum of consciousness they experienced even though they would have divided it up in their own way, and for their own purposes, and therefore would have "created their own version of human consciousness" (2002a:131).

A Theory of Religion: 'Rooted in the Brain'

For Lewis-Williams, then, religion, like art, is fundamentally "rooted in the brain" (2004:129), and in its original form is "inevitably intertwined with the origins of art" (2007:9). A closer look at the specific elements of the human nervous system is essential to understanding Lewis-Williams's theory of religion and this is most fully developed in *Inside the Neolithic Mind: Consciousness, Cosmos, and the Realm of the Gods* (co-authored with D. G. Pearce, 2005) which aims at accounting for the transition in thought-life from the Upper Paleolithic to the Neolithic and how the new belief-systems meshed with daily life. Given that persons do not live alone but in communities, understanding will require more than just understanding the brain.

Lewis-Williams and Pearce explicitly espouse the view that there must have been some kind of "social contract" (*ala* Rousseau) that held the Neolithic together and allowed Neolithic society to function relatively smoothly. But they also quickly point out that this cannot be the whole story (2005:39). As they see it, there must have been a concomitant contract – the consciousness-contract – that allowed the Neolithic to cope with daily mental experiences. As they point out: "*Whatever influence the material environment may have on human behaviour, and we do not underestimate its power, all people have to live with and to accommodate the products of their brains in a society of other brains and bodies*" (36; emphasis added). What is common to all societies, moreover, is the mental states produced by the brain, and understanding its "activity", therefore, is to understand what is common to (universal in) all societies. Nevertheless, as already noted, the brain/mind cannot by itself account for the particularities of specific cultures. Thus, in anticipation of their study of the Neolithic, they point out that the generality of the claims they make about Neolithic experience and belief

is founded on the working of the human brain that, in all its electro-chemical complexity, creates what we call our minds. The neurological functioning of the brain, like the structure and functioning of other parts of the body, is a human universal. The specific content of individual minds, their thought, images and memories, are another matter altogether; content is largely, but not entirely, provided by cultures as they are, or were, at a specific time in human history. Content is therefore always changing. The way in which brain structure and content interact to produce unique life-patterns and belief systems is a key issue that we explore (2005:6-7).

While the social sciences treat the specific expressions of these universals, neuropsychology deals with universal characteristics of human thought. It is, therefore, the altered states of consciousness – from light to deep trance – that constitute the foundation for religious experience and, subsequently, religious belief. These stages of consciousness are wired into the human nervous system, although the meanings given to them are cultural interpretations of the altered states of consciousness experienced. According to Lewis-Williams and Pearce, then, it is out of the varied states of consciousness produced by the brain/mind – that is, out of the neurologically based "consciousness contract" – "that notions of the ineffable and the three dimensions of religion [experience, belief, and practice] flow..." (2005:39).

According to Lewis-Williams, the spectrum of human consciousness ranges along a continuum moving from what he calls the consciousness of rationality or "alert consciousness" to "deep trance"; a spectrum that moves from outward-looking states of mind to inward-directed states of mind. The outward-looking states of mind include both the waking state as well as "light" altered states such as daydreaming, lucid dreaming (and dreaming states), while the inward-directed states are generated wholly in and by the brain. The inward-directed trance states are all similar, despite how they are effected or caused (that is, whether they are induced by migraines, dancing, chanting, rhythmic clapping of hands, drums, meditation, sensory deprivation, hunger, fatigue, flickering light, extreme pain, near-death experiences, drugs, temporal lobe epilepsy, or schizophrenia), and pass through three stages from light to deep trance that are not discreet but rather grade into one another. That potential for the human nervous systems to enter into such altered states of consciousness and to generate hallucinations is of great antiquity, not only wired into the human nervous system but the mammalian nervous system more generally.

Although Lewis-Williams is aware that many archaeologists are uneasy about invoking theories of human consciousness in attempting to account for the thought-life and beliefs of prehistoric human communities, he is nevertheless convinced that it is only through the adoption of such new approaches to interpreting the data that research in the discipline can be advanced, and particularly so "for piecing together humankind's first 'religious' concepts and experiences ..." (2002b [1988]:189). Using a neurological model of the kind he proposes – one that focuses attention less on human intelligence and more on the marginalized elements of human consciousness – Lewis-Williams argues, is a reasonable avenue of research in seeking an understanding of both the ancient human mind and of the mind of the San. It is reasonable because modern human beings (and therefore all anatomically modern human beings) have brains that function in scientifically discoverable ways and modern neuropsychological research involving laboratory subjects has provided us with information about the ways human consciousness affects us and, therefore, how it could have influenced people's minds in prehistoric times. He admits that we do not know how the functioning of the brain produces consciousness, but he also points out that this is not the stumbling-block some might consider it to be, because all that is necessary in this context is that we are able to describe some aspects of consciousness. And on the basis of the results of modern neuropsychological research he points out that there exists a broad range of human psychological experiences beyond what has been called the "consciousness of rationality" – that is, of "waking problem-oriented thought" – upon which he constructs a neuropsychological model of human experience that he believes will better account for both San rock art paintings and the art of Upper Paleolithic peoples. Such a model, he maintains, will show the brain to be the root of all (at least) archaic spirituality/religion which, given the ethnographic evidence, is best described as shamanistic.

The altered states of consciousness that humans experience follow two trajectories (normal and intensified), can be visual, aural, somatic, or psychological and, especially in the later stages of the intensified trajectory of inward-directed states of consciousness, can give rise to felt encounters of an alternative reality. Neuropsychological research, as Lewis-Williams points out, reveals three stages in trance experience among human beings. The first stage in this trajectory involves only what he refers to as "light trance" and includes experiences that are common to all (or at least most) humans across all cultures. This level includes hypnagogic experiences – states of consciousness intermediate between wakefulness and sleep – that include vivid mental imagery generated by the optic system (entoptic phenomena) and a transfer of sensation from one sensory modality to another (synesthesia). The entoptic phenomena generally consist of geometric mental images that include a diversity of forms such as grid or lattice designs, sets of parallel lines, bright dots and short flecks, zigzag lines, nested catenary curves, filigrees and thin meandering lines, and spirals. Given their ubiquity it is surmised that they are hard-wired into brains and therefore inescapable and must be dealt with by individuals and the communities they comprise. Given the particularities of the social and cultural settings involved, some of these basic experiences may be ignored and others valued and cultivated, and may even be transformed into systems of symbols.

A second, deeper, stage of altered consciousness involves attempting "to make sense of the entoptic forms", as Lewis-Williams puts it, "by construing them as objects with emotional or religious significance" (2005:50). This is often connected with an experience of passage through a vortex, descent into an underworld, entering a portal into another world, and something like a near-death experience. These experiences are clearly hallucinatory.

Lewis-Williams (and Pearce) describe the third and deepest level of trance state as follows: "Emerging from the vortex, subjects enter a bizarre, ever-changing world of hallucinations" (2005:54). The individual's experience here is "autistic": wholly closed-off from the outside world. Hallucinations have an increased vividness, are experienced in all senses, are associated with powerful emotions and feelings, and involve processes of fragmentation and integration that produce compound images in which the subjects participate in their own imagery. Such deep trances also create the illusion of dissociation from one's body that produces sensations of possession or a sense of extra-corporeal travels in an alternate world, and puts one in touch with spirits and other supernatural powers, giving one control of the weather, or of spirit-animals (and actually feeling themselves becoming animals), or power to heal the sick.

Understanding the nature of the human brain/mind, and the mercurial type of consciousness it creates, suggests to Lewis-Williams (and his several collaborators) the possibility of deriving principles of human behaviour – including social behaviour – that might account for the variety of ways in which humans have expressed themselves through time and around the world, and particularly so in ancient cultures. This becomes clear in reviewing his later work on the San, his interpretation of "the mind" of our Upper Paleolithic forebears, and, as already noted, in his analysis of the transition in thought from the Upper Paleolithic to the Neolithic.

Shamanism: From 'Spiritual' to 'Religious' Experience

In *San Spirituality: Roots, Expression, and Social Consequences* (2004) Lewis-Williams, in collaboration with D. G. Pearce, elaborates the ideas about the San set out in his doctoral dissertation. In their account of the Howieson's Poort stone industry of the Middle Stone Age in South Africa, they insist that any attempt to explain the use of quartz in that industry cannot simply rest either on grounds of technological innovation or on the attractiveness of the materials used. They claim, rather, that the Howieson's Poort people (among others) appear to have perceived a connection between the crystals and "some sort of spirituality" (2004:19). By "spirituality" they mean to refer to "the luminous geometric forms seen in early [entoptic] states of altered consciousness and they argue that the luminous character of quartz and its capacity to trigger electrochemical experiences in the human brain makes sense of the use of quartz among our early human ancestors. For Lewis-Williams and Pearce, therefore, there is no understanding of San spirituality without an understanding of neuropsychology, and San "religion" is simply its socialized form. Religion, that is, is therefore the product of the psychological and somatic experiences of altered states of consciousness that our prehistoric ancestors used as a resource for "negotiating" their world (2004:30). Thus, according to Lewis-Williams and Pearce, "[prehistoric] [r]eligion is not so much an attempt to explain the natural world, its regularities and catastrophes and to cope with death, as [it is] a way of coming to terms with [and making use of] the electrochemical functioning of the brain" (2004: xxiv). It is also important to keep in mind here that Lewis-Williams recognizes that this "negotiation of consciousness", as he and Pearce refer to it, is not only the foundation for spirituality and religion but also the foundation of social differentiation (2004:25). Different communities, that is, made use of this resource in diverse ways, since cultures can mold the brain wiring even if they cannot eliminate it (2004:29). Thus they contend: "All attempts to make sense of the products of our brains – the normal and the intensified trajectory – must necessarily be social, not simply personal and individual; they must be fashioned in and accepted by the community" (2004:34). Religion (spirituality), therefore, derives from "the development and harnessing of a type of consciousness that is unique to anatomically modern *Homo sapiens*" (2004:25) but the distribution of the experiences of those altered states of consciousness also made possible significant social distinctions. As Lewis-Williams argues in *The Shamans of Prehistory: Trance and Magic in the*

Painted Caves" (jointly authored with Jean Clottes): "By drawing on this resource, some people were able to lay claim to special knowledge and special insights that set them apart from and probably above others" (1998:114).

For Lewis-Williams (and Jean Clottes) then, the religion – that is, the Ur-religion – that emerges from altered states of consciousness in *Homo sapiens* is (and possibly must be) a form of shamanism. As they point out, the art of the Upper Paleolithic communities of Europe provides considerable evidence for the existence of an early (if not the earliest) form of shamanism and the ethnographic record suggests that some form of shamanism universally characterizes hunter-gatherer societies. It is clear, he and Clottes argue, that the beliefs that derive from altered states of consciousness, on being institutionalized, give rise to beliefs in an alternative (supernatural) reality which, consequently, constitute humankind's Ur-religion – a religion created by "the first attempt to deal with functions of the brain" (2002b:190).

In a more recent article Lewis-Williams and Clottes argue that shamans were people with special powers gained by virtue of their peculiar (mental) experiences that were interpreted as giving them access to different levels of (supernatural) reality and put them in touch with special powers which they used for the benefit of their people (and, no doubt, to their own advantage). Shamanism, therefore, was not simply a reflection of the private experiences of individuals but was, rather, embedded in the social fabric of early human societies. The universality of shamanism among hunter-gatherers, therefore, is not the result of diffusionism but rather the product of a "universal neurological inheritance that includes the capacity of the nervous system to enter altered states of consciousness and the need to make sense of the resulting experiences and hallucinations within a foraging community" (2007:33). Interpreting the cave art of the Upper Paleolithic as the expression of the experience of the shaman and shamanic community, he and Clottes then claim, makes possible the most coherent account of it that we possess. As they put it: "We have every reason to believe: (a) that the people of the period would have had to make sense of altered states within the context of their hunting and gathering economy and social relations and (b) that the human behaviour that this process entailed left physical traces" (2007:35); and they believe that allowing these "hypotheses" to interact with the data will confirm that judgment. Lewis-Williams's theory of religion, then, might best be summarized as beliefs in the supernatural that result from the efforts of our Upper Paleolithic ancestral communities "to make sense of and to socialize the products of the nervous system" (2007:34). And understanding that theory, therefore, requires consideration of both "the behaviour of the nervous system and the social contexts in which the behaviour takes place..." (2007:34).

For Lewis-Williams and his collaborators, then, understanding the nature of the human brain/mind and the mercurial type of consciousness it creates is essential for understanding (explaining) religion. They do not, however, think religion is wholly explicable simply in neuropsychological terms because, as Lewis-Williams and Pearce point out, "[h]uman brains exist in societies" (2005:55) which means that there is interaction between the mental – neurologically generated – experiences described above, and the culturally specific content that is incorporated into those experiences. As they put it: "We argue that it is out of the socially situated spectrum of consciousness ... that religion develops. One could say that religious experience is, in the first instance, a result of taking the introverted end of the spectrum at face value within a given cultural context" (2005:55). That is, people harness the mental states created by the brain (in either natural or induced conditions) by interpreting them "as witnesses to the existence of cosmological realms ... and supernatural beings that can impinge on daily material life ..." (2005:26). Such influence on society, therefore, comes from shared rather than individual or idiosyncratic (neurologically generated) beliefs which then undergo elaboration which, as they put it, "takes on a life of its own within a social structure that has its own tensions and divisions" (2005:27). They acknowledge that the system of beliefs that is produced does not need to correspond in every detail to the mental experience from which it derives, although they nevertheless insist on the essential character of the mental experience.

Testing the Theory

The explanatory potential of the neuropsychological theory, as Lewis-Williams points out, can only be determined by letting it intersect with empirical data. And this is what he allows it to do in relation to achieving an understanding: of the origins of art in the Upper Paleolithic in *The Mind in the Cave (2002a)*; of San rock art and San spirituality and religious beliefs in *San Spirituality* (2004) and in *The Mind in the Cave*; of the incredible similarities of images in the rock art and cave paintings of communities around the world and throughout time and of the interest in deep underground chambers and passages in *The Mind in the Cave* and *A Cosmos in Stone* (2002b); of the widespread existence and similarity of shamanic systems of thought and practice including belief in a multi-level (tiered) cosmos and in trans-cosmological travel in the sources already mentioned and in *Shamans in Prehistory* (1998); and of the great transition from Upper Paleolithic to Neolithic religion in his *Inside the Neolithic Mind* (2005).

It is not possible here to examine all the empirical data that Lewis-Williams and his collaborators account for in terms of altered states of consciousness. It will suffice to bring to attention just a few of the sets of data they see as explicable in terms of the inescapable experiences generated by the human nervous system common to us and our ancient ancestors. They argue, for example, that the neuropsychological theory is successful in accounting not only for the existence of some form of shamanism in the Upper Paleolithic period, but also for its surprising similarities worldwide (2002b [1997]:222). One sees this clearly, he argues, in relation to shamanic altered states of consciousness found in African San rock art, the rock art of the Cosos in North America, the art of the Tikano in South America, and Huichol art in Central America. He writes (2002b [1997]:198):

> In the iconography of each of these four arts there are geometric motifs isomorphic with the geometric forms of stage 1, construed geometrics referable to stage 2, and stage 3 iconic images that are sometimes therianthropic and sometimes combined with geometric forms (Lewis-Williams and Dowson, 1988). Arts known independently to be associated with shamanic altered states of consciousness thus display a complex set of features – not merely geometric motifs – that fits the set of features established by laboratory research on the mental imagery of altered states of consciousness. If the model is tested against an art not associated with altered states of consciousness, say Rembrandt's work, no such fit will be found....

In his treatment of the San he points out "that human neurology also explains the ubiquity of the tiered cosmos [of shamanism]" and that "[c]osmological travel [by shamans] is paralleled by mental travel along the spectrum of consciousness and its attendant transformations" (2004:35). These "realities", as well as the access to unseen realms of power are, in his view therefore, "verified by altered states of consciousness" (2004:35). Comparing the San material to the similarities found elsewhere in the world, as well as to the art of the Upper Paleolithic he writes (2002a:170):

> I can think of no other way to explain the similarities: they are the product of the universal human nervous system in altered states of consciousness, culturally processed but none the less still recognizably deriving from the structure and electro-chemical functioning of the nervous system.

There is simply no evidence in the natural world, he maintains, that can empirically account for these similarities (2004:82). And the shamanic world view is persuasive to the "common person" in these cultures, he argues, because all persons have "some inkling of the spirit world in their own dreams. In the absence of any knowledge of human neurology, dreams provide them with personal, indeed, incontrovertible, evidence of the existence of the spirit realm" (2002a:177). It also helps in elucidating the paintings in that they make perfect sense as "reified visions of the spirit world" experienced by the seer. In applying the model/theory to Upper Paleolithic art, for example, Lewis-Williams notes that he and T. A. Dowson found evidence in that material for all three stages of altered consciousness (2002b [1988]:181):

> Many (but not all) of the so-called signs are similar to stage I entoptic forms; some seem to have been construed as animals in the manner of stage 2 hallucinations, though this is more difficult to demonstrate; and there are monsters, therianthropes, and other features of the third and deepest stage of trance. The puzzling co-occurrence of geometric and representational depictions throughout the Upper Paleolithic is thus explained. They are not two evolutionary categories or two parallel kinds of art. Rather, they are two kinds of universally experienced mental percepts. As neuropsychological research shows, we should expect to find both of them in an art derived from altered states of consciousness.

Religion: The Expression of Spiritual/Religious Experience

The explorations and analyses of ancient rock art and prehistoric cave paintings undertaken by Lewis-Williams (and his colleagues and coauthors) in search of an explanation of their meaning and signifycance, both for the individuals who created them and the societies in which they thrived, has ultimately provided us with a plausible scientific theory of religion and a scientifically credible account of the origin, (or at least one of the origins), of religion. The earliest form of religion (Ur-religion), according to Lewis-Williams, emerged with our *Homo sapiens* ancestors and was essentially shamanistic in character. And shamanism, he further argues, can only be accounted for in terms of the (universal) neurological inheritance of the anatomically modern human brain (mind) which harbours the potential for altered states of consciousness, including a wide variety of hallucinatory experiences that came to bear significantly on the social group. Lewis-Williams's theory of religion, therefore, is essentially a neuropsychological account of religion. However, unlike many other attempts to account for religion in terms of the cognitive or other neurological resources of the human brain/mind, Lewis-Williams's theory is not reductionistic in the sense that it attempts to account for religion (rather than mere spirituality) wholly in neurological terms. He agrees that the human mind exhibits a form of consciousness that goes beyond simply paying attention to the contingencies of everyday life and that persons in deep trance-states (subjectively) experience an "otherworldly" reality. But he does not equate that bare experience (which is wholly the product of the human nervous system) with religion. In *The Mind in the Cave* he refers to that aspect of the consciousness spectrum as producing a "primordial spirituality" (2002a:289) which he differentiates from "religious experience" which involves interpretation of such apparent "encounters". Religion, moreover, also involves organization (socialization) of such "spiritual" (otherworldly) experiences; this transforms the merely spiritual into the religious (religion).

In *Inside the Neolithic Mind* Lewis-Williams (and Pearce) again argues that "religion" is more than just "primordial spirituality"; religion is an integrated whole of three interlocking dimensions of human experience that ultimately derive from the brain *which*, it must be remembered, is always historically/contextually situated. The three interlocking dimensions are religious experience, religious beliefs, and religious practice. When a trance state (that is, "spirituality", or what Jean Clottes calls "an awakened consciousness" [2006]) is *interpreted* as contact with a supernatural world not of one's own making, a religious experience results. And when such experiences are codified with respect to specific social contexts, they maintain, we have the production of religious beliefs. Religious rituals then can be seen as "practices" that both draw upon religious experience and manifest religious belief. Thus he (and Pearce) argue "that it is out of the socially situated spectrum of consciousness ... that religion develops" (2005:55). Consequently, even though religion as so described is ultimately embedded in neurology, neurology is simply one of many resources available to human manipulation in the service of existence; alone it is incapable of accounting for religion. As Lewis-Williams and Pearce put it: "Human beings are sentient. They know what is going on around them. They are Marionettes of neither their environment, nor their cultures, nor their neurology" (2005:285). As has been noted above, an adequate theory of religion must take into consideration the "social contract" as well as the "consciousness contract". And this is what sets anatomically modern human beings apart from animals and pre-humans (1998:114). In making use of their shared continuum of altered states of consciousness as a resource for ordering their lives (in their art and religion), they differentiated themselves from their hominid forebears (and cousins) and gave birth to society as we know it today (1988:114).

For Lewis-Williams, then, religion is a wholly human phenomenon, even though the essence of religion necessarily includes a "supernatural" component "known" in an act or acts of revelation. And this is true, he maintains, not only for ancient or Ur-religion, but of modern religion as well. As he and Jean Clottes note (1998:114):

> [The shamanistic framework ultimately] gave shape to the so-called 'higher' religions that still postulate a realm above and a realm below, visions of those worlds and prominent personages who, symbolically or in reality, move between those realms. The world has not escaped from the trammels of the Upper Paleolithic.

In the eyes of contemporary religious devotees, therefore, his theory is reductionistic, despite his protestations to the contrary, in that it, in some sense at least, reduces it to nothing more than an altered state of consciousness and social elaborations of the experiences so generated. This charge of reductionism, however, is not one Lewis-Williams intends to dispute for he cannot envisage any possibility of a rapprochement between science and religion (2005:290). Religion, he admits, deals with the products of the introverted end of the consciousness spectrum but he does not see this as giving it immunity from science. Science, on the other hand, deals with the external world and all that it contains, including consciousness and its products, and this exposes the supernatural world (as a product of that consciousness) to scientific study.

In *The Mind in the Cave* Lewis-Williams asks whether "we are in danger of losing something supremely valuable" (2002a:291) if we account for the supernatural aspect of life, whether in ancient or modern form, in this scientific fashion. He thinks not: "We can catch our breath when we walk into the [centuries-old] Hall of Bulls", he writes, "without wishing to recapture and submit to the religious beliefs and regimen that produced them" (2002a:291). The historically (rather than the merely anatomically and behaviourally) modern person, he argues, has been emancipated (even if not completely) "from the imperatives of voices and visions ... in a slow 'stop-go-retreat' process" (2002a: 288), making it possible to distinguish what occurs in the head (otherworldliness) from what lies beyond us (worldliness or, simply, the external world). Belief in supernatural realms and influences, then, is anachronistic (2002a:290; see also 2006:163); it is mystical atavism (2002a:291) whether expressed in traditional religious terms or a form of what he calls "urban shamanism" (2002a:288).

CAN THE STUDY OF RELIGION BE SCIENTIFIC?

DIMITRIS XYGALATAS

If you are in a company of people of mixed occupations, and somebody asks you what you do, and you say you are a college professor, a glazed look comes into his eye. If you are in a company of professors from various departments, and somebody asks what your field is, and you say philosophy, a glazed look comes into his eye. If you are at a conference of philosophers, and somebody asks you what you are working on, and you say philosophy of religion...

- Nelson Pike, as quoted in Bambrough 1980: 289.

Religion is both a fascinating and an important subject of study. People often do truly amazing things for their faith. They fast, they abstain from sex, they walk on fire, perform self-flagellation, or nail themselves on crosses. They take great risks and spend time, money, and all kinds of valuable resources for their religion. These are all very exciting behavioural expressions, not only because of the sensationalist imagery that they provide, but mainly because of their effects on people's lives. Religious incentives and prohibitions, religious fundamentalism, sectarian violence, religion's influence on politics, education, and even science, have major consequences for practically all humans, religious and non-religious alike. However, despite of religion's tremendous importance, the study of religion has a marginal place in the academia, and departments of religion are usually underfunded and attract fewer and fewer students. Why is this so? It is certainly not because there is lack of public interest in the subject which is attested to by the millions of books on religion, atheism, and 'spirituality' (whatever that may mean) sold every year. I argue that the problem lies in the study of religion itself.

The truth is that the study of religion has a bad reputation, due to the way it has been – and largely is still being – conducted. Unlike other disciplines, like psychology or sociology, that have fully embraced scientific methods and criteria from the natural and the health sciences, the study of religion remains fragmented and isolated. It is isolated because it is conducted almost exclusively within the humanities, without much dialogue with other disciplines, and largely ignoring the extraordinary advances in the explanation of human behaviour that emerging from the life sciences. To make matters worse, most religion departments to this day either share roof with or are part of theological or 'Divinity' schools and faculties. This fact alone suggests a blurred line between the study of religion and the practice of religion, between knowledge on religion and religious knowledge, and deters people from taking the field seriously. Imagine that you are a freshman at a university and you have to choose your courses. How would you feel about signing up for an introductory course in Political Science offered by the School of Communism, or Evolution 101 in a department of Creationism?

In addition, the study of religion is also fragmented, because the directly observable behaviour and the explicit beliefs on which it almost exclusively relies are the beginning and not the end of the story when it comes to the understanding of religious – or any other – behaviour. On the other hand, the few natural scientists who are concerned with the study of religious behaviour usually focus exclusively on implicit processes in the brain, in decontextualized laboratory settings. Unfortunately, more often than not the two parties never share their insights and findings.

Despite these problems, there is real progress. During the last decades, a new paradigm for the study of religion has been proposed and is slowly but steadily gaining ground. I am referring to the emerging field of the Cognitive Study of Religion, which attempts to bridge the gap between the study of religion and the natural sciences by shifting the focus from mere description and interpretation to more explanatory, testable models of religious behaviour.

The Study of Religion: A Troubled History

In 1859, Charles Darwin published the *Origin of Species*, where he described the process of evolution by natural selection, a theory that was to become the foundation of all life sciences. During the next half-century, Darwinian theory had a profound influence on the study of religion, and provided a guiding theoretical framework for its coherent scientific study, allowing for the first time for the real focus of this study 'to be located, not in transcendental philosophy, but in... this-wordly categories' (Sharpe 1986: 24).

Meanwhile, the cultural evolutionist theories of such scholars as Auguste Comte, Thomas Malthus, Herbert Spencer, and Lewis H. Morgan, were developed independently. Certain forms of this early cultural evolutionism were used to defend contradictory ideological perspectives including racist, colonialist and imperialist practices, laissez-faire economics, slavery, eugenics, and Nazi propaganda. These simplistic notions of social progress sought support in the Darwinian theory of evolution; they did not, however, rest on scientific evidence but emerged from strongly held political and ideological commitments. After World War I, when the critique of cultural evolutionism became widely accepted, social scientists radically changed their approach to the study of human societies, placing an emphasis on cultural relativism, in an attempt to avoid the ethnocentrism of past approaches.

Scholars of religion turned to less rigorous phenomenological treatments of religion, which averted theory and favoured hermeneutic over explanatory approaches. Religion was considered as a *sui generis* domain, neither connected nor reducible to any other level of human behaviour. (e.g. see Otto 1923; Eliade, 1991; Sharma, 2001; Pals 1987). According to this view, the study of religion requires special and unique methods that are not available to the other disciplines. As Wilfred Cantwell Smith put it (1959: 42), 'no statement of religion is valid unless it can be acknowledged by that religion's believers'. Explaining religion would be to 'explain it away', or would miss its 'essence' (whatever that may mean); therefore, scholars of religion should try to understand and describe rather than explain the object of their study (Pyysiäinen 2004). Consequently, this view considered the study of religion as the exclusive privilege of theologians or – at best – as a pursuit strictly limited within the humanities. By denying the authority and the right of scientific theoretical thought to examine religious phenomena and rejecting the usefulness of science for the explanation and interpretation of religious action and behaviour, (McCutcheon 1997), it has effectively isolated the field and alienated scholars from other disciplines. Thus, religion generally remained strictly a subject matter of the humanities and either embraced the hermeneutic stance or was mainly concerned with the social functions of religion and ignored psychological processes and the properties of the individual brain/mind. Psychologists, on the other hand, either abstained altogether from entering a domain in which they considered that mediocre work was being done, or focused on specific religious experiences, such as mysticism or ecstasy, (James 1958) as if they were paradigmatic of all forms of religiosity. (Lawson 2000).

The Cognitive Turn

During the last decades, there has been an attempt to reconcile the study of religion with the study of the mind, an attempt from which the interdisciplinary field of the Cognitive Study of Religion has emerged. The origins of this field can be found in the emergence of Cognitive Science, which was proposed in the 1950s as the new paradigm for the study of mental processing and gradually replaced the previously dominant behaviourist paradigm (Skinner 1938). The 'cognitive revolution' began from research in Artificial Intelligence (Turing 1950; McCarthy et al. 1962, 1969; Minsky 1954; Newell et al. 1958, 1972; Simon 1969) and the study of memory (Miller 1956; Broadbent 1958). In the study of culture, the most influential player was linguist Noam Chomsky (1957), who argued for a shift of focus 'from behaviour and its products to the system of knowledge that enters into behaviour' (1986: 28). By showing that important aspects of language acquisition are the product of universal capacities of the human brain, he delivered the final blow to Skinner's radical behaviourism, which saw language as a learned habit (1959). Chomsky argued for the existence of an innate human linguistic capacity, a 'universal grammar' that underlies every language's particular grammar. He went on to propose his famous theory of *transformational-generative grammar* to model the innate knowledge that underlies the human ability to speak and understand language. For Chomsky, then, linguistics is ultimately a *psychological* pursuit, aiming to specify the biologically based principles of the mind that constrain the form of all natural languages. Soon thereafter, growing attention began to be directed to studies that focused on underlying commonalities rather than on surface variability of human traits and searched for a 'universal grammar' underlying the particular semantics of cultural phenomena. (Berlin and Kay 1969; Hays 1983; Brown 1984; Atran 1990)

The first calls for a cognitive study of religious behaviour appeared in the 1970's, (Lawson 1976; Staal 1979a, 1979b) and Dan Sperber's seminal book on symbolism (1975) was the work that had the greatest impact. Sperber criticized the existing semiotic approaches to symbolism and the dominant structuralist anthropological approach of his time, and proposed that symbolism is best understood not as a system of abstract signs and their meanings with its own rules, but rather as part of our normal mental processes of reasoning about the world. In 1996, he stressed the role of the cognitive constraints that bias the distribution of cultural representations, and shortly thereafter drew an analogy between the spreading of ideas and the spreading of viruses in populations. He proposed an 'epidemiological approach' to cultural phenomena, which provides a framework to explore how certain ideas are selected for transmission while others simply fade away (1996). Opposing Richard Dawkins' theory of transmission of cultural representations as *memes* (Dawkins 1976), Sperber maintained that representations get transmitted not through precise replication, but through a process of constant transformation and re-representation. The aim of the study of culture is to explain which psychological dispositions underlie the formation and distribution of representations and under which circumstances a relative stabilization of form or content occurs in the process of their generation. Representations move, as they are continuously transformed, towards attractors around which transformation tends to be limited and relative stability occurs. These attractors can be cross-cultural and panhuman or culture-specific and unsteady. The selective stabilization brought about by these attractors is the main force driving cultural evolution.

Towards the end of the twentieth century, cognitive scientists turned to investigate religious phenomena. Some of the scholars that established the field were Stewart Guthrie, with his work on anthropomorphism (1980 1993); Pascal Boyer, with his work on the naturalness of religious concepts (1994); David Lewis-Williams (1981), Merlin Donald, (1991) Steven Mithen (1996) and Terrence Deacon (1997), with their theories regarding the evolution of the human mind, symbolic representations, and the origins of religion; E. Thomas Lawson and Robert McCauley (Lawson and McCauley 1990; McCauley and Lawson 2002), with their *Ritual Form Hypothesis* on the relationship between ritual structure and people's intuitions about the efficacy of a ritual; and Harvey Whitehouse (1992, 1995, 2004), with the *Modes Theory* on the role of memory in religious transmission. On the basis of the pioneering work of these scholars, a plethora of publications appeared in the cognitive study of religion at the dawn of the new millennium.

The proponents of this paradigm argue that despite contextual differences, the seemingly boundless variability of cultural forms and expressions recorded by comparative scholars or religion is subject to panhuman constraints. The properties of the human brain (e.g. memory, theory of mind), the interaction between individuals and that between groups as manifested in historical forms and developments (e.g. political changes, wars, etc.), as well as the physical world in which we live and its particular geographies (e.g. available sources of food, geographic isolation), actively shape and constrain human expression and behaviour (Whitehouse 2004). Religion is no exception; religious behaviour is shaped by the same universal mental capacities of the human brain, while the representation of religious actions draws on the same panhuman cognitive mechanisms used for the representation of any kind of action.

The Debate on Reductionism

The cognitive trend constitutes the most important approach to the study of religion that has appeared over the last decades. However, despite its auspicious beginning, it has not yet managed to produce the paradigm shift that it promised, and it has met strong resistance within the field of the Study of Religion. One of the most common objections to the cognitive study of religion is that it is a 'reductionist' enterprise, which aims to 'explain religion away'. These, however, are two different claims, and one does not necessarily imply the other.

First of all, the cognitive study of religion *is* indeed reductionist, involving the use of a variety of methods and tools. Reductionism implies that theories or things of one sort can account for theories or things of another (McCauley 2001). This reduction can be intra-theoretic, involving competing theories within the same level of analysis (e.g. nativist and connectionist) or inter-theoretic, implying that the nature of complex things may be reduced to the nature of simpler or more fundamental things, e.g. the reduction of psychological states to neurological procedures (See McCauley 1996). Both kinds, but particularly latter, are often fiercely rejected by many scholars within the study of religion. These scholars often embrace the *sui generis* view of religion (religion cannot be reduced to anything non-religious), which leads to an anachronistic, dualist way of thinking. Others mistakenly identify reductionism with cultural eliminativism; they consider that reducing mental states to a material (brain) states is to deny the role of culture, history, and ecology in the shaping of our representations. This stance I call naïve anti-reductionism. Although more sophisticated versions of anti-reductionism have certainly been proposed[1], the following two examples of naïve anti-reductionism are in fact very useful in showing what reductionism is certainly *not* about:

> With reductionism comes the conviction that a court proceeding to try a man for murder is 'really' nothing but the movement of atoms, electrons, and other particles in space, quantum and classical events, and ultimately to be explained by, say, string theory. (Kauffman 2006)

*

> Reductionism implies that any apparently higher levels, such as life, mind and religious ideas of 'God', can all be fully explained in terms of the lower level sciences of chemistry and physics. (Haught 1995: 73)

Of course, no one of sane mind would try to explain the intentions of a judge or the concept of God in terms of atoms and electrons (that would constitute a case of what Daniel Dennett has termed 'greedy reductionism' [1995]). What reductionism actually means is that complex systems can be described and explained within the framework of a multi-level, hierarchical scientific organisation, where each level can be explained in terms of what is going on *one* level down (Dawkins 1986). Therefore, reductionism claims that social phenomena can in principle be reduced to psychological procedures,

---

1. For example, Ernst Mayr has argued for the autonomy of biology, and showed why it cannot be reduced to physics (2004). Also, Fodor (1974; 1975) and Putnam (1967) have famously argued for the autonomy of psychology and insisted that psychological phenomena are not necessarily reducible to a lower level of analysis, because higher-level phenomena are multiply realized by an extremely broad and not well delineated set of lower-level realizers. However, also see Richardson 1979, Shapiro 2004, and Bechtel 2007, for some of the criticisms against this position. In particular, on multiple realizability as an argument *for* and *not against* reducibility, see Bechtel and McCauley 1999; McCauley 2007a.

which can be explained in terms of biology, which can be further reduced to chemistry, which can ultimately be reduced to physics.

For example, if we wanted to study drug addiction, we could conduct a sociological study, which would perhaps reveal a correlation between drug use and crime rates. The causes of this correlation could, of course, be explained on a sociological level, e.g. lack of economic resources, lower educational levels, etc. But if we want to have a deeper understanding of what causes this pattern in a population, we need to explain how drugs affect behaviour on an individual level (psychology). If we want to know why *that* is, we would need to study the way our brain works, and the working of various neurotransmitters, receptors, and synapses in the brain (biology). But if we ask why these neurotransmitters have the ability to trigger specific receptors and what effect this has on the brain, we need to know the properties of the drug molecules, the chemical synthesis of each neurotransmitter, etc. (chemistry). Finally, if we wish to know about the very nature of particular molecules, we must understand the forces that cause atoms to be attracted or repelled by other atoms, as well as the forces that keep the particles of every atom together (physics).

As I said, this reduction is done *one level at a time*, and this is how we get from complex to simpler phenomena. But now imagine that we try to explain drug-related crime by referring to the movement of protons and electrons. This notion is absurd because it leaves a tremendous gap between the higher and lower levels of analysis, telling us nothing about the relation between the two, which is the aim of reductionism in the first place. But again, this naïve type of reductionism exists only in the minds of some of those who oppose it.

Neither does this mean that in order to explain religion we need to move all the way down to a physical level of explanation. We can study a phenomenon on various levels, by identifying the causal mechanisms that operate at this level, according to which aspect of the phenomenon we wish to explain (see Craver and Bechtel 2007). It is unfounded to claim that a lower-level explanation would somehow 'explain away' the top-level interpretation or even the phenomenon itself. To return to the example of drug addiction, explaining the role of neurotransmitters does not reject sociological explanations – if anything, it strengthens their own explanatory potential – and it certainly does not 'explain away' addiction. Similarly, discussing the role of neurotransmitters in ritual trance does not negate the social effects of rituals (e.g. their role in social bonding and group cohesion); to the contrary, it provides a much more powerful understanding of these effects and their causes. Without this kind of analysis, there can be no progress in science.

Any kind of account of an event worth the term 'explanation' involves some sort of reductionism (Slingerland 2008). When we study 'religion', we study religious practices, behaviours, texts, artefacts, social groups and institutions and their particular structures and histories. If we strip religion of the historical, the social, the psychological level, and so forth, nothing is left in the end. Thus, 'it is the attempt to be a true anti-reductionist that ironically leads one to explain religion away in the sense that one no longer knows what it is that one is studying' (Pyysiäinen 2008).

Finally, there are those who argue that a reductionist study of religion would alienate believers by doubting the truth value of religious propositions. It is understandable that those who seek funding or are employed by theological institutions or wish to recruit religious participants in their research programmes may have such preoccupations. This, however, is not a scientific concern, but a political one. The usefulness of any given method or the validity of any hypothesis must be evaluated on the basis of evidence and logical coherence, not on the basis of ideological agendas (Pyysiäinen 2008).

In addition to reducing phenomena to lower states of analysis, a cross-disciplinary study can also resort to theoretical and methodological guidance from higher levels of analysis. For example, psychological causal explanations play an important role in generating and testing neurobiological hypotheses (Bickle 2003: 178). Human activities involve several levels of complexity - cognitive processes, individual actions, short- and long-term interactions between individuals, groups, and societies (Hinde 2005). Each of these levels interacts dynamically with other levels, as well with environmental factors and established socio-political structures. A study of human behaviour should take all these levels into account. Of course, no single study or discipline could ever comprehensively deal with such an enormous number of variables. Therefore, what we need is not a grand unified theory of human behaviour, but rather a range of scientific theories that target specific aspects of this behaviour. 'All scientific explanation is partial explanation from the perspective of some analytical level or other. Scientific theories and the explanations they inspire are selective. They neither explain everything nor wholly explain anything' (McCauley 2007a). Thus, a scientific study of religion involves an *explanatory pluralism*, borrowing tools, methods, theoretical strategies and experimental techniques across explanatory levels. In such an integrative model of study, lower levels of analysis can help refine or re-adjust higher level approximations, while resorting to upper levels of analysis can play a role in justifying lower level proposals (McCauley 1996; 1998; 2007b).

The need for integration

Although scholars of religion are increasingly getting interested in cognitive theories, they are still largely detached from empirical and experimental work and frequently fail to acknowledge the universal features of the human brain, which constrain cultural expressions. A *radical methodological relativism* often plagues the field, considering that each culture and each religion can only be understood in its own terms. This stance in effect weakens the scientific study of religion, as it denies the very constants that would make the variables more meaningful. Such an approach ignores the universal psychological mechanisms that condition human behaviour. Furthermore, there is a general reluctance to move beyond the explicit representations and into the implicit workings of the mind. The mental concepts and processes related to religiosity are much more complex than what subjects report, or even believe, about their own beliefs. As important as it may be, it is not enough to describe what people do and what they say; it is also important to detail the extent to which these explicit representations are informed by implicit cognitive processes.

On the other hand, although many cognitive scientists have now turned to the study of religious representations, for the most part they study their subjects within the narrow confines of the laboratory, without paying much attention to the environmental and socio-cultural variables that shape human behaviour. Furthermore, certain cognitive scientists advocate a *radical methodological materialism* that argues that mind and brain are one and the same thing, and therefore we can know everything we need to know about the mind by studying the brain in the lab. Although I agree with the first proposition, that mind states *are* essentially brain states, and despite the progress that has been made in the brain sciences, we are still very far from understanding the brain sufficiently at the physical level, both epistemologically and technologically, and therefore we must incorporate other levels of analysis. Also common among certain cognitive scientists is a *methodological individualism that* sees groups of people as mere collections of individuals, therefore arguing that individual behaviour is all there needs to be explained in order to understand cultural phenomena. (Thagard, 2005) I find this view simply naïve, as it fails to appreciate the fact that individuals do not exist outside societies (except for certain rare pathological conditions). Finally, cognitive scientists have largely placed an exclusive emphasis on implicit processes, discarding the importance of explicit ideas and attitudes. (Whitehouse, 2004) But this seems contradictory. Starting from the premise that all normal human minds function on the same principles, and thus form implicit notions in the same ways, then how can we account for the variability that we have set out to explain in the first place if we reject the role of explicit knowledge?

It seems, then, that despite the pioneering efforts of the scholars mentioned above, the two sides have not managed to achieve genuine integration. We are rather left with two different approaches of religion, and the proponents of each are suspicious of and refuse to collaborate with the advocates of the other. To use Thomas E. Lawson's apt metaphor (referring to psychology, ethnography, and history), this situation reminds of a methodological bigamy, when it would ideally be a tripartite free love.

Bigamy is a triangular relationship in which at the apex of the triangle is the bigamist and at the base of the triangle are the objects of bigamy. Essential to this triangular relationship is that one member at the base is ignorant of the existence of the other... This ignorance spells trouble in the short or the long run when the bigamy is discovered. And the consequent knowledge typically alienates all concerned. In tripartite free love, each member of the relationship knows about the other. This does not necessarily spell trouble as long as mutual respect is present. (Lawson, 2004: 2).

The relationship of the study of religion with cognitive science has been a bigamous one. The two members at the base of the triangle either (pretend to) ignore or to treat each other with jealously and contempt. It is obvious that such a relationship cannot be productive.

A Synthetic Approach

What is needed is an approach that will be not only *analytical* but also *synthetic*. In such an approach, the traditional methods of the humanities and those of cognitive science are not antagonistic but complementary. Acknowledging the importance of cognitive factors is consistent with the existence of social factors, and vice versa. All humans share the same basic perceptual and neurological apparatus, which provides a biological commonality to cultural variability. (Thagard, 2005) Religious behaviour does not exist outside culture and cannot exist without individuals. It is constrained by the properties of the human brain as much as it is by historical and socio-political developments. Cognition and knowledge does not exist in isolation in neuronal states inside the brain, nor is it trapped inside individual minds, but is also realized through interaction with other individuals. (Hutchins 1995) A comprehensive study of religion (and all cultural phenomena) should take into account three levels of analysis: the biological/neurological (what goes on in individual brains), the psychological (what goes on in the mind), and the social (what goes on between minds). The common denominator of all three of those domains is the brain, which sets the constraints within which our minds operate and the parameters for socio-cultural forms.

Thus, a Cognitive Study of Religion will be multidisciplinary and methodologically integrative. The traditional textual, historiographic and ethnographic methods should by no means be replaced, but rather complemented by methods, theories, and technology from the natural sciences. Obviously, such an intellectually demanding task must necessarily involve cross-disciplinary collaboration. The anthropologist who works in the field will inform and consult the psychologist working in the lab in order to make better sense of the behaviours they are studying, just like, for example, archaeologists need to consult molecular physicists for the carbon dating of their findings or psychologists, sociologists, mathematicians and computer scientists work together to develop models of social behaviour.

We therefore need a holistic, synthetic and collective approach to the study of religion, which will combine the strengths of various disciplines in order to overcome their weaknesses. Since the goal of such an approach is to move from a hermeneutic to a scientific study of religion, it will have to produce empirically testable and cross-culturally verifiable theories. The study of the universal properties of the human brain can offer a non-ethnocentric framework for the understanding of cultural variability, and consequently provide a common basis for comparative studies of religion. (Martin, 2005) A Cognitive Study of Religion will offer a much more robust theoretical framework of comparative religion, based on theories that can be cross-culturally tested in experimental settings, in the field, and by historical assessment. I believe that its achievements will be commensurate with the combined strengths of its constituting disciplines.

Luther H. Martin has been one of the first scholars of religion to recognize the potential and importance of cognitive studies for the field, and has consistently advocated the need for a scientific, interdisciplinary and explanatory study of religion. Through his teaching, his publications, and his support and guidance to students and young scholars, he has been one of the most inspiring and influential figures in the field. As I am among those who have been lucky to know Luther as a scholar, an author, a teacher and a friend, it is my great privilege to contribute to this volume in his honour.

GENERAL BIBLIOGRAPHY

Adriani, A. 1961. *Repertorio d' arte dell' Egitto greco-romano*. Ser. A, vols I-II. Palermo.

Akira, H., 1990. *A history of indian buddhism: From sakyamuni to early mahayana* (P. Groner, Trans.). Honolulu. University of Hawaii Press.

Alcock, J., 2001. *Triumph of Sociobiology*. New York: Oxford University Press.

Aldridge, A., 2000. *Religion in the Contemporary World: a Sociological Introduction*. Cambridge: Polity Press.

Alexander, P.S., 1992. "'The Parting of the Ways' from the Perspective of Rabbinic Writings". In: J. Dunn (ed.), *Jews and Christians. The Parting of the Ways A.D. 70 to 135. The Second Durham-Tübingen Research Symposium on Earliest Christianity and Judaism (Durham, September, 1989)*. Tübingen, 1-25.

Alanus de Insulis, 1855. *De fide Catholica contra hæreticos sui temporis, praesertim Albigenses*. In: *Patrologia cursus completus*. Ed. Jacques-Paul Migne. *Series Latina* 210: 305—430. Paris: Garnier.

------------------,1966. *Der Anticlaudian oder die Bücher von der himmlischen Erschaffung des Neuen Menschen. Ein Epos des lateinischen Mittelalters*. Trans. William Rath. Stuttgart: Mellinger.

Alföldi - Rosenbaum E., 1976. "Alexandriaca. Studies on Roman Game Counters III", *Chiron* 6: 205-239.

Alles, Gregory, 2008. "Introduction". In: Alles, G. (ed.), *Religious Studies: A Global View*. London, New York: Routledge, 1-13.

Almond, Gabriel A. - Appelby, R. - Scott, Sivan, Emmanuel, 2003. *Strong Religion: The Rise of Fundamentalism around the World*. Chicago: The University of Chicago Press.

Alt, K., 1990. *Philosophie gegen Gnosis. Plotins Polemik in seiner Schrift II 9*. Abhandlungen der geistes- und sozialwissenschaftlichen Klasse. Akademie der Wissenschaften und der Literatur. Jahrgang 1990 no. 7. Stuttgart: Franz Steiner Verlag.

--------, 1990. *Philosophie gegen Gnosis. Plotins Polemik in seiner Schrift II 9*. Abhandlungen der geistes- und sozialwissenschaftlichen Klasse. Akademie der Wissenschaften und der Literatur. Jahrgang 1990 no. 7. Stuttgart: Franz Steiner Verlag.

Ameling, W. (ed.), 2004. *Inscriptiones Judaicae Orientis*. vol. 2: *Kleinasien*, Tübingen.

------------------------, 2007. "Die jüdische Diaspora Kleinasiens und der "epigraphic habit". In: J. Frey - D. Schwartz - S. Gripentrog (eds.), *Jewish Identity in the Greco-Roman World / Jüdische Identität in der griechisch-römischen Welt*. Leiden- Boston, 253-182.

Ammerman, Nancy T., 2002. "Connecting Mainline Protestant Churches with Public Life". In: Robert Wuthnow /John H. Evans, (ed.), *The Quiet Hand of God. Faith-Based Activism and the Public Role of Mainlaine Protestantism*. Berkeley: University of California Press, 129-158.

------------------------, 2005. *Pillars of Faith. American Congregations and Their Partners*. Berkeley: UCP.

Amundsen, L., 1928. "A *Magical Text on an Oslo Ostracon*", Symbolae Osloenses 7: 36-37.

Anderson, Benedict, 1994. "Exodus", *Critical Inquiry* 20: 314-27.

Anderson, C., 1999. *Pain and its ending: The four noble truths in the Theravda buddhist canon*. Richmond: Curzon Press.

Anderson, Hugh, 1976. *The Gospel of Mark* (New Century Bible Commentary). Grand Rapids: Eerdmans.

Ando, Cl., 2008. *The Matters of the Gods. Religion and the Roman Empire*. Berkeley-London.

Angelis, Fr. De - Garstad, B., 2006. "Euhemerus in Context", *Classical Antiquity* 25, 4 (October): 211-242.

Antes, P. - A. W. Geertz - R. R. Warne (ed.), 2004. *New approaches to the study of religion*. Berlin: Walter de Gruyter.

Anthropology. 1966. *Harvard Crimson*. Retrieved 6 August 2008, from http://www.thecrimson.com/article.aspx?ref=247740.

Anttonen, Veikko, 1996. "Rethinking the Sacred: The Notions of the Human Body and Territory in Conceptualizing Religion". In: T. Indinopulos - E. Yonan (eds.), *The Sacred and its Scholars*. Leiden: Brill, 36-64.

------------------------, 1999. "Does the Sacred Make a Difference? Category Formation in Comparative Religion". In: T. Ahlbäck (ed.), *Approaching Religion, Part 1.*, Åbo, Finland: The Donner Institute, 9-24.

------------------------, 2000. "Toward a Cognitive Theory of the Sacred: An Ethnographic Approach". *Folklore: Electronic Journal of Folklore* 14: 41-48.

Appadurai, A., 1996. *Modernity at large: Cultural dimensions of globalization*. Minneapolis: University of Minnesota Press.

Arnal, William, 2007. "The Gospel of Mark as Reflection on Exile and Identity". In: W. Braun - R. T. McCutcheon (eds.), *Introducing Religion: Essays in Honor of Jonathan Z. Smith*. London: Equinox, 57-67.

Armstrong, A. H., 1978. "Gnosis and Greek Philosophy". In: B. Aland (ed.), *Gnosis. Festschrift für Hans Jonas.* Göttingen: Vandenhoeck and Ruprecht, 87-124.

Asad, Talal, 1993. *Genealogies of Religion: Discipline and Reasons of Power in Christianity and Islam.* Baltimore: The Johns Hopkins Press.

Ashton, S.-A., 2001. *Ptolemaic royal sculpture from Egypt: the interaction between Greek and Egyptian traditions.* Oxford.

------------------, 2002. "Ptolemaic royal images". A review of Stanwick, P. E. Portraits of the Ptolemies. Greek kings as Egyptian Pharaohs. *Journal of Roman Archaeology* 17: 543-550.

------------------, 2003. *The last queens of Egypt.* Harlow.

------------------, 2004. "The Egyptian tradition", in: A. Hirst and M. Silk (eds.), *Alexandria Real and Imagined* (Centre for Hellenic Studies, King's College Londomn, Publications 15). Aldershot, HA – Burlington, VT., 15-40.

------------------, 2005. *Roman Egyptomania.* London.

Assmann, J. 2003. *Die mosaische Unterscheidung oder der Preis des Monotheismus.* Munich and Vienna: Carl Hanser Verlag.

Atran, Scott, 1990. *Cognitive Foundations of Natural History: Towards an Anthropology of Science* Cambridge. Cambridge University Press.

------------------, 2002. *In Gods We Trust: The Evolutionary Landscape of Religion.,* Oxford- New York: Oxford University Press.

-------------- - Norenzayan, A., 2004. "Religion's Evolutionary Landscape: Counter-intuition, Commitment, Compassion, Communion". *Behavioral and Brain Sciences* 27: 713-770.

Babha, H., 1994. *The Location of Culture,* London.

Bachelor, S., 1997. *Buddhism without beliefs: A contemporary guide to awakening.* New York: Riverhead Books.

Bagnall, R., 1988. "Greeks and Egyptians: Ethnicity, Status, and Culture", in: Fazzini, R. and Bianchi, R. S. (eds.), *Cleopatra's Egypt. Age of the Ptolemies.* New York: 21-27

Balch, David L. "The Suffering of Isis/Io and Paul's Portrait of Christ Crucified (Gal. 3:1): Frescoes in Pompeian and Roman Houses and in the Temple of Isis in Pompeii". *Journal of Religion* 83, no. 1 (2003): 24–55.

Bambrough, Renford, 1980. "Editorial: Subject and Epithet". *Philosophy* 55 (213): 289-290.

Bandaranayake, Senake, 1974. *Sinhalese Monastic Architecture.* Leiden: Brill.

Bandura, Albert, 1990. Mechanisms of Moral Disengagement In: W. Reich (Ed.), *Origins of Terrorism: Psychologies, Ideologies, Theologies, States of Mind.*, Cambridge: Cambridge University Press.

Barkow, J. H. - Cosmides, L. - Tooby, J., 1992. *The adapted mind: Evolutionary psychology and the generation of culture*. New York: Oxford University Press.

Baron-Cohen, S., 1995. *Mindblindness: An Essay on Autism and Theory of Mind.* Cambridge,MA: MIT Press.

Barrett, J. L. - Frank C. Keil, 1996. "Conceptualizing Nonnatural Entity: Anthropomorphism in God Concepts". *Cognitive psychology* 31: 219-247.

Barrett, J. L., 1999. "Theological Correctness: Cognitive Constraints and the Study of Religion". *Method and Theory in the Study of Religion* 11, 325-339.

------------------, 2000. "Exploring the Natural Foundations of Religion". *Trends in Cognitive Science* 4: 29-34.

------------------, 2001. "How Ordinary Cognition Informs Petitionary Prayer". *Journal of Cognition and Culture* 1: 259-269.

------------------, 2002. "Dumb Gods, Petitionary Prayer and the Cognitive Science of Religion". In: I. Pyysiäinen - V. Anttonen (eds.), *Current Approaches in the Cognitive Science of Religion*. London - New York: Continuum, 93-109.

------------------, 2004. *Why Would Anyone Believe in God?* Walnut Creek, CA: Alta Mira Press,

------------------ - Orman, B. van, 1996. "The Effect of the Use of Images in Worship on God Concepts". *Journal of Psychology and Christianity* 15: 38-45.

------------------ - Nyhof, M.A., 2001. "Spreading Non-natural Concepts: The Role of Intuitive Conceptual Structures in Memory and Transmission of Cultural Materials". *Journal of Cognition and Culture* 1,1: 69-100.

Beard, M., 1991. "Writing and Religion: Ancient Literacy and the Function of the Written Word in Roman Religion". In: J. H. Humphrey (ed.), *Literacy in the Roman World.*, Ann Arbor, MI: The University of Michigan, 35-58.

Barton, Stephen C., 1992. "The Communal Dimension of Earliest Christianity: A Critical Survey of the Field", *Journal of Theological Studies* 43 : 399-427.

Baumgarten, Helga, 2006. *Hamas. Der politische Islam in Palästina*. München: Diederichs.

Becher, T., Trowler, P. R., 2001. *Academic tribes and territories: Intellectual enquiry and the culture of disciplines*, second edition. Open University Press, Buckingham, UK.

Bechtel, William, 2007. "Reducing Psychology while Maintaining its Autonomy via Mechanistic Explanations:. in: M. Schouten - H. Looren de Jong (eds.), *The Matter of the Mind: Philosophical Essays on Psychology, Neuroscience and Reduction*, Oxford: Basil Blackwell, 172-198.

------------------ - Robert McCauley, 1999. "Heuristic Identity Theory (or Back to the Future): the Mind-Body Problem Against the Background of Research Strategies in Cognitive Neuroscience". *Proceedings of the 21st Annual Meeting of the Cognitive Science Society*. Mahwah, NJ: Lawrence Erlbaum Associates, 67-72.

Beck, H. G. 1993. *Vom Umgang mit Ketzern. Der Glaube der kleinen Leute und die Macht der Theologen*. Munich: Verlag C.H.Beck.

Beck, R., 2004. *Beck on Mithraism: Collected Works, with New Essays* (Ashgate Contemporary Thinkers on Religion Series. Ashgate). Aldershot.

------------, 2006. *The Religion of the Mithras Cult in the Roman Empire: Mysteries of the Unconquered Sun*. Oxford - New York: Oxford University Press.

Becker, J. O., 2004. *Deep listeners: Music, emotion, and trancing*. Bloomington: Indiana University Press.

Behr C. A. 1981. *Aelius Aristides. The Complete Works, II, Orations XVII-LIII*. Leiden.

Bell, C. M., 1997. *Ritual: Perspectives and dimensions*. New York: Oxford University Press.

Bendlin, A., 2000. "Looking Beyond the Civic Compromise: Religious Pluralism in Late Republican Rome". In: E. Bispham - C. Smith (eds.), *Religion in Archaic and Republican Rome and Italy: Evidence and Experience*. Edinburgh: Edinburgh University Press,. 115-135, and 167-171.

Bentley, G. E. Jr. 2004. *Blake Records*. 2nd edition. New Haven and London: Yale University Press.

Benzon, W., 2001. *Beethoven's anvil : Music in mind and culture*. New York: Basic Books.

Berkwitz, Stephen, 2007. *The History of the Buddha's Relic Shrine. A Translation of the Sinhala Thūpavamsa*. New York: Oxford University

Press.

Berlin, Brent - Paul Kay, 1969. *Basic Color Terms: Their Universality and Evolution.* Berkeley: University of California Press.

Bernabé, A., 1997. 'Elementos orientales en el orfismo', in: *Actas e/el Congreso Español de Antiguo Oriente Próximo, El Mediterráneo en la Antiguedad, Oriente y Occidente, Madrid,octubre de 1997.*

---------------, 2003. *Hieros logos. Poesiá órfica sobre dos dioses et alma y el más aleá.* Madrid.

Berner, U., 1981. "Universalgeschichte und kreative Hermeneutik. Reflexionen anhand des Werkes von Mircea Eliade". *Saeculum* 32: 221-41

---------------, 1992. "Religio und Superstitio. Betrachtungen zur römischen Religionsgeschichte". In: Theo Sundermeier (ed.), *Den Fremden wahrnehmen. Bausteine für eine Xenologie.* Gütersloh: Gütersloher Verlagshaus Gerd Mohn, 45-64

---------------,1997. Auslegung der Genesis im Spannungsfeld zwischen Monotheismus und Dualismus". In: E. Sedlmayer (ed.), *Schlüsselworte der Genesis II. Polaritäten - Kräfte Gleichgewichte.* Berlin: Dreieck Verlag der Guardini-Stiftung, 279-294.

---------------, 2006. "'Wahr' oder 'unwahr' – die Mosaische Unterscheidung. Die Diskussion um den Monotheismus des Mose und die Thesen Jan Assmanns". In *Welt und Umwelt der Bibel* 41/3: 46-49.

---------------, 2007. "Antike und Christentum im Mittelalter. Alanus ab Insulis als Dichter und Theologe". In: Christoph Auffarth (ed.) *Religiöser Pluralismus im Mittelalter? Besichtigung einer Epoche der europäischen Religionsgeschichte.* Berlin: LIT Verlag, 25-37.

---------------, 2008. "Verlauf und Ertrag der Monotheismusdebatte ausreligionswissenschaftlicher Sicht". In: Lukas Bormann (ed.), *Schöpfung, Monotheismus und fremde Religionen.* Neukirchen-Vluyn: Neukirchener Verlag, 37-61.

Bernstein, Alan E., 1992. *The Formation of Hell. Death and Retribution in the Ancient and Early Christian Worlds.* Ithaka, London: Cornell University Press.

Betz, H.D. (ed.), 1992. *The Greek Magical Papyri in Translation Including the Demotic Spells*^{2}, London.

------------ (ed.), 2003. *The "Mithras Liturgy". Text, Translation and Commentary,* Tübingen: Mohr Verlag.

Bianchi, Ugo, 1961. "Après Marbourg (Petit discours sur la méthode)". *Numen,* 8 (1), 64-78 = Repr. in: id., 1979. *Saggi di metodologia della storia delle religioni.* Roma: Ateneo & Bizzarri, 229-246.

----------------, 1964. "Lo studio delle religioni non cristiane", in: *Le missioni e le religioni non cristiane*, Milano, 121-137 = In: id., 1979. *Tra mondo e salvezza: Problemi del Cristianesimo di oggi* (Vita e pensiero). Milano, 69-96.

----------------, 1965. "L'XI Congresso di storia delle religioni", *Rivista di storia e letteratura religiosa* 1: 529-530.

----------------, 1975. *La religione greca.* Torino: UTET.

----------------, 1975. *The History of Religions.* Leiden: Brill

----------------, 1980. "Iside dea misterica. Quando?". In: *Perrenitas. Studi in onore di A. Brelich. Promossi della Cattedra di Religioni del mondo classico dell' Università di Roma.* Roma, 9-36.

----------------, 1987. "History of Religions". In: M. Eliade (ed.), *Encyclopedia of Religion*, vol. 6. New York: Macmillan, 399-408.

Bickle, John, 2003. *Philosophy and Neuroscience: A Ruthlessly Reductive Account (Studies in Brain and Mind).* Boston: Kluwer Academic Publishers,.

Bij de Vaate, Alice J. - Henten, J.W. van, 1996. "Jewish or Non-Jewish? Some Remarks on the Identification of Jewish Inscriptions from Asia Minor", *BO* 53: 16-28.

Bilabel, C. F., 1929. 'Die gräko-ägyptische Feste', *N. Heidelb. Jahrbuch*: 1-51.

Blackmore, S., 1999. *The meme Machine.* New York: Oxford University Press.

Blake, William. 1989. *The Complete Poems.* ed. W.H.Stevenson. 4th ed. London: Longman.

Bleeker, Claas Jouco, 1960. "The Future Task of the History of Religions". *Numen* 7: 221-234.

Bloch, Maurice, "Why Religion is Nothing Special But is Central", *Philosophical Transactions of the Royal Society* 363: 2055-2061.

Bloch, M., Parry, J. P., 1982. *Death and the Regeneration of life.* Cambridge- New York: Cambridge University Press.

Bludau, A., 1925. *Die Schriftfälschungen der Häretiker: Ein Beitrag zur Textkritik der Bibel.* Münster: Aschendorff.

Boas, F., 1938. *Mind of primitive man* (2nd Ed.). New York: Macmillan.

Boffo, L., 1996. *Iscrizioni greche e latine per lo studio della Bibbia.* Brescia.

Böhl, E., 1881, "Alte christliche Inschriften nach dem Text der Septuaginta", *Theologische Studien und Kritiken*, 54: 692-713.

Böhlig, A. 1989. "Zur Struktur des gnostischen Denkens". In: *Gnosis und*

Synkretismus. Gesammelte Aufsätze zur spätantiken Religionsgeschichte. Pt. 1, Wissenschaftliche Untersuchungen zum Neuen Testament 47. Tübingen: Mohr, 3-24.

Bollas, Christopher, 1995. *Cracking Up: The Work of Unconscious Experience*. New York: Hill and Wang.

Bonacursus ex hæretico Catholicus. 1855. *Vita hæreticorum*. In: *Patrologia cursus completus*. Ed. Jacques-Paul Migne. *Series Latina* 204: 775-92. Paris: Garnier.

Bonner: C., 1950. *Studies in Magical Amulets Chiefly Graeco -Egyptian*, Ann Arbor – London.

Bonner, Michael – Mine, Ener – Amy, Singer (eds.), 2003. *Poverty and Charity in Middle Eastern Contexts*. New York: SUNY.

Boon,James A., 1972. "Further Operations of Culture in Anthropology: A Synthesis of and for Debate", *Social Science Quarterly* 52: 221-52 = Louis Schneider and Charles M. Bonjean (eds.), 1973. *The Idea of Culture in the Social Sciences*. Cambridge University Press.

Borchers, K., 1980. *Mythos und Gnosis im Werk Thomas Manns. Eine religionswissenschaftliche Untersuchung*. Freiburg/Br.: Hochschulverlag.

Borst, A., 1990. "Die dualistische Häresie im Mittelalter". In: *Barbaren, Ketzer und Artisten. Welten des Mittelalters*. 2nd ed. Munich: R.Piper, 199-231.

-----------, 1992. *Die Katharer*. Freiburg: Herder.

Bosworth, Brian, 1999. "Augustus, the *Res Gestae* and Hellenistic Theories of Apotheosis". *Journal of Roman Studies*, 89, 1 : 1–18.

Bottéro, J., 1991. "U anthropogonie mésopotamienne ct 1' élément divin en 1' homme". In: Ph. Borgeaud (ed.), *Orphism et Orphée. En l' honneur de Jean Ruchardt/ texts réunis et éditès*. Genève, 221-225.

Bottomore, T. B., 1987. *Sociology: a guide to problems and literature*. London: Allen & Unwin.

Bouché-Leclercq A., 1903-1907. *Histoire des* Lagides, vols. I- IV. Paris.

Bowersock, G.W. *Hellenism in Late Antiquity*. Cambridge: Cambridge University Press, 1990.

Bowie, Ewen, 1994. "The Readership of Greek Novels in the Ancient World". In: James Tatum (ed.), *In Search of the Ancient Novel*. Baltimore - London: Johns Hopkins University Press, 435–459.

Boyd, P., Richerson, P. J., 1985. *Culture and the evolutionary process*. Chicago: University of Chicago Press.

Boyer, P., 1990. *Tradition as truth and communication: a cognitive description of traditional discourse*. Cambridge: Cambridge University Press.

------------, 1994. *The naturalness of religious ideas: a cognitive theory of religion*. Berkeley - London: University of California Press.

------------, 2000. "Functional Origins of Religious Concepts: Ontological and Strategic Selection in Evolved Minds". *Journal of the Royal Anthropological Institute* (N.S.) 6: 195-214.

------------, 2001. *Religion Explained. The Evolutionary Origin of Religious Thought*. Basic Books, New York = 2002. *Religion explained: the human instincts that fashion gods, spirits and ancestors*, London: Vintage.

------------, 2002. "Why do gods and spirits matter at all?". In: I. Pyysiäinen - V. Anttonen (Eds.), *Current approaches in the cognitive science of religion*. London - New York: Continuum, 68-92

------------ - Ramble, C., 2001. "Cognitive Templates for Religious Concepts: Cross-Cultural Evidence for Recall of Counter-Intuitive Representations:. *Cognitive Science* 25, 4: 535-564.

------------ - Lienard, P., 2006a. "Why ritualized behavior? Precaution systems and action parsing in developmental, pathological, and cultural rituals". *Behavioral and Brain Sciences* 29: 595-650.

------------, 2006b. "Whence collective ritual? A cultural selection model of ritualized Behavior". *American Anthropologist*, 108, 4: 814-827.

Boynton, S., 2000. Training for the liturgy as a form of monastic education. In: J. Ferzeco and C. Muessig (Eds.), *Medieval monastic education*. Leicester University Press, London, 7-20.

------------, 2007. "Prayer as liturgical performance in eleventh and twelfth-century monastic psalters". *Speculum* 82, 4: 896-931.

------------ - Cochelin, I., 2006. "The sociomusical role of child oblates at the abbey of Cluny in the eleventh century". In: S. Boynton - R.-M. Kok (eds.), *Musical childhoods & the cultures of youth*. Middleton, CT.: Wesleyan University Press, 3-24.

Braun, Willi – McCutcheon, Russell (eds.), 2000. *Guide to the Study of Religion*. London: Cassell Academic Press.

Brelich, Angelo, 1960. "Ai margini del 10° Congresso Internazionale di Storia delle Religioni (Marburgo, 11 e 17-9-1960)". *Studi e materiali di storia delle religioni* 31: 121-128.

Breytenbach, J. Cilliers, 1990. "Paul's Proclamation and God's Thriambos. (Notes on 2 Corinthians 2:14–16b)". *Neotestamentica* 24: 257–71.

Bricault, L., 2001. *Atlas de la diffusion des culte isiaques (Ive s. av. J. - C. – IVe apr. J. - C.)* (Mémoires de l'Académie des Inscriptions et Belles-Lettres). Paris.

-------------- (ed.), 2005. *Recueil des inscriptions concernant les cultes isiaques* (Mémoires de l'Académie des Inscriptions et Belles-Lettres 31), vols. I-III, Paris.

--------------------, 2006. *Isis, Dame des flots* (Aegyptiaca Leodiensia, 7). Liége.

-------------- (ed.), 2008. *Bibliotheca Isiaca*, I. Bordeux.

Brilliant, Richard, 1999. "'Let the Trumpets Roar!' The Roman Triumph". In: Bettina Bergmann and Christine Kondoleon (eds.), *The Art of Ancient Spectacle*. Studies in the History of Art 56. New Haven - London: Yale University Press, 221–229.

Britton, Ronald, 1998. *Belief and Imagination: Explorations in Psychoanalysis*. London: Routledge.

Broadbent, Donald, 1958. *Perception and Communication*. Oxford: Pergamon.

Brodie, Th. L., 2001. "Towards Tracing the Gospel's Literary Indebtedness to the Epistles". In: D. R. MacDonald, *Mimesis and Intertextuality in Antiquity and Christianity*, Harrisburg, 104-116.

Bromberg, Philip M., 2006. *Awakening the Dreamer: Clinical Journeys*. New Jersey: The Analytic Press.

Brown, Cecil H., 1984, *Language and Living Things: Uniformities in Folk Classification and Naming*. New Brunswick, NJ.: Rutgers University Press.

Brown, D.E., 1991. *Human universal*. New York: McGaw-Hill.

Brown, Peter, 1981. *The Cult of the Saints: Its Rise and Function in Latin Christianity*. Chicago: The University of Chicago Press.

Brown, Tr. S., 1964. "Euhemerus and the Historians", *HThR* 39, (October): 259-274.

Brown, W. E., 1955. "Some Hellenisitc Utopias", *The Classical Weekly* 48, 5 (January): 57-62.

Brugger, Winfried, 2001. *Einführung in das öffentliche Recht der USA*. 2. A. München: C. H. Beck.

Buser, T., 1968. "Gauguin's Religion." *Art Journal* 27: 375-80.

Bryson, B., 1996. "Anything but Heavy Metal: Symbolic exclusion and musical dislikes." *American Sociological Review* 61: 884-899.

Bulbulia, J., 2008. "Meme infection or religious niche construction? An adaptationist alternative to the cultural maladaptationist hypothesis". *Method and Theory in the Study of Religion* 20: 67-107.

Burkert, Walter, 1983. *Homo Necans: The Anthropology of Ancient Greek Sacrificial Ritual and Myth*. Translated by Peter Bing. Berkeley: University of California Press.

----------------------, 1985. *Greek Religion*. Cambridge, MA: Harvard University Press.

----------------------, 1987. *Ancient Mystery Cults*. Cambridge, MA: Harvard University Press,

----------------------, 2003. "Orpheus und Ägypten". In: *id., Die Griechen und die Orient Von Homer bis zu den Magiern*. München: C. H. Beck Verlag, 79-106.

Burris, John P., 2005. "Comparative-historical Method [Further Considerations]". In: L. Jones, (ed.), *Encyclopedia of Religion*, second edition. , Detroit: Macmillan, 1871-1873.

Burton, A., 1972. *Diodorus of Siculus, Book I. A Commentary* (EPRO, 29). Leiden: Brill.

Buss, D., 1999. Evolutionary Psychology: The New Science of the Mind. Boston: Allyn and Bacon.

Cadge, Wendy, 2008. "De Facto Congregationalism and the Religious Organizations of Post-1965 Immigrants to the United States: A Revised Approach", *JAAR* 76: 344-374.

Calder, William M., 1994. "Epilogue". In: Calder W. M. III, (ed.), *Usener und Wilamowitz. Ein Briefwechsel 1870-1905*. Stuttgart-Leipzig, 71-78.

Calderini, Ar., 1935. *Dizionario dei nomi geografici e topografici delll' Egitto greco-romano* , vol. I. Cairo.

Carruthers, M., 1998. The Craft of Thought: Meditation, Rhetoric, and the making of Images, 400-1200. New York: Cambridge University Press.

Carlson-Thies, Stanley W., 2003. "Charitable Choice: Bringing Religion Back into American Welfare". In: H. Heclo/ W.M. McClay (eds.), *Religion Returns to the Public Square: Faith and Policy in America*. Washington, D. C.: Woodrow Wilson Press; Baltimore, Md: John Hopkins University Press, 269-297.

Carter, J. R., 1993. "Music in the Theravāda Buddhist heritage: In chant, in song, in Sri Lanka". In: *On understanding Buddhists: Essays on the*

Theravāda tradition in Sri Lanka., Albany, NY: State University of New York Press, 133-152.

Carus, P., 1882. *Lieder eines Buddhisten.*, Dresden: R. von Grumbkow.

Carus, P., 1886. *The Principles of Art, from the Standpoint of Monism and Meliorism.* Boston: Industrial Art Teacher's Association.

------------, 1888. "Spiritism and immortality". *Open Court* 2, 68: 1360-1362.

-----------,1895a. *The gospel of Buddha, according to old records.* Retrieved September 19, 2008, from http://books.google.com/ books?id=yNbvuMZNRpoC&dq=carus+gospel+buddha&source=gbs_summary_s&cad=0.

------------, 1895b. "The significance of music". *Monist* 5: 401-407.

------------, 1896a. "Goethe a Buddhist". *Open Court* 10, 445: 4832-4837.

------------, 1896b. "Buddhism and the Religion of Science". *Open Court* 10, 446: 4844-4845.

------------, 1896c. *The religion of science.* Retrieved November 12, 2008, from http://books.google.com/books?id= vPFqRlCDyMMC& printsec= frontcover&dq=carus+religion+science#PPP1,M1.

------------, 1899. *Sacred tunes for the consecration of life: Hymns of the religion of science.* Chicago: Open Court.

------------, 1900. "Popular music". *Open Court* 14, 2: 122-123.

------------, 1904. "Three buddhist stanzas, done into English verse and set to music". *Open Court* 18, 10: 625-628.

------------, 1905. "The three characteristics". *Open Court* 19: 563-567.

------------, 1906. "Music in education". *Open Court* 20: 311-313.

------------, 1907. "A new system of notation for violin music". *Open Court* 21, 10: 584-591.

Casadio, Giovanni, 2005. "Historiography: Western Studies [Further Considerations]". In: L. Jones, (ed.), *Encyclopedia of Religion*, second edition. Detroit: Macmillan, 4042-4052.

----------------------, 2007. "Comparative Religion Scholars in Dialogue: New Evidence from Letters addressed by Mircea Eliade to Ugo Bianchi'. *Euresis. Cahiers roumains d'études littéraires et culturelles* 4 (3-4), 121-135.

----------------------, 2009. "Ancient Mystic Religion: The Emergence of a New Paradigm from A. D. Nock to Ugo Bianchi". *Archaeus* 13: 35-94.

Casanova, José, 1994. *Public Religions in the Modern World.* Chicago: University of Chicago Press.

----------------------, 2007. "Die religiöse Lage in Europa". In: Hans Joas & Klaus Wiegandt (eds.), *Säkularisierung und die Weltreligionen.* Frankfurt/M.: Fischer, 322–357.

Casarico, I., 1981. "Note su alcune feste nell'Egitto tolemaico e romano", *Aegyptus* 61: 121-142.

Casebeer, William D., 2003. "Moral Cognition and Its Neural Constituents". *Nature Reviews – Neuroscience* 4: 841-847.

Cassian, J., 1997. *The conferences.* Mahwah, N.J.: Newman Press

Chamoux, Fr., 2003. *Hellenistic Civilization.* transl. by M. Roussel-Marg. Rousssel. Malden, Ma-Oxford.

Chaniotis, A., 2004 (2003). 'The Divinity of Hellenistic Rulers'. In: A. Erskine (ed.), *A Companion to the Hellenistic World.* Blackwell Companions to the Ancient World. Maldem-Oxford (reprint), 431-445.

Chaves, Mark, 2004. *Congregations in America.* Cambridge (Mass.): Harvard UP. 2004.

Chehab, Zaki, 2007. *Inside Hamas. The Untold Story of Militants, Martyrs and Spies* . London: Tauris.

Chausse, M. de La, 1700. *Le gemme antiche figurate di Michelangelo Causeo de la Chausse.* Roma.

Chomsky, Noam, 1957. *Syntactic Structures.* The Hague: Mouton.

----------------------, 1959. "A Review of B. F. Skinner's *Verbal Behavior*". *Language* 35, 1: 26-58.

----------------------, 1986. *Knowledge of Language: Its Nature, Origin, and Use.* New York: Praeger.

Clark, A. L., 2004. "Testing the two modes theory: Christian practice in the later middle ages". In: H. Whitehouse - L. H. Martin (eds.), *Theorizing religions past: Archaeology, History, and Cognition.* Waltnut Creek, CA: AltaMira Press, 125-142.

----------------, 2007. "Why all the fuss about the mind? A medievalist's perspective on cognitive theory". In: R. Fulton, R. - B. Holsinger (eds.), *History in the comic mode: Medieval communities and the matter of the person.* New York: Columbia University Press, 170-181.

Clark, Gillian, 2004. *Christianity and Roman society.* New York: Cambridge University Press.

Clarysse, W., 1980. "Philadelpheia and the Memphites in den Zenon Archive". In: *Studies in Ptolemaic Memphis* (Studia Hellenistica, 24). Leuven, 91-122.

----------------- - Vandorpe, K. 1995. *Zenon, un homme d'affaires grec à* l' ombre des Pyramide. Leuven.

Clasquin, M., 1992. "Gnosticism in Contemporary Religious Movements. Some Terminological and Paradigmatic Considerations". *Journal for the Study of Religion* 5: 41-55.

Cohen, A.,1989. *The Symbolic Construction of Community*. London-New York.

Coleman, John, S. S.J., 2003. "American Catholicism, Catholic Charities, and Welfare Reform". In: Hugh Heclo -Wilfred M. McClay (eds), *Religion Returns to the Public Square. Faith and Policy in America*. Washington, D.C.: Woodrow Wilson Center Press, 239-267.

Collins, R., 2004. *Interaction Ritual Chains*. Princeton, N.J.: Princeton University Press.

Collins, S., 1990. "On the very idea of the pali Canon". *Journal of the Pali Text Society*, XV: 89-126.

-------------, 2003. "What is Literature in Pali". In: S. Pollock (ed.), *Literary Cultures in History: Reconstructions from South Asia*. Berkeley, University of California Press, 649-688.

Cooke, D., 1927. "Euhemerism: A Mediaeval Interpretation of Classical Paganism". *Speculum* II 4 (October): 396-410.

Cosmides, L. - J. Tooby, 1997. "Evolutionary psychology: A Primer. Retrieved November 20, 2008, from Center for Evolutionary Psychology".Web site: http://www.psych.ucsb.edu/ research/cep/ primer. html

Couliano, I.P., 1984. "The Gnostic Revenge. Gnosticism and Romantic Literature". In: J. Taubes (ed.), *Gnosis und Politik*. Religionstheorie und politische Theorie 2. Munich: Wilhelm Fink Verlag 290-306.

---------------, 1992. *The Tree of Gnosis. Gnostic Mythology from Early Christianity to Modern Nihilism*. Trans. H.S. Wiesner and the author. San Francisco: Harper Collins.

Cozolino, Louis J., 2002. *The Neuroscience of Psychotherapy: Building and Rebuilding the Human Brain*. New York & London: W. W. Norton & Company, 2002.

Craver, Carl - William Bechtel, 2007. "Top-Down Causation Without Top-Down Causes". *Biology and Philosophy* 22: 547–63.

Croitoru, Joseph, 2007. *Hamas. Der islamische Kampf um Palästina*. München: C. H. Beck.

Curtius, E. R., 1938. "Zur Literarästhetik des Mittelalters II". *Zeitschrift für romanische Philologie* 58: 129-232.

----------------------,1993. *Europäische Literatur und lateinisches Mittelalter*. 11th ed. Tübingen and Basel: Francke Verlag.

Damasio, Antonio R., 1994. *Descartes' Error: Emotion, Reason, and the Human Brain.* New York: Avon Books.

Damon, S. F., 1988. *A Blake Dictionary. The Ideas and Symbols of William Blake.* Rev. ed. Hanover/London: Brown University Press.

Damrosch, L. Jr., 1980. *Symbol and Truth in Blake's Myth.* Princeton: Princeton University Press.

Daniel – Maltomini 1990-1992: R.W. Daniel, F. Maltomini (eds.), Supplementum Magicum, 2 voll., Papyrologica Coloniensia 16, 1-2, Opladen 1990.

Davie, Grace, 1994. *Religion in Britain since 1945: Believing without Belonging.* Oxford: Blackwell.

----------------, 2000. *Religion in Modern Europe. A Memory Mutate.* Oxford: Oxford University Press.

--------------, 2006. "Vicarious Religion: A Methodological Challenge". In: Nancy Ammerman (ed.), *Everyday Religion: Observing Modern Religious Lives.* New York: Oxford University Press, 21–37.

Dattari, G., 1901. Monete Imperiali Greche. Numi Augg. Alexandrini. Catalogo della Collezione G. Dattari. Compilato dal Propietario. Cairo.

Davies, Stevan, 1996. "Mark's Use of the Gospel of Thomas", *Neotestamentica* 30: 307-334.

Dawkins, R., 1976. *The selfish gene.* New York: Oxford University Press.

----------------, 1982. *The extended phenotype: The gene as the unit of selection.* Oxford, San Francisco: Freeman.

----------------, 1986. *The Blind Watchmaker.* New York: W. W. Norton & Company.

----------------, 2006a. *The God Delusion.* Boston: Houghton Mifflin Co.

----------------, 2006b. *The God delusion.* London: Bantam Press.

Deacon, Terrence, 1997. *The Symbolic Species: The Co-Evolution of Language and the Human Brain.* London: Penguin.

Decaroli, R., 2004. *Haunting the Buddha: Indian popular religions and the formation of Buddhism.* New York: Oxford University Press.

DeConick, April D., 2005. *Recovering the Original Gospel of Thomas. A History of the Gospel and its Growth.* London - New York: T. and T. Clark Publishers.

----------------------, 2007. *The Original Gospel of Thomas in Translation.* London - New York: T. and T. Clark Publishers.

Deen Schildgen, Brenda, 1999. *Power and Prejudice: The Reception of the Gospel of Mark*. Detroit: Wayne State University Press.

Deissmann, A., 1905, "Verkannte Bibelzitate in syrischen und mesopotamischen Inschriften", *Philologus* 64 (N.F. 18): 475-478.

DeLoache, Judy- Alma Gottlieb (eds.), 2000. *A World of Babies: Imagined Childcare Guides for Seven Societies*, Cambridge: Cambridge University Press.

Dennett, D. C., 1987. *The Intentional Stance*. CambridgemMA: The MIT Press.

——————————, 1992. *Consciousness explained*. London: Allen Lane.

——————————, 1995. *Darwin's Dangerous Idea*. New York: Simon & Schuster.

——————————, 1991. *Consciousness Explained*. Boston, Toronto, London: Little, Brown and Company.

——————————, 2006. *Breaking the Spell: Religion as a Natural Phenomenon*. New York: Viking.

——————————, 2007. *Breaking the spell: religion as a natural phenomenon*, London: Penguin.

——————————, 2008. *Public Education, Knowledge, and Visions for non-toxic Religion*. Chicago, Il.: *American Academy of Religion*

Detienne, M., 1994. *The Gardens of Adonis. Spices in Greeek Mythology*. transl. by J. Lloyd, with an Introduction of J.-P. Vernant. Princeton, N.J.: Princeton University Press.

Dharmapala, A., 1897. "Is there more than one Buddhism? In reply to the Rev. Dr. Ellinwood". *The Open Court* 11: 82-84.

Diez de Velasco, F. – Molinero Polo, M. A., 1994. "Hellenoaegyptiaca 1: Influences égyptiennes dans l' imaginaire gree de la mort: quelques exemples d' un emprunt supposé (Diodore 1, 92, 1-4: 1, 96,4-8)", *Kernos* 7: 75-93.

Donadoni, S., 1964. "Due testi oracolari copti". In: AA.VV., *Synteleia* V, Napoli: . Arangio Ruiz, 286 ss. = Id., 1989. *Cultura dell' Antico Egitto. Scritti di Sergio F. Donadoni*, Roma, 531 ss.

Donahue, John R., 1992. "The Quest for the Community of Mark's Gospel". In: Frans van Segbroeck et al. (eds.), *The Four Gospels, 1992: Festschrift Frans Neirynck* . vol. 2. Leuven: Leuven University Press, 1992), 819-34.

——————————, 1995. "Windows and Mirrors: The Setting of Mark's Gospel", *Catholic Biblical Quarterly* 57 (1995): 1-26.

Donovan, J., 1990. *Gnosticism in Modern Literature. A Study of the Selected Works of Camus, Sartre, Hesse, and Kafka*. New York and London:

Garland Publishing.

Douglas, M., 1986. *How institutions think*. Syracuse, NY: Syracuse University Press.

Drees, W., 1996. *Religion, Science and Naturalism.*, Cambridge: Cambridge University Press.

Drescher J., 1950. "A Coptic Amulet". In: AA.VV. *Coptic Studies in Honor of Walter Ewing Crum*, Boston, MA, 265-270.

Dumoulin, Heinrich, 1988. *Zen Buddhism: A History*. Vol. 1. New York: Macmillan Publishing Company.

Dunand, F. 1973a: *Le culte d' Isis dans la bassin oriental de la Méditerranée*. vol. I: *Le culte d'Isis et les Ptolémées* (EPRO, 26). Leiden: Brill.

---------------, 1973b. *Le culte d' Isis dans le bassin oriental de la Méditerranée*. vol. II: *Le culte d' Isis en Grèce* (EPRO, 26). Leiden: Brill.

---------------, 1979. *Religion populaire en Égypte romaine* (EPRO, 76). Leiden: Brill.

---------------, 1983a. "Culte royal et culte imperial en Égypte. Continuités et ruptures". In: G. Grimm - H. Heinen - E. Winter [Hrsg.], *Das Römisch-Byzantinische Ägypten. Akten des internationalen Symposions 26-30 September 1978 in Trier* (Aegyptiaca Treverensia. Trierer Studien zum Griechisch- Römischen Ägypten, vol. 2). Mainz am Rhein, 47-58.

---------------, 1983b. "Cultes Égyptiens hors d' Égypte". In: E. Van't Dack-P. Van Dessel-W. Van Gucht (eds.), *Egypt and the Hellenistic World. Proceedings of the International Colloquium, Leuven 24-26 May 1982* (Studia Hellenistica, 27). Leuven, 75-98.

---------------, 1991. "L' Égypte Ptolémaïque et Romaine". In: *ead.* – Chr. Zivie-Coche (eds.), *Dieux et Hommes en Égypte (3000 av. J.- C. 395 apr. J.-C.). Anthropologie religieuse*. Paris, 197-329.

---------------, 2000 (1992). 'The Factory of Gods". In: Chr. Jacob-Fr. de Polignac, *Alexandria, Third Century B. C. The Knowledge of the World in a Single City*. Alexandria, 158-161.

---------------, 2004 (2003). 'The Religious System in Alexandria". In: A. Erskine (ed.), *A Companion to the Hellenistic World*, (Blackwell Companion to the Ancient World). Malden, Ma-Oxford, 253- 263.

Dunbar, R., 1996. *Grooming, Gossip, and the Evolution of Language*. Cambridge, MA: Harvard University Press.

---------------, 2003. "Social Brain: Mind, Language and Society in evolutionary Perspective". *Annual Review of Anthropology* 32: 163-181.

------------- - Barrett, L. - Lycett, J., 2005. *Evolutionary Psychology: A beginners Guide*. Oxford: Oneworld Publications.

Durand, J.-L., Scheid, J., 1994. " 'Rites' et 'Religion'. Remarques sur certains préjugés des historiens de la religion des Grecs et des Romains". *Archives de sciences sociales des religions* 85: 23-43.

Durkheim, Émile, 1954. *The Elementary Forms of the Religious Life*. trans. J.W. Swain. Glencoe, Il.: The Free Press.

Ebach, H., 1990. "Euhemerismus". In: H. Cancik-B. Gladigow - M. Laubscher [eds.], *Handbuch religionswissenschaftlichen Grundbegriffe*, vol. II. Stuttgart-Berlin-Köln, 365-368.

Ebertz, Michael N. ,1997. *Kirche im Gegenwind. Zum Umbruch der religiösen Landschaft*. Freiburg: Herder.

Edgar, C. C. (ed.), 1931. *Zenon Papyri*, vols I-IV (Catalogue général des auites égyptiennes du monde du Caïre,). Cairo.

Edmondson, Jonathan C., 1999. "The Cultural Politics of Public Spectacle in Rome and the Greek East, 167–166 B.C.E.". In: Bettina Bergmann and Christine Kondoleon (eds.), *The Art of Ancient Spectacle* (Studies in the History of Art 56). New Haven- London: Yale University Press, 77–95

Edsman, Carl-Martin, 1974. "Theology or Religious Studies?" *Religion* 4: 59-74.

Eickelman, Dale F. - Jon W. Anderson (eds.), 2003. *New Media in the Muslim World. The Emerging Public Sphere*. Bloomington: Indiana UP.

Empereur, J.-Y., 2002. *Alexandria : past, present and future*. London.

--------------------, 2002. *Alexandrina 2*. Le Caire : Institut francais d'archeologie orientale. Cairo.

Eliade, M., 1958. *Patterns in comparative religion*. trans. R. Sheed. New York: World.

---------------, 1959. *Sacred and the profane*. Trans. by W. R Trask. New York: Harcourt Brace Jovanovich.

-------------, 1968. *Myth and reality*. Trans. by W. R. Trask. New York: Harper & Row.

------------, 1987. *The Sacred and the Profane*. London: Harcourt Brace & Company.

------------, 1989. *The myth of the eternal return*, or, *Cosmos and History*. London: Arkana.

----------, 1991. *Images and Symbols: Studies in Religious Symbolism*. Trans. by Philip Mairet. Princeton, NJ.: Princeton University Press.

---------- - Kitagawa, J. M (eds.), 1959. *The History of Religions: Essays in methodology*. Chicago: University of Chicago Press.

Elm, Susanna, 1994. *Virgins of God: The Making of Asceticism in Late Antiquity*. Oxford: Oxord Clarendon Press.

Elliott, J. K., 2000. "Mark 1.1-3 - A Later Addition to the Gospel?" *New Testament Studies* 46: 584-88.

Ellis, E. E., 1988. "Biblical Interpretation in the New Testament Church". In: M. J. Mulder - H. Sysling (eds.), *Mikra. Text, translation, Reading and Interpretation of the Hebrew Bible in Ancient Judaism and Early Christianity*. Vol. II 1. Assen - Maastricht: Van Gorcum, 691-725

Elster, J., 1983. *Explaining Technical Change: A Case Study in the Philosophy of Science*. Cambridge: Cambridge University Press.

Euthymius Zigabenus. 1865. *Panoplia dogmatica*. In: *Patrologiae cursus completus*. Ed. Jacques-Paul Migne. *Series Graeca*, vol. 130. Paris: Migne.

Faber, R., 1984. "Eric Voegelin. Gnosis-Verdacht als polit(olog)isches Strategem". In: J. Taubes (ed.), *Gnosis und Politik*. Religionstheorie und politische Theologie, vol. 2. Munich: Wilhelm Fink Verlag, 230-248.

Fauth W. , 1995. *Helios Megistos. Zur synkretistischen Theologie der Spätantike* (EPRO 125). Leiden: Brill.

Fazwi, I. - I. Lübben, 2004. *Die ägyptische Jama'a al-Islamiya und die Revision der Gewaltstrategie.*. Berlin: Deutsches Orient-Institut.

Feeney, D., 1998. *Literature and Religion at Rome: Cultures, Contexts and Beliefs*. Cambridge: Cambridge University Press.

Feissel, D., 1984. "La Bible dans les inscriptions grecques". In: C. Mondésert (ed.), *Le monde grec ancient et la Bible*, (Bible de sous temps), Paris: Beauchesne, 223-231.

Feldherr, Andrew. *Spectacle and Society in Livy's* History. Berkeley, Calif./London: University of California Press, 1998.

Felle, A.E., 2006. *Biblia epigraphica. La sacra scrittura nella documentazione epigrafica dell' orbis christianus antiquus (III-VIII secoli)*. Bari.

Felle, A.E., 2007. 'Judaism and Christianity in the Light of Epigraphic Evidence ([3rd] – [7th] cent. C.E.)', *Enoch* 29: 354-377.

Ferguson, J., 1973. *Heritage of Hellenism*. London.

Fichtenau, H., 1992. *Ketzer und Professoren. Häresie und Vernunftglaube im Hochmittelalter*. Munich: Verlag C. H. Beck.

Fields, R., 1992. *How the swans came to the lake: A narrative history of Buddhism in America*, third edition. Boston, Mass.: Shambhala Publications.

Filoramo, G., 1990. *A History of Gnosticism*. Oxford: Basil Blackwell.

Fine, S. - Rutgers, L.V., 1996. 'New Light on Judaism in Asia Minor During Late Antiquity: Two Recently Identified Inscribed Menorahs', *JSQ* 3: 1-23.

Fish, Stanley, 1980. *Is There a Text in This Class? The Authority of Interpretive Communities*. Cambridge: Harvard University Press.

Fischer, F. W., 1972. *Max Beckmann. Symbol und Weltbild. Grundriß zu einer Deutung des Gesamtwerks*. Munich: Wilhelm Fink Verlag.

------------------, 1977. "Geheimlehren und moderne Kunst. Zur hermetischen Kunstauffassung von Baudelaire bis Malewitsch". In: R. Bauer et al. (ed.), *Fin de Siècle. Zu Literatur und Kunst der Jahrhundertwende*. Studien zur Philosophie und Literatur des neunzehnten Jahrhunderts 35. Frankfurt: V. Klostermann, 344-77.

Fleck, L., /1981 (1935). *Genesis and development of a scientific fact* (. Second edition. T. J. Trenn & R. K. Merton (eds.). Transl. by F. Bradley - T. J. Trenn. Chicago: Chicago University Press,

Fodor, Jerry A., 1974. "Special Sciences (Or: The Disunity of Science as a Working Hypothesis?)". *Syntheses* 28: 97-115.

------------------, 1975. *The Language of Thought*. New York: Thomas Y. Crowell Company.

Foley, Robert, 1995. *Humans Before Humanity*, Oxford & Cambridge: Blackwell Publishers Ltd.

Fonagy, Peter, 2001. "The Psychoanalysis of Violence. Paper presented to the Dallas Society for Psychoanalytic Psychotherapy". Accessed online www.psychematters.com/papers/fonagy4.htm Cited with permission of the author.

------------------ - Gergely, Gyorgy - Jurist, Elliot L - Target, Mary, 2002. *Affect Regulation, Mentalization, and the Development of the Self*. New York: Other Press.

------------------ - Target, Mary, 2003. *Psychoanalytic Theories: Perspectives from Developmental Psychopathology*. New York: Brunner-Routledge

Fontaine, P. F. M., 1992. *The Light and the Dark. A Cultural History of Dualism*, vol.7. Amsterdam: J.C. Gieben.

------------------, 2005. *The Light and the Dark. A Cultural History of Dualism*, vol. 20. Groningen: Gopher Publishers.

Frankfort, H., 1978 (1948). *Kingship and the Gods. A Study of Ancient Near eastern Religion as the Integration of Society and Nature* (with a new Preface of S. N. Kramer). Chicago-London.

Frankfurter, D., 1998. *Religion in Roman Egypt : assimilation and resistance.* Princeton; Chichester.

Franko, G. F., 1988. "Sitometria in the Zenon archive: Identifying Zenon's Personal Documents", *BASP* 25: 13-98.

Fraser, P. M., 1972. *Ptolemaic Alexandria*, Vols. I-IIb. Oxford.

---------------, 1998 (1972). *Ptolemaic Alexandria*. Vols. I-III. Oxford.

Freeman, Walter J., 1999. *How Brains Make Up Their Minds.* London: Weidenfeld & Nicolson.

Freud, Sigmund, 2001 (1927). *The Future of An Illusion. Standard Edition.* volume XXI. transl. by James Strachey. London: Vintage.

Frey, P. J. - B., 1975. *Corpus of Jewish Inscriptions. Jewish Inscriptions from the Third Century BC to the Seventh Century AD,* vol. I: *Europe. With a prolegomenon by Baruch Lifshitz.* New York.

Frilingos, Christopher A., 2004. *Spectacles of Empire. Monsters, Martyrs, and the Book of Revelation.* Divinations: Rereading Late Ancient Religion. Philadelphia: University of Pennsylvania Press.

Frith, U., 2003. *Autism: Explaining the Enigma.* Second edition. Oxford: Blackwell Publishers.

Früchtel, E. 1994. "Weltflucht und Weltentfremdung". In: R. Berlinger and W. Schrader (eds.), *Gnosis und Philosophie: Miscellanea.* Elementa 59. Amsterdam - Atlanta: Rodopi, 175-196.

Fry, T. (ed.), 1981. (Rb 1980). *The Rule of St. Benedict in latin and english, with notes and thematic index.* Collegeville, MN.: Liturgical Press.

Fujiwara, Satoko, 2008. "Japan". In: Alles, G. (ed.), *Religious Studies: A Global View.* London, New York, Routledge, 191-217.

Gager, J. G. (Ed.), 1992. *Curse Tablets and Binding Spells from the Ancient World.* Oxford University Press, New York, Oxford.

Galbreath, R., 1981. "Problematic Gnosis. Hesse, Singer, Lessing, and the Limitations of Modern Gnosticism". *The Journal of Religion* 61: 20-36.

Gama, A., D', 2008. "Folk and Myth breaks Harvard mold. *Harvard Crimson*". Retrieved 12 October 2008, from http://www.thecrimson.com/article.aspx?ref=524524.

Garstad, B. 2004. "Belus in the Sacred History of Euhemerus", *CP* 99: 246-257.

Geary, Patrick, 1978. *Futra Sacra: Thefts of Relics in the Central Middle Ages*. Princeton University Press, Princeton.

Geertz, Armin W., 1999. "Definition as Analytical Strategy in the Study of Religion", *Historical Reflections/Réflexions Historiques* 25 (3): 445-475.

Geertz, Clifford, 1973. *The Interpretation of Cultures: Selected Essays*. New York: Basic Books, Inc.

Geffken, J., 1925. "Leon von Pella", *RE* 14, 2012- 2014.

Gehrke, H. - J., 2000. Ιστορία του Ελληνιστικού κόσμου. μετ. Α. Χανιώτης. Επιμ. Κ. Μυραζέλης. Athens.

Geisen, R. 1992, *Anthroposophie und Gnostizismus. Darstellung, Vergleich und theologische Kritik*. Paderborner Theologische Studien 22. Paderborn: Ferdinand Schöningh.

Geissen A., 1974-1982. *Katalog Alexandrinischer Kaisermünzen der Sammlung des Instituts für Altertumskunde der Universität zu Köln*, vols I-V. Opladen.

Gensichen, J., 1910. *De Scripturae Sacrae vestigiis in inscriptionibus latinis christianis*, Greifswald.

Gentili, G. V., 1954. 'Siracusa. Saggi di scavo a sud del vilae Paolo Orsi, in predio Salerno Alerta', *NSc* S. VIII 8: 306-312.

Gentili, G. V., 1959-1960. "I busti futili di Demetra e Kore a Siracusa", *ASS* 5/6: 5-20.

Germano, D. and K. Trainor (eds.), 2004. *Embodying the dharma: Buddhist relic veneration in Asia*. Albany, NY: State University of New York Press.

Gernet, L. - Boulanger, A., 1932. *La génie Grec dans la religion*. Paris.

Gibson, L.H., 1999. *The Jewish Manumission Inscriptions of the Bosporus Kingdom*, Tübingen.

Giddens, Anthony, 2006. *Sociology*. Cambridge: Polity Press.

------------------------, 1996. *The Consequences of Modernity*. Cambridge: Polity Press.

Gilhus, Ingvild Sælid, 2006. *Animals, Gods and Humans. Changing views of animals in Greek, Roman and Early Christian ideas*. London - New York: Routledge,

----------------------------, 2008 (in press). Sacred Marriage and Spiritual Knowledge: Relations between Carnality and Salvation in the *Apocryphon of John*. In: Uro, R. and M. Nissinen (Eds.), *Sacred Marriages: Divine Human Sexual Metaphor*. Eisenbrauns, 487-510.

Gilovich, T., Griffin, D. and Kahneman, D., 2002. *Heuristics and Biases:*

The Psychology of Intuitive Judgment. Cambridge: Cambridge University Press.

Goddio, F., 1998. *Alexandria : the submerged royal quarters.* London.

----------------, - Clauss, M., 2006. *Egypt's Sunken treasures.* Munich- Berlin- London- New York.

Goodger, J., 1982. "Judo Players as a Gnostic Sect". *Religion* 12: 333-344.

Gonce, L.O. - Upal, M.A. - Slone, D.J. - Tweney, R.D., 2006. "Role of Context in the Recall of Counterintuitive Concepts". *Journal of Cognition and Culture* 6. 3-4: 521-547.

Gordon – Marco Simón 2010: R.L. Gordon, F. Marco Simón (eds.), *Magical Practice in the Latin West. Papers from the Intarnational Conference Held at the University of Zaragoza, 30 sept.-1 Oct 2005,* Religions in the Greco–Roman World 168, Leiden – Boston 2010.

Gould, S., 1999. *Rocks of Ages: Science and Religion in the Fullness of Life.* New York: The Ballantine Publishing Group.

Graf, F., 1991. "Prayer in Magic and Religious Ritual". In: C. A. Faraone - D. Obbink (eds.), *Magika Hiera. Ancient Greek Magic and Religion.,* London - New York: Oxford University Press, 188-213.

-----------, 1997. *Magic in the Ancient World.* Cambridge, MA.: Harvard University Press.

---------- — Johnston, S. Illes, 2007. *Ritual Texts for the afterlife. Orpheus and the Bacchic Gold Tablets.* London-New York.

Graf, Friedrich-Wilhelm, 1999. " 'In God we Trust.' Über mögliche Zusammenhänge von Sozialkapital und kapitalistischer Wohlfahrts- ökonomie". In: Friedrich Wilhelm Graf/Andreas Platthaus/ Stephan Schleis- sing (eds.), *Soziales Kapital in der Bürgergesellschaft,* Stuttgart: Kohlhammer, 93-130.

Gragg, D., 2004. "Old and New in Roman Religion: A Cognitive Approach". In: L. H. Martin – H. Whitehouse, (eds.), *Theorizing Religions Past.* Walnut Creek: AltaMira Press, 69-86.

Green, P., 1990. *From Alexander to Actium. The Hellenistic Age,* London.

Griffiths, J. G. (ed.), 1970. Plutarch, *De Iside et Osiride, edited with an Introduction, Translation and Commentary.* Cambridge.

---------------------, 1980. *The Origins of Osiris Cult and his Cult* (Studies in the History of Religions (Supplement to *NUMEN*). Leiden: Brill .

Grimes, Ronald, 1999. "Jonathan Z. Smith's Theory of Ritual Space".

Religion 29: 261-73.

Grose, S., 1923. *Catalogue of the MsLean Collection of Greek Coins*, vol. I. Cambridge.

Guillelmus abbus Sancti Theodirici. 1855. *De erroribus Guillelmi de Conchis ad Sanctum Bernardum*. In: *Patrologia cursus completus*. Ed. Jacques-Paul Migne. *Series Latina* 180: 333-40. Paris: Garnier.

Guthrie, Steward, 1980. "A Cognitive Theory of Religion". *Current Anthropology* 21: 181-203.

----------------------, 1993. *Faces in the Clouds: A New Theory of Religion*. Oxford, New York: Oxford University Press.

----------------------, 1996. "Religion: What is it?", *Journal for Scientific Study of Religion* 35: 412-419.

----------------------, 2002. "Animal Animism: Evolutionary Roots of Religious Cognition". In: I. Pyysiäinen - V. Anttonen (eds.), *Current Approaches in the Cognitive Science of Religion*. Continuum, London, New York, 38-67.

----------------------, 2007. "Anthropology and Anthropomorphism in Religion". In: H. Whitehouse - and J. Laidlaw (eds.), *Religion, Anthropology, and Cognitive Science*, Durham, NC: Carolina Academic Press, 37-62.

Haardt, R., 1967. *Die Gnosis. Wesen und Zeugnisse*. Salzburg: Otto Müller Verlag.

Habermas, Juergen, 2002. *Religion and Rationality: Essays on Reason, God, and Modernity*. Cambridge, MA: The MIT Press.

Habicht, C., ^{2}1970. *Gottmenschtum und Griechische Städte*, München.

Hägg, Tomas, 1994. . "Orality, Literacy, and the 'Readership' of the Early Greek Novel". In: Roy Eriksen (ed.), *Contexts of Pre-Novel Narrative. The European Tradition* (Approaches to Semiotics 114). Berlin- New York: Mouton De Gruyter, 47–81.

Hamman, A., 1980. "La prière chrétienne et la prière païenne: formes et différence". *Aufstieg und Niedergang der römischen* Welt II. 23.2, 1190-1247.

Hani, J., 1976. *La religion Égyptienne dans la pensée de Plutarque* (Collection d' Études Mythologiques). Paris.

Hanratty, G., 1980a. "Gnosticism and Modern Thought I". *The Irish Theological Quarterly* 47: 3-23.

----------------, 1980b. "Gnosticism and Modern Thought II". *The Irish Theological Quarterly* 47: 119-32.

Häring, N., 1977. "Die Rolle der Heiligen Schrift in der Auseinandersetzung des Alanus de Insulis mit dem Neumanichaismus". In: A. Zimmerman (ed.), *Die Mächte des Guten und Bösen. Vorstellungen im XII. und XIII. Jahrhundert über ihr Wirken in der Heilsgeschichte*. Miscellanea Mediaevalia 11. Berlin and New York: Walter de Gruyter, 315-343.

Harle, J. C., 1994. *The Art and Architecture of the indian Subcontinent.*, New Haven - London: Yale University Press.

Harmon, D. P., 1978. "The Family Festivals of Rome". *Aufstieg und Niedergang der römischen* Welt II.16.2, 1592-1603.

Harnack, A.v., 1912. *Über den privaten Gebrauch der heiligen Schriften in der Alten Kirche*, Leipzig.

Harper, C. L., 2005. *Spiritual Information: 100 Perspectives on Science and Religion*. Philadelphia: Templeton Foundation Press.

Harrauer Ch., 1987. *Meliouchos. Studien zur Entwicklung religiöser Vorstellungen in griechischen synkretischen Zaubertexten* (Wiener Studien Supplementband 11). Wien.

Harris, P. L., 1994. "Thinking by Children and Scientists: False Analogies and Neglected Similarities". In: L. A. Hirschfeldand - S. A. Gelman (eds.), *Mapping the Mind: Domain Specificity in Cognition and Culture*. Cambridge: Cambridge University Press, 294-315.

Harris, E. J., 1998. *Ananda Metteyya: The first British emissary of Buddhism*. Kandy: Buddhist Publication Society.

Hartmann, K. 1982. *Die Rechnung mit Gott. Gnostische Strömungen in Kirchengeschichte und Gegenwart*. Stuttgart: Quell-Verlag.

Harvey, David, 1990. *The Condition of Postmodernity*. Oxford: Blackwell.

Hassan, Nasra, 2001. "An Arsenal of Believers". *The New Yorker*, November 19.

Haught, John F., 1995. *Science and Religion: from Conflict to Conversation*. New York: Paulist Press.

Hays, Terrence, 1983. "Ndumba folkbiology and general principles of ethnobotanical classification and nomenclature". *American Anthropologist* 85: 592-611.

Hazzard, R. A. 2000. *Imagination of a Monarchy: Studies in Ptolemaic Propaganda* (Phoenix, tome supplementaire, XXXVII). Toronto - Buffalo - London.

Heever, Gerhard A. Van den, 2005. "'Loose Fictions and Frivolous

Fabrications'. Ancient Fiction and the Mystery Religions of the Early Imperial Era". Ph. D. diss. Pretoria: University of South Africa.

Heiler, Friedrich, 1959. "The History of Religions as a Preparation for the Co-operation of Religions". In: M. Eliade - J. M. Kitagawa (Eds.), *The History of Religions. Essays in Methodology.*, Chicago: University of Chicago Press, 132-160.

Heinen, H., 1981. 'Alexandrien – Weltstadt und Residenz'. In: N. Hinske (Hrsg.), *Kulturbetrachtungen dreier Jahrtausende in Schmelztiegel einer mediterranen Großstadt*. Mainz, 3-12.

Henderson, H., 1993. *Catalyst for controversy: Paul Carus of Open Court*. Carbondale, Il.: Southern Illinois University Press.

Hengel, M., 1992. "Die Septuaginta als von den Christen beanspruchte Schriftensammlung bei Justin und den Vätern vor Origenes". In: J. Dunn (ed.), *Jews and Christians; The Parting of the Ways A.D. 70 to 135. The Second Durham-Tübingen Research Symposium on Earliest Christianity and Judaism (Durham, September, 1989)*. Tübingen, 39-84.

Henrichs, A. 1984. 'The Sophists and Hellenistic Religion: Prodicus as the Spiritual Father of the Isis Aretalogies', *HSCPh* 88: 139-158.

Herrmann, J. J. Jr, (1999) (2000). "Demeter-Isis or the Egyptian Demeter?", *JDAI* 114: 65-123.

Hertz, R., 1960. "A contribution to the study of the collective representation of death". In: R. Needham - C. Needham (transl.), *Death and the right hand*. Cohen and West, London, 197-212.

Herzog-Hauser, G., 1943. 'Tyche', *RE* 7A 2, 1643-1689.

Hinde, R., 1999. *Why Gods Persist*. New York: Routledge.

Hinde, Robert A., 2005. "Modes Theory: Some Theoretical Considerations". In: H. Whitehouse- R. McCauley (eds.), *Mind and Religion: Psychological and Cognitive Foundations of Religion*. New York: Altamira Press, 31-55.

Hölbl, G., 2001. *A History of the Ptolemaic Empire*. transl. by T. Saavedra. London-New York.

Hoexter, Miriam, 2005. 'The 'Waqf' and the Public Sphere'; Jan-Peter Hartung, 'Die fromme Stiftung [*waqf*]. Eine islamische Analogie zur Körperschaft?. In: H. G. Kippenberg/ Schuppert (Hrg.), *Die verrechtlichte Religion: Der Öfentlichkeitsstatus von Religionsgemeinschaften*. Tübingen, Mohr Siebeck, 287-314.

-------------------- - Shmuel N. Eisenstadt -Nehemia Levtzion (eds.), 2002. *The Sphere in Muslim Societies*. Albany: Suny

Hoheisel, K. 1993. "Heil und Erlösung durch Gnosis und Mysterien heute". In: H. Kochanek (ed.), *Heil durch Erfahrung und Erkenntnis. Die Herausforderung von Gnosis und Esoterik für das frühe Christentum und seine Gegenwart*. 71-89. Veröffentlichungen des Missionspriesterseminars St. Augustin bei Bonn, Nr. 429. Nettetal: Steyler Verlag, 71-89.

Holland, J. L., 1997. *Making vocational choices: A theory of vocational personalities and work environments*. third edition. Odessa, FL: Psychological Assessment Resources.

Horbury, W. - Noy D. (eds.), 1992. *Jewish Inscriptions of Graeco-Roman Egypt; with and Index of the Jewish Inscriptions of Egypt and Cyrenaica*, Cambridge.

Hordern, J., 2000. "Notes on the Orphik Papyrus from Gurôb (P. Gurôb 1; Pack^2 2464)", *ZPE* 129: 131-140.

Horn, W. D., 1987. "Blake's Revisionism: Gnostic Interpretation and Critical Methodology". In: D. Miller et al. (ed.), *Critical Paths. Blake and the Argument of Method*. Durham and London: Duke University Press, 72-97.

Horsley, G. H. R. - Lee, G. A., 1993, "A Preliminary Checklist of Abbreviations of Greek Epigraphic Volumes", *Epigraphica* 55: 129-170.

Horst, P., Van der., 1994. "Silent Prayer in Antiquity". *Numen* 41: 1-25.

------------------------, 1991. *Ancient Jewish Epitaphs; An Introductory Survey of a Millenium of Jewish Funerary Epigraphy (300 BCE – 700 CE)*, Kampen.

Huber, C. 1988. *Die Aufnahme und Verarbeitung des Alanus ab Insulis in mittelhochdeutschen Dichtungen*. Zurich and Munich: Artemis Verlag.

Humphrey, Hugh M., 2006. *From Q to "Secret" Mark: A Composition History of the Earliest Narrative Theology*. London: T & T Clark.

Humphrey, Nicholas, 1986 (2002). *The Inner Eye*, London: Faber and Faber Ltd.; Oxford: Oxford University Press, reprint.

Hunter, David G., 2007. *Marriage, Celibacy, and Heresy in Ancient Christianity. The Jovinianist Controversy*. Oxford: Oxford University Press.

Hunter, James Davison, 1991. *Culture Wars. The Struggle to Define America. Making Sense of the Battles over the Family, Art, Education, Law and Politics*. New York: Basic Books.

Hutchins, Edwin, 1995. *Cognition in the Wild*. Cambridge, MA: MIT Press.

Jacoby, F., 1907. 'Euhemeros', *RE* VI 1, 957-972.

Jalabert, L., 1914. "Citations bibliques dans l' epigraphie grecque", in: *DACL*, vol. V,1,Paris, 1731-1756.

Jambet, Christian, 1999. "Préface", in: S. Schmidtke (ed.), *Correspondance Corbin-Ivanow. Lettres échangées entre Henry Corbin et Vladimir Ivanow de 1947 à 1966.* Leuven-Paris: Diffusion Peeters, 5-10.

James, Williams, 1958. *The Varieties of Religious Experience.* New York: Mentor.

----------------------, 1985. *The Varieties of religious experience.* New York: Penguin.

Jenkins, R., 1996. *Social Identity,* London - New York.

Jensen, J.S., 2003. *The Study of Religion in a New Key - Theoretical and Philosophical Soundings in the Comparative and General Study of Religion.* Aarhus: Aarhus University Press.

----------------- - Luther H. Martin (eds.), 2002. *Rationality and the Study of Religion.* London: Routledge.

Johnson, K., 2004. "Primary emergence of the doctrinal mode of religiosity in prehistoric southwestern Iran". In: H. Whitehouse, L. H. Martin, (eds.), *Theorizing religions past: Archaoelogy, history, and cognition.* Walnut Creek: AltaMira Press, 45-66.

Jonas, H., 1963. *The Gnostic Religion. The Message of the Alien God and the Beginnings of Christianity.* 2nd ed. Boston: Beacon Press.

------------, 1964 (1934). *Gnosis und spätantiker Geist 1. Die mythologische Gnosis.* 3rd ed. Göttingen: Vandenhoeck and Ruprecht.

Juergensmeyer, Mark (ed.), 1992. *Violence and the Sacred in the Modern World.* London: Frank Cass.

-----------------------------------, 2003. *Terror in the Mind of God: The Global Rise of Religious Violence.* Berkeley. University of California Press.

Interrogatio Iohannis. 1929. In *Die Vorgeschichte der christlichen Taufe,* by R. Reitzenstein, Leipzig and Berlin: B.G. Teubner, 297-311.

Ismael, Jacqueline S. - Tareq Y. Ismael, 1995. "Cultural Perspectives on Social Welfare in the Emergence of Modern Arab Social Thought". *The Muslim Word* 85: 82-106.

Kajanto, I, 1972. "Fortuna", *Aufstieg und Niedergagand der Römischen Welt* II 17,1, 502-558.

Kakar, Sudhir, 1996. *The Colors of Violence: Cultural Identities, Religion, and Conflict.* Chicago: The University of Chicago Press.

Kant, L.H., 1987, "Jewish Inscriptions in Greek and Latin", *Aufstieg und Niedergagand der Römischen Welt* II, 20. 2, 671-813.

Kauffman, Stuart A., 2006. "Beyond Reductionism; Reinventing the Sacred". *Edge* 197, Web edition.

Kaufmann, Franz-Xaver, 2003. *Varianten des Wohlfahrtstaates*. Frankfurt: Suhrkamp[1]. Elmar Rieger/Stephan Leibfried, *Grundlagen der Globalisierung. Perspektiven des Wohlfahrtsstaates*. Frankfurt: Suhrkamp 2001, p. 205.

Keil, B., (ed.), 1958. *Aelii Aristidis Smyrnaei quae supersunt omnia*, vol. II. Berlin.

Kelemen, D., 1999a. "The Scope of Teleological Thinking in Pre-School Children". *Cognition* 70: 241-272.

Kelemen, D., 1999b. "Why are Rocks Pointy?: Children's Preference for Teleological Explanations of the Natural World". *Developmental Psychology* 33: 1440-1452.

Kelhoffer, James A., 2000. Miracle *and Mission: The Authentication of Missionaries and Their Message in the Longer Ending of Mark* (WUNT 2/112). Tübingen: Mohr-Siebec.

------------------------------. 2004. "How Soon a Book Revisited: EUAGGELION as a Reference to 'Gospel' Materials in the First Half of the Second Century". *Zeitschrift für die neutestamentliche Wissenschaft* 95: 1-34.

Keller, C.-A., 1985. "Gnostik als religionswissenschaftliches Problem". *Theologische Zeitschrift* 41: 59-73.

----------------. 1987. "Christliche Gnosis und Gnosisversuche der Neuzeit. Was ist Erkenntnis?". In: J. Müller et al. (ed.), *New Age - aus christlicher Sicht*. Freiburg: Paulusverlag; Zurich: Theologischer Verlag, 51-94.

----------------------. 1994. "Gnostik. Urform christlicher Mystik". In: R. Berlinger and W. Schrader (eds.), *Gnosis und Philosophie: Miscellanea*. Elementa 59. Amsterdam - Atlanta: Rodopi, 197-225.

Kent, Eliza F., 2005. "Representing Caste in the Classroom: Perils, Pitfalls and Potential Insight". *Method and Theory in the Study of Religion* 17: 231-241.

Kern, O., 1922. *Orphicorum Fragmenta*. Berlin.

Kilner, M. F., 1977. *The Shoulder Bust in Sicily and South and Cental Italy. A Catalogue and Materials for Dating*. Göteborg.

Kinnard, Jacob, 2004. "The Field of the Buddha's Presence". In: D. Germano - K. Trainor (eds.), *Embodying the Dharma: Buddhist Relic Veneration in Asia*. Albany: SUNY Press, 117-43.

Kippenberg, H. G., 1997. "Magic in Roman Civil Discourse: Why Rituals Could Be Illegal". In: P. Schäfer - H. G. Kippenberg (Eds.), *Envisioning Magic: A Princeton Seminar and Symposium*. Leiden: Brill, 137-163

----------------------, 2005. "Consider that it is a Raid on the Path of God: The Spiritual Manual of the Attackers of 9/11". *Numen* 52: 29-57.

----------------------, 2008. *Gewalt als Gottesdienst. Religionskriege im Zeitalter der Globalisierung.* München: C.H.Beck.

Kirkpatrick, Lee A. - Shaver, Phillip R., 1990. "Attachment Theory and Religion: Childhood Attachments, Religious Beliefs, and Conversion". *Journal for the Scientific Study of Religion* 29, 3: 315-334.

Kirkpatrick, Lee A., 2005. *Attachment, Evolution, and the Psychology of Religion.* New York: The Guilford Press.

Kitchen, J., 2009. "Variants, Arians and the Trace of Mark: Jerome and Ambrose on '*neque filius*' in Matthew 24:36". In: I. van't Spijker (ed.), *The Multiple Meaning of Scripture: The Role of Exegesis in Early Christian and Medieval Culture* (Commentaria 2). Leiden: Brill, 15-40.

Klinghardt, M., 1999. "Prayer Formularies for Public Recitation: Their Use and Function in Ancient Religion". *Numen* 46 : 1-52.

Knott, Kim, 2005. *The Location of Religion: A Spatial Analysis.* London: Equinox.

Köenen, L., 1983. "Die Adaptation ägyptischer Königsideologie am Ptolemäerhof". In: *Hellenistic World. Proceedings of the International Colloqiuim, Leuven 24-26 May 1982.* Louvain, 143-190.

----------------, 1993. 'The Prolemaic King as a Religious Figure', in: A. Bulloch-E. Gruen-A. Long-A. Stewart (eds.), *Images and Ideologies: Self-Definition in the Hellenistic World.* Berkeley, 25-113, with a response by Fr. Walbank, 116-129.

Koester, Helmut, 1983. "History and Development of Mark's Gospel (From Mark to Secret *Mark* and 'Canonical' Mark)". In: B. C. Corley (ed.), *Colloquy on New Testament Studies: A Time for Reappraisal and Fresh Approaches.* Macon: Mercer University Press, 35-57.

Koester, Helmut, 1990. *Ancient Christian Gospels: Their History and Development.* Philadelphia: Trinity Press International; London: SCM.

Kolta, K.S., 1968. *Die Gleichsetzung ägyptischen und griechischen Götter bei Herodot.* Diss. Tübingen.

Koester, Helmut, 1983. "History and Development of Mark's Gospel (From Mark to Secret *Mark* and 'Canonical' Mark)". In: B. C. Corley (ed.), *Colloquy on New Testament Studies: A Time for Reappraisal and Fresh Approaches.* Macon: Mercer University Press, 35-57.

Koslowski, P.,1988a. "Einleitung. Philosophie, Mystik, Gnosis". In: *idem* (ed.), *Gnosis und Mystik in der Geschichte der Philosophie.* Zurich and Munich: Artemis Verlag, 9-12.

----------------, 1988b. "Gnosis und Gnostizismus in der Philosophie. Systematische Überlegungen". In: *idem* (ed.), *Gnosis und Mystik in der Geschichte der Philosophie*. Zurich and Munich: Artemis, 368-399.

----------------, 1992. "Christliche Gnosis als andere Aufklärung. Überlegungen zur christlichen Philosophie". In: *Russische Religionsphilosophie und Gnosis. Philosophie nach dem Marxismus*. Hildesheim: Bernward Verlag, 87-129.

----------------, 2006. "Philosophische Religion als Form postmodernen Denkens". In: *idem* (ed.), *Philosophische Religion. Gnosis zwischen Philosophie und Theologie*. Munich: Wilhelm Fink Verlag, 223-238.

Kotansky; R., 1994. *Greek Magical Amulets. The inscribed Gold, Silver, Copper and Bronze* lamellae. Part I: *Published Texts of Known Provenance* (Papyrologica Coloniensia 22, 1). Opladen.

Kövecses, Zoltán, 2003. *Metaphor and Emotion. Language, Culture, and Body in Human Feeling*. Cambridge: Cambridge University Press.

Krämer, Gudrun, 2005. "Aus Erfahrung lernen? Die islamische Bewegung in Ägypten", in Clemens Six - Martin Riesebrodt -Siegfried Haas (eds.), *Religiöser Fundamentalismus. Vom Kolonialismus zur Globalisierung*. Innsbruck: Studien Verlag, 185-200.

Kraemer, R. S., 1991, "Jewish Tuna and Christian Fish: Identifying Religious Affiliation in Epigraphic Sources", *HThR* 84: 141-162.

Kretschmar, G., 1953. "Zur religionsgeschichtlichen Einordnung der Gnosis". In: K. Rudolph (ed.), *Gnosis und Gnostizismus*. Wege der Forschung 262. Darmstadt: Wissenschaftliche Buchgesellschaft 1975, 426-437.

Kristeva, J., 1967, 'Le mot, le dialogue et le roman', *Critique* 23: 438-465.

Kuntzmann, Raymond, 1986. *Le Livre de Thomas (NHII, 7)*. (Bibliothèque Copte de Nag Hammadi, 'Textes,' 16). Québec: Les presses de L'Université Laval.

Kuttner, Ann., 1999. "Hellenistic Images of Spectacle, from Alexander to Augustus". In: Bettina Bergmann and Christine Kondoleon (eds.), *The Art of Ancient Spectacle*. (Studies in the History of Art 56). New Haven- London: Yale University Press, 97–123.

La' da, C. A., 2003. "Encounters with Egypt: The Hellenistic Experience", in: R. Mathews – C. Roemer (eds.), *Ancient Perspectives on Egypt*. London, 157-169.

La gloire: (1998). *La gloire d'Alexandrie: 7 mai-26 juillet 1998*. Paris.

Laidlaw, J., 2007. "A well-disposed social anthropologist'sproblems with the 'cognitive science of religion". In: H. Whitehouse, J. Laidlaw (eds.), *Religion, anthropology, and cognitive science*. Durham: Carolina Academic Press, 211-246.

Lakoff, George, 1987. *Women, Fire, and Dangerous Things. What Categories Reveal about the Mind*. Chicago, London: The University of Chicago Press.

Laland, K.N., 2007. Niche construction, human behavioural ecology and evolutionary psychology. In: Dunbar, R.I.M. and L. Barrett (eds.), *Oxford handbook of evolutionary psychology*. Oxford University Press, New York, pp. 35-48.

------------------, - Brown, G. R., 2002. *Sense and nonsense: evolutionary perspectives on human behavior*. Oxford: Oxford University Press.

Lambert, M., 1991. *Ketzerei im Mittelalter. Eine Geschichte von Gewalt und Scheitern*. Freiburg: Herder.

Lanman, J. A., 2007. "How 'Natives' Don't Think: The Apotheosis of Overinterpretation". In: H. Whitehouse - J. Laidlaw (Eds.), *Religion, Anthropology, and Cognitive Science*. Durham: Carolina Academic Press, 105-132.

Laquer, Thomas, 1992. *Making Sex. Body and Gender from the Greeks to Freud*. Cambridge. MA.- London: Harvard University Press.

Laqueur, R., 1938. s.v. "Timotheos", *RE* VI A, 1338.

Lausberg, Marion - Tinnefeld Franz. s.v. "Martialis". In: Hubert Cancik, Helmuth Schneider (Antike), and Manfred Landfester (Rezeptions- und Wissenschaftsgeschichte) (eds.). *Der Neue Pauly*. Edited Brill Online, http://0-www.brillonline. nl.oasis. unisa.ac.za: 80/ subscriber/ entry?entry= dnp_e725280, accessed 11 November 2007.

Lawson, E. Thomas, 1976. "Ritual as Language". *Religion. A Journal of Religion and Religions* 6: 123-139.

------------------------------, 1984. "Functionalism Reconsidered". *History of Religions* 23: 372-381.

------------------------------, 2000. "Cognition". In: W. Braun – R. T. McCutcheon (eds.), *Guide to the Study of Religion*. London: Cassell, 75-84.

------------------------------, 2004. "The Wedding of Psychology, Ethnography, and History: Methodological Bigamy or Tripartite Free Love?". In: H. Whitehouse – L. H. Martin (eds.), *Theorizing Religions Past: Arcaeology, History, and Cognition,* Oxford: Altamira, 1-5.

——————————————, McCauley, R. N., 1990. *Rethinking Religion: Connecting Cognition and Culture*. Cambridge: Cambridge University Press.

Layton, Bentley, 1987. "The Book of Thomas the Contender Writing to the Perfect". In: *The Gnostic Scripture. A New Translation with Annotations and Introductions*. Doubleday, New York, 400-409.

Leclercq, H., 1914. "Citations bibliques dans l' épigraphie latine". In: *DACL*. Vol. V, 1. Paris, 1756-1780.

Lee, P. J., 1987. *Against the Protestant Gnostics*. New York and Oxford: Oxford University Press.

Leeuw, G Van der., 1963. *Religion in essence and manifestation: A Study in phenomenology*. vols I-II. Trans. by J. E. Turner. New York: Harper & Row.

Leman, Nicholas, 1996. "Kicking in Groups". In: *The Atlantic Monthly* 277: 22-26.

Leslie, A., 1994. "ToMM, ToBY, and Agency: Core Architecture and Domain Specificity". In: L. Hirschfeld - S. A. Gelman (eds.), *Mapping the Mind*. New York: Cambridge University Press, 119-148.

Leslie, M., 2000. "Divided they stand". *Stanford Magazine*. Retrieved 6 August 2008, from http://www.stanfordalumni.org/ news/ magazine /2000/ janfeb/ articles /anthro.html.

Levitin, D. J., 2006. *This is your brain on music: The science of a human obsession*. New York, N.Y.: Dutton.

Lévy-Bruhl, Lucien, 1979. *How Natives Think*. Trans. L. Claire. New York: Alfred A Knopf.

Lévy-Strauss, Claude, 1966. *The Savage Mind*. London: Weidenfeld and Nicolson.

Lewis, T. T., 2000. *Popular Buddhist texts from Nepal: Narratives and rituals of Newar Buddhism*. Albany: State University of New York Press.

Lewis-Williams - James David, 1981. *Believing and Seeing: Symbolic Meanings in Southern San Rock Painting*. London: Academic Press.

Lietzmann, D.H., 1934 (ed.), *Griechische Papyri*[4], Berlin.

Lieu, J.M., 2004. *Christian Identity in the Jewish and Greco-Roman World*. Oxford.

Lincoln, Br., 1986. *Myth, Cosmos, and Society: Indo-European Themes of Creation and Destruction*. Cambridge.

——————————, 1989. *Discourse and the Construction of society. Comparative Studies of Myth, Ritual and Classification*. New York-Oxford.

----------------, 1996. "Theses on Method". *Method & Theory in the Study of Religion* 8: 225-227.

----------------, 2003. *Holy Terrors: Thinking about Religion after September 11.*, Chicago: The University of Chicago Press.

----------------, 2004. "Epilogue". In: Sarah I. Johnson (ed.), *Religions of the Ancient World: A Guide*. Cambridge: The Belknap Press of Harvard University Press, 657-667.

Linder, M. - John Scheid, 1993. "Quand croire c'est faire: le problème de la croyance dans la Rome ancienne". *Archive des sciences sociales des religions* 81 : 47-61.

Ling, T. (ed.), (1981). *The Buddha's philosophy of man.* Everyman, London - Rutland.

Lisdorf, Anders, 2004. "The Spread of Non-natural Concepts: Evidence from the Roman Prodigy Lists". *Journal of Cognition and Culture* IV 1: 151-73.

----------------, 2005. "The conflict over Cicero's house: An analysis of the ritual element in 'De Domo Sua," *Numen* 52: 445-464.

Littleton, C. Scott, 1965. "A Two-dimensional Scheme for the Classification of Narratives". *Journal of American Folklore* 78: 21-27.

Littlewood, R., 1998. "Living gods: In [Partial] Defence of Euhemerus", *Anthropology Today* 14, 2 (April): 6-14.

Livre des deux principes. 1973. Ed. and trans. Christine Thouzellier. Sources chrétiennes 198. Paris : Éditions du cerf.

Lloyd, A. B., 1976. *Herodotus Book II. Commentary 1-98* (EPRO, 43). Leiden: Brill.

----------------, 1988. *Herodotus Book II. Commentary 99-192* (EPRO, 43). Leiden: Brill.

Lohlker, Rüdiger, 2006. *Islamisches Völkerrecht. Studien am Beispiel Granada.* Bremen: Kleio Humanities.

Loos, M., 1974. *Dualist Heresy in the Middle Ages.* Academia: Prague.

Lüderitz, G., 1983. Corpus jüdischer Zeugnisse aus der Cyrenaica mit einem Anhang von Joyce M. Reynolds, Wiesbaden.

Mack, Burton, 1988. *A Myth of Innocence: Mark and Christian Origins*. Philadelphia: Fortress.

MacMullen, R., 1982, 'The Epigraphic Habit in the Roman Empire', *AJP* 103: 233-246.

Madigan, Kevin, 2007. *The Passions of Christ in High-Medieval Thought: An Essay in Christological Development* .Oxford Studies in Historical Theology; Oxford: Oxford University Press.

Malunowicz, L., 1982. 'Citations bibliques dans l'épigraphie grecque', in: E. A. Livingstone (ed.), *Studia Evangelica* VII. Berlin, 333-337.

Manganaro, G., 1965. "Ricerche di antichità e di epigrafia siceliote, I: Per il culto di Demeter in Sicilia", *RAC* 17: 183-189, tav. LXIV-LXV.

Manning, J. G., 2003. *Land and Power in Ptolemaic Egypt. The Structure of Land Tenure*. Cambridge.

Marcus, Joel, 2000. "Mark – Interpreter of Paul", *New Testament Studies* 46: 473-487.

Marjanen, A., 2005. "What Is Gnosticism? From the Pastorals to Rudolph". In: A. Marjanen (ed.), *Was There a Gnostic Religion?*. Göttingen: Vandenhoeck and Ruprecht, 1-53.

Markschies, C., 2001. *Die Gnosis*. Munich: C.H. Beck.

Markschies, Chr., 2006. *Das antike Christentum; Frömmigkeit, Lebensformen, Institutionen*. München.

Martial. *De Spectaculis*. Translated by A. S. Kline. http://www.tonykline.co.uk/PITBR/Latin/Martial.htm#_Toc123798956, accessed 11 November 2007.

Martin, L. H., 1987. *Hellenistic Religions. An Introduction*. New York, Oxford: Oxford University Press.

----------------, 1990. "Greek Goddesses and Grain: The Sicilian Connection", *Helios* 17, 2: 251-261.

----------------, 1994. 'The Anti-individualistic Ideology of Hellenistic Culture', *Numen* 41: 117-190.

----------------, 1995a. "Fortuna". In: K. van der Toorn - B. Becking – P.W. van der Horst (eds.), *Dictionary of Deities and Demons in the Bible*, Leiden, 635-638.

----------------, 1995b. "Tyche". In: K. van der Toorn-B. Becking – P.W. van der Horst (eds.), *Dictionary of Deities and Demons in the Bible*, Leiden, 1653-1656.

----------------, 1996. 'The Post-Eliadean Study of Religion and the New Comparativism". *Method & Theory in the Study of Religion* (id. [ed.]. Introduction to a symposium on 'The new comparativism in the study of religion] 8,1: 1-3.

----------------, 1997. "Rationality and Relativism in History of Religions Research". In: Jeppe S. Jensen – id (eds.), *Rationality and the Study of*

Religion. Aarhus: Aarhus University Press = 2003 paperback, ed. London: Routledge.

-------------- (ed.), 1999. *The Definition of Religion in Social-scientific/Historical Research*". *Historical Reflections/Réflexions Historiques*, 25. 3.

--------------, 2000a. "Secular Theory and the Academic Study of Religion". In: T. Jense - M. Rothstein (eds.), *Secular Theories on Religion. Current Perspectives.*, Copenhagen: Museum Tusculanum Press, 137-148.

--------------, 2000b. "Comparison". In: W. Braun –R. McCutcheon (eds.), *Guide to the Study of Religion*. London-New York, 45-56.

--------------, 2001a. "Comparativism and sociobiological Theory". *Numen* 48: 290-308.

--------------, 2001b. "The Academic Study of Religion during the Cold War: The Western Perspective". In: *id.* - I. Dolezalová - D. Papoušek (eds.). *The Study of Religion during the Cold War, East and West*. New York: Peter Lang Press.

--------------, 2002. "Marcel Gauchet: The disenchantment of the world". *Method & Theory in the Study of Religion* 14, (1): 114-120.

--------------, 2003a. "Cognition, Society and Religion: a new approach to the study of culture". *Culture and Religion*. IV 2: 207-231.

--------------, 2003b. "Kingship and the Hellenistic Consolidation of Religio-Political Power". In: *idem* – P. Pachis (eds.), *Theoretical Frameworks for the study of Graeco-Roman Religions. Adjunct Proceedings of the XVIIIth Congress of the International Association for the History of Religions, Durban, South Africa, 2000*, Thessaloniki: University Studio Press, 89-96.

--------------, 2004a. "Toward a Scientific History of Religions". In: id. - H. Whitehouse (eds.), *Theorizing Religions Past: Arcaeology, History, and Cognition*, Walnut Creek: Altamira Press,. 7-14.

--------------, 2004b. "Toward a Cognitive History of Religions". In: *Unterwegs. Neue Pfade in der Religionswissenschaft. Festschrift für Michael Pye zum 65*. München: Biblion, 73-80.

--------------, 2004c. "Ritual Competence and Mithraic Ritual". In: T. Light – B.C. Wilson (eds.), *Religion as a Human Capacity: A Festschrift in Honor of E. Thomas Lawson*. Leiden: Brill, 245-263.

--------------, 2004d. "History, Historiography, and Christian Origins: The Case of Jerusalem". In: R. Cameron and M. Miller (eds.). *Redescribing Christian Origins*, Leiden: Brill, 263-273.

----------, 2004e. "Redescribing Christian Origins: Historiography or Exegesis?". In: R. Cameron and M. Miller (eds.). *Redescribing Christian Origins*, Leiden: Brill, 475- 481.

----------, 2004f. "The Very Idea of Globalization: The Case of Hellenistic Empire". In: *id.* - Panayotis Pachis (eds.). *Hellenisation, Empire and Globalization: Lessons from Antiquity*. Thessaloniki: Vanias Press, 123-139.

----------, 2004g. "Petitionary Prayer: Cognitive Considerations". In: B. Luchesi - K. von Stuckrad (eds.), *Religion in kulturellen Diskurs. Festschrift für Hans G. Kippenberg zu seinen 65. Geburtstag*. 2004, Berlin: Walter de Gruyter. 115-126.

----------, 2004h. Toward a scientific history of religions. In: H. Whitehouse, Martin, L. H. (Eds.), *Theorizing religions past: Archaeology, history, and cognition*. AltaMira Press, Walnut Creek.

----------, 2005a. "Comparative Religion". In: J. Hinnells (ed.), *The Routledge Companion to the Study of Religion*. London – New York . 205-222.

----------, 2005b. "Cognitive Science of Religion". In: John Hinnells (ed.). *The Routledge Companion to the Study of Religion*. London – New York : Routledge, 473-488.

----------, 2005c. "The Promise of Cognitive Science for the Historical study of Religions, with Reference to the Study of Early Christianity", Lecture held in Helsinki, 1 September in: www.helsinki. fi/collegium/events/Luther_Martin. pdf, 1-35, accessed 16 June 2009.

----------, 2005d. "Performativity, Narrativity, and Cognition: Demythologizing the Roman Cult of Mithras". In: W. Braun (ed.), *Persuasion and Performance, Rhetoric and Reality in Early Christian Discourses*. Waterloo: Wilfrid Laurier University Press, 187-217.

----------, 2008a. "Can Religion Really Evolve? (And What Is It Anyway?)". In: J. Bulbulia - R. Sosis - E. Harris -R. Genet - C. Genet - K. Wyman (eds.). *The Evolution of Religions: Studies, Theories, and Critique.*, Santa Margarita, CA: The Collin Foundation Press.. 349-355.

----------, 2008b. "What Do Religious Rituals Do? (And How Do They Do It?): Cognition and the Study of Religion". In: R. McCutcheon - W. Braun (eds.). *Introducing Religion: Festschrift for Jonathan Z. Smith*, London: Equinox, 311-325.

----------, 2008c. "Daniel Dennett's *Breaking the Spell:* A Response to Its

Critics". *Method & Theory in the Study of Religion* 20,1: 61-66.

-----------, 2008d. "The Academic Study of Religion: A Theological or Theoretical Undertaking?". In: P. Pachis - P. Vasiliadis - D. Kaimakis (eds.). *PHILIA KAI KOINONIA. Cultures in Contact: Essays in Honor of Professor Gregorios D. Ziakas*, Thessaloniki: Vanias Publications, 333-345.

----------- - Paden, W. 1970. "Beyond Osiris: The evolution of a religion studies program". *American Academy of Religion conference*, New York.

--------- - Goss, J. (eds.), 1985. *Essays on Jung and the study of religion*. Los Angeles: University Press of America.

---------- - Gutman, H. - Hutton, P. H. (eds.), 1998. *Technologies of the self: A seminar with Michel Foucault*. Boston: University of Massachusetts Press.

------------- - Whitehouse, H. (eds.), 2004. *Theorizing Religions Past: Archaeology, History, and Cognition*. Walnut Creek, CA: AltaMira Press.

Marvin Olasky,1992. *The Tragedy of American Compassion*. Washington D.C.: Regnery.

Matlin, M. W., 2002. *Cognition*. fifth edition. Stamford, CT: Thomson Wadsworth.

Mayr, Ernst, 2004. *What Makes Biology Unique? Considerations on the Autonomy of a Scientific Discipline*. Cambridge: Cambridge University Press.

Mazzoleni, D., "Patristica ed epigrafia", 1989. In: A. Quacquarelli (ed.), *Complementi interdisciplinary di patrologia*, Roma, 319-365.

McCarthy, John, 1969. "Some Philosophical Problems from the Standpoint of Artificial Intelligence". In: B. Meltzer - D. Michie (eds.), *Machine Intelligence.*, Edinburgh: Edinburgh University Press, 463-502.

-------------------------, et al. 1962. *LISP 1.5. Programmer's Manual*. Cambridge, MA: MIT Press.

McCauley, R.N., 1986. "Intertheoretic Relations and the Future of Psychology". *Philosophy of Science* 53: 179-199.

-------------------, 1988. "Epistemology in an Age of Cognitive Science". *Philosophical Psychology* 1: 143-52.

-------------------, 1996. "Explanatory Pluralism and the Coevolution of Theories in Science". In: *idem* (ed.), *The Churchlands and Their Critics.*, Oxford:. Blackwell, 463-502.

-------------------, 1998. "Levels of Explanation and Cognitive Architectures". In: W. Bechtel - G. Graham (eds.), *Blackwell Companion to Cognitive Science*. Oxford: Blackwell, 611-624.

―――――, 2000. "The Naturalness of Religion and the Unnaturalness of Science". In: F. Keil and R. Wilson (eds.). *Explanation and Cognition*. Cambridge, MA: MIT Press, 61-85.

―――――, 2001. "Reduction". In: Wilson, Robert A. - Frank C. Keil (Eds.), *The MIT Encyclopedia of the Cognitive Sciences (MITECS)*. Cambridge, MA: MIT Press, 712-714.

―――――, 2007a. "Reduction: Models of Cross-Scientific Relations and Their Implications for the Psychology-Neuroscience Interface". In: P. Thagard (ed.), *Handbook of the Philosophy of Science: Philosophy of Psychology and Cognitive Science*. Amsterdam: Elsevier, 105-158.

―――――, 2007b. "Enriching Philosophical Models of Cross-Scientific Relations: Incorporating Diachronic Theories". In: M. Schouten – H. Looren de Jong (eds.), *The Matter of the Mind: Philosophical Essays on Psychology, Neuroscience and Reduction*, Oxford: Blackwell Publishers, 199-223.

―――――, in progress. *The Naturalness of Religion and the Unnaturalness of Science: A Comparison of Their Cognitive Foundations*.

―――――, - Thomas E. Lawson, 2002. *Bringing Ritual to Mind; Psychological Foundations of Cultural Forms*. Cambridge: Cambridge University Press.

McCloskey, M., 1983. "Intuitive Physics". *Scientific American* 248: 122-30.

McCorkle Jr., W. W., 2007. *From corpse to concept: A cognitive theory on the ritualized disposal of dead bodies*. Unpublished Dissertation., Belfast: Queen's University of Belfast.

―――――, 2008. "Memes, genes, and dead machines: Evolutionary anthropology of death and burial". In: J. Bulbulia - R. Sosis - E. Harris - R. Genet - C. Genet – K. Wyman (eds.), *The evolution of religion: Studies, theories, and critiques*. Santa Margarita, CA: Collins Foundation Press, 287-292.

―――――, forthcoming. *The evolution of ritual*.

McCutcheon, Russell T., 1997. *Manufacturing Religion: The Discourse on Sui Generis Religion and the Politics of Nostalgia*. New York: Oxford University Press.

―――――, 2000. "Myth". In: W. Braun – id. (eds.), *Guide to the Study of Religion*. London-New York: Continuum, 190-208.

——————————————, 2001. *Critics not Caretakers. Redescribing the Public Study of Religion.* New York: State University of New York Press.

——————————————, 2003. *The Discipline of Religion. Structure, Meaning, Rhetoric.* London: Routledge.

——————————————, 2007. "They Licked the Platter Clean': On the Co-Dependency of the Religious and the Secular", *Method & Theory in the Study of Religion* 19/3&4: 173-199.

——————————————, 2008. "A Gift with Diminished Returns", *Journal of the American Academy of Religion* 76/3: 748-765.

McKenzie - J Gibson- S. Reyes. A. T., 2004. "Reconstructing the Serapeum in Alexandria from the Archaeological Evidence", *Journal of Roman Studies* 94:73–114.

McKeown, Elisabeth, 2005. *Catholic Studies*. Unpublished paper presented at the meeting of the ACHA in Seattle, session *Theologians and Historians*.

Merkelbach 1984: R. Merkelbach, *Mithras*, Königstein/Ts. 1984.

Merkelbach., R., 1999. "Die goldenen Totenpässe: ägyptisch, orphiscb, bakchisch", *ZPE* 128: 1-13.

Merlin, Donald, 1991. *Origins of the Modern Mind: Three Stages in the Evolution of Culture and Cognition*. Cambridge, MA.: Harvard University Press.

——————————————, 2001. *A Mind So Rare: The Evolution of Human Consciousness*. New York & London: W. W. Norton & Company.

Merrifield, Andrew, 1993. "Place and Space: A Lefebvrian Reconciliation". *Transactions of the British Institute of Geographers* 18: 516-531.

Merton, R. K., 1972. "Institutional imperatives of science". In: B. Barnes (ed.), *Sociology of Science*. Baltimore: Penguin Books, 65-79.

——————————————, 1973. *Sociology of science: Theoretical and empirical investigations*, Storer, N. W. (Ed.). University of Chicago Press, Chicago.

Meyer, E.A., 1990, "Explaining the Epigraphic Habit in the Roman Empire: the Evidence of Epitaph", *JRS* 80: 74-96.

Michalski, Kryzstof (ed.), 2006. *Religion in the New Europe*. Budapest: Central European University Press.

Mikalson, J. D., 1998. *Religion in Hellenistic Athens*, (Studies in Hellenistic Civilisation, XXIX). Berkeley-Los Angeles-London.

——————————————, 2006. "Greek Religion: continuity and change in the Hellenistic Period", In: Gl. R. Bugh [ed.], *The Cambridge Companion to the*

Hellenistic World. Cambridge - New York – Melbourne - Madrid-Cape Town – Singapore - Sao Paolo, 208-222.

Miller, George, 1956. "The Magical Number Seven, Plus or Minus Two: Some Limits on Our Capacity for Processing Information". *The Psychological Review* 63: 81-97.

Milne, J. G., 1933. Catalogue of Alexandrian Coins. Oxford.

Minsky, Marvin, 1954. *Neural Nets and the Brain Model Problem*, Ph.D. dissertation, Princeton University.

Miro, E. De, 1963. "I recenti scavi sul progetto di S. Nicolo in Agrigento", *CASA* 2: 57-63.

Mishal, Shaul - Avraham Sela, 2000. *The Palestinian Hamas. Vision, Violence, and Coexistence.* New York: Columbia UP.

Mitchell – Van Nuffelen 2010a: S. Mitchell, P. Van Nuffelen (edd.), *Monotheism Between Pagans and Christians in Late Antiquity*, ISACR 12, Leuven –Walpole [MA] 2010.

Mitchell – Van Nuffelen 2010b: S. Mitchell, P. Van Nuffelen (edd.), *One God. Pagan Monotheism in the Roman Empire*, Cambridge – New York – Melbourne et al. 2010.

Mithen, S. J., 1996a. *The prehistory of the mind: A search for the origins of art, religion, and science.* London: Thames and Hudson.

----------------, 1996b. *The Prehistory of the Mind: The Cognitive Origins of Art and Science.* London: Thames and Hudson.

Mneimneh, Hassan - Makiya, Kanan, 2002. "Manual for a Raid". *The New York Review of Books* 49,19 (January 17).

Moll, Jorge - Roland Zahn - Ricardo de Oliveira-Souza - Frank Krueger - Jordan Grafman, 2005. "The Neural Basis of Human Moral Cognition", *Nature Reviews – Neuroscience* 6: 799-809.

Montanari, E., 1974. "L' episodio delle pelegrinazioni di Demetra delle fonti di Ovidio, *Fast.* IV 502-62 e *Met.* V 446-61", *ASNP*, S. III 4,1: 109-137.

Moose, P., Von, 1998. "Die Begriffe *öffentlich* und *privat* in der Geschichte und bei den Historikern". *Saeculum* 49: 161-192.

Mora, F., 1986. *Religione e Religioni nelle Storie di Erodoto* (Le Edizioni Universitarie Jaca, 18). Milano.

----------, 1994. *Arnobio e i culti di mistero. Analisi storico-religiosa del V libro dell' 'Adversus Nationes'* (Storia delle Religioni, 10). Roma.

Morgan, C. L., 1894. "Three aspects of Monism". *The Monist* 4, 3: 321-332.

Morley E. J. (ed.), 1967. *Henry Crabb Robinson on Books and Their*

Writers. References in Robinson's Diary, Travel Journals and Reminiscences. 1, New York: AMS Press.

Morley, N., 2004. *Theories, Models, and Concepts in Ancient History.* London: Routledge.

Müller, F. M., 1872. *Lectures on the science of religion.*, New York: Charles Scribner and Co.

Murphy, Tim, 2003. "Toward a Semiotic Theory of Religion", *Method & Theory in the Study of Religion* 15: 48-67.

Ndebele, Njabulo, 1994. *South African Literature and Culture.* Manchester: Manchester University Press.

Nestle, E., 1883. 'Die alten christlichen Inschriften nach dem Text der Septuaginta', *Theologische Studien und Kritiken* 56: 153-154.

Neusner, Jacob, 2005. "Review of Jonathan Z. Smith, *Relating Religion: Essays in the Study of Religion*". *Religion* 35: 128-130.

Newell, Allen, et al., 1958. Elements of a Theory of Human Problem Solving. *Psychological Review* 65, 151-166.

----------------- - Herbert A. Simon, 1972. *Human Problem Solving.* Eaglewood Cliffs: Prentice-Hall.

Nguyen, V. Henry T., 2007. "The Identification of Paul's Spectacle of Death Metaphor in 1 Corinthians 4.9" *New Testament Studies* 53, 4: 489–501.

Nietzsche, Fr., 1967. *On the Genealogy of Morals: A Polemic* (trans. W. Kaufmann and R. J. Hollingdale). New York: Vintage.

Nikolaides, Th. 2001. «Ιστορία και Ανθρωπολογία. Η Δημιουργία της Ύστερης Αρχαιότητας. Εισαγωγή του Μεταφραστή», in: P. Brown, *Η Δημιουργία της Ύστερης Αρχαιότητας.* Transl. by Th. Nikolaides. Athens, 13-19.

Nilsson, M.P., 1957, *The Dionysiac Mysteries of the Hellenistic and Roman Age.* Lund.

----------------, ^{3}1974 - ^{3}1976. *Geschichte der Griechischen Religion,* (Handbuch der Altertumswissenschaft V2, 1-2), vols. I-II. München.

Nock, A. D., 1933. *Conversion. The Old and the New in Religion from Alexander the Great to Augustine of Hippo.* London - Oxford-New York.

Nock, A. D., 1972a. "Notes on Ruler Cult I-IV". In: *idem, Essays on Religion and the Ancient World* (Selected and Edited, with an Introduction by, Bibliography, Notes of Nock's Writings and Indexes by Z. Stewart). Vol. I. Oxford, 134- 159.

----------------, 1972b. 'Soter and Euergetes'. In: idem, *Essays on Religion and the Ancient World, op. cit.*, vol. II, 720-735.

Norris, Pippa- Ronald Inglehart, 2004. *Sacred and Secular Religion and Politics Worldwide*. Cambridge: U.P.

Noy, D. (ed.), 1993. *Jewish Inscriptions of Western Europe, vol.1: Italy (excluding the City of Rome), Spain and Gaul*, Cambridge.

---------- (ed.), 1995. *Jewish Inscriptions of Western Europe, vol. 2: The City of Rome*. Cambridge.

---------- - H. Bloedhorn (eds.), 2004. *Inscriptiones Judaicae Orientis, vol. 3: Syria and Cyprus*, Tübingen.

---------- - Panayotov A. - Bloedhorn H. (eds.), 2004. *Inscriptiones Judaicae Orientis*, vol. 1: *Eastern Europe*. Tübingen.

Obolensky, D., 1948. *The Bogomils. A Study in Balkan Neo-Manichaeism*. Cambridge: University Press.

Oldopp, Birgit - Rainer, Prätorius, 2002. " 'Faith Based Initiative': Ein Neuansatz in der U.S. – Sozialpolitik und seine Hintergründe". In: *Zeitschrift für Sozialreform* 48: 28-52.

Orr, D. G., 1978. "Roman Domestic Religion". *Aufstieg und Niedergang der römischen Welt* II.16.2, 1557-1591.

Orrieux Cl., 1983. *Les Papyrus de Zenon. L' horizon d' un grec en Égypte au IIIe siècle avant J.C.* (Préface d' Eduard Will). Paris.

----------------, 2000 (1992). "The Network", in: Chr. Jacob-Fr. de Polignac, *Alexandria, Third Century B. C. The Knowledge of the World in a Single City*, Alexandria, 175-186

Otter, M., 2008. "Entrances and exits: Performing the psalms in Goscelin's *Liber Confortatorius*". *Speculum* 83, 2: 283-302.

Otto, Rudolf, 1923. *The Idea of the Holy*. New York, Oxford: Oxford University Press.

Pace, B., 1946. *Arte e civiltà della Sicilia antica*. Vol. III. Geneva-Roma-Napoli – Città di Castello.

Pachis, P., 1988. *Το νερό και το αίμα στις μυστηριακές λατρείες της ελληνορωμαϊκής εποχής*. Diss. Thessaloniki.

-------------, 1998. *Δήμητρα Καρποφόρος. Θρησκεία και αγροτική οικονομία του αρχαιοελληνικού κόσμου* (Θρησκειολογία, 17). Athens.

-------------, 2002. 'Η έννοια της περιπλάνησης κατά τη διάρκεια των Ελληνιστικών χρόνων', *Επιστημονική Επετηρίδα Θεολογικής Σχολής*. Νέα

Σειρά, Τμήμα Θεολογίας (Τιμητικό Αφιέρωμα στον Ομότιμο Καθηγητή Νικόλαο Αθ. Ματσούκα), 12: 273-323.

----------, 2003a. Ἴσις αρποτόκος, τομ. Α΄: Οἰκουμένη, Προλεγόμενα στὸν συγκρητισμὸ τῶν ἑλληνιστικῶν χρόνων, Thessaloniki.

----------, 2003b. 'The Hellenistic era as an Age of Propaganda: The Case of Isis' cult', in: L.H. Martin-*idem* [eds.], *Theoretical Frameworks for the study of Graeco-Roman Religions*. Thessaloniki, 97-125.

----------, 2004. "'Manufacturing Religion' in the Hellenistic Age: The Case of Isis-Demeter Cult", in: L. H. Martin-*idem*, *Hellenisation, Empire and Globalization: Lessons from Antiquity. Acts of the panel held during the 3rd Congress of the European Association for the Study of Religion, Bergen, Norway, 8-10 May 2003*. Thessaloniki, 163-207 = id. 2010. "'Manufacturing Religion' in the Hellenistic Age: The Case of Isis-Demeter Cult", in: id., *Religion and Politics in the Graeco-Roman World. Redescribing the Isis – Sarapis, Cult*. Thessaloniki: Barbounakis Editions, 21-74.

----------, 2008. "*Hominibus vagis vitam*: The Wandering of *Homo Hellenisticus* in an Age of Transformation". In: W. Braun-R. McCutcheon (eds.), *Relating Religion. Essays in Honor of Jonathan Z. Smith*, London-Oakville, 388-405.

----------, 2009. "Imagistic Modes of Religiosity and the study of the Cults of Graeco-Roman World". In: L. H. Martin, P. Pachis (Ed.), *Imagistic Traditions in the Graeco-Roman World: A Cognitive Modeling of History of Religious Research*, Thessaloniki: Vanias, 15-34.

Paden, William E., 1999. "Sacrality and Worldmaking: New Categorical Perspectives". In: Ahlbäck, T. (ed.), *Approaching Religion*, Part 1, Åbo, Finland: The Donner Institute, 165-180.

----------, 2000a. "Sacred order". *Method and Theory in the Study of Religion* 12: 207-225.

----------, 2000b. Elements of a New Comparativism. In: Patton, K. and B. Ray (Eds.), *A Magic Still Dwells: Comparative Religion in the Postmodern Age*. Berkeley: University of California Press, 182-192.

----------, 2008. "Connecting With Evolutionary Models: New Patterns in Comparative Religion". In: Braun, W. and R. McCutcheon (Eds.), *Introducing Religion: Essays in Honor of Jonathan Z. Smith*. Equinox, London, 406-17.

Pals, Daniel L., 1987. "Is Religion a *Sui Generis* Phenomenon?" *Journal of the American Academy of Religion* 55: 259-282.

Papini L., 1985. "Biglietti oracolari in copto dalla necropoli di Antinoe". In: T. Orlandi – F. Wisse (eds.), *Acts of the Second International Congress of Coptic Studies, Roma, 22-26 September 1980*. Roma, 245-256.

Papini, L., 1992. *Domande oracolari: elenco delle attestazioni in greco e in copto*, Analecta Papyrologica 4: 21-28.

Parker R., 2002. s.v. 'Timotheos'. *Neue Pauly*, XII 1, 598-599.

Patzig, G., 1983. *Ethik ohne Metaphysik*. Göttingen: Vandenhoeck & Ruprecht.

Pauen, M. 1994. *Dithyrambiker des Untergangs. Gnostizismus in Ästhetik und Philosophie der Moderne*. Berlin: Akademie Verlag.

Pearson, B. A., 2005. "Gnosticism as a Religion". In: A. Marjanen (ed.), *Was There a Gnostic Religion?* Göttingen: Vandenhoeck and Ruprecht, 81-101.

------------------, 2007. *Ancient Gnosticism. Traditions and Literature*. Minneapolis: Fortress Press.

Pedahzur, Ami - Perliger, Arie - Weinberg, Leonard, 2003. "Altruism and Fatalism: the Characteristics of Palestinian Suicide Terrorists". *Deviant Behaviour: An Interdisciplinary Journal* 24: 405-423.

Peronne, L., 2003. "Prayer and the Constitution of Religious Identity in Early Christianity". *Proche-Orient Chrétien* 53: 260-288.

Perpillou-Thomas, Fr., 1993. *Fêtes d' Égypte Ptolémaïque et Romaine d' après la documentation papyrologique Grecque* (Studia Hellenistica, 3,1). Louvain.

Pestman, P. W. (ed.), 1980. *A Guide to the Zenon Archive* (with Contributions by W. Clarysse, M. Korver, M. Muszynski, A. Shutgens, W. J. Tait, J. K. Winnicki) (Papyrologica Lugdono-Batava, vols 21 A and B). Leiden.

Peterson, E., 1926. *Eis theos. Epigraphische, formgeschichtliche und religionsgeschichtliche Untersuchungen*, Göttingen.

Petrou, Ioannis S., 2004. *Christianity and Society. Sociological analysis of relations between Christianity and Society and Culture*. Thessaloniki: Vanias (in Greek).

------------------, 2005a *Multiculturality and Religious Freedom*. Thessaloniki: Vanias (in Greek).

------------------, 2005b. *Social Theory and Modern Culture*. Thessaloniki: Vanias (in Greek).

Petrus Siculus. 1860. *Sermo primus adversus Manichaeos, qui et Pauliciani dicuntur*. In: *Patrologiae cursus completus*. Ed. Jacques-Paul Migne. *Series Graeca* 104. Paris: Migne,1305-1330.

Piatelli-Palmarini, M., 1994. *Inevitable Illusions: How Mistakes of Reason Rule Our Minds*. transl. by Piatelli-Palmarini and K. Botsford. New York: John Wiley and Sons.

Pietri, Ch., 1985. "La Bible dans l' épigraphie de l' Occident latin". In: J. Fontaine and Ch. Pietri (eds.), *Le monde latin antique et la Bible*. Paris, 189-205.

Pinker, S., 2002. *The blank slate: The modern denial of human nature*. New York:Viking.

Piranomonte – Marco Simón 2012: M. Piranomonte, F. Marco Simón (eds.), *Contesti magici Contextos mágicos. Atti del Convegno Internazionale, Roma, Palazzo Massimo, 4-6 novembre 2009*, Roma 2012.

Plantzos, D., 1996. "Ptolemaic Cameos of the second and first centuries BC". *Oxford Journal of Archaeology*, 15, 1: 39 – 61.

Plantzos, D., 1999. *Hellenistic engraved gems*. Oxford.

Polanyi, M., 1962. "Republic of Science: Its political and economic theory". *Minerva* 1: 54-74.

Pollard, J. - Reid, H., 2006. *The Rise and Fall of Alexandria: Birthplace of the Modern Mind*. New York- London.

Poole, Fitz - Porter J., 1986. "Maps and Metaphors: Towards Comparison in the Anthropology of Religion". *Journal of the American Academy of Religion* 54, 3: 411-498.

Poole, R.S., 1865. *Catalogue of the Coins of the Ptolemies*. London.

------, R.S., 1892. *Catalogue of the Coins of Alexandria and the Nomes*. London.

Porter, S.E., 1997. "The Use of the Old Testament in the New Testament: A Brief Comment on Method and Terminology". In: C.A. Evans and J. A. Sanders (eds.), *Early Christian Interpretation of the Scriptures of Israel. Investigations and Proposals*, Sheffield, 79-96.

--------------, 2006. "Further Comments on the Use of the Old Testament in the New Testament". In: Th.L. Brodie, D. R. MacDonald and S.E. Porter (eds.), *The Intertextuality of the Epistles. Explorations of Theory and Practice*. Sheffield, 98-112.

Portes, Alejandro, 1998. "Social Capital: Its Origins and Applications in Modern Sociology". *Annual Review of Sociology* 24: 1-24.

---------------------- - Patricia Landolt, 1996. "The Downside of Social Capital". *The American Prospect* 26: 18-21.

Post, Jerrold M., 1990. "Terrorist psycho-logic: Terrorist Behaviour as a product of psychological forces". In: W. Reich (ed.), *Origins of Terrorism: Psychologies, Ideologies, Theologies, States of Mind*. Cambridge: Cambridge University Press, 25-40.

Post, Jerrold M., 2004. "Cited in George Hough, Does Psychoanalysis have anything to offer an Understanding of Terrorism? Panel Report of the American Psychological Association Meeting in Boston, June 22, 2003". *Journal of the American Psychoanalytic Association* 52, 3: 813-828.

----------------------, 2006. "The Psychology of Terrorism. All in the Mind". Australian Broadcasting Corporation. April 15, 2006. Accessed online www.abc.net.au/rn/science/mind/stories/s1614281.htm

Prätorius, Rainer, 2003. *In God we Trust. Religion und Politik in den USA*. München: C.H. Beck.

Preisendanz, K. (ed.),³1972-1974. Papyri Magicae Graecae. Leipzig

Price, S. R. F., 1998 (1984). *Rituals of Power. The Roman Imperial Cult in Asia Minor*, Cambridge: Cambridge University Press (reprint).

Provitera, G. A. 1980. "Politica religiosa dei Dinomenidi e ideologia dell' *optimus rex*". In: G. Piccaluga (ed.), *Perenitas. Studi in onore di A. Brelich*. Roma, 393-411.

Puschmann, Th. (ed.), 1879. *Alexander von Tralles. Ein Beitrag zur Geschichte der Medizin*. Wien.

Putnam, Hilary Whitehall, 1967. The Nature of Mental States. In: Capitan, W. and D. D. Merrill (Eds.), *Art, Mind and Religion*. Pittsburgh: University of Pittsburgh Press, 37-48.

Putnam, Robert D., 1993. *Making Democracy Work. Civic Traditions in Modern Italy*. Princeton: U.P.

-------------------------------, 1995. "Bowling Alone: America's Declining Social Capital". In: *Journal of Democracy* 6: 65-78

-------------------------------, 2000. *Bowling Alone. The Collapse and Revival of American Community* New York: Simon & Schuster.

-------------------------------, 2001. *Gesellschaft und Gemeinsinn. Sozialkapital im internationalen Vergleich*. Gütersloh: Bertelsmann.

Pyysiäinen, Ilkka, 2002. "Mind and Miracles". *Zygon* 37.3, 729 - 740 = id., 2004. *Magic, Mind and Miracles. A Scientist's Perspective*. Cognitive Science of Religion Series. Walnut Creek, CA: Altamira Press, 729-740.

-------------------------, 2008. "After Religion: Cognitive Science and the Study of Human Behaviour". Published online at: http://e-religions.net/page.php?id=27&comments=1.

Quispel, G., 1973. "Gnosis und hellenistische Mysterienreligionen". In: U. Mann (ed.), *Theologie und Religionswissenschaft. Der gegenwärtige Stand ihrer Forschungsergebnisse und Aufgaben im Hinblick auf ihr gegenseitiges Verhältnis*. Darmstadt: Wissenschaftliche Buchgesellschaft, 318-331.

---------------, 1975. "Hesse, Jung und die Gnosis". In: *Gnostic Studies* 2. Uitgaven van het Nederlands Historisch-Archaeologisch Instituut te Istanbul 34/1. Istanbul: Nederlands Historisch-Archaeologisch Institut in het Nabije, 241-258.

-------------, 1978. "Hermann Hesse and Gnosis". In: B. Aland (ed.), *Gnosis. Festschrift für Hans Jonas*. Göttingen: Vandenhoeck and Ruprecht, 492-507.

-------------, 1980. "Gnosis and Psychology". In: B. Layton (ed.), *The Rediscovery of Gnosticism* 1. *The School of Valentinus*. Studies in the History of Religions. Supplements to Numen 41. Leiden: E.J. Brill,17-31.

-------------,1981. "Gnosis". In: M. J. Vermaseren (ed.), *Die Orientalischen Religionen im Römerreich*. Études préliminaires aux religions orientales dans L'Empire romain 93. Leiden: E.J. Brill, . 413-435

--------------------, 1995. *Gnosis als Weltreligion*. 3rd ed. Lehre und Symbol 38. Bern: Origo Verlag.

Rapoport, David. C., 1993. "Comparing Militant Fundamentalist Movements and Groups". In: B. Appleby - R. Scott – M. E. Marty (eds.), *The Fundamentalism Project*. Vol. 3. Chicago: The University of Chicago Press, 429-461.

-------------------------, 1999. *Ritual and religion in the making of humanity*. Cambridge: Cambridge University Press.

Raine, K., 2002. *Blake and Tradition*. Vol. 2. London: Routledge and Kegan Paul.

Ramasway, Krishnan - Antonio de Nicolas - Aditi Banerjee (eds.), 2007. *Invading the Sacred: An Analysis of Hinduism Studies in America*. Delhi, India: Rupa & Co.

Rawls, John, 2001. *Justice as Fairness. A Restatement*. edited by Erin Kelly. London: Belknap Press.

Reardon, B. P., (ed), 1989. *Collected Ancient Greek Novels*. Berkeley - Los Angeles/London: University of California Press.

Reed, Annette Yoshiko, 2002. "ΕΥΑΓΓΕΛΙΟΝ: Orality, Textuality, and the Christian Truth in Irenaeus' *Adversus Haereses*", *Vigiliae Christianae* 56: 11-46.

Reekmans, T., 1983. "Archives de Zénon: Situation et comportment des entrepreneurs indigènes". In: E-Van't Dack–P. Van Dessel-W. Van Gucht (eds.), *Egypt and the Hellenistic World*. Studia Hellenistica, 27. Leuven, 325-350.

Renaut, Allain, 2004. *La Fin de l' Autorité*. Paris: Editions Flammarion.

Rennie, Bryan (ed.), 2006. *Mircea Eliade: a Critical Reader*. London: Equinox.

Reynolds, C. W., 1987. "Flocks, herds, and schools: A distributed behavioral model". *Computer Graphics*, 21,4: 25-34.

Rhys-Davids, T. W., Oldenberg, H., 1996. *Vinaya texts* . Vol. 13). Delhi: Motilal Banarsidas.

Richardson, Robert C., 1979. "Functionalism and Reductionism". *Philosophy of Science* 46, 4: 533-558.

Richter, R. 1996. "Politische Ordnung und Fremde Kultur im Bild der hellenistischen Utopie". In: B. Funk [ed.], *Hellenismus, Beiträge zur Erforschung von Akkulturation und politischen Ordnung in den Staaten des hellenistischen Zeitalter. Akten des Internationalen Hellenismus-Kolloqiums 9-14 März 1994 in Berlin*, Tübingen, 629-652.

Rizzolatti,1. Giacomo - Luciano Fadiga - Leonardo Fogassi - Vittorio Gallese, 2002. "From Mirror Neurons to Imitation: Facts and Speculations". In: Andrew N. Meltzoff & Wolfgang Prinz (eds.), *The Imitative Mind: Development, Evolution, and Brain Bases*. Cambridge & New York: Cambridge University Press, 247-266.

Rochat, P., Morgan, R., and Carpenter, M., 1997. "Young Infants' Sensitivity to Movement Information Specifying Social Causality". *Cognitive Development* 12: 441-465.

Rose, Susan, 1993. "Christian Fundamentalism and Education in the United States". In: M. E. Marty – R. Scott. Appleby (eds.), *The Fundamentalism Project*. Volume 2. Chicago: The University of Chicago Press, 452-489.

Rostovtzeff, M., 1941. *Social and Economic History of the Hellenistic World*, vols I-III. Oxford (reprint).

Roussel, P., 1916. *Les cultes ègyptiennes à Dèlos du IIIe au Ier siècle av. J. - C. (Annales d'Est)*. Nancy.

Rottenwöhrer, G. 1986. *Unde malum? Herkunft und Gestalt des Bösen nach heterodoxer Lehre von Markion bis zu den Katharern.* Bad Honnef: Bock and Herchen Verlag.

RPC= The Roman Provincial Coinage project: www.rpc.ashmus.ox.ac.uk.

Rudolph, K., 1980. *Die Gnosis. Wesen und Geschichte einer spätantiken Religion.* 2nd ed. Göttingen: Vandenhoeck and Ruprecht.

---------------, 1992. "Erkenntnis und Heil. Die Gnosis". In: C. Colpe et al. (eds.), *Spätantike und Christentum. Beiträge zur Religions- und Geistesgeschichte der griechisch-römischen Kultur und Zivilisation der Kaiserzeit.*. Berlin: Akademie Verlag, 37-54.

Runciman, S., 1960. *The Medieval Manichee. A Study of the Christian Dualist Heresy.* Cambridge: Cambridge University Press.

Rutgers, L. V., 1998. "The Importance of Scripture in the Conflict Between Jews and Christians: the Example of Antioch". In: L.V. Rutgers, P.W. van der Horst, H.W. Havelaar, L. Teugels (eds.), *The Use of Sacred Books in the Ancient World,* Louvain, 287-304.

Ruyer, R., 1977. *Jenseits der Erkenntnis. Die Gnostiker von Princeton.* Vienna and Hamburg: Paul Zsolnay Verlag.

Sabottka, M., 1989. Das Serapeum in Alexandria : Untersuchungen zur Architektur und Baugeschichte des Heiligtums von der freuhen ptolemaischen Zeit bis zur Zersterung 391 n. Chr. PhD Dissertation. Technischen Universität Berlin.

Sacks, O. W., 2007. *Musicophilia: Tales of music and the Brain.* New York: Alfred A. Knopf.

Sahlins, Marshall, 1985. *Islands of History* .Chicago: University of Chicago Press.

---------------------, 2005. "Structural Work: How Microhistories Become Macrohistories and Vice Versa", *Anthropological Theory* 5: 5-30.

Saler, Benson, 1993. *Conceptualizing Religion: Immanent Anthropologists, Transcendent Natives, and Unbounded Categories.* E.J. Brill, Leiden.

-----------------, 1997 (2003). "Lévy-Bruhl, Participation, and Rationality". In: J. Jensen - L. H. Martin (eds.), *Rationality and the Study of Religion.*, Aarhus: Aarhus University Press, 44-64.

Salomon, R., 1998. *Indian epigraphy: A guide to the study of inscriptions in sanskrit, prakrit, and the other indo-aryan languages.* New York: Oxford University Press.

---------------- - Allchin, F. R. - Barnard, M., - British Library, 1999.

Ancient buddhist scrolls from gandhara: The british library kharosthi fragments., London: British Library.

Sanzi E., 2002. Mithras: a *deus invictus* among Persia, Stars, Oriental Cults and Magical Gems". In: AA.VV., *Charmes et sortilèges. Magie et magiciens* (Res Orientales, 14). Bures-sur-Yvette, 209-230.

------------, 2008. "Il santo martire Colluto: archiatra del copro e dell'anima. Osservazioni storico-religiose su alcune testimonianze copte". In: *Asclepio e Cristo. Culti terapeutici e taumaturgia nel mondo mediterraneo antico fra pagani e cristiani. Convegno internazionale, Agrigento, 20-21 novembre 2006*. Roma, 189-202

Sauer, R. 1992. "Die Waldorfpädagogik. Eine gnostische Erziehungslehre?". In: W. Eckermann et al. (eds.), *Erlösung durch Offenbarung oder Erkenntnis? Zum Wiedererwachen der Gnosis.* Ed. Kevelaer: Verlag Butzow and Bercker Kevelaer, 185-206.

Scheid, J., 2003. *An Introduction to Roman Religion*. Edinburgh: Edinburgh University Press.

Schenke, H. M., 1965. "Hauptprobleme der Gnosis". *Kairos* 7: 114-33.

---------------------, 1985. "Radikale sexuelle Enthaltsamkeit als Hellenistisch-jüdisches Vollkommenheitsideal im Thomas-Buch (NHC II,7)". In: U. Bianchi (ed.), *La tradizione dell' enkratéia. Motivazioni ontologiche e protologiche. Atti del Colloquio Internationale–Milano, 20-23 aprile, 1982.* Roma: Edizione dell' Atheneo, 263-291.

------------, 1989. *Das Thomas-Buch: Nag-Hammadi-Codex II, 7/neu herausgegeben, übersetzt und erklärt.* (Texte und Untersuchungen zur Geschichte der altchristlichen Literatur138). Berlin: Akademie-Verlag.

Schiff, 1905. s.v. 'Eleusis' (4). *RE* 5, 2339-2342.

Schildgen, Brenda Deen, 1980. *Cliffs of Fall* (New York: Seabury).

Schimmel, Annemarie, 1960. "Summary of the Discussion". *Numen* 7: 235-239.

Schmidt, M. G.,2000. *Demokratietheorien. Eine Einfuerung.* Opladen: Verlag Leske & Budrich.

Schneider, C., 1967-1969. *Kulturgeschichte des Hellenismus.* vols I-II, München.

Schoenborn, Ulrik, 1994. "Vom Weinstock. Die Gleichnisrede in *LibThom* (NHC II,7) 144, 19-36". In: H. Preissler, et al. (eds.), *Gnosisforschung und Religionsgeschichte. Festschrift für Kurt Rudolph zum 65. Geburtstag.* Marburg: Diagonal-Verlag, 267-285.

Scholz, U. W., 1980. "Zur Erforschung der römischen Opfer (Beispiel die Lupercalia)". In: J. Rudhardt - O. Reverdin (eds.), *La sacrifice dans l'antiquité*. Fondation Hardt, XXVII. Genève: Vandœvres, 289-340.

Schopen, Gregory, 1996. "The Suppression of Nuns and the Ritual Murder of Their Special Dead in Two Buddhist Monastic Texts". *Journal of Indian Philosophy* 24: 563-592.

------------------------1997a. *Bones, stones, and buddhist monks: Collected papers on the archaeology, epigraphy, and texts of monastic buddhism in india.*, Honolulu: University of Hawaii Press.

------------------------, 1997b. "Relic". In: M. Taylor, (ed.), *Critical Terms for the Study of Religion*. Chicago: The University of Chicago Press, 256-268.

------------------------, 2004. *Buddhist monks and business matters: Still more papers on monastic buddhism in india*. Honolulu: University of Hawaii Press.

Schore, Alan, 2003. *Affect Regulation and the Repair of the Self*. New York: W.W. Norton.

Schroeder, Fr. M. 1989. "The Self in Ancient Religious Experience:, in: A.H. Amstrong (ed.), *Classical Mediterranean Spirituality. Egyptian, Greek, Roman* (World Spirituality, 15), New York, 336-359.

Schuppert, Gunnar Folke, 2005. "Skala der Rechtsformen für Religion: vom privaten Zirkel zur Körperschaft des öffentlichen Rechts. Überlegungen zur angemessenen Organisationsform für Religionsgemeinschaften". Sullivan, Denis J., 11–35.

Scott, James, 1977. "Patronage or Exploitation?". In: Ernest Gellner - John Waterbury (eds.), *Patrons and Clients in Mediterranean Societies*. London: Duckworth, 21-39.

Schwabe, M. - Lifshitz, B., 1974. *Beth She'arim, vol. II: The Greek Inscriptions*, Jerusalem.

Schwarz, G., 1987. *Triptolemos. Ikonographie einer Agrar- und Mysteriengottheit* (Gräzer Beiträge, Suppl. 2). Horn.

Sfameni Gasparro, G., 1985/1986. *I Culti egiziani nel mondo ellenistico-romano: destino dell' uomo e religiosità isiaca* (Università di Messina Facoltà di Lettere e Filosofia) (Lezioni di Storia delle Religioni raccolte dalla Dr. Concetta Aloe Spada). Messina.

------------------------, 1986. *Misteri e culti di Demetra* (Storia delle Religioni, 3). Roma.

------------------------, 1997. "Daimon and Tychē in the Hellenistic Experience". In: P. Bilde-T. E. Pedersen-L. Hannestad-J. Zahle [eds.],

Conventional Values of the Hellenistic (Studies in Hellenistic Civilization, VIII). Aarhus, 67-109 = ead, 2002. *Oracoli, Profeti, Sibille. Revelazione e salvezza nel mondo antico*. Bibliotheca di Scienze Reliigose, 171. Roma, 255-301.

---------------------------, Sfameni Gasparro 2010: G. Sfameni Gasparro, *Dio unico, pluralità e monarchia divina. Esperienze religiose e teologie nel mondo tardo-antico*, Brescia 2010.

Shapiro, Lawrence A., 2004. *The Mind Incarnate*. Cambridge, MA.: MIT Press.

Sharf, Robert, 1999. "On the Allure of Buddhist Relics". *Representations* 66: 75-99.

Sharma, Arvind, 2001. *To the Things Themselves: Essays on the Discourse and Practice of the Phenomenology of Religion.*, Berlin - New York: Walter de Gruyter.

Sharpe, E. J., 1986. *Comparative religion: A history*. Second edition. La Salle, Il.: Open Court.

Sheridan, J. F., 1957. *Paul Carus: A study of the thought and work of the editor of the Open Court Publishing Company*. Unpublished Ph.D. Thesis, University of Illinois, Urbana, IL.

Shweder, Richard A., 1991. *Thinking Through Cultures: Expeditions in Cultural Psychology*. Cambridge: Harvard University Press.

Sider, Ronald J. - Heidi Rolland, Unruh, 2001. "Evangelism and Church-State Partnerships". *Journal of Church and State* 43: 267-296.

Siegel, Daniel J., 2001. "Toward an Interpersonal Neurobiology of the Developing Mind: Attachment Relationships, 'Mindsight,' and Neural Integration," *Infant Mental Health Journal* 22 (1-2): 67-94.

Silk, J. A., 1994. "The victorian creation of buddhism". *Journal of Indian Philosophy, 22,* 171-196.

Silk, M., 2006. "Hold the Prayers". *Religion in the News* 9, 1.

Silverstein, T., 1948. "The Fabulous Cosmogony of Bernardus Silvestris". *Modern Philology* 46: 92-116.

Simon, Herbert, 1969. *The Sciences of the Artificial.*, Cambridge, MA.: MIT Press.

Skinner, Burrhus Frederic, 1938. *The Behaviour of Organisms: An Experimental Analysis*. New York: Appleton-Century-Crofts.

Skowronek, St. –Tkaczow, B., 1979. "Le culte de la déesse Déméter à Alexandrie". In: L. Kahil – Chr. Augé (eds.), *Mythologie gréco-romaine. Mythologies péripheriques. Étude d' iconographie. Paris 17 mai 1979*

(Colloques international du C.N. R. S., 593). Paris, 131-144, fig. I.

Slingerland, Edward, 2008. "Who's Afraid of Reductionism? The Study of Religion in the Age of Cognitive Science". *Journal of the American Academy of Religion* 76.2: 375-411.

Singerman, Diane, 2004. "The Networked World of Islamist Social Movements". In: Quintan Wiktorowicz (ed.), *Islamic Activism. A Social Movement Theory Approach.* Bloomington: Indiana UP 2004, pp. 143-163.

Slone, D. J. - Gonce, L. - Upal, A. - Edwards, K. - Tweney, T., 2007. "Imagery Effects on Recall of Minimally Counterintuitive Concepts". *Journal of Cognition and Culture* 7: 355-367.

Slone, J., 2004. *Theological Incorrectness: Why Religious People Believe What They Shouldn't.* New York: Oxford University Press.

Sloterdijk, P., 1991. "Die wahre Irrlehre. Über die Weltreligion der Weltlosigkeit". In: P. Sloterdijk and T. H. Macho (eds.), *Weltrevolution der Seele. Ein Lese- und Arbeitsbuch der Gnosis von der Spätantike bis zur Gegenwart* 1. Munich: Artemis and Winkler Verlag.

Smail, D. L., 2008. *On deep history and the brain.* Berkeley: University of California Press.

Smith, J. Z., 1982a. "The Bare Facts of Ritual", in: id., 1982. *Imagining Religion: From Babylon to Jonestown*, Chicago: The University of Chicago Press, 53-65.

----------------, 1982b. *Imagining Religion. From Babylon to Jonestown.* Chicago: Chicago University Press.

----------------, 1987. *To Take Place: Toward Theory in Ritual.* Chicago: The University of Chicago Press.

----------------, 1990. *Drudgery Divine: On the Comparison of Early Christianities and the Religions of Late Antiquity.* Chicago: The University of Chicago Press.

----------------, 1992. "Scriptures and Histories" *Method & Theory in the Study of Religion* 4: 97-105

----------------, 1993 (1979). *Map is Not Territory.* Chicago: The University of Chicago Press.

----------------, 1998. "Canons, Catalogues and Classics". In: A. Van der Kooij and K. Van der Toorn (eds.), *Canonization and Decanonization: Papers Presented to the International Conference of the Leiden Institute for the Study of Religions (LISOR), Held at Leiden 9-10 January 1997* . Leiden: Brill, 295-311.

----------------, 2004a. *Relating Religion: Essays in the Study of Religion,* Chicago - London: University of Chicago.

----------------, 2004b. "The Topography of the Sacred", in: id. *Relating Religion: Essays in the Study of Religion*. The University of Chicago Press, Chicago, 101-116.

Smith, R., 1988. "The Modern Relevance of Gnosticism". In: J. M. Robinson (ed.), *The Nag Hammadi Library in English*. 3rd ed. San Francisco: Harper and Row, 532-549.

Smith, Wilfred Cantwell, 1959. "Comparative religion: Wither – and why?". In: M. Eliade and J. M. Kitagawa (Eds.), *The History of Religions. Essays in Methodology*, Chicago, 31-58. = Id., 1976. *Religious Diversity*, 1976. Repr. in abridged form and an introduction by the editor, W. G. Oxtoby. New York: Harper & Row, 138-157.

SNG= *Sylogge Nummorum Graecorum*: www.sylloge-nummorum-graecorum.org.

Snyder, Graydon, 1985 (2003, rev. edit.). *Ante Pacem: Archaeological Evidence of Church Life Before Constantine*. Macon: Mercer University Press.

Solmsen, F., 1979. *Isis among the Greeks and Romans*. Cambridge.

Speckhard, Anne, 2006. "The Psychology of Terrorism. All in the Mind". Australian Broadcasting Corporation. April 15.

Sperber, D., 1975. *Rethinking symbolism*. Cambridge- New York: Cambridge University Press.

----------------, 1985. "Anthropology and Psychology: Towards an Epidemiology of Representations". *Man* 20: 73-89.

----------------, 1996. *Explaining culture: A naturalistic approach*. .Oxford - Cambridge, MA.: Blackwell.

----------------, 2000. *Metarepresentations: A multidisciplinary perspective.*, Oxford - New York. Oxford University Press.

Spuerri, W., 1979. 'Leon von Pella', *Kleine Pauly* 3, 565.

Spyridakis, S., 1968. "Zeus is Dead: Euhemerus and Crete", *The Classical Journal* 63, 8 (May): 337-340.

Staal, Frits, 1979a. "Ritual Syntax", in: M. Nagatomi, et al. (eds.), *Sanskrit and Indian Studies: Essays in Honour of Daniel H. H. Ingalls*. Dordrecht: D. Reidel, 119-142.

Staal, Frits, 1979b. "The meaninglessness of ritual". *Numen* 26, 1: 2-22.

Stambaugh, J. E., 1972. *Sarapis under the Early Ptolemies*. Leiden.

Stammkötter, F.-B., 1992. "Wir wollen sie mit der Vernunft widerlegen. Origenes in der Auseinandersetzung mit der Gnosis". In: W. Eckermann et al. (eds.), *Erlösung durch Offenbarung oder Erkenntnis?*. Keverlaer: Verlag

Butzow und Bercker, 69-83.

Stanley, C.D., 1997. "The Rhetoric o Quotation: An Essay on Method". In: C.A. Evans and J.A. Sanders (eds.), *Early Christian Interpretation of the Scriptures of Israel. Investigations and Proposals.* Sheffield, 44-58.

Stanwick, P. E., 2002. *Portraits of Ptolemies. Greek Kings as Egyptian Pharaoh.* Austin.

Stein, Ruth, 2003. "Evil as Love and as Liberation: The Mind of a Suicidal Religious Terrorist". In: D. Moss, (ed.), *Hating in the First Person Plural: Psychoanalytic Essays on Racism, Homophobia, Misogyny, and Terror.* New York: Other Press, 281-310.

Stephens, S. A., 1994. "Who Read Ancient Novels?". In: James Tatum (ed.), *The Search for the Ancient Novel.* Baltimore/ London: Johns Hopkins University Press, 405–18.

--------------------, 2003. *Seeing Double. Intercultural Poetics in Alexandria* (Hellenistic Culture and society, XXXVII). Berkeley-Los Angeles-London.

Stern, Jessica, 2003. *Terror in the Name of God: Why Religious Militants Kill.* New York: HarperCollins.

Steward, A., 1993. Faces of Power. Alexander's Image and Hellenistic Politics. Berkeley.

Stewart, Z., 1977. "La religione". In: R. Bianchi Bardinelli [ed.], *Storia della civiltà dei Greci.* Vol. VIII. Milano, 503-616.

Stowe, D. W., 2004. *How sweet the sound: Music in the spiritual lives of Americans.* Cambridge, MA.: Harvard University Press.

Stowers, Stanley K., 2004. "Mythmaking, Social Formation, and Varieties of Social Theory". In: Ron Cameron - Merrill P. Miller (eds.), *Redescribing Christian Origin.* SBLSS, 28. Atlanta: Society of Biblical Literature; Leiden: Brill, 489-495.

Straten, F. T., Van, 1981. "Gifts for the Gods". In: H. S. Versnel, (ed.), *Faith, Hope and Worship: Aspects of Religious Mentality in the Ancient World..* Leiden: Brill, 65-151.

Strauss, C., Quinn, N., 1997. *Cognitive theory of cultural meaning.* Cambridge: Cambridge University Press.

Strenski, Ivan, 2003. "Sacrifice, Gift and the Social Logic of Muslim 'Human Bombers' ". *Terrorism and Political Violence* 15, 3: 1-34.

Strohm, H. 2005. *Die Gnosis und der Nationalsozialismus. Eine religionspsychologische Studie.* Aschaffenburg: Alibri Verlag.

Strong, John S., 2004. *Relics of the Buddha.* Princeton: Princeton University Press.

----------------------, 2007. "Buddhist Relics in Comparative Perspective: Beyond the Parallels". In: D. Germano - K. Trainor (eds.), *Embodying the Dharma: Buddhist Relic Veneration in Asia.*, Albany: SUNY Press, 27-50.

Strozier, Charles B., 2002. *Apocalypse: On the Psychology of Fundamentalism in America.* Eugene, OR.: Wipf and Stock.

Stroumsa, G. G., 1985. "Die Gnosis und die christliche 'Entzauberung der Welt'". In: W. Schluchter (ed.), *Max Webers Sicht des antiken Christentums. Interpretation und Kritik.* Frankfurt: Suhrkamp, 486-508.

----------------------, 1992. "Anti-Cathar Polemics and the *Liber de duobus principiis*". In: B. Lewis - F. Niewöhner (eds.), *Religionsgespräche im Mittelalter.* Wolfenbütteler Mittelalter-Studien 4. Wiesbaden: Otto Harrassowitz, 169-183

Sullivan, A., 2001. "This is a Religious War." *The New York Times Magazine.* 7 (October 7): 44-47, and 52-53.

Sullivan, Denis J., 1994. *Private Voluntary Organizations in Egypt. Islamic Development, Private Initiative, and State Control.* Gainesville: University Press of Florida.

---------------------- - Sana Abed-Kotob, 1999. *Islam in Contemporary Egypt. Civil Society vs. The State.* Boulder (Color.) : Lynne Rienner.

Svoronos, 1904, Ta Nomismata ton Ptolemaion. Athens.

Swearer, Donald K., 2004. *Becoming the Buddha: The Ritual of Image Consecration in Thailand.* Princeton: Princeton University Press.

Szemler, G. J., 1972. *The Priests of the Roman Republic. A Study of Interactions Between Priesthoods and Magistracies.* (Collection Latomus). Bruxelles.

Tabor, J., 2006. *The jesus dynasty: Stunning New Evidence about the Hidden History of Jesus.* London: Harper Element.

Tambiah, Stanley, 1990. *Magic, Science, Religion, and the Scope of Rationality.* Cambridge: Cambridge University Press.

Taubes, S. A., 1954. "The Gnostic Foundations of Heidegger's Nihilism". *The Journal of Religion* 34: 155-72.

Taylor, L. R., 1975. *The Divinity of Roman Emperor.* Papers and Monographs of the American Philological Association, 1. New York (reprint).

Thagard, Paul, 2005. *Mind: Introduction to Cognitive Science.* Cambridge, MA.: MIT Press.

Thomas, Christine, 2008. "Place and Memory: Response to Jonathan Z. Smith on *To Take Place,* On the Occasion of its Twentieth Anniversary".

Journal of the American Academy of Religion 76, 3: 773-780.

Thomas, R. K., 2001. *Lloyd Morgan's Canon: A history of misrepresentation* [Electronic Version]. History & Theory of Psychology Eprint Archive. Retrieved November 12, 2008 from http://htpprints.yorku.ca/archive/ 00000017/.

Thompson, D. B., 1972. Ptolemaic oinochoai and portraits in faience : aspects of the ruler-cult. Oxford.

Thompson, D. J., 1988a. *Memphis under the Ptolemies*. Princeton.

—————————, 1988b. "Demeter in Graeco-Roman Egypt". In: Willy Clarysse–Antoon Schoors–Harco Willems (eds.), *Egyptian Religion. The Last Thousand Years. Studies Dedicated to the Memory of Jan Quaegebeur* (Orientalia Lovaniensia Analecta/ Department Orientalisstick, Kath. Univers. Leuven, 1995). Leuven, 699-707.

Tinbergen, N., 1951. *The Study of instinct.*, Oxford: Oxfrod Clarendon Press.

Tomasello, M., 1999. *The Cultural Origins of Human Cognition*. Cambridge, MA.: Harvard University Press.

Tondrieu J., 1948. "Princesses ptolémaïques comparées ou identifiès à des déesses (IIIe-Ier siècle avant J.-C", *BSRA* 37: 12-33.

Tooby, J. - Cosmides, l., 1992. "Psychological foundations of culture", in: J. H. Barkow, ead. – id. (eds.), *Adapted mind: Evolutionary psychology and the generation of culture*, Oxford: Oxford University Press, 19-136.

Totti, M., 1986. *Ausgewählte Texte der Isis-und Sarapis Religion*. Hildesheim-Zürich- New York.

Trainor, Kevin, 1992. "When is Theft not a Theft? Relic Theft and the Cult of the Buddha's Relics in Sri Lanka". *Numen* 39, 1: 1-26.

—————————, 1997. *Relics, Ritual, and Representation in Buddhism: Rematerializing the Sri Lankan Theravada Tradition*. Cambridge: Cambridge University Press.

————————— (forthcoming). "Buddhism in a nutshell: The uses of *Dhammapada* 183". In: C. Anderson et al. (Eds.), *Embedded religions: Essays in honor of W. S. Karunatillake*. S. Godage and Brothers, Colombo.

Tremlin, T., 2006. *Minds and Gods - the Cognitive Foundations of Religion*. New York: Oxford University Press.

Tröger, K. W., 2001. *Die Gnosis. Heilslehre und Ketzerglaube*. Freiburg and Basel: Herder.

Troianovski, A. S., 2005. "Bio-Anthro profs seek own dept. *Harvard*

Crimson". Retrieved 6 August 2008, from http://www.thecrimson.com/article.aspx?ref=508049.

Tsing, A. L., 1993. *In the realm of the diamond queen.* Princeton: Princeton University Press.

Turcan, R., 2000. *The Gods of Ancient Rome: Religion in Everyday Life from Archaic to Imperial Times.* Edinburgh: Edinburgh University Press.

Turing, Alan M., 1950. "Computing machinery and intelligence". *Mind* 50: 433-460.

Turner, John D., 1975. *The Book of Thomas the Contender. Coptic Text with Translation, Introduction and Commentary.* (Society of Biblical Literature Dissertation Series 23). Missoula, mont.: Scholars Press.

----------------------, 1989. "The Book of Thomas the Contender writing to the Perfect". In: B. Layton (ed.), *Nag Hammadi Codex II, 2-7.* Volume II. Leiden: Brill, 171-205.

----------------------, 2007. "The *Book of Thomas* and the Platonic Jesus". In: L. Painchaud - P.-H Poirier (Eds.), *L'évangile selon Thomas et les textes de Nag Hammadi,' Bibliothèque Copte de Nag Hammd*i, '. Québec – Paris: Etudes.' Les Presses de l´Université Laval/Peeters.

Tyson, Joseph B., 2006. *Marcion and Luke-Acts: A Defining Struggle.* Columbia: University of South Carolina Press.

Tweed, T. A., 2000. *The American encounter with Buddhism, 1844-1912: Victorian culture and the limits of dissent.* Chapel Hill, NC: University of North Carolina Press.

Tweney, R. D. - Upal, M. A. - Gonce, L. O. - Slone, D. J. - Edwards, K., 2006. "The Creative Structuring of Counterintuitive Worlds". *Journal of Cognition and Culture* 6: 483-498.

Tylor, E. B., 1871. *Primitive culture.* London: Murray.

Uro, Risto, 2003. *Thomas. Seeking the Historical Context of the Gospel of Thomas.*, London - New York: T. and T. Clark Publishers

Vanderlip, V. Fr., 1972. *The Four Greek Hymns of Isidorus and the Cult of Isis* (American Studies in Papyrology, vol. 12). Toronto.

Verhoeven, M. J., 2004. Introduction: The dharma through Carus's lens. In: *The gospel of Buddha according to old records, by Paul Carus.*, Chicago: Open Court, 1-101.

Vermaseren, M.J. (ed.), 1956-1960. Corpus Inscriptionum et Monumentorum Religionis Mithriacae. Vols. I-II., The Hague.

Vernière, Y, 1990. "L'expédition mythique d'Osiris-Dionysos en Asie et ses prolongements politiques". In: J. François – A. Motte (eds.), *Mythe et Politique. Actes du Colloque du Liège, 14-16 septembre 1989 organisé par le Centre de recherches Mythologiques de l' Universitè de Liège*. Bibliothèque de la Faultè de Philosophie et Letrres de l' Universitè de Liège, fasc. CCLVII. Paris, 279-285.

Versnel, H. S., 1981. "Religious Mentality in Ancient Prayers", in: id. (ed.), *Faith, Hope and Worship: Aspects of Religious Mentality in the Ancient World*. Leiden Brill, 1-64.

––––––––––––––––––, 1990 (1998). *Ter Unus. Isis, Dionysus, Hermes. Three Studies in Henotheism* (Inconsistencies in Greek and Roman Religion, 1). Leiden-New York-Køhenhavn-Köln.

––––––––––––––––––, 1991. "Beyond Cursing: The Appeal to Justice in Judicial Prayers" In: C. A. Faraone - D. Obbink (Eeds.), *Magika Hiera: Ancient Greek Magic and Religion*. New York - Oxford, Oxford University Press, 60-106.

Viellefond, J-R., 1972. "ΔΑΜΑΤΕΡ ΦΟΙΝΙΚΟΠΕΖΑ". In: *Studi classici in onore di quintilio Catandella*, vol. I. Catania, 141-148.

Visser, C. E., 1938. *Götter und Kulte im ptolemäischen Alexandrien*. Amsterdam.

Voegelin, E., 1960. "Religionsersatz. Die gnostischen Massenbewegungen unserer Zeit". *Wort und Wahrheit* 15: 5-18.

Volkmar, Gustav, 1857. *Die Religion Jesu*. Leipzig: Brockhaus.

Vozo, G., 1973. "Eloro". In: P. Pelagatti - *idem* (eds.), Archeologia nella Sicilia sud-oriente (Centre Jean Bénard, Napoli). Siracusa, 117-126, tav. XLI.

–––––––––––, 1980-1981. 'L' attività della Sicilia orientale. Parte I, *Kokalos* 26-27: 685-687, tav. CXXVIII – CXXIX).

Waardenburg, Jacques, 2007. *Muslims as Actors: Islamic Meanings and Muslim Interpretation in the Perspective of the Study of Religions*. Berlin - New York. Walter de Gruyter.

Wachsmuth, D., 1980. "Aspekte des antiken mediterranean Hauskults". *Numen* 27: 24-75.

Walbank, Fr. W., 1993. *Ο Ελληνιστικός κόσμος*. Transl. By T. Βαρβέρης. Επιμ. Λ. Μανωλόπουλος-Π. Νίγδελης). Thessaloniki: Vanias

Wallace, A., 1966. *Religion: An anthropological view.*, New York, Random House.

Warner, R. S., 2008. (2007). "Presidential address: Singing and solidarity". *Journal for the Scientific Study of Religion* 47, 2: 175-190.

Weber, Max, 1978. *Economy and Society.* Ed. by Guenther Roth and Claus Wittich. 2 vols. Berkeley: University of California Press.

---------------, 2001a. *Wirtschaft und Gesellschaft.* Volume 1: *Gemeinschaften,* ed. by Wolfgang Mommsen together with Michael Meyer. Max Weber Gesamtausgabe I/22-1. Tübingen: Mohr Siebeck 2001.

---------------, 2001b. *Wirtschaft und Gesellschaft.* Vol. 2. *Religiöse Gemeinschaften.* Ed. by Hans G. Kippenberg together with Petra Schism and Jutta Niemeier. Max Weber Gesamtausgabe I/22-2. Tübingen: Mohr Siebeck.

Weber, R., 1983. *Gottfried Benn. Zwischen Christentum und Gnosis.* Europäische Hochschulschriften I. Deutsche Sprache und Literatur 568. Frankfurt/Main: Peter Lang.

Wegener, F., 2007. *Alfred Schuler, der letzte deutsche Katharer. Gnosis, Nationalsozialismus und mystische Blutleuchte.*Gladbeck: KFVR Verlag.

Wegner, M., 1982. "Terrakoten einer Frau mit einem Ferkel", in: *APARCHAI. Nuove ricerche e studi sulla Magna Grecia e la Sicilia antica in onore di Paolo Enrico Arias* (Pisa, I, 201-219; vol III, tav. 57-58.

Weinrich, O., 1912. "Theoi Epekooi". *Mitteilungen des Deutschen Archäologischen Instituts. Athenische Abteilung* 37: 1-68.

Weise, G., 1997. "Zur Spezifik der Intertextualität in literarischen Texten". In: J. Klein - U. Fix (eds.), *Textbeziehungen. Linguistische und literaturwissenschaftliche Beiträge zur Intertextualität.* Tübingen: 39-48.

Werblowsky, R. J. Zwi, 1960. "Marburg – And After?". *Numen* 7: 215-220.

Wesseling, Berber.1988. "The Audience of the Ancient Novels". In: H Hofmann (ed.0 *Groningen Colloquia on the Novel.* Groningen: Egbert Forsten, 67–79.

West, M. L., 1994. "Ab ovo. Orpheus, Sanchuniaton, and the origins of the Ionian world model", *CQ* 44: 289-307.

-------------, 2000 (1997). *Tite East Face of Helicon.* Oxford.

Werner, Martin, 1923. *Der Einfluss paulinischer Theologie im Markusevangelium: Eine Studie zur neutestamentlichen Theologie.* BZNW 1. Giessen: Töpelmann.

West Eberhard, Mary Jane, 2003. *Developmental Plasticity and Evolution,* Oxford: Oxford University Press.

White, D. 1964. "Demeter's Sicilian cult as a Political Instrument", *GRBS*

5,4: 261-279.

Whitehouse, H., 1992. "Memorable Religions: Transmission, Codification and Change in Divergent Melanesian Contexts". *Man* (N. S.) 27, 4: 777-797.

---------------------, 1995. *Inside the Cult: Religious Innovation and Transmission in Papua New Guinea*. Oxford: Oxford University Press.

-------------------, 2000. *Arguments and icons: Divergent modes of religiosity.*, Oxford - New York: Oxford University Press.

------------------, 2004a. *Modes of Religiosity: A cognitive Theory of religious Transmission.*, Walnut Creek, CA: AltaMira Press.

------------------, 2004b. "Theorizing religions past". In: H. Whitehouse – L. H. Martin (Eeds.), *Theorizing religions past: Archaeology, history, and cognition.* Walnut Creek: AltaMira Press, 215-232.

------------------ - McCauley, R. (eds.) 2005. *Mind and religion: psychological and cognitive Foundations of Religiosity.* Walnut Creek, Calif.- Oxford: AltaMira Press.

----------------- - J. Laidlaw (eds.) 2007. *Introduction: Religion, Anthropology, and Cognitive Science.* Durham, N.C.: Carolina Academic Press.

Wiebe, Donald, 1991. *The Irony of Theology and the Nature of Religious Thought.* Montreal – Kingston – London - Buffalo: McGill-Queen's University Press.

------------------, 1999.*The Politics of Religious Studies. The Continuing Conflict with Theology in the Academic.* New York: St. Martin's Press.

Wilkins, E. Gr., 1980 (1917). *Know Thysel in Greek and Latin Literature,* Chicago.

Williams, Michael A., 1996. *Rethinking 'Gnosticism.' An Argument for Dismantling a Dubious Category.* Princeton: Princeton University Press.

Williams, M. A., 2005. "Was There a Gnostic Religion? Strategies for a Clearer Analysis". In: A. Marjanen (ed.), *Was There a Gnostic Religion?.* Göttingen: Vandenhoeck and Ruprecht, 1-53

Willis, I. (ed.), 1970. *Ambrosii Theodosii Macrobii Saturnalia.* Leipzig.

Willis 1970: I. Willis (ed.), *Ambrosii Theodosii Macrobii Saturnalia,* Lipsiae 1970.

Wilson, D.S., 2002. *Darwin's cathedral: Evolution, religion and the nature of society.* University of Chicago Press, Chicago.

Wilson, E. O., 1987. "Kin recognition: An Introductory synopsis". In: Fletcher, D.J.C. and C.D. Michener (eds.), *Kin recognition in animals.* John

Wiley & Sons, New York, 7-18.

----------------, 1999. *Consilience: The Unity of knowledge*. New York: Random House.

Wilson, R. M., 1994. "Gnosis and Gnosticism. The Messina Definition". In: G. Sfameni Gasparro (ed.), *Agape Elpis. Studi Storicoreligiosi in Onore di Ugo Bianchi*. Storia delle Religioni 11. Rome: "L'Erma" di Bretschneider, 539-551.

Winiczyk, K., 1991. *Euhemeri Messenii reliquiae*. Stuttgart.

--------------------, 2002. *Euhemeros von Messene: Leben, Werk, und Nachwirkung*. München-Leipzig.

Winkler, John J., 1990. *The Constraints of Desire. The Anthropology of Sex and Gender in Ancient Greece*. London - New York: Routledge.

Winnicott, D. W., 1965. *Morals and Education. In The Maturational Process and the Facilitating Environment*. London: Karnac.

Wiseman, Timothy Peter, 2004. *The Myths of Rome*. Exeter: University of Exeter Press.

Wittgenstein, L. - Anscombe, G. E. M., 2003. *Philosophical Investigations - the German Text, with a Revised English Translation*. Malden, MA: Blackwell.

Woodward, J., 2000. "Explanation and Invariance in the Special Sciences". *British Journal for the Philosophy of Science* 51: 197-254.

Woolf, G., 1996, "Monumental Writing and the Expansion of Roman Society in the Early Empire:, *JRS* 86: 22-39.

"World Islamic Front Statement", February 23, 1998. http:// www. fas. org/irp/world/para/docs/980223-fatwa.htm

Wuthnow, Robert, 2004. *Saving America? Faith-Based Services and the Future of Civil Society*. Princeton: U.P.

Wycherley, R.E., ^{2}1962. *How the Greeks Built Cities*. London.

Yardley, A. B., 2006. *Performing Piety: Musical Culture in medieval English Nunneries*. New York: Palgrave Macmillan.

Young, F.M., 1997. *Biblical Exegesis and the Formation of Christian Culture*, Cambridge.

Zaidman Bruit, L. - Pantel Schmitt, P., 1992. *Religion in the Ancient Greek City*. Transl. by P. Cartledge. Cambridge, MA. Cambridge University Press.

Zaman, Muhammad, 2002. *The Ulama in Contemporary Islam: Custodians of Change*, Princeton, NJ: Princeton University Press.

Zanker, Paul. 1988. *The Power of Images in the Age of Augustus*. transl. by Alan Shapiro. Ann Arbor, Mich.: University of Michigan Press.

Ziad Abu-Amr, 1997. "Shaykh Ahmad Yasin and the Origins of Hamas". In: R. Scott Appleby (ed.), *Spokesmen for the Despised. Fundamentalist Leaders of the Middle East.* Chicago/London: University of Chicago Press, 225-256.

Zielinski, Th., 1923. "Les origines de la religion hellenistique", *RHR* 88: 173-192.

Zumschlinge, M., 1976. *Euhemeros. Staatstheoretische und staatsutopische Motive.* Diss. Bonn.

Zuntz, G. 1963 (1972). "Once more the So-Called 'Edict of Philopator on the Dionysiac Mysteries' (BGU 1211)", *Hermes* 91: 228-239 = *idem*, 1972. *Opuscula Selecta*. Manchester- Cambridge, 88-101.

INDEX OF AUTHORS

Achilles Tatius	137,138, 140, 143, 147, 148
Aeschylus	72
Alanus ab Insulis	16,18-23
Apuleius	72, 284
Athenaeus	146
Bernardus Silvestris	23
Cicero	135,136,n.10,146
Claudius Ptolemy	271
Clement of Alexandria	14, 40,104
Dante	29
Dio Chrysostom	1, 4, 6-8
Diodorus Siculus	269-272, 275-287
Dionysius of Halicarnassus	122,123,128,
Euhemerus of Messene	270,272-274, 277, 282
Euthymios Zigabenos	17,22
Hecataeus of Abdera	272,273,278
Herodotus	284-286
Homer	5
Horatius	119, 124, 126, 130
Ignatius of Antioch	260
Irenaeus	39,399
Julius Cassianus	103
Justin, Martyr	39,398
Kongfuzi (Confucius)	237
Leon of Pella	273
Livy	120,n.2, 122, 125, 127 -129

Lucian	143, 144,n.22
Macrobius	340
Mark	31-46
Martial	133-134
Matthew	35,37-38, 41,74, 190, 398
Melito	399
Milton	26
Nonnos	143,144, 148, 286
Origen	14, 39,108,399
Paul	40,42-45, 106, 252, 253,n.2, 256
Petrus Siceliotes	16
Pliny (the Elder)	69,70
Plotinus	14
Plutarch	122, 139-143, 284
Polybius	125,126
Seneca	69,n.72
Sophokles	71
Suetonius	237
Tacitus	137, 238
Tertullian	39
Thucidides	136
Valerius Maximus	122,n.3, 126,

INDEX OF WRITERS

Anderson, C.	246
Arendt H.	154
Arnal, W.	33, 34,44,260
Asad, T.	295-296
Assmann, J.	11,n.5,15
Atran, Sc.	83,89,90-92, 95,97,120,229,462
Balangangadhara, S.N.	236,258
Balch, D.	143,146
Barrett, J. L.	75-77,119,-121,169,230
Barth, Fr.	414-422,427,430-436
Becker, J.	374,375, 378,382-384,387,388,391,394
Benn, G.	12
Benzon,W.	373,388,398,393,n.23
Bianchi, U.	10,15,n.17, 49,50-53, 55-58,60,n. 19, 61,63-65, 67, 281, 282, 342
Blake, W.	12,16,24-27
Bleeker, C. J.	53,55, 56,n.11, 58, 59,n.16,61,63
Bloch, M.	244,265
Boas, Fr.	178,182
Bonner, C.	345
Bowersock, Gl.	143,144
Boyer, P.	12,16,24-27,n.7,83,89,90-92,95-97, 117,118, 123,219, 220, 222, 232, 242,n.1, 245,246,n.4, 247
Brandon, S.G.F.	53,59
Bromberg, Ph.	160
Brown, P.	272,288,326-327
Bulbulia, J.	296,298
Camus, Alb.	12
Carus, P.	376, -380, 382-388, 394
Casanova, J.	191,n.2, 192,193,254, 255, 259
Chomsky, N.	462

Cloninger, R.   93
Clottes, J.   451,455,456
Collin, R.   373, 392,n.20
Colpe, C.   53
Comte, A.   461
Cosmides, L.   7, 184,241,296
Courtright, P.   256
Culianu, I.P.   53
Damasio, Ant.   155, 392,n.9
Darwin, Ch.   238,460
Davie, Gr.   215-216
Dawkins, R.   239,240,242,308,312, 463,465
Deacon, T.   90,-92,96,463
Dennett, D.   85,221,228,229,238,240,265,n.13,308,313,465
Donahue, J.   31,32
Donald, M.   86,172,463
Doniger, W.   256,257
Douglas, M.   183,184
Duchesne-Guillemin, J.   59
Dumézil, G.   53

Eliade, M.   50,53,54,60,182,185,301,303,307,n.5,
                                     314,315,461

Foley, R.   88-90
Fonagy, P.   151,152,155,157,158,159,168,169
Foucault, M.   188,238,251,260,266,294
Freud, S.   153,154,164,170,294

Gauguin, P.   12,26
Geertz, Cl.   83-85,261
Goodenough, E.R.   59
Goody, J.   242,414,438,439
Gottschalk, L.   326
Guthrie, St.   75,464

Hamer, D.	93
Habermas, J.	153,167
Hassan, N.	159,160,166,167
Haviland, B.	179
Hertz, R.	244
Hesse H.	12,17
Holm, N. G.	53
Honko, L.	53
Horn, W. D.	25
Hultkrantz, Å.	53
Humphrey, N.	83,93
James, W.	301,335,462
Jonas, H.	9,26
Juergensmeyer, M.	151,155,160,161,163,165
Kafka, Fr.	12
Kakar, S.	166
Kandinsky, W.	12
Kant, Larry H.	313,n.7, 335
Keil Fr.	76,340
Kippenberg, H.	73,168
Kishimoto, H.	60
Kitagawa J.M.	54,59,314
Koslowski P.	10,n.3,13
Kotansky, R.	347
Kripal, J.	256
Kristeva, J.	407
Laidlaw, J.	244-246
Laine, J.	256
Lawson E. Th.	221,n.1,224,462,463,469
Leeuw, G. van der	300
Lerner, Y.	158

Lewis, T.	239,461
Lincoln, Br.	55,151,172,275,276, 288
Lisdorf, A.	119
Long, J.M.	59
Malthus, Th.	461
Mann, Th.	12
Martin, L. H.	45,54,81,99,103,113,133,175,185-188,209,217,237,248,251,269,270,271,273-274, 281, 284,288,293,n.1,302,307,311,314, 315,318, 322, 326,335, 336, 413, 414, 470,471
McCauley, R.	224,231,233
McCloskey, M.	232
McCutcheon, R.	276, 288, 308, 461
McNeill, W. H.	388
Meslin, M.	53
Mithen, St.	229,241,463
Morgan, C. L.	379
Morgan, Kenneth W.	53,56, 61,-63
Morgan, Lewis H.	461
Müller, M.	182,238
Nasr, Seyyed H.	50, 54,n.8
Ndebele, N.	163,164
Nikolaides, Th.	163,164
Nock, A.D.	53,274
Nyberg, S.	53
Nyhof, M.	119-121
Olcott, H. S.	384
Pentikäinen, J.	53
Persinger, M.	93
Petersen, J.	178
Pettazzoni, R.	53,58

Piatelli-Palmerini, M.   217, 234
Portes, A.   200
Post, J.   162
Pratchett, T.   81
Putnam, R.   193-196,200,201,n.27
Pye, M.   53
Pyysiäinen, I.   117,119, 123, 461,467

Quispel G.   1o,12,15,16, 23,24,27

Raine, K.   25
Rappaport, R.A.   299
Rochat, Ph.   228
Rose, S.   161,170
Rowling, J. K.   3
Rudolph, K.   10, 11,n.37, 44

Sack, Ol.   373
Sahlins, M.   31,n.2, 32,43,n.5, 111-112
Sartre, J.-P.   12
Schenke, H.-M.   9,n.1,102.,n.5, 104, 111, 112
Schneider, H. W.   36,n.16, 55, 279
Schore, All.   155
Sharf, R. H.   326, 328, 334, 441
Sharpe, E. J.   53,182,461
Siegel D. J.   93-94
Simon, M.   59,195,n.12,339,462
Smart, N.   650,53.75
Smith, J. Z.   11,32, 33,n.10, 45, 46,n.41, 51, 152, 165, 178, 365, 283, 314, 325, 327, 328, 330, 331, 333, 335, 336
Smith, W. C.   46,n.41, 53,-55,57, 58,65, 66, 265, 461
Sorensen, J. P.   26
Spencer, H.   461
Sperber, D.   1-2, 4, 5, 6, 179, 183,239,393,463
Stauss, Cl. L.   53,179,255,294,333,335

Stein, R.	155,162,163,167,168
Stern, J.	155,158,161,163, 166,168,169
Tolkien, J. R. R.	3, 238
Tooby, J.	3,n.2, 7, 183,241,296
Trainor, K.	377,n.5, 383,n.12
Tröger, K. W.	10,13,14,n.13,15,n.19
Tweed, Th.	376, 377, 384
Volkmar, G.	32
Waardenburg, J.	41,42,n.34
Wallace, Ant.	300
Warner, R.St.	12,201,202
Weber, M.	12,201,202
Werner, M.	41
West-Eberhard, M. J.	87
Whitehouse, H.	76,n.7, 237,242,243, 263, 264, 308, 413,n.1,414, 415, 433, 437,n. 6, 463, 464, 469
Widengren, G.	53
Williams, D.-L.	441-457, 463
Winnicott, D.M.	169
Wittgenstein, L.	212
Zaehner, R.C.	59
Zaman, M.	262

GENERAL INDEX

Abilities, human	71, 74, 75, 77, 78
Abraham	20, 343
Activities, church	193, 195
------------, religious	70, 73, 191
------------, ritual	70
------------, social	195 n. 12
Activity, magical	72
Adam	261, 423
Adonai	343-344
Adultery	139, 141
Aegean islands	405
Aemilia	130 n. 9
Agathon Daimonion	342
Agency	
---------, individual	158, 165, 166
---------, supernatural	244, 245, 246
Agent(s), superhuman	264, 301
----------, supernatural	245, 246 n. 4
Agriculture	276-278, 280
Alexander	146, n. 25, 238, 270, 274, 285, 286, 351-354, 364
-------------, of Tralles	347
------------, *Romance*	352 n. 2, 364
Alexandreia ad Issum	144
Alexandria	36, 37, 40.45, 138, 273, 279-280, 281, 286, 352, 354, 355, 357, 358, 359, 360, 361, 362, 363, 364, 365, 366, 398, n. 5, 403
Alexandrianisation	365, 366
Altered states of consciousness	302, 384, 443-444, 447, 448, 449, 450, 451, 453, 454, 455, 456
American Anthropological Association	179
---------------, International for the History of Religions	61, 67

American Society for the Study of Religion	55, 61, 62
Ammon	352, 353, 356 n. 13, 360, 364
--------- -Zeus	352
Amphitheatre(s)	134, 146
Amulet, magical	342, 343, 345, 347
Anagraphê	41
Analogical notion	49, 239, 243 n. 3, 269, 293, 300, 301, 344, 348, 404
Analytical, approach	376
Anamnesis	303-304
Ananda Maitreyya	378, 385-386
Ancestor(s)	167, 170, 246, 297, 329, 330, 353, 416, 417, 418, 422, 423, 450, 453, 455
Anfushi triad	356, 368
Angels	20, 198, 199, 201, 343, 390
Anger	108, 156, 166
Animal(s)	6, 82, 84, 85, 89, 99, 100, 103, 104, 105, 106, 108, 112, 113, 114, 146, 230, 240, 241, 278, 298, 356, 379, 420, 421, 432, 449, 454, 356
Amowkam, temple	416
Anthropology, cognitive	245
-----------------, cultural	177, 231
-----------------, psychological	83
Army of God	169
Anti-Petrine	40
Anti-reductionism	465
Anticlaudianus de Antirufino	21
Antiochia	274
Anxieties	22
------------,existential	95
Aphrodite	140, 147, 347
------------ / Cleopatra	147

Apocalyptic literature
(Christian) 108
Apollonios 347
Arabs 17, 56, 62
Archaeology 85, 177, 238, 243 n. 3, 442
Archangels 343
Arsenal of Believers 159
Arsinoe II 356 n. 12-13, 358, 359, 369
Art, prehistoric 443
Artemis/Isis 140, 141
Artifacts 2, 137, 238, 295, 296, 394, 408
Artificial Intelligence 463
Arunta 329
Ascetic, Christianity 99
---------, ideals 99, 100
---------, life 100, 101
---------, message 109
---------, setting 101
---------, teachings 101
---------, tractates 101
Asceticism 100, 101, 113, 114
Asia Minor 17, 132 n. 2, 142 n. 17, 146 n. 29, 400
Asian Religions 54
Astarte, of Sidon 137
Atheism 191, 311, 459
Attus Navius 119, 128, 129
Audiens 69
Authoritarianism 309, 320
Authority, institutional 441
-----------, Ptolemaic 355, 358
-----------, religious 181, 296, 357, 361-362, 493
-----------, personal 442

Baktaman, of inner New Guinea 415-439
Balkan 404, 405

Benefaction(s)	271, 272, 282, 283
Behavior	73, 76, 77, 81, 84, 88, 90, 94, 95, 96, 106, 112, 157, 161, 178, 179, 184, 213, 220, 224, 229, 237, 239, 245, 246, 247, 254, 260
Belief(s)	13, 14, 20, 21, 22, 154, 155, 157, 158, 161, 164, 165, 166, 169, 170, 175, 297, 298, 299, 315, 442, 443, 445, 446, 449, 453
-----------, religious	153, 154, 157, 157, 159, 160, 165, 318, 443, 444, 455, 456, 457
-----------, idiosyncratic	452
Believer	55, 57, 58 n.15, 75, 76, 154, 154, 155, 157, 159, 170, 193, 202, 203, 247, 284, 310, 320, 330, 348, 462, 464
Believing	16, 152, 155, 165, 191, 192, 443
Belonging	2, 15, 21, 44, 114, 191, 192, 199, 200, 358, 364
Benefactors	275, 276
Bible	20, 22, 172, 377, 397, 398, 394, 404, 407, 408, 409
-------, Hebrew	224, 397, 399, 401, 402, 406
Biblical criticism	39
Bigamy	469
Biology	87, 90 n. 9, 93, 106, 188, 220, 295, 295, 376, 379, 466 n. 1
---------, evolutionary	87, 308, 312
Blood	49, 151, 247, 253 n. 2, 329, 353, 423
Body	20, 21, 76, 85, 95, 99, 103, 104, 106, 107, 109, 110, 113, 114, 118, 127, 139, 170, 184, 244, 246, 247, 286, 287, 293, 326, 327, 330, 331, 334, 335
--------, language	423, 430, 435, 436, 438
Bogomils	10, 13, 17, 18, 22
Bonacursus	19, 20 n. 42
Book of Thomas the Contender	99, 100-104, 106-109, 111-114,

Brain(s)	1-3, 7, 74, 82-90, 92-94, 96, 123, 179, 219, 211, 220, 227, 240, 241, 293, 295, 296, 319, 336, 374, 388, 444, 445, 446-453, 455, 460, 462, 464-466, 468, 470
Brotherliness	202
Buddha	54, 244-247, 327, 328, 329, 330-334, 337, 380-382, 384, 386 n. 11, 387
Buddhism	13, 54, 180, 200, 202, 244-248, 259, 376, 377, 378, 380, 384, 385, 386 n. 11,
Buddhist, Sri Lankan	385
Byblos	139, 148, 361
Byzantines	17
Caesarion	358
Cain	18
Callixenos	146
Camillus	121-122
Canonization	45
Canopus	360
Capital, social	191, 193-195, 200, 300
Caracalla	342, 361
Carthage	129
Categories (Jungian)	11
Cathars	10, 13, 16, 18, 19-21, 23 n. 45
Catholic, education	50
----------, outlook	50
Causa efficiens	20
Causa formalis	20
Cemetery temple	417, 418, 422
Center	43, 65, 81, 129, 196, 197, n. 18, 243, 332, 383, 358
Ceremony	70, 332
Chalkis	409
Chastity	119, 129
Choir, monastic	381, 392

Christianity
-------------, Early   34-35 n. 11, 36, 46
Chronotopes   328

Church Fathers   29, 461
Church, Fathers   26, 398-399
------------, Catholic   21, 310
---------, Episcopal   50
Churinga   329, 330
Civilization(s)   139, 168, 238, 275, 276-278, 280, 281, 285,
                                      311, 313, 414
Claremont   55, 56, 62, 67, 293 n. 1
Classification   87, 252, 263, 265, 266, 328, 378
Claudia Qiuinta   129, 130
Clea   136, 141
Cleopatra I   359
------------, III   359
------------, VII   351, 357, 358, 359
-------------, as «Nea Isis»   359
Codification   96, 413, 418, 412, 430, 433, 434, 437, 438
Cognition
-------------, historical   237, 240
-------------, human   6, 69, 75, 76, 88, 89, 90, 214, 234
-------------, social   91, 96
Cognitive poetics   99
------------, revolution   243, 462
------------, Sciences   74, 231, 243, 244, 245, 247, 248, 315, 335
------------, Study of Religion
(CSR)   3 n.2, 7, 246, 461-464, 470
Cognitivist(s)   4, 84, 264
------------, approaches   1
------------, cult-titles   5
Colosseum   134
Comitium   125, 128
Commodus   361, 370

Communication	72, 74, 76, 89, 118, 156, 161, 166, 178, 207, 248, 275, 284, 288, n. 4, 415, 420, 422, 426, 427, 436, 437, 438, 439
----------------, interpersonal	74, 76, 77, 156, 157
Communities (religious)	58, 191, 194-197, 200-202, 207, 229, 318, 321, 394

----------------, Christian	35, 398, 399, 491
Community	31-34, 37, 70, 95, 151, 152, 155, 159, 160, 192, 195 n. 12, 197, 201, 203-205, 244, 245, 247, 299, 310, 313, 317, 318, 369 n. 3, 375, 382, 384, 387, 388, 389, 393, 398, 404, 409, 416, 418, 419, 424, 444-446, 450, 451
Comparative religion(s)	49, 54, 57, 58, 185, 293, 294, 295, 300, 303, 470
-------------- - historical method	49
Comparativism	84, 301
Comparison(s)	9, 18, 26, 49, 50 n. 1, 57, n. 14, 84, 86, 128, 211, 217, 221, 223, 225, 230, 233, 234, 284, 294, 300, 302, 326, 262, 392, 400, 402, 416

----------------, cognitive	225
Competence, symbolic	82, 92
Confucianism	54
Congregation	197-200, 202, 203, 207, 252, 254, 382
Consciousness	81, 82, 83, 85-88, 144, 179, 220, 296, 302, 382 n. 9, 384, 408, 415, 443, 444, 445-457
Consilience	214, 294, 295, 300
Constitutions, National	309
Constraints, panhuman	464
Context, ceremonial	136
-----------, religious	399
------------, socio-historical	399
Continuity	15, 17, 18, 66, 303, 417, 423-426
Contra hereticos	18, 20 n. 33, 22
Conventions, International	309
Correctness, theological	76

Cosmographia	23
Cosmos, heavenly	340
Counter-prayers	71
Creation	14, 17, 18, 19, 21, 23, 26
Creative Explosion	445
Creativity	17, 82, 426
Crete	137, 403, 405
Crown, Hathoric	359
Cult(s), public	70, 78, 274
Cultural genetics	414, 438
---------, heroes	278
---------, phenomena	462, 463, 469, 470
---------, representations	237, 463
---------, theatrical of the Graeco-Roman world	148
---------, transformation	414, 415, 416, 430, 434
---------, variability	296, 462
Culture(s)	
---------, local	211, 375
---------, modern	308, 309, 316, 317, 319
Curiosity	86, 123, 253
Custom(s)	70, 72, 73, 74, 75, 77, 313, 316, 402
Daemones	140
Daphne	144, 145
Day of Lord (apokalyptic)	38
de fide Catholica contra haereticos	19 n. 33, 22
Declamation, Sophistic	140
Deep history	294, 303
Defining struggle	43
Definition (phenomenological)	11
Delos	405
Demarcation, religious	397, 399, 408, 409

Demeter	140, 277 n. 1, 279, 281, 283, 284, 285, 287
Demiurge	13, 22, 25
------------, Gnostic	22
Democracy	161, 193, 194, 309, 319, 320, 322
Department, of Anthropology	177, 180
------------------, Religion	176, 178, 182-188
Depression	156
Description(s)	9, 14, 15, 82, 112, 136, 137, 143-144, 145 n. 24, 155, 160, 166, 248, 257, 259, 271, 272, 279-280 n. 2, 281, 282, 284, 286, 287, 319, 380, 390, 417, 460
Destiny, astrological	343
Deuteronomy	397, 403, 406, 408, 409
Development, evolutionary	84
-----------------, ontogenetic	83
Diaspora	397, 400, 405, 406, 408
-----------, Jewish	405, 406, 408
Diffusionism	451
Dimension, cosmic	303, 340
Dionysos	140
Dionysus	140, 143, 144, 146, 148, 282, 284, 285, 286, 287
--------- / Antony	147
Discourse	
------------, on religions	258, 260, 262, 264, 265
------------, scholarly	51
------------, theological	53, 398
Dismemberment	139, 141, 286
Dimension(s)	308, 310, 312, 315, 317, 321, 322
Dispositions, imperial	148
Dissociation	155, 159, 160, 168, 449
Divine	
--------, Office	388, 389, 390, 391, 392
Documents, Gnostic	12, 15, 27
Dogmatism	55

Domestic rituals, Roman	70
Domitian	361
Double exile	33
Dreams	296, 384, 454
Drugs	161, 196, 447
Dualism(s)	13-20, 22, 23, 25, 91, 155
-----------, Cathar	19
Dumb gods	69, 75
Ears, of the divinity	70
Ecclesia Gnostica	24
Ecclesiastical, welfare	156
Ecstasy	95, 384, 391, 393, 394, 462
Education	50, 57, 70 n. 3, 76, 81, 94, 161, 164, 169, 170, 185, 194, 203, 204, 207, 310, 311 n.3, 357, 383, 390, 459, 466
Efessos	373
Egypt	
-------, Christian	341
-------, (Upper)	100, 114
Egyptomania	271
Ekphrasis	136, 137, 138, 140
Elements, visual	136
Eleusis	280
Eliminativist, cultural	88
Emancipation, human	309
Emic	54, 326, 328
Emotion(s)	73, 84, 91, 100, 102, 106, 109, 231, 296, 301, 374, 375, 378, 379, 383, 391, 427, 429, 431, 449
Empathy	83, 93
Enargeia	134, 135, 136 n. 10
Encratic ascetic life	101
Endemic	151, 160, 179

Enemies of God	153, 171
Enlightenment	39, 44, 111, 172, 380
epékoos	70
epicharis	347
Epidemics	1-2, 240
Epidemiological approach	463
Epigramme(s)	134
Epigraphic texts, Christian	400
Epithumia	105, 106
Erinyes	71
Eros	107, 137
Erotes	147
Euhemerism	271, 272, 273
Euphemia	344
Europa	137, 140
European History of Religions	9, 10, 11, 12, 16, 24, 27
------------, Religious History	9, 15, 27
Euthenia	362
Evacuation	160
Eve	262, 423
Events, counter-intuitive	117, 118, 121, 123
--------, miraculous	117, 118, 121, 122, 123, 130
Evil(s)	13, 17, 20, 21, 22, 25, 96, 103, 110, 118, 138, 152, 154, 159, 162, 163, 166, 169, 317, 343
Evolution	
------------, cultural	183, 294, 297, 463
------------, Darwinian	218, 294
Evolutionism, cultural	461
Exegesis of the Lord's Oracles	39
Experience(s), hallucinatory	455
------------, moral	96
------------, musical	372, 378, 379, 380, 383, 387
------------, religious	93, 154, 157, 160, 221, 257, 300, 420, 422, 434, 438, 441, 442, 450, 452, 456, 462
------------, spiritual	450, 455

Experiment(s)	75, 94, 96, 119, 120, 214, 215, 438
Extra-corporeal travels	449
Factor(s), cultural	308, 312, 317, 373, 374, 375, 378, 383
Faith-Based and Community Initiatives	197
Father	5, 159, 163, 167, 170, 198, 202, 340, 343
Fatwa	171
favete linguis	70
Fear	156, 157, 168, 171, 213, 303, 379, 390, 419, 422, 433
Feelings of shame	156
Female(s)	104, 119, 159, 296, 388
Fertility cult	416, 421
---------, goddess	354
Festival(s)	77, 145 n. 25, 280, 393
Fire	83, 100, 104-108, 110, 111, 113, 126, 130, 241, 242, 339, 343, 346, 418, 460
Formulae, Jewish	402
Forum, Roman	125, 128
Four Noble Truths, the	246, 385
Freedom, human	309, 313, 314, 317, 320
Fundamentalism(s)	95, 161, 459
----------------, religious	161, 459
Fundamentalists, Christian American	157
Games, gladiatorial	154
Gem(s)	331, 345, 346, 359
Generalization(s)	
------------------, invalid	326, 335
------------------, valid	326, 335
Globalization	191
Glossolalia	233
Gnosis	9, 10, 13, 14, 15, 16, 24, 26, 27, 55 n. 8, 392

Gnosis, antique	9, 10, 11, 13, 14, 15, 16, 24, 25, 26, 27
---------, Christian	15
---------, Jewish	15
Gnosticism	9, 10, 12, 13, 14, 24, 25 n. 57
Gods, Graeco-Roman	85
Good	13, 14, 14, 17, 19, 20, 21, 21, 25, 25
Gorgonia	345, 346
Gospel of Mark	31, 32, 33 n. 9-10, 34, 35, 37 n.19, 38, 40, 41, 42, 45
-----------, Peter's	38, 41
----------, *Thomas*	34, 36 n. 17
Graces	147
Guardian of the Race	5-6
Guru Complex	431, 434, 438
Hadra	358, 359
Hadrian	361
Hallucinations	447, 449, 451, 454
Hamas	161
------- (zeal)	206
Heis Zeus Sarapis	341, 342
Helios	141, 339, 340, 341, 342, 344, 345, 346
Hellenistic, era	271, 281, 285, 286
--------------, world	269, 271, 274, 283
Henotheistic, sensibility	342
Heracles	139
Hercules	353
Heresy	13, 19, 23
Hermoupolis	345
Heroes	272, 278, 353, 435
Heterotopias	328
Hinduism	50, 54, 200, 256 n. 6, 257, 259
Historians, ancient	209, 237
Historical-comparative approach	49
Historiography, cognitive	209, 210, 211, 215

History of religions	9, 10, 11, 12, 16, 18, 24, 27, 49, 52, 57, 59, 61, 66, 248, 293, 294, 295, 300, 303
Homily	252, 253, 262
Hominid(s)	89, 445
homo faber	336
------, *fabricatus*	336
------, *sapiens sapiens*	445
Horatius Cocles	119, 124, 126, 130
Howieson's Poort people	450
Human Rights	192, 310, 322
Humanity	6, 14, 17, 18, 19, 88 n. 8, 89, 143, 271, 276, 277, 278, 281, 282, 284, 285, 304, 379
Humiliation	163, 165, 168, 171
Hymns, Buddhist	373, 376, 377, 380 n. 8, 385, 387, 394
Hypékoos	69
Hypotyposes	40
Iao	339, 343, 344, 345
Identity, Egyptian	356, 358, 365
-----------, group	95, 229
Ideology	146 n. 29, 147, 275, 276, 283, 288, 351, 354, 357, 358, 360, 365, 424, 425, 426, 434, 435
Images, prehistoric	444
---------, sacred	422
Immigrant(s)	200, 377
Impereality	134
Increase	56, 62, 76, 78, 93 n. 13, 97, 217, 283, 343, 421, 422, 428, 430, 440
Information, genetic	84
Initiation	94, 137, 284, 413, 415, 416, 419, 420, 422, 427, 428, 429, 430, 431, 432, 434, 435, 436, 437, 438
inlustre monumentum	136

Inscriptions, Jewish	357, 399, 400, 401, 402, 403, 404, 405, 406, 407, 408, 409, 410, 411
Insiders	51 n. 3, 55, 298, 327
Institutions, religious	169, 313, 318, 319, 320, 426
Integrated Causal Model (ICM)	7, 184
Integration	16, 93,14, 94, 96, 193, 196, 215, 294, 408, 415, 419, 420 n. 4, 423, 433, 434, 449, 468, 469
Intellectuals, Christian	36, 38
Interaction ritual	373
International Association for the History of Religions (IAHR)	55, 56, 61, 62, 67
interpretatio Graeca	285
Interreligious dialogue	53, 57
Interrogatio Iohannis (Bogomil text)	17
Intifada	205
Intuition(s), ontological	220
Intuitive Psychological System	241
Isara	346
Isis	138-143, 146-148, 277-278, 281-285, 287, 354, 356-357 n. 15, 358-360, 362, 365, 370
-----, *benefactress*	284
-----, Euploia	361
-----, *Galaktotrophousa*	361
----, Lochias	362
-----, of Akra Lochiados	362
-----, Pharia	361
------, Therenouthis	361
Islam	10 n.4, 13, 50 n. 2, 54, 161, 162, 167, 171, 192, 200, 201, 203, 205, 206, 223, 261, 262, 322
Islamic Mission (dawa)	203
Israel	56, 57, 67, 199, 205, 206, 244, 397, 398

Jahiliyya	168
Janiculum Hill	124
Jerusalem, special IAHR Congress	55, 67
Jesus	33-34 n.10, 35, 36, 37, 42, 57, 57, 99, 100, 103, 111, 113, 200, 202, 235, 343, 398, 399, 409
------, Christ	17, 199, 252, 256, 398
------, David	343
------, son of Abraham	343
Jihad	163, 203, 205, 206
Johannine Corpus	34
Judaism	151, 171 n. 4, 192, 200, 201, 259, 261
---------, ancient	73
Julius Caesar	128
Juno	121, 123, 124
Jupiter Capitolinus	145
--------, Optimus Maximus	3
Justification(s), religious	169
kata thymon	71
Katiam, temple	417, 418
King	5, 198, 199, 272, 274, 275, 282, 331, 332, 334-339, 364
Kings mother	359
Kingship	272, 274, 300
Kivung religion in East	415-420, 422-430, 433-436, 438-439
Knowing	14, 36, 91,102, 220, 254, 264, 302, 429
Knowledge	
------------- (Erkenntnis)	14
------------- (Wissen)	14
Koine	285
Kore	281
Koriam	435, 436,438
Kosmokrator	342

Lagawop	25
Language	89, 90, 100, 102, 106, 113, 180, 212, 237, 238, 246, 259, 265, 288, 294, 295, 296, 298, 301, 3012, 303, 327, 328, 336, 377, 378, 382, 386, 392, 393, 404, 422, 430, 435, 436, 438, 463
Lar Familiaris	129
Lars Porsenna	124
Laskar Jihad group	163
Late Antiquity	9, 10, 11, 13, 15, 16, 17, 20, 24, 281, 288, 327, 397, 399, 401, 402, 405, 409
Law, roman	73
Leucippe and Cleitophon	137-141, 143-144, 147-148
Liber de duobus principiis	19
Liberation	164, 205, 309, 318
Life (bodily)	103, 105
Linguistics	61, 177, 239, 421, 463
Literature, canonical	385
Local	32, 33, 44, 58, 67, 192, 193, 197, 200, 204, 207, 211, 253, 254, 257, 262, 265, 266, 271, 274, 283, 285, 296, 326, 352, 353, 357, 404, 422, 435, 436
logoumena	142
Luna	345
Magic, Graeco-Roman	72
Magical Papyri	106, 339, 343, 345, 347, 348, 443
Magician(s)	72, 106
Magna Mater	129
Male(s)	104, 109, 111, 113, 142, 159, 163, 296, 388, 427, 431, 432, 433, 438
Manichaeism	15, 16, 100
Marburg	57-59, 67
Marcion and Luke-Acts	43
Marcionism	17

Marcionite, clan	43
Marriage(s), spiritual	111
Martyrdom	160, 167
Mechanism(s), generative	327, 328, 334
Meditation	44, 384-386, 391 n. 19
Medicine men	443
Megalensia	129
Meligenetor	339, 346
Melikertes	339
Meliorism	378, 379
Meliouchos	339
Memes (memeplexes)	240
Memorability	118, 130
Memory	2, 6, 45, 51, 86, 87, 88, 117, 220, 287, 298, 302, 303, 344, 366, 414, 423, 424-427, 428 n. 5, 430, 432, 434, 437, 438, 462, 464
Menelaos	138
Mental representations	156, 157
Mentalization	155-159, 162, 169
Messina Gnosis-Colloquium (1966)	10
Metalanguage	12, 14
Metaphor(s)	10, 11 n. 5, 13 n. 10, 15, 100, 102, 106-113, 157, 179, 263, 270, 389, 417, 421-423, 429, 444, 469
Method(s)	31, 32 n. 6, 46 n. 41, 49, 52, 53 n. 7, 59, 60 n. 19, 166, 175, 176, 187, 188, 211, 221, 243, 270, 272, 277, 474, 376 n. 2, 380, 385, 442, 460, 461, 464, 467-470
Methodological, bigamy	469
------------------, individualism	468
------------------, guidance	467
------------------, integrative	470
------------------, materialism	468
------------------, relativism	468

569

Methodologies	84, 155, 239, 294
Methodology, empirical	155, 294
Metiochus	143
Metonymies	103
Mickey Mouse (problem)	3
Middle Ages (European)	10, 13, 15, 16, 17, 20, 21, 22
---------, Paleolithic mind	445
Militancy, religious	152, 155, 162, 163, 165, 166, 169
Mind(s)	
---------, human	1-3, 6-8, 87, 90, 211, 212, 215, 218, 219, 229, 230, 233, 240, 242, 243, 248, 413, 445, 449, 452, 455, 463, 469
Miracle(s)	
------------, in the strong sense	119
------------, in the weak sense	117, 119
Miraculum	118
Mirari	118
Mithraism	133, 345, 347
Mithras	133, 339, 342, 343-348
---------, Liturgy	343
Mobilization	95
Models, evolutionary	293, 299
Modernity	152, 161, 170, 258, 261, 309, 311-312, 317, 318, 321, 322
Modes of thinking, dehumanizing	153
Monarchs	273-277, 281, 283, 285, 286
Monism	377, 380
Moral(s)	74, 91, 94-97, 130, 153, 154, 158, 161-164, 166, 169, 171, 218, 222, 223, 224, 232, 276, 302, 381, 382, 383, 417, 423
Morality	81 n.1, 82, 90 n. 8, 92, 95, 176, 196, 202, 221
Murders, mythic	140
Music	12, 81, 188, 237, 373-375, 377-380, 382-383, 385-391, 393-394

Muslim Brothers	204, 205, 206
---------, faithful	168
Mysterion	36
Mystery	
---------, cults	78, 133, 142, 147, 284, 414
--------, religions	135 n. 3
Mysticism	13, 380, 462
Mystics	13
Mystification	432, 433
Myth of Innocence	33 n. 10, 34
Myth(s), Roman	117-121, 123
----------, popular Roman	123, 129
Mythos	14, 22, 26
Nag Hammadi, papyri	24
Narrative	14, 17, 33-35, 39, 40, 41, 85, 94, 95, 118, 120, 123, 129, 130, 133, 135, 136, 137, 138, 140-144, 146-148
-------------, mythic	133, 142-144, 146
National Socialism	11
Naturalness of religion	1, 463
------------, cognitive	227, 233
Natural selection, Darwinian	238, 239, 295, 460
Neanderthals accounts	445
Needs, social	199
Neo-Marxism	294
Neolithic society	446
Neoplatonism	9, 11, 14
Neuroscientists, hardcore	84
Neurology, human	453, 454, 456
Neurons	2, 93, 263, 293
Neuropsychological research	444, 448, 452, 454
Neuroscience	93 n. 14, 96 n. 17, 97, 294, 373
Neurotransmitters	466
New Guinea	413, 415, 435, 437

New Testament	36 n. 14, 41 n. 32, 43
Nicaea in Bithynia, Asia Minor	403
Nihilism (modern)	9
Nike	341
Nilogenia	346
Nilometer	356
Nilus	360, 362
Novel	81, 137, 138 n. 13-14, 139, 140, 144, 146, 149, 179
Numen	58-60, 152
Nymph(s)	147
Nysa, in Arabia	282
Occupation, Syrian	355
Odysseus	71
Offerings	246, 332, 416, 418, 422, 429
Officials, religious	308, 311, 315 n. 11, 320
Oinochoai	356, 359
Old Testament	20, 71 n. 5, 73, 398, 405, 406
Olympian gods	273
Olympic Games	1, 4
----------, Zeus	4, 6
Omnipotence	24, 75, 340
Ontogeny	87, 93
Ontological commitment	3, 219
Oracle of Sarapis	339, 346
Orations	4
Order, moral	91, 95
-------, sacred	299
Organizations, religious	195 n. 13, 196
Orpha	343
Orphamiel	343
Orthodoxy, Christian	50
Orthodox, centre	43
Osiris	138-142, 147, 276, 278, 280, 282, 287, 354, 360-361

ouroboros	345
Outsiders	221, 257, 436
Ownership, private	389
Palestine	108, 397, 404, 257, 436
Pallas Athena	3
Panegyrists, Roman	137
Paradigm(s)	2, 176, 183, 184, 243, 299, 328, 408, 461, 462, 464
Paradise	18, 160, 161
Paranatural beings	3
Parasitic	90
Parthenope	143
pater familias	70
Patristic period	38, 39
Paul VI, Pope	55
Paulicians	13, 16, 17
pax deorum	73
Peace	102
Pelusium	139, 141
Perception(s)	160, 164, 175, 219, 227, 233, 310, 311 n. 4, 312 n. 6, 315, 316, 319-321, 336, 360, 365
Performance	119, 129, 133, 134, 143, 147, 148, 194, 227, 275, 373, 374, 385, 390, 392, 393, 424, 425, 426, 429, 430, 434, 437, 438, 443
Performance, «autopilot»	393
----------------, ritual	75, 298, 390
Pergamos	273
Periphery	45, 182, 230
Pharaohs	285, 286
Pharaonica	352 n. 2, 363, 364, 366
Pharos	138, 351, 355, 358, 359, 361, 362, 363, 365, 369
Phenomena, counter-intuitive	117, 118, 123, 130
Phenomenological	

understanding	57
Phenomenology	49, 57, 294, 295, 296
Phidias	1, 4-6
Phillip II	353
Philomela	138, 140
Philosophy (existential)	9
Phrygia	403, 409
Phytomorphic	114
Pietas	124
Place	18, 33, 35 n.12, 36 n.17, 37, 39 n. 26, 42, 44, 73, 82, 84, 109, 139, 141, 159, 161, 172, 184, 186, 202, 203, 210, 242, 244, 256, 260, 261, 265, 295, 301, 303, 325, 326, 328, 329, 330, 332, 334-336, 341, 346, 348, 352, 355-385, 390, 393 n. 24, 397, 398, 402, 403, 406, 415, 416, 418, 420 n. 4, 424, 423, 424, 436
Plant (s)	99, 100, 108, 110, 112-114
plasmata	142
Pleistocene, late	414
Poetry, elegiac	72
Pollution	244, 443
pompa triumphalis	146
Pompei	143, 144 n. 20, 147
Pons Sublicius	124
Pontus	404
Port, Alexandrian	362
Positivism	11
Post-Enlightenment Gospel	44
Power(s)	14,18, 22, 94, 103,104, 111,119, 134, 137, 140, 142, 145,n.25, 158, 160, 163, 168, 169, 191, 193, 202, 205, 207, 252, 260 n. 8, 261 n. 9, 270, 273, 274, 281, 288, 299, 307 n. 1, 309, 310, 316, 318, 327, 334, 336, 339, 340, 341, 343, 347, 353, 354, 357, 374, 377, 381, 382, 385, 388, 391, 419, 422, 435, 437,

	446, 449, 451, 453
--------, imperial	134
Powers, counter-intuitive	119
Practices, hierarchical	153
praesentia	327
Prayer(s)	
----------, magical	72, 342
----------, petitionary	75
----------, silent	70-74, 77-78
Priest(s)	70, 146, 252, 271, 273, 279, 282, 355, 357 n. 19, 414
Problems, biological	219
-----------, mechanical	219
-----------, social	219
Process, cognitive	77, 90, 117, 235, 239, 387, 467, 468
---------, psychic	154
---------, Phylogenetic	83
Prodicus	272
Profane	153, 159, 315
prohibition(s)	197, 231, 317, 334, 389, 459
Propaganda, political	351, 354, 355, 360, 361, 365
--------------, royal	355
Prophet	168, 203, 340, 381, 382, 384, 390, 417, 426, 435
Protagoras	273
Protector of Cities	5
------------, of Suppliants	5
Providence	7
Prudence	21, 22
Psychoanalysis	11, 153, 154, 172
Psychologists, experimental	188
Psychology	
--------------, developmental	155
--------------, evolutionary	214, 294, 336
--------------, outdoor	83
Ptah-Hephaistos	360

Ptolemaic coinage	352, 353
------------ period, early	351
--------------------, late	351, 354, 357- 356, 362
Ptolemies	272, 274, 279, 280, 286, 287, 352, 353, 354, 355, 356, 357, 358, 359, 363-365
Ptolemy I the Soter	279, 351, 353, 367
---------, II Philadelphus	148, 356 n. 13, 358, 367
---------, IV	354, 356 n. 12, 367
---------, V	351, 357, 359, 370
---------, VI	355, 356, n. 12
--------, XII	359
--------, XIII	370
Purification, spiritual	159
Radical methodological materialism	468
Ramakrishna Mission Ashrams	53
Ranks, hierarchical	389
Rapes, mythic	140
Rationalism	319
Rationalization(s)	77, 121
Reality, political	136, 270
Relationships, human	309
Redemption	67, 203
Redescription	259, 336
Reductionism	457, 464, 465, 461, 467
-----------------, historicist	57
-----------------, sociological	57
References, cultural	134
Reflection on exile and identity	32 n. 7, 33
Relationship, horizontal	238
----------------, vertical	238
Relativism	461, 468
Relic(s)	325, 326-329, 331-336

Religion
----------, Greek   151
----------, natural   185, 255
----------, popular   219, 221, 223, 225, 226, 232 - 235
----------, Roman   69, 70 n. 2, 73, 76, 77
Religions, memorable   414
----------, monotheistic   75
----------, public   192 n. 5, 254
Religionswissenschaft   50 n. 2, 58, 183
Religiosity, Gnostic   15, 27
----------, Graeco-Roman   69, 70, 72, 74, 75, 77, 78
Religious, Roman life   70, 78
----------, Studies   59, 60 n. 18, 239, 298, 301, 302
----------, thought   3, 81, 90, 158, 167, 426
Remus   119, 126, 128, 129, 130
Representations, counterintuitive   117, 220, 246 n.4, 247
Rhea Silvia   128
Ritual(s)
-------, Form Hypothesis   463
----------, mystery   284
Rock art   441-444, 448, 452, 453, 455
Roman Curia   56
--------, Empire   32, 148, 360, 363, 366, 400, 401
Rome   41, 53, 63, 66, 119, 120, 121, 122 n. 3, 123, 124, 134, 136, 143, 144 n. 24, 45 n. 25, 146 n. 27, 340, 362, 398 n. 1, 400
Romulus   119, 126
Rosicrucianisn   11
Rulers   238, 280, 351, 355, 358 n. 2, 364

Sabaoth   343, 344
Sacrality   299, 33, 334, 516
Sacred   2, 39, 44, 129, 130, 142, 153, 157, 159, 160, 165-167, 171, 195 n. 13, 233, 253, 255, 256,

	258, 272, 282, 298, 299, 300, 303, 315, 316, 317, 325, 326, 329, 330, 333- 336, 341, 344, 356, 360, 362, 364, 373, 374, 386, 387, 391, 418, 419, 422, 424, 428, 430, 436, 474, 484, 493, 494, 500, 515, 520, 522, 537
Saint Colluthus	342
Samaritans	402
Sangha, assemble	382
Sarapeion	354 n. 6, 355, 356, 358, 363, 367, 368
Sarapis	138, 147, 339, 340-343, 346, 356, 358, 360, 361, 363, 365, 370
-------- - Pantheos	361, 370
Sardis	405
Sarousin	341
Satan	13, 17, 18, 22, 162
Satanael	17, 22
Schizophrenia	447
Science(s)	3, 11, 27, 36, 58, 60 n. 18, 74, 75, 78, 81, 90 n. 9, 93 n. 14, 97, 164, 176, 177, 181, 182, 183, 188, 209, 211, 213, 214, 215, 217- 226, 228, 230, 231, 232, 233, 234, 237, 238, 243-245, 247, 248, 293, 294, 295, 297, 300, 302, 304, 307, 308 n. 1, 309, 315, 317, 319, 320, 326, 335, 376-379, 384, 386-388, 394, 414 n. 2, 413, 447, 460, 461, 462, 465, 467, 468, 470
---------, reflective	155
Scientific study of religion	57, 59, 269, 370, 307, 309, 315, 321, 373, 441, 468, 470
Scipio Africanus	129
Scriptures	19, 46 n. 41, 321, 399, 407-409
Second Hellenism	340, 342, 346, 347
Secularism	161, 163, 165, 170, 255, 259, 261
Seleucids	274, 355
Self-awareness	165, 170

-----, -consciousness	88
-----, -flagellation	459
-----, -incitement (Selbstentzündung)	18
----- -organization, societal	192
------, -sacrifice	95
Semi-gods	353
Semiramis	144
Sensibilities, religious	93
Servius Tullius	119, 120, 123, 129, 130
Seven Arts	21
Shamanism	450, 451, 453, 455, 457
Shamans	92, 443, 450, 451, 453
Shii Muslim	50
Shipwreck(s)	139, 141
Sign(s)	118, 124, 138, 187, 299, 329, 340, 397, 408, 442, 454, 463
Sin	157, 171, 259, 262, 264, 416
Sobek	360
Social and ideological formations, of early Christians	41
Social science	36 n.16, 176, 177, 423, 447
Socialization	94, 211, 455
Socio-anthropological work on religion	298
Sol	341
Sophia	345, 346, 346, 403 n. 7
Sorcerers	82, 443
Soul(s)	17, 20, 21, 22, 63, 100 n. 3, 104, 106, 109, 110, 114, 175, 330, 346, 397
Space, public	73, 143
-------, sacred	2, 39 n. 27, 44, 129, 130, 142 n. 17, 153, 157, 159, 160, 165, 166, 167, 171, 195 n. 13, 233, 253, 255, 256, 258, 272, 282, 287, 297-300, 303, 315-317, 325, 326, 329, 330, 333-336,

	341, 344, 356, 360, 361, 373, 374, 386, 387, 391, 418, 419, 422, 424, 428, 430, 436
Spectacle	147
----------, public	134, 136
Sphere, public	27, 204
Sphinxes	356, 360, 368
Spiritualism	384
Spirituality, Graeco-Roman	77
Stabilization	77, 463
Standard Social Scientific Model (SSSM)	7
Stars	25
Structuralism	420
Structures, cognitive	156
Studies, empirical	151, 152
Studio delle Religioni	65, 66
Suffering of the gods	140
Supernatural agents	95, 245, 246, 248
------------, Beings	2, 3, 96, 245, 452
Svarga Ahram, at Rishikesh	54
Symbol(s)	84, 86, 142, 165, 187, 239, 281, 297, 303, 334, 353, 359, 360, 401, 408, 421, 422, 427-430, 432, 433, 442, 443, 445, 449
Syncretic version	352
Synthetic, approach	469, 470
Syria	100 n. 1, 145 n. 25, 404
Syriac Christianity (Early)	16
Taboo(s)	416, 419, 433
Tale-telling, mythical	134
Tales, memorable	123
Taoism	54
Taxonomy, epistemological	176

Temptation-theme   20
Temporal lobe epilepsy   447
Texts
------- (biblical)   109, 398, 401, 403, 407
--------, epigraphic   400, 401, 403, 404, 407
---------, magical   339, 400
The Book of Urizen   25
theologoumena, Egyptian   269
Theology   21, 36, 41, 50, 66, 140, 178, 217, 223-226,
                                    230, 232, 233, 234, 255, 259, 260, 294

-----------, reflective   226
Theory
------------, memory   415
------------, mind   224, 228-229, 230, 231, 232, 241, 242, 248,
                                    464

--------, of knowledge
(Aristotelian)   14
-----------, rhetorical   135
Theosophical Society   384
Theriomorphic   113
Thesmophoros (Law-giver)   276
Thessalonica   408
Thessaloniki   276, 283, 347
Third Temple in Jerusalem   158
Thomas   12, 34, 36 n. 17, 99, 100-104, 106-109, 111,
                                    113, 114, 280, 325, 377, 461, 463, 469

Thought (sexual)   101, 104
Thought, religious   3, 81, 90, 91 n. 10, 158
Thematization   294, 295, 300
Tiber   124, 125, 126, 129
Time   5, 11, 14, 17, 24, 26, 34, 38, 44, 45, 49, 62,
                                    66, 67, 75, 84, 91, 96, 97, 100, 102, 110, 111,
                                    138, 139, 153, 155, 158, 162, 168, 169, 170,
                                    176, 178, 185-187, 192, 195, 196-203, 205,
                                    206, 219, 222, 224, 227-230, 237, 239, 245,

	247, 259, 264, 270, 274, 276, 278, 280, 282-286, 293, 294, 296, 300, 303, 308, 311, 318, 320, 322, 330, 345, 352, 354, 355, 357, 362, 377, 385, 386, 388, 389, 390, 391, 399, 398, 401, 408, 409, 413, 415, 425, 427, 431, 434, 439, 443, 444, 447, 449, 459, 461, 463, 466
Timotheus, Eumolpides priest	279
Tokyo	60, 67
Torah	397, 398, 404
Tractates (ascetic)	101
Tradition, Attic	281
-----------, Gnostic	15, 25
-----------, oral	415, 425, 426, 428, 438, 439
-----------, religious	14, 54, 55-58, 96, 129, 153, 161, 162, 163, 171 n. 4, 172, 221, 247, 262, 271, 293, 297, 322, 415, 433, 434
-----------, Sicilian	281
Traditions, non-literate	414, 417, 437, 438
-----------, literate	416, 428, 438, 439
Trajectory	34, 148, 335, 448, 450
Trance states	374, 447, 455
Transempirical, powers or beings	92
Transformation	121, 164, 166, 167, 207, 284, 414, 415, 416, 428 n. 5, 430, 433, 434, 435, 439, 453, 462, 463
Transformational-generative grammar	462
Transmission	
----------------, horizontal	239
----------------, religious	426, 463
Trinitarian and Christological Dogma	22

Triptolemus	279-281, 286
Trisagion hymn	406
Triumph(s)	21, 54, 134, 144, 145, 146 n. 28-29
------, Miracle	332, 223
-----, -motif	111
Typhon	139, 141
Typology, historical	50
Umma	297
Unbeliever	168, 205
Understanding, intuitive	76
UNESCO	62
Universals	50, 447
Università Urbaniana Propaganda Fide	53
University of Chicago	182, 185
Upper Paleolithic art	444, 445
--------------------, people	444, 445, 448
-----------------------, period	444, 445, 453
Urizen	25
Value(s)	10, 27, 36, 39, 40, 41, 44, 57, 59, 88, 91, 94, 96, 123, 153, 158, 161, 166, 169, 170, 172, 178, 180, 185, 195 n.13, 196, 199, 209, 220, 221, 237, 256, 266, 271, 295, 297, 303, 313, 315, 320, 321, 322, 335, 365, 375, 383, 385 n.10, 397, 420, 427, 429, 432, 441, 443, 449, 452, 476
verba praeire	70
Vestal, virgins	119, 121
Vespasian	361
Vicaria Dei	22
Vicissitude(s)	141
Victory	146, 147, 171 n. 5

Vietnam War Memorial,
in Washington D.C.   325
Villa of the Mysteries,
in Pompei   147
Violence   156, 159, 160-163, 166-171, 206, 328, 459
-----------, sectarian   459
Viruse(s)   239, 463
Visual arts   4
voces magicae   339, 340, 344- 347

Wealth   134 n. 2, 203, 204
World Congress of Faith   58
World's Parliament of Religion 385

Xenophanes   313

Yoga, Brahmanical   385
Yolam, temple   417, 418, 431
Zeus   1-6, 137, 340, 342, 352, 353, 360
------ / Sarapis   138, 342
--------, Helios, Mithras,
Sarapis   339, 340, 341, 346
------, Homognios   5
-------, Xenios   5
Zionist(s)   163

CPSIA information can be obtained at www.ICGtesting.com
Printed in the USA
BVOW02s1233111214

378532BV00005B/13/P